Stability Regions of Nonlinear Dynamical Systems

Theory, Estimation, and Applications

This authoritative treatment covers theory, optimal estimation, and a range of practical applications.

The first book on the subject, written by leading researchers, this clear and rigorous work presents a comprehensive theory for both the stability boundary and the stability regions of a range of nonlinear dynamical systems, including continuous, discrete, complex, two-time-scale, and non-hyperbolic systems, illustrated with numerical examples. The authors also propose the new concepts of quasi-stability regions and relevant stability regions and their complete characterizations.

Optimal schemes for estimating stability regions of general nonlinear dynamical systems are also covered, and finally the authors describe and explain how the theory is applied in areas including direct methods for power system transient stability analysis, nonlinear optimization for finding a set of high-quality optimal solutions, stabilization of nonlinear systems, ecosystem dynamics, and immunization problems.

Hsiao-Dong Chiang is Professor of Electrical and Computer Engineering at Cornell University, Ithaca, NY, Founder of Bigwood Systems, Inc. (BSI), Ithaca, NY and Founder of Global Optimal Technology, Inc. (GOTI). He is a Fellow of the IEEE.

Luís F. C. Alberto is Professor at the School of Engineering of São Carlos, University of São Paulo, Brazil, and was a director of the SBA (Brazilian Society of Automation) 2013–2014.

Stability Regions of Nonlinear Dynamical Systems

Theory, Estimation, and Applications

HSIAO-DONG CHIANG
Cornell University

LUÍS F. C. ALBERTO
University of São Paulo

CAMBRIDGE
UNIVERSITY PRESS

CAMBRIDGE
UNIVERSITY PRESS

University Printing House, Cambridge CB2 8BS, United Kingdom

Cambridge University Press is part of the University of Cambridge.

It furthers the University's mission by disseminating knowledge in the pursuit of education, learning and research at the highest international levels of excellence.

www.cambridge.org
Information on this title: www.cambridge.org/9781107035409

First published 2015

A catalogue record for this publication is available from the British Library

Library of Congress Cataloguing in Publication data
Chiang, H. (Hsiao-Dong)
Stability regions of nonlinear dynamical systems : theory, estimation, and applications / Hsiao-Dong Chiang, Cornell University, Luís F. C. Alberto, University of São Paulo.
pages cm
ISBN 978-1-107-03540-9 (hardback)
1. Stability. 2. Dynamics. 3. Nonlinear control theory. I. Alberto, Luís Fernando Costa.
II. Title.
QA871.C52 2015
003′.85–dc23

2014049631

ISBN 978-1-107-03540-9 Hardback

Contents

Preface

There are several books available dedicated to the broad subject of stability of nonlinear dynamical systems. There are, however, just a few books available covering the important subject of stability regions (or regions of attraction or domains of attraction) of nonlinear systems. This subject is usually covered in books on nonlinear systems at an introductory level and in a fragmented manner. The motivation for writing this book was prompted by the critical need to treat the subject of stability regions in a clear, concise and comprehensive way for a wide range of audiences.

Knowledge of stability regions is fundamental in general nonlinear dynamical systems. It is of equal, if not more, importance to the notion of stability in nonlinear systems. The problem of determining stability regions of nonlinear dynamical systems plays an important role in many disciplines in engineering and the sciences. For instance, this problem appears in the areas of direct methods for power system stability analysis, stabilization of nonlinear systems, design of manipulators in robotics, design of nonlinear controllers, analysis and synthesis of power electronics, associative memory in artificial neural networks, solution methods for nonlinear optimization problems, ecosystem dynamics, immunization problems, economics and so forth. Hence, there is a broad range of professionals who can benefit from knowledge of current developments in the area of stability regions.

Currently, the practice of studying the stability of an equilibrium point or operating state is insufficient for the characterization of nonlinear dynamical systems since stable operating points of physical systems are rarely globally stable. In other words, the stability region is usually only a subset of the state space, and determination of this region is essential in many applications. In fact, the problem of determining the stability region is old and yet so difficult that breakthroughs in the complete characterization of stability regions of nonlinear dynamical systems were achieved only in the last twenty-five years. The development of the theory of stability boundaries and its applications has reached a level of sufficient maturity that the publication of a book exclusively dedicated to this important subject is justified.

The main objective of this book is to provide a comprehensive treatment of the theory of stability regions of nonlinear dynamical systems and effective and yet scalable methods for estimating stability regions. The book is divided into four parts. The first part devotes nine chapters to the theoretical treatment of stability regions. The theory of the stability region is developed for several classes of nonlinear dynamical systems, including continuous, discrete-time, complex and constrained nonlinear dynamical systems. The second part is devoted to the development of effective, yet scalable, numerical methods for estimating stability regions. This part is composed of six chapters, which contain methods for optimally estimating stability regions for continuous, discrete-time, complex and constrained nonlinear dynamical systems. Methods for estimating relevant stability boundaries are also presented. It is also shown that by exploring the special structure of the study system model, improved and more powerful results can be obtained. Some developments in this direction are presented in this part, in which the structure of second-order systems is explored to obtain more powerful results on the structure of the stability boundary and to develop more efficient methods for estimating stability regions.

It is our belief that sound theoretical development should be combined with practical implementation and application to solve real-world problems. The fourth part of this book presents two practical applications of stability region theory and estimation methods to real-world problems. The on-line direct stability analysis of large-scale power grids is presented in Chapter 20. The theory of stability regions plays a key role in this practical application. The development of stability-regions-based methodologies to overcome the drawbacks of iterative local search methods and that of modern heuristics in the search for the global optimum is another application presented in Chapter 21. The new paradigm for solving nonlinear optimization problems has several distinguishing features achieved by transforming the search space into a state space composed of the union of stability regions.

In summary, this book offers, in a rigorous manner, a comprehensive theory for both the stability boundary and the stability regions of general nonlinear dynamical systems. Several topological and dynamical characterizations of the stability boundaries for large classes of nonlinear dynamical systems, including both continuous and discrete-time dynamical systems, are presented. Several effective and scalable numerical methods for optimally estimating stability regions, along with their practical applications, are presented to illustrate the theoretical developments presented in the book.

We hope that this book can contribute to inspire others to apply the theory and estimation methods of stability regions to relevant problems arising in the sciences and in engineering and to promote further developments in this fascinating area of research.

Acknowledgements

We started to work on the general subject of stability regions of nonlinear dynamical systems as graduate students at the University of California, Berkeley, in the mid 1980s and at the University of São Paulo, São Carlos, Brazil, in the mid 1990s, respectively. The need to develop fast on-line direct methods for power system stability analysis re-triggered the development of the theory of stability regions in the 1980s, and we were lucky to work on this exciting subject and to contribute to this area with theoretical and numerical developments, and practical applications.

We acknowledge our research advisors, Professor Felix Wu (EECS) and Professor Morris Hirsch (Mathematics), from University of California at Berkeley, and Professor Newton G. Bretas, from University of São Paulo, for introducing us to the world of stability regions. We carry the importance of their advice with us to this day. We also acknowledge those who patiently taught us about nonlinear systems, nonlinear circuits and dynamical systems in the beginning of our careers as researchers. Our thanks go to Professor Morris Hirsch, Professor Shankar Sastry, Professor Pravin Varaiya and Professor Leon Chua of University of California at Berkeley and Professor Hildebrando M. Rodrigues of University of São Paulo.

For many years, several former PhD students contributed to the development of the material presented in this book. In particular, we would like to acknowledge Dr. Lazhar Fekih-Ahmed, Dr. Chia-Chi Chu, Dr. Jaewook Lee, Dr. Ian Dobson, Dr. Mathew Varghese and Dr. Warut Suampun of Cornell University. We also acknowledge Dr. Ana Paula Mijolaro, Dr. Josaphat R. R. Gouveia and Dr. Taís R. Calliero of University of São Paulo.

We are greatly indebted to Dr. Lazhar Fekih Ahmed for joint work on quasi-stability regions and constrained stability regions, to Dr. Chia-Chi for joint work on the BCU method and on the constructions of energy functions. Joint work with Dr. Jaewook Lee on the stability regions of non-hyperbolic dynamical systems and on the closest UEP method has been very rewarding. Dr. Fabiolo M. Amaral, has been working with us for many years on the theory of stability region bifurcations and stability regions of discrete systems; his insightful discussions were fundamental to the development of this theory. We would like to express our special thanks to Dr. Edson A. R. Theodoro whose work was relevant to the development and application of the theory of stability regions for two-time-scale systems.

Our academic colleagues have also been a guiding source of support and encouragement. We are very thankful to our colleagues at Cornell University and University of São

Paulo. A special thanks goes to Professor James S. Thorp, Professor Robert J. Thomas, Professor Rodrigo A. Ramos and Professor João B. A. London Jr.

Last but not least, we would like to thank our families, for their full-fledged support and understanding over the past several years.

Hsiao-Dong Chiang
Luís F. C. Alberto

1 Introduction

Stability has been regarded by many as a fascinating and difficult problem of human culture. Stability problems are present in all our lives. The act of standing is an example of maintaining stability that we learn to control at a very early age. Stability of physical systems and of social systems, such as economies and ecosystems, are frequently discussed on the news. Stability is present everywhere and is a fundamental subject that permeates engineering and the sciences.

Stability is a very broad subject, and the concept of stability can be formulated in a variety of ways depending on the intended use of stability analysis and design. As such, at least 50 different terms for stability concepts appear in the literature. An important subject closely related to stability is the *stability region* (i.e. *region of attraction* or *domain of attraction*) of nonlinear dynamical systems, the main subject of this book.

Many nonlinear physical and engineering systems are designed to be operated at an equilibrium state (i.e. equilibrium point). A first and foremost requirement for successful operation of these systems is to maintain stability of this equilibrium state. Stability requires robustness of the equilibrium point to small perturbations, i.e. the system state returns to the equilibrium state after small perturbations. Since most physical and engineering systems are not globally stable, equilibrium states can only be restored under a limited amount (or size) of perturbation. Intuitively, one can state that a system sustaining a larger size of perturbation is "more" stable or "more" robust than another system. We will see that this "degree" of stability is related to the concept of the stability region of nonlinear dynamical systems, which will be thoroughly explained and explored in this book.

1.1 Degree of stability

In order to illustrate the concept of stability and the "degree" of stability or "robustness" of stability, we consider two solid bodies of different size lying on the ground as shown in Figure 1.1. Their respective positions are stable. The stable position of the solid A is recovered after a small perturbation is applied to the solid, as indicated in Figure 1.2. However, if the perturbation is large enough, then the body will reach another equilibrium position, as seen in Figure 1.3, and will not return to the original (stable) equilibrium point. This system possesses multiple equilibrium states, some of which are stable but not globally stable. If one needs to determine which solid has a more

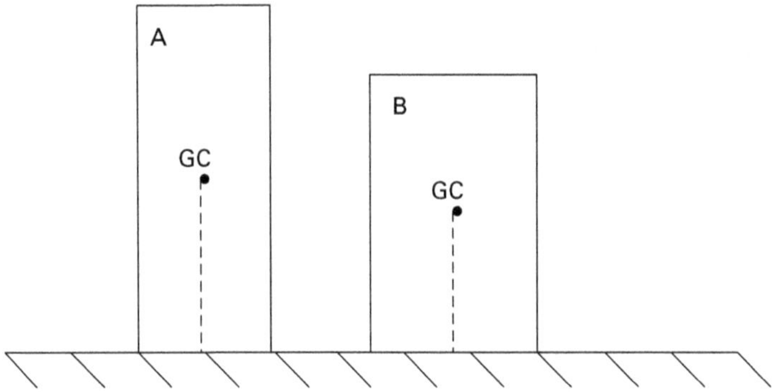

Figure 1.1 Solids A and B have the same weight and volume. The point GC indicates the center of gravity of the solid. Their positions are both stable but the stability of the position of solid B is more robust to perturbations than that of solid A. The stability region of B is larger than the stability region of A.

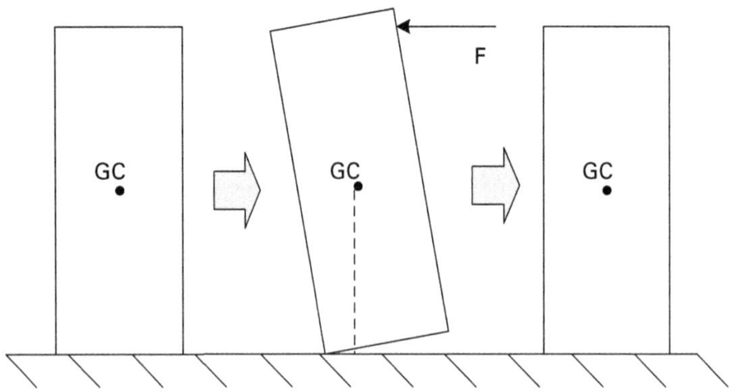

Figure 1.2 The solid is in a stable position. A small perturbation F is applied to the solid. After the removal of that perturbation, the solid returns to its original stable position.

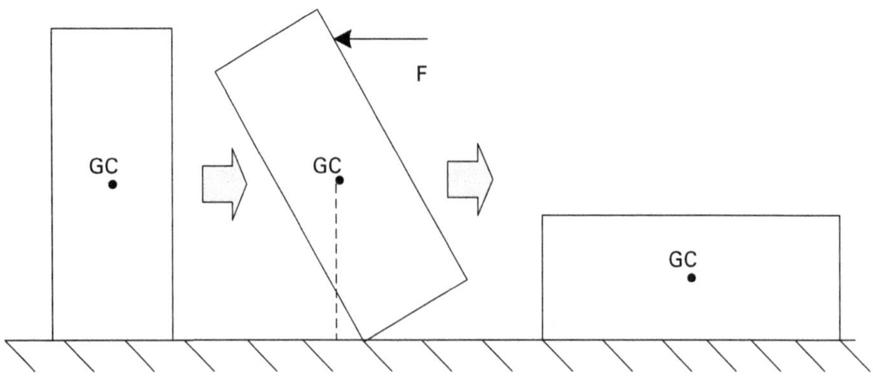

Figure 1.3 The solid is in a stable position. A sufficiently large perturbation F is applied to the solid to make the solid settle down into another stable position, showing that the stable positions of this system are not globally stable.

"stable" equilibrium position, then it is obvious that solid B is "more stable" than solid A. More precisely, the "amount" of disturbance (i.e. perturbation) needed to push solid B away from its stable position is much larger than the amount of disturbance that is needed to push solid A away from its stable position. This "degree of stability" is related to the concept of stability region. It is clear that the stable position of solid B has a larger stability region than the stable position of solid A.

Generally speaking, the concepts of both stability and asymptotic stability are local and do not provide information regarding how robust the system is with respect to disturbance and/or model uncertainty. The concept of stability region, on the other hand, is a global one and gives a complete picture of "the degree of stability" with respect to noise or perturbations or model uncertainty. More specifically, knowledge of the stability region provides, for a given initial condition or a specified amount of perturbation, information on whether or not the system will settle down to a desirable steady state condition.

1.2 Stability regions

We consider the following (autonomous) nonlinear dynamical system

$$\dot{x} = f(x), \quad x \in R^n. \tag{1.1}$$

It is natural to assume the function (i.e. the vector field) $f: R^n \rightarrow R^n$ satisfies a sufficient condition for the existence and uniqueness of the solution. The solution of (1.1) starting at x_0 at time $t = 0$ will be denoted $\phi(t, x_0)$, or $x(t)$ when it is clear from the context.

For an asymptotic stable equilibrium point $\hat{x}$, there exists a number $\delta < 0$ such that $\|x_0 - \hat{x}\| < \delta$ implies $\phi(t, x_0) \rightarrow \hat{x}$ as $t \rightarrow \infty$. In other words, there exists a neighborhood of the equilibrium point such that every solution starting in this neighborhood is attracted to the equilibrium point as time tends to infinity. If δ can be chosen arbitrarily large, then every trajectory is attracted to $\hat{x}$ and $\hat{x}$ is called a *global asymptotically stable equilibrium point.* There are many physical systems containing asymptotically stable equilibrium points but not globally stable equilibrium points. A useful concept for these kinds of systems is that of the **stability region** (also called the **region of attraction** or **domain of attraction**). The stability region of a stable equilibrium point x_s is the set of all points x such that

$$\lim_{t \rightarrow \infty} \phi(t, x) = x_s. \tag{1.2}$$

In words, the stability region of x_s is the set of all initial conditions x whose trajectories tend to x_s as time tends to infinity. We will denote the stability region of x_s by $A(x_s)$, and its closure by $\bar{A}(x_s)$, respectively; hence

$$A(x_s) := \{x \in R^n : \lim_{t \rightarrow \infty} \phi(t, x) = x_s\}. \tag{1.3}$$

When it is clear from the context, we write A for $A(x_s)$. From a topological point of view, the stability region $A(x_s)$ is an open, invariant and connected set. The boundary of the

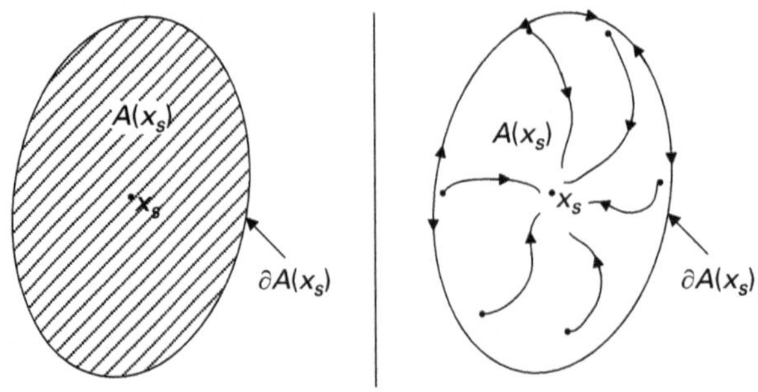

Figure 1.4 As time increases, every trajectory in the stability region $A(x_s)$ converges to the asymptotic stable equilibrium point (SEP) x_s and every trajectory on the stability boundary evolves on the stability boundary.

stability region $A(x_s)$ is called the **stability boundary** (also called the **separatrix**) of x_s and will be denoted by $\partial A(x_s)$. Figure 1.4 illustrates the concept of stability region and stability boundary.

The concept of stability region of an asymptotic stable equilibrium point can be extended to that of other types of attractors. For example, the stability region of an asymptotic stable closed orbit, γ, is defined as follows:

$$A(\gamma) = \{x \in R^n : \lim_{t \to \infty} d(\phi(t,x),\gamma) = 0\}$$

where $d(.,.)$ is a distance function. The stability regions of other types of attractors, such as asymptotically stable quasi-periodic solutions and asymptotically stable chaotic trajectories, are similarly defined. Global stability (i.e. stability in the large) rarely occurs in physical and engineering systems due to physical and operational limits, such as saturations and control actions. In addition, the cost of designing systems to be globally stable, when feasible, is usually very high. These factors make the task of determining stability regions of nonlinear systems of great importance for practical application. It is fair to state that knowledge of the stability region of an attractor is equally important to verification of the stability of the attractor itself.

Knowledge of the stability region is essential in a variety of application areas such as direct methods for power system transient stability analysis [51,65,81,121,207,224,257,264,280], stabilization of nonlinear systems [101,120,147,179,183,227,250,253,279], decentralized control design for nonlinear systems, power system voltage collapse problems [121,126], schemes for choosing manipulator specifications and parameters in robotics [26,196], the design of associative memory in artificial neural networks [71,83,127], solution methods for nonlinear optimization problems [48,49,56,62,162,163], the effect of discretized feedback in a closed loop system on stability [155], and so forth. Knowledge of the stability region of a stable limit cycle is critical in areas such as biology [19] and robotics [274,275].

Determining stability regions of nonlinear dynamical systems is an old problem, and yet it remains challenging. Recent advances in theoretical developments of stability regions and in the development of effective methods for estimating stability regions offer promising results to meet the challenges. These advances can be classified as follows.

Theoretical development:

- characterization of a subset of stability regions of general nonlinear dynamical systems,
- complete characterization of stability boundaries of equilibrium points and attractors of a fairly large class of nonlinear dynamical systems,
- complete characterization of stability boundaries of fixed points of a class of time-discrete nonlinear dynamical systems,
- characterization of stability boundaries of two-time-scale systems,
- complete characterization of stability boundaries of a class of nonlinear non-hyperbolic dynamical systems,
- characterization of relevant stability boundaries of a class of nonlinear dynamical systems.

Methods for estimating the stability region:

- optimal estimation of stability regions of a fairly large class of continuous nonlinear dynamical systems,
- optimal estimation of stability regions of second-order systems,
- optimal estimation of stability regions of a large class of discrete dynamical systems,
- constructive approach to iteratively improve estimations of stability regions,
- optimal estimation of relevant stability regions of a class of nonlinear dynamical systems.

In this book, a comprehensive treatment of these contributions is presented in a structured way, followed by advanced topics on stability regions and by illustrations of stability-region-based practical applications.

1.3 Characterization and estimation of stability regions

It is not easy to trace the first attempt to address the issue of stability regions in the literature, but the history of development of the concept of stability sheds some light on it. The subject of stability has attracted a significant, if not the most, amount of research and development in the area of nonlinear system analysis, design and control. The theory of stability and, in particular, methods to estimate stability regions are related to energy function theory. The mathematicians Leonhard Paul Euler and Joseph-Louis Lagrange, in the eighteenth century, established relations between the equilibrium point and stability with maxima and minima of energy functions. Aleksandr Mikhailovich **Lyapunov** (1857–1918), inspired by his master's thesis on fluid dynamics, developed a general theory of stability in his PhD thesis: *The general problem of the stability of motion* (1892). Lyapunov derived sufficient conditions for stability based on the existence of a scalar function, nowadays called the Lyapunov function in his honor.

George David Birkhoff (1884–1944) made an important contribution by developing stability theory for the asymptotic behavior of trajectories of differential equations.

Krasovskii was probably the first to prove an invariance principle [149]. This principle explores the concept of limit sets introduced by Birkhoff. Joseph P. LaSalle, who received his PhD in 1941, developed a similar invariance principle in 1960 [158]. LaSalle's invariance principle is probably the first tool proposed for estimating stability regions in a systematic way. The work of Krasovskii and LaSalle led to the development of the expression of stability region estimates in the form of level sets of Lyapunov-like functions.This pioneering work has sparked the development of a great number of methods for estimating stability regions of high-dimensional nonlinear systems.

The existing methods for estimating stability regions proposed in the literature can be classified into Lyapunov-function-based methods and non-Lyapunov-function-based methods. Unfortunately, the vast majority of methods always offer rather conservative estimations of stability regions; in other words, these methods offer estimated stability regions which are only a (small) subset of entire stability regions. This conservative estimation can lead to serious consequences. For example, conservative estimates of the stability region may result in unnecessary interruptions in the operation of power systems and in expensive over-design of control systems. Thus, there is a serious need for the development of effective methods for accurately estimating stability regions of high-dimensional nonlinear dynamical systems.

Lyapunov-function-based methods have been popular for estimating the stability regions of stable equilibrium points. This class of methods is applicable to large-scale nonlinear systems but may suffer from overly conservative estimation of stability regions. The degree of conservativeness of Lyapunov-function-based methods in estimating stability regions depends on the underlying Lyapunov function and the associated value of the critical level. Finding a good Lyapunov function for estimating stability regions is not an easy task. There is no systematic way to derive a Lyapunov function for general nonlinear dynamical systems. Recent advances along this line of research and development include the proposal of maximal Lyapunov functions [256], the optimal estimation of stability regions based on a given Lyapunov function [53], LMI optimization techniques for constructing Lyapunov functions, and estimating stability regions for polynomial dynamical systems [41,76,137,164,247,250,252] and for non-polynomial systems [39]. These optimization-based methods entail high computation effort, tending to grow rapidly with the system dimension, making them unsuitable for estimating stability regions of high-dimensional nonlinear dynamical systems. The LMI-based estimation methods generally yield (overly) conservative estimations of stability regions.

Another advance is the constructive Lyapunov function methodology for determining optimal Lyapunov functions to estimate stability regions [60]. The constructive methodology yields a sequence of estimated stability regions which form a strictly monotonic increasing sequence, and yet each of them is contained in the stability region of the system under study. The constructive methodology can either stand by itself or be used with existing methods serving as the input to the constructive methodology. This methodology can reduce the conservativeness in estimating stability regions via the Lyapunov function approach.

There are very few methods which are able to estimate the entire stability region. They all entail serious computational problems, making their application impractical. For example, Zubov's method [106,175] offers a technique for computing the entire stability region via the "optimal" Lyapunov function. However, constructing this "optimal" Lyapunov function requires solving a set of nonlinear partial differential equations (PDEs) which are difficult, if not impossible, to solve. Because of this problem, several techniques have been proposed which attempt to approximate the solution of the PDEs, but with limited success. Furthermore, it has been found that even for some second-order systems the use of approximated solutions to construct the "optimal" Lyapunov function still results in rather conservative estimations.

Another method which attempts to reduce the conservativeness in estimating the stability boundary was proposed in [97], in which the estimated stability boundary was synthesized from a number of system trajectories obtained by backward integration. This method, known as the trajectory reversing method [97,99,168], is suitable only for low-dimensional systems due to its excessive computational burden. Other methods for estimating the stability region include approximating the stability boundary by polytopes [208] and the cell-to-cell mapping method [128].

A majority of the existing proposed methods for estimating stability regions are only applicable to low-dimensional nonlinear systems. Very few of them are applicable to high-dimensional nonlinear systems (say several hundred or thousands of dimensions). To be practical, estimation methods need to be able to deal with very large-scale systems (say, tens of thousands of state variables).

An important breakthrough in the development of the theory of stability regions occurred in the 1980s with the formulation of a comprehensive theory of stability regions and a complete characterization of the stability regions of a class of nonlinear dynamical systems [54]. This class of nonlinear dynamical systems is characterized by its ω-limit set being composed of equilibrium points and limit cycles. A conceptual method based on the complete characterization derived in [54], when feasible, can find the *entire* stability region. This method is based on a complete characterization of the stability boundary via the stable manifolds of unstable equilibrium points and unstable limit cycles on the stability boundary. For low-dimensional nonlinear dynamical systems, the derivation of the stable manifolds can be achieved by numerical methods. However, current computational methods are inadequate for computing the stable manifolds of unstable equilibrium points and/or of unstable limit cycles of high-dimensional nonlinear dynamical systems.

Knowledge of the complete characterization of a stability boundary has led to the development of several effective methods for accurate estimation of stability regions. By exploring the complete characterization of the stability boundary, several computational schemes for optimally estimating the stability regions have been developed, see for example [53]. These complete characterizations have also led to the development of effective methods to estimate relevant parts of the stability boundary. These methods have been fundamental to advancing a variety of important practical applications.

1.4 Practical applications of stability regions

Estimating or determining stability regions is central to solving many problems arising in sciences and engineering. The following list is a non-exhaustive enumeration of applications where knowledge of stability regions plays an important role.

1. Biology:
   - micromolecules and macromolecules;
   - dynamics of ecosystems.
2. Biomedicine:
   - dynamics of the immune response;
   - human respiratory models.
3. Control:
   - sliding control systems;
   - control of nonlinear systems;
   - linear systems with saturated controls;
   - control of polynomial systems.
4. Economics:
   - economic growth rate;
   - carrying capacity of the human population.
5. Robotics:
   - asymptotically stable walking cycle of a bipedal robot;
   - regulator design of robot manipulators.
6. Power grids and power systems:
   - direct power system transient stability analysis;
   - dynamic voltage stability analysis.
7. Power electronic circuits:
   - power-electronic-based converters;
   - DC-to-DC converters.
8. Neural networks:
   - dynamic recurrent neural networks;
   - associative memory;
   - Hopfield neural network models.
9. Optimization problems:
   - unconstrained optimization problems;
   - constrained optimization problems.

We briefly describe some of the above applications based on knowledge of stability regions.

1.4.1 Immune response

Stability regions provide insight into the analysis of the immune response where interaction among lymphocyte populations, antigens and antibiotics occurs [110,139,225,267]. The coexistence of multiple equilibriums has been reported in these models and the outcome of a treatment depends not only on how antigens are administered but also on the initial condition of the system when the treatment began. Depending on the initial condition, the dynamics of the treatment will converge to the attractor whose stability region contains the initial condition.

The immune system is primarily composed of a large number of cells called lymphocytes. Lymphocytes produce antibodies that bind to invading organisms (antigens) in order to eliminate them. An animal can produce a very large number of different antibodies (10^6 to 10^7) [110]. The presence of an antigen stimulates the proliferation of cells (lymphocytes) with the specific antibody for that invader. The following set of differential equations models the dynamics of positive and negative cells of the immune system:

$$\dot{x}_+ = x_+[R(x_-, x_+) - D(x_-, x_+) - k_4] + k_5$$
$$\dot{x}_- = x_-[R(x_+, x_-) - D(x_+, x_-) - k_4] + k_5$$

where x_+ and x_- respectively represent the concentrations of positive and negative cells in the organism. The constant k_5 is an influx rate term while k_4 is a death rate term. The function R is a replication rate that describes the proliferation of the positive cells, while the function D is a death rate due to killing by antibodies or anti-antibodies. A typical phase portrait of the immune dynamical system is shown in Figure 1.5. Typically, immunization systems have four asymptotically stable equilibrium points: the virgin state, indicated as

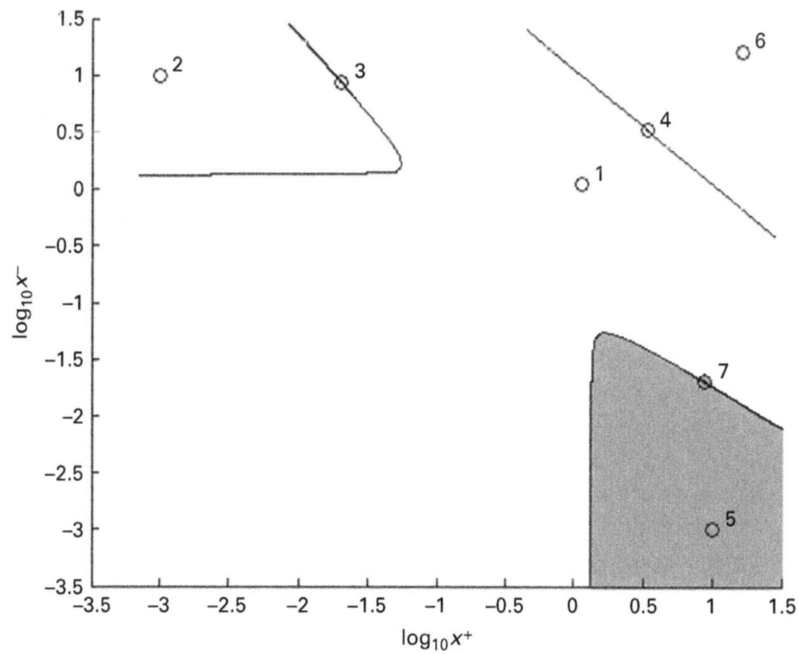

Figure 1.5 The stability region of the immune state (equilibrium #5) of a typical immunization system is highlighted. The immune system contains four asymptotically stable equilibrium points (SEPs): equilibriums #1 (virgin state), #2 (anti-immune state), #5 (immune state) and #6 (suppressed state). Their stability boundaries are depicted in this figure as black lines. Equilibriums #3, #4 and #7 are unstable equilibrium points lying on the stability boundaries of these four SEPs.

equilibrium #1 in Figure 1.5, where the concentrations of both positive and negative cells are low, the immune state, depicted as equilibrium #5, where the concentration of positive cells is much larger than that of negative cells, the anti-immune state, equilibrium #2, where the concentration of negative cells is high compared to the concentration of positive cells, and a suppressed state, equilibrium #6, where both concentrations are high. The immunization problem consists of designing a proper perturbation to the immune system, an immunization shot for example, such that the initial condition of the system after perturbation lies inside the stability region of the immune state equilibrium point, highlighted in gray in Figure 1.5. As a result, the immune system of the organism will abandon the virgin state equilibrium #1 and settle down in the immune state equilibrium #5.

1.4.2 Ecosystems

In biology, knowledge of stability regions is relevant to the problems of ecosystem dynamics in which different species coexist, in particular the coexistence of predators and prey species [19,148,176]. For example, the influence of commercial exploitation of a population of salmon on the size of stability regions was investigated in [206]. The resilience of an ecosystem can be viewed as the problem of determining whether or not a certain initial population lies inside the stability region of an attractor. Large stability regions are usually related to high-level "resilience" of the ecosystems.

1.4.3 Micromolecules and macromolecules

The task of finding saddle-points lying on a potential energy surface plays a crucial role in understanding the dynamics of micromolecules as well as in studying the folding pathways of macromolecules such as proteins. This task has been a topic of active research in the field of computational chemistry for more than two decades. Several methods have been proposed in the literature based on the Hessian matrix at the saddle-points, see for example [21,141]. These methods are not applicable to high-dimensional problems because the computational cost increases tremendously as the system dimension increases. Due to the scalability issue, several first-derivative-based methods for computing the saddle-points have been proposed, see for example [189,210]. A detailed description of such methods, along with their advantages and disadvantages, can be found in a survey paper [119]. A stability-region-based method has been developed for finding saddle-points of high-dimensional problems [213,214]. This method is based on the transformation of the task of finding the saddle-points into the task of finding the dynamic decomposition points lying on the stability boundary of two local minima (i.e. two stable equilibrium points). This method does not require that the gradient information starts from a local minimum (i.e. a stable equilibrium point). It finds the stability boundary in a given direction and then traces the stability boundary until the dynamic decomposition point (i.e. the saddle-point) is reached. This tracing of the stability boundary is far more efficient than searching for saddle-points in the entire search space. This again illustrates that knowledge of the stability boundary plays a key role in the development of this effective computational method.

1.4.4 Respiratory model and economics

Multiple equilibrium points exist in the respiratory model derived for humans [268] and in that derived for bacteria [91]. The presence of multiple equilibriums indicates that global stability is not possible in these models and knowledge of the stability region of each stable equilibrium point can provide a complete picture of the dynamic behavior of these respiratory models. The asymptotic behavior of a respiratory model will converge to the stable equilibrium point whose stability region contains the initial condition.

Nonlinear analysis of problems arising in economics can also benefit from knowledge of stability regions and how the stability region changes relative to parameter variations [16]. For example, an investigation of the impact of changes in technological level and saving rate on the relationship between the carrying capacity of the human population and economic growth rate was conducted in [34]. Multiple equilibriums as well as abrupt changes in their stability regions due to parameter variations have been reported and analyzed.

1.4.5 Nonlinear control systems

Characterization of the stability region is of great importance in modern nonlinear control systems [101,120,147,179,183,227,250,253]. It has been recognized that high feedback control gain can destabilize closed loop nonlinear systems [147,253]. Contraction of the stability region due to high feedback gain contributes to this destabilization. The task of maximizing the size of the stability region has become a design objective in nonlinear system analysis and controller design [270,230], including the design of linear system controllers with saturation [77,101,120,183].

The characterization of stability regions of nonlinear control systems is important in practical applications, see for example [3,43,84,133,154,287]. Local stability analysis is based on a linearized model and does not provide information regarding the nature of the stability region. The stability region can only be characterized by taking into account the nonlinearity of the underlying nonlinear system. We present one example of controller design: one where the stability region shrinks with the increase in controller gain. This example serves to illustrate that overlooking both the nonlinear terms and the stability region characterization in the controller design can result in a closed loop nonlinear system with a very small size stability region and, consequently, with a small degree of robustness with respect to both state and parameter perturbations.

Consider the following nonlinear system from [147]:

$$\begin{aligned} \dot{x}_1 &= x_2 + ax_1 - \frac{1}{3}u^3 \\ \dot{x}_2 &= bx_2 + u \end{aligned} \tag{1.4}$$

where $u \in R$ is an input, a and b are unknown parameters satisfying $|a| < c$ and $|b| < c$, with $c > 0$ as an upper bound for the uncertain parameters a and b. The objective is to design a state feedback controller that stabilizes the origin of this system. For this purpose, consider the linearization of system (1.4)

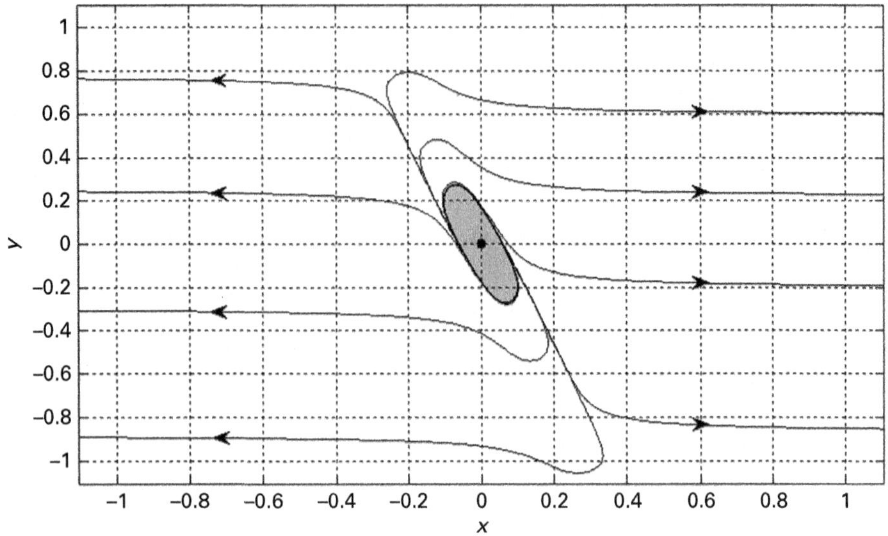

Figure 1.6 Illustration of the stability region of the nonlinear closed loop system (1.4) and (1.6) for $a = b = 0$, $c = 1$ and $\gamma = 4$ and some system trajectories which serve to verify the position of the stability boundary.

$$\begin{aligned} \dot{x}_1 &= x_2 + ax_1 \\ \dot{x}_2 &= bx_2 + u \end{aligned} \tag{1.5}$$

with the following feedback control:

$$u = -\gamma^2 x_1 - \gamma x_2. \tag{1.6}$$

By substituting (1.6) into (1.5) and computing the eigenvalues of the closed loop system, we conclude that the equilibrium point (0.0, 0.0) (i.e. the origin) of the linearized closed loop system is asymptotically stable, for any $|a| < c$ and $|b| < c$, if $\gamma > 2c$. Assuming the same state feedback (1.6) is employed to stabilize the nonlinear system (1.4), Figure 1.6 shows the stability region of the origin of the nonlinear closed loop system for $a = 0, b = 0$, $c = 1$ and $\gamma = 4$. Since the eigenvalues of the Jacobian matrix evaluated at the origin are $-2 \mp j2\sqrt{3}$, the origin of the closed loop system is indeed asymptotically stable. In spite of that, it is clear from Figure 1.6 that the stability region is too small to ensure robustness with respect to perturbations.

There is a nonlinear relationship between the stability boundary and the controller. Numerical studies show that the stability boundary of (0.0, 0.0) (the origin) is composed of an unstable limit cycle. In addition, the stability region shrinks as the controller gain γ increases, shown in Figure 1.7. This simple example illustrates that the linearization approach for controller design is completely independent of the size of the stability region. A design for a good-performance controller based on the linearized model can lead to very small size stability regions. As a consequence, it is important to adopt the size of the stability region as one criterion in designing linear controllers for controlling nonlinear systems.

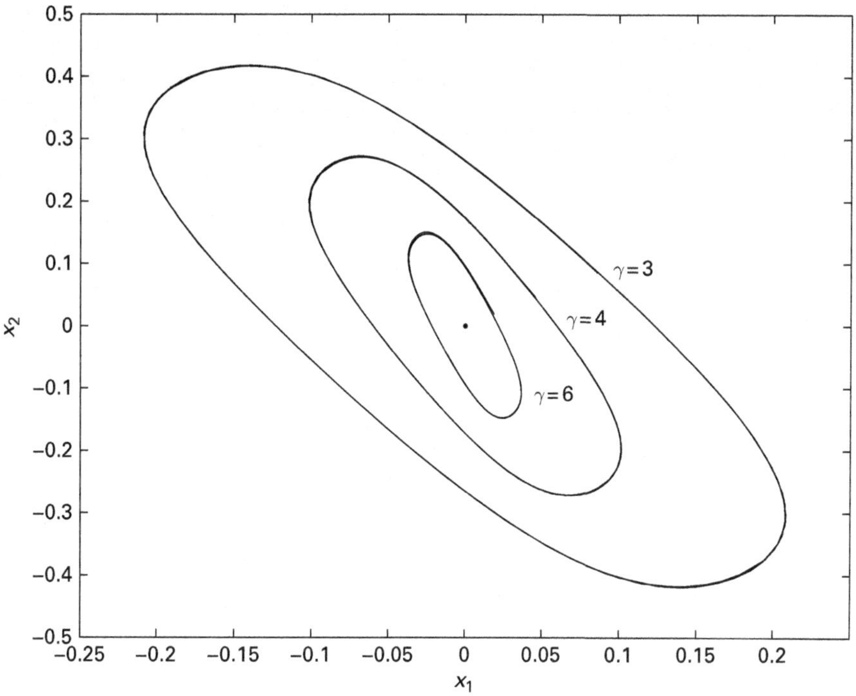

Figure 1.7 The stability region of the origin of the nonlinear closed loop system (1.4) and (1.6) shrinks as the controller gain γ increases. As the feedback gain increases from a magnitude of 3 to 6, the corresponding stability boundary significantly shrinks.

1.4.6 Power systems and power electronics

Characterizations of the stability of an operating point and its stability region are very important in many engineering problems. For instance, they play a key role in several applications in electric power systems [51,65,81,121,207,224,257,264]. Nowadays, on-line transient stability assessment of power systems is a reality due to the development of comprehensive stability region theory and of fast computational methods for estimating stability regions. In particular, this on-line transient stability analysis was made possible due to the development of the controlling UEP method and the BCU method, which were developed by exploring the stability boundaries of a class of nonlinear power system models. These methods have been applied to direct stability analysis of very high-dimensional systems, of more than 50 000 state variables [59,241].

Determination of the stability region is also essential in the area of power electronics. It is common to employ analysis of linearized power electric models and numerical simulation of nonlinear power electronic models to gain insight into the dynamical behavior of power electronic circuits. However, knowledge of the stability region can complement the nonlinear simulation approach to provide information regarding the robustness of the stable states and a comprehensive reliability assessment of the power electronic circuit under design. Power electronic based converters, for example, can be designed to operate nearly as ideal regulators. These power converters behave nearly as constant power loads and are a source of destabilization in power distribution systems. As a consequence, stability regions of power distribution systems connected to these constant power loads are a subset of the entire space [100]. Voltage-regulated boost DC-to-DC converters are also an example of power electronic circuits in which knowledge of stability regions is important [177,217]. For both power electronic circuits, determining stability regions is important to ensure the robust performance of the designed power electronic circuits.

1.4.7 Artificial neural networks

Complete stability is a desired property for recurrent artificial neural networks [71,83,127]. Complete stability ensures that every trajectory converges to an equilibrium point. The property of complete stability can be translated into the problem of checking whether or not the union of the stability regions of all stable equilibrium points (SEPs) covers almost the entire state space. The coexistence of multiple SEPs indicates that the output of a recurrent neural network depends not only on the input signal but also on the history of the dynamics or, more precisely, on which stability region the initial condition of the network lies when the input signal is injected into the recurrent artificial neural network [71,83,127].

1.4.8 Robotics

Stability regions of limit cycles are relevant in the analysis, design and monitoring of robotics [166]. Enlarging the stability region of the asymptotically stable walking cycle of a bipedal robot is a criterion in the design of robust controllers [229,274,275]. Keeping the robot inside the stability region of the walking cycle prevents its falling. The design of controllers with a constraint on the minimum size of the stability region is of great interest in the problem of regulator designing for robot manipulators [26,179,196]. The problem of robot navigation can be formulated in terms of stability regions. More precisely, it can be translated into the problem of designing a nonlinear system with a vector field which has the final destination as an asymptotically stable equilibrium point and the initial position of the robot lying inside the stability region of the stable equilibrium point [32,145].

1.4.9 Applications to nonlinear optimization

Knowledge of stability regions can provide an effective approach to solving nonlinear optimization problems. This approach to solving optimization problems was recently proposed and demonstrated [48,49,56,62,162,163]. Its potential to become one of the mainstream approaches to solving optimization problems is very promising. Indeed, optimization technology has practical applications in almost every branch of science, economics, engineering and technology.

One popular method for solving nonlinear optimization problems is to use an iterative local search procedure which can be described as follows: start from an initial vector and search for a better solution in its neighborhood. If an improved solution is found, repeat the search procedure starting from the new solution as the initial solution; otherwise, the search is terminated. Local search methods usually become trapped at local optimal solutions and are unable to escape from them. In fact, the great majority of existing nonlinear optimization methods for solving optimization problems usually come up with local optimal solutions but not the global optimal solution.

The drawback of iterative local search methods has motivated the development of a number of more sophisticated local search methods, termed modern heuristics, designed to find better solutions via introducing some mechanisms that allow the search process to escape from local optimal solutions. The underlying "escape" mechanisms use certain search strategies to accept a cost-deteriorating neighborhood to make escape from a local optimal solution possible. These sophisticated local search algorithms include simulated annealing, genetic algorithm, Tabu search, evolutionary programming and particle swarm operator methods. However, these sophisticated local search methods, among other problems, require intensive computational effort and usually cannot find the globally optimal solution. To overcome the difficulties encountered by the majority of existing optimization methods, the following two important and challenging issues in the course of searching for multiple high-quality optimal solutions need to be fully addressed:

(C1) how to effectively move (escape) from a local optimal solution and move toward another local optimal solution;
(C2) how to avoid revisiting local optimal solutions which are already known.

A stability-region-based methodology which is a new paradigm for solving nonlinear optimization problems will be presented in this book. This new methodology has several distinguished features, to be described in this book, and can address the two challenges (C1) and (C2).

1.5 Purpose of this book

The main purpose of this book is to present a comprehensive framework for developing the theory of stability regions of different types of nonlinear dynamical systems. Advanced topics on stability regions will also be presented. In addition, this book presents a unified framework to estimate stability regions of nonlinear dynamical systems. Optimal schemes for estimating stability regions of nonlinear dynamical systems will be presented for different types of these systems. Practical applications of stability regions to several topics of engineering are demonstrated. The transformation of nonlinear optimization problems into the stability region problem is also presented and shown to be a fruitful research and development area.

Nonlinear dynamical systems arising from different engineering disciplines and the sciences are vast and appear in a variety of different forms. This book covers the following types of nonlinear analysis:

- continuous nonlinear dynamical systems,
- discrete-time nonlinear dynamical systems,
- complex continuous nonlinear dynamical systems,
- constrained continuous nonlinear dynamical systems,
- two-time-scale nonlinear dynamical systems,
- interconnected nonlinear dynamical systems,
- non-hyperbolic nonlinear dynamical systems.

From computational viewpoints, effective schemes for estimating the stability regions of the above different types of nonlinear dynamical systems are described and analyzed. In addition, a constructive methodology for optimally estimating stability regions will be presented and analyzed.

From a practical viewpoint, knowledge of stability regions has significant applications. This book describes practical applications of stability regions to the following areas:

- direct stability analysis in electric power systems,
- stability-region-based nonlinear optimization methods.

We believe that solving challenging practical problems efficiently can be accomplished through a thorough understanding of the underlying theory, in conjunction with exploring the special features of the practical problem under study to develop effective solution methodologies. This book covers both comprehensive theoretical developments of stability regions and comprehensive solution methodologies for estimating stability regions.

There are 22 chapters contained in this book. These chapters can be classified as shown in Figure 1.8.

In summary, this book presents theoretical developments of stability regions of various types of nonlinear dynamical systems as well as solution methodologies for estimating these stability regions. The interplay between stability regions and bifurcations is also described. On the application side of the knowledge of stability regions, the long-standing research and development of fast stability-region-based methods for power system transient stability analysis and the emerging area of stability-region-based nonlinear optimization methods are presented in detail. In particular, this book

- develops a comprehensive theory of stability regions of nonlinear dynamical systems (including continuous and discrete systems),
- presents a complete characterization of stability boundaries of a class of nonlinear dynamical systems,
- presents a complete characterization of stability boundaries of a class of complex nonlinear dynamical systems,

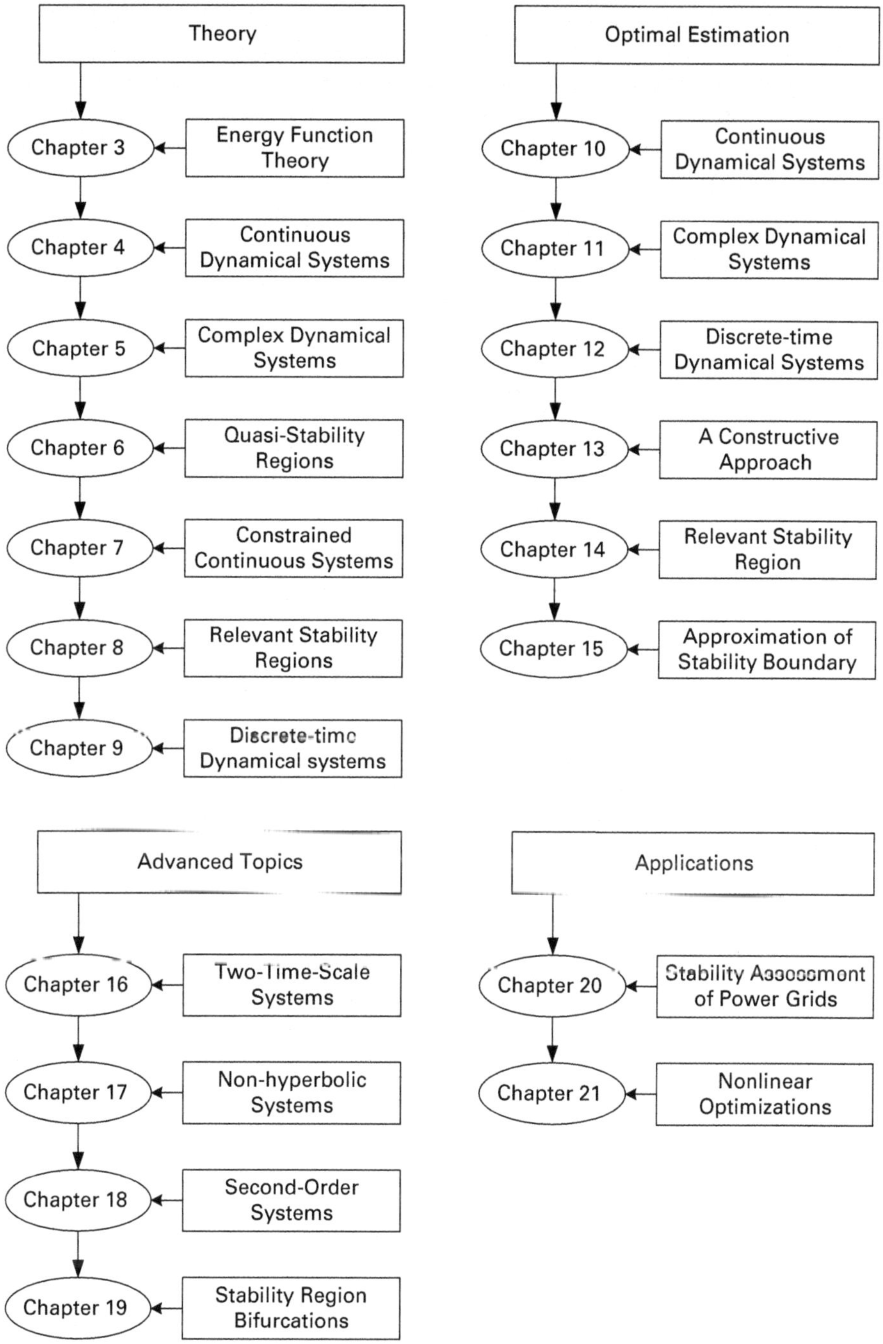

Figure 1.8 An overview of the organization and content of this book.

- develops a comprehensive theory of stability regions of two-time-scale nonlinear dynamical systems,
- develops a comprehensive theory of stability regions of a class of non-hyperbolic nonlinear dynamical systems,
- develops schemes for estimating stability regions of general nonlinear dynamical systems,
- presents optimal schemes for estimating stability regions of a class of nonlinear dynamical systems,
- presents a complete characterization of relevant stability regions of a class of nonlinear dynamical systems,
- presents optimal schemes for estimating relevant stability regions of a class of nonlinear dynamical systems.

It is expected that the reader has some familiarity with the theory of ordinary differential equations and nonlinear dynamical systems. This book can be read sequentially, chapter to chapter, or the reader may jump to chapters of interest. The minimum suggested path would be Chapters 1, 2, 3, 4, 6, 8, 10, 13, 18 and 22. This sequence will give to the reader a very comprehensive understanding of the theory of stability regions for continuous dynamical systems and the methods for estimating stability regions. Readers may want to include Chapters 5 and 11 to gain some insight into the characterization of the stability boundaries of a more general class of continuous dynamical systems, i.e. systems that exhibit complex behavior such as closed orbits and chaos.

Chapters 9 and 12 present a complete characterization of stability boundaries and methods to estimate the stability region of nonlinear discrete-time systems. They can be read separately but the contents of Chapters 4 and 10 would be useful background. Chapter 7 extends the characterization of stability boundaries developed in Chapter 4 to the class of nonlinear constrained systems, i.e. systems that are modeled by a set of differential and algebraic equations. It is suggested that Chapter 4 should be read before Chapter 7. The advanced topic chapters require some working knowledge of the theory chapters and of the optimal estimation chapters. The same is true for the application chapters.

Part I

Theory

2 Stability, limit sets, and stability regions

The study of stability region theory requires some knowledge of stability theory and dynamical systems. In this chapter, we review relevant concepts of stability and dynamical systems theory. The chapter is not intended to present a comprehensive treatment of stability and dynamical systems. Some of these concepts, which are relevant for the development in this book, and their implications will be discussed in detail. Other concepts, which are needed to prove some of the analytical results presented in this book, will be introduced and referenced in the literature when appropriate. More advanced readers can skip this chapter. A more comprehensive treatment of the contents of this chapter can be found, for example, in the books [122,193] on topology, [220] on analysis, [24,117] on stability and [107,216,276] on dynamical systems.

2.1 Mathematical preliminaries

We shall require some definitions and notation from set theory. Most of our problems are posed in ***n*-dimensional Euclidean space**, R^n; in particular $R^1 = R$ is just the set of real numbers or the real line, R^2 is the (Euclidean) plane, and R^3 is the usual (Euclidean) space. Points in R^n are printed in lower case type x, y, etc. and we will sometimes use the coordinate form $x = (x_1, \ldots, x_d)$. If x and y are points of R^n, the (Euclidean) **distance** between them is $|x - y| = \left(\sum_{i=1}^{d} |x_i - y_i|^2\right)^{1/2}$. A **vector space**, such as R^n, embedded with a distance is called a **metric space**.

Sets, which will generally be subsets of R^n, are denoted by capital letters (e.g., E, F, K, etc.). In the usual way, $x \in E$ means that the point x is a member of the set E, and $E \subset F$ means that E is a subset of F. We write $\{x$: condition$\}$ for the set of x for which the "condition" is true. The **empty set**, which contains no elements, is written $\varnothing$. We sometimes use a superscript $^+$ to denote the positive elements of a set (e.g., R^+ is the set of positive real numbers).

The **closed ball** of center x and radius r is defined by $\{y: |y - x| \leq r\}$. Similarly, the **open ball** is $B_r(x) = \{y: |y - x| < r\}$. Thus the closed ball contains its bounding sphere, but the open ball does not. Of course, in R^2 a ball is a disk, and in R^1 a ball is just an interval. If $a < b$, we write $[a, b]$ for the **closed interval** $\{x: a \leq x \leq b\}$ and (a, b) for the **open interval** $\{x: a < x < b\}$.

We write $E \cup F$ for the **union** of the sets E and F (i.e., the set of points belonging to either E or F). Similarly, we write $E \cap F$ for their **intersection** (i.e., the points in both E and F). More generally $\cup_i E_i$ denotes the **union** of an arbitrary collection of sets $\{E_i\}$ (i.e., those points in at least one E_i) and $\cap_i E_i$ denotes their **intersection**, consisting of the points common to all of the sets E_i. A collection of sets is **disjoint** if the intersection of any pair is the empty set. The **difference** $E\backslash F$ or $E-F$ consists of those points in E that are not in F, and $R^n\backslash E$ is called the **complement** of E.

If E is any set of real numbers, the **supremum**, sup E, is the least number m such that $x \leq m$ for every x in E. Similarly, the **infimum**, inf E, is the greatest number m such that $m \leq x$ for every x in E. Roughly speaking, we think of inf E and sup E as the minimum and maximum of the numbers in E, though it should be emphasized that inf E and sup E need not themselves be in E.

The **diameter**, diam E, of a subset E of R^n is the greatest distance apart of pairs of points in E; thus diam $E = \sup\{|x-y|: x, y \in E\}$. A set A is **bounded** if it has finite diameter or, equivalently, if it is contained in some sufficiently large ball.

We have already used the terms "open" and "closed" in connection with intervals and balls, but these notions extend to much more general sets. Intuitively, a set is closed if it contains its boundary and open if it contains none of its boundary points. More precisely, a subset E of R^n is **open** if, for every x in E, there is some ball $B_r(x)$ of positive radius r, centered at x and contained in E. A set E is **closed** if its complement is open; equivalently E is closed if every sequence x_j in E that is convergent in R^n, say to a point x of R^n, converges in E, i.e. x in E. In other words, every **accumulation point** or **limit point** of E is a point of E. A point p is a limit point of the set E if every neighborhood of p contains a point $q \neq p$ such that $q \in E$. The empty set $\emptyset$ and R^n are regarded as both open and closed. The union of any collection of open sets is open, as is the intersection of a *finite* collection of open sets. The intersection of any collection of closed sets is closed, as is the union of a *finite* number of closed sets.

The smallest closed set containing a set E, more precisely, the intersection of all closed sets that contain E, is called the **closure** of E. Equivalently, the closure of a set is the union of the set with all its accumulation points. Similarly, the **interior** of a set E is the largest open set contained in E, that is the union of all open subsets of E. The **boundary** of E is defined as the set of points in the closure of E but not in its interior. The boundary of a set E will be denoted by ∂E and

$$\partial E = \overline{E} \cap \overline{(R^n - E)}.$$

A subset of R^n is **compact** if it is closed and bounded. For other spaces, such as infinite dimension spaces, this characterization of compact sets as closed and bounded sets is not valid, see [220] for a comprehensive discussion about compact sets. A set E is thought of as connected if it consists of just one "piece"; formally E is **connected** if there do not exist open sets U and V such that $U \cup V$ contains E and with $E \cap U$ and $E \cap V$ disjoint and non-empty. A subset E of R^2 is termed **simply connected** if both E and $R^2\backslash E$ are connected.

A function $f: R^n \rightarrow R^m$ is of **class** C^r, or simply a C^r-function, if its derivatives up to order r exist and are continuous. $Df(\cdot)$ denotes the derivative of function f and $D^r f(\cdot)$ denotes the derivative of order r. A function is of **class** C^∞ if it is of class C^r for all $r > 0$. Let U and V be two open sets in R^n. A C^r-function $f: U \rightarrow V$ is a **diffeomorphism** of class C^r if the inverse of f exists, i.e. there exists a C^r-function $g: V \rightarrow U$, such that $g^o f$ is the identity function in U. A **homeomorphism** is a diffeomorphism of class C^0. The preimage or inverse image of a set B under function f will be denoted $f^{-1}(B)$ and is defined by $f^{-1}(B) = \{x{:}f(x) \in B\}$.

Occasionally we need to indicate the degree of smoothness of a curve or surface. We say that such a set is C^k ($k = 1, 2, \ldots$) if it can be locally defined, with respect to suitable coordinate axes, by a function that is k times differentiable with a continuous kth derivative. A curve or surface is C^∞ if it is C^k for every positive integer k. Objects in a metric space X that can be locally defined by a continuous and invertible function $h{:}R^n \rightarrow X$, whose inverse is also continuous, of coordinates in a Euclidean space R^n in X are called manifolds of dimension n. Examples of manifolds are smooth curves, spheres, ellipsoids and smooth surfaces.

2.2 Autonomous nonlinear dynamical systems

We consider the following (autonomous) nonlinear dynamical system

$$\dot{x} = f(x), \tag{2.1}$$

where $x \in R^n$ is a vector of state variables. It is natural to assume the function (i.e. the vector field) $f: R^n \rightarrow R^n$ satisfies a sufficient condition for the existence and uniqueness of a solution. A sufficient condition for this is to assume that f is a C^r-function, with $r \geq 1$, i.e. a function which is r times differentiable and has continuous derivatives. This condition guarantees, for each initial condition x_0, the existence of a **maximal interval of existence** $I = (\omega_-, \omega_+) \subset R$, containing the origin, and a unique continuously differentiable function $x(t){:}I \rightarrow R^n$, which is a solution of the differential equation (2.1), satisfying the initial condition $x(0) = x_0$.

The theory of the existence and uniqueness of solutions of differential equations is beyond the scope of this book, nevertheless the following result, which is an important consequence of this theory, will be important for the developments in this book.

THEOREM 2-1 **(Maximal interval of existence)**
Let $x(t)$ be a solution of (2.1) and let $[0, \omega_+]$ be the maximal interval of existence (to the right) of this solution. If there exists a compact set $K \subset R^n$ such that $x(t) \in K$ for all $t \in [0, \omega_+]$, then $\omega_+ = \infty$, that is, the solution exists and is defined for all $t \geq 0$.

Theorem 2-1 establishes that bounded solutions of ordinary differential equations have to be defined for all positive times. It is obvious that trajectories of physical systems are defined for all positive times, however, our mathematical models attempt to describe (or capture) behaviors of these physical systems. Consequently, it is necessary to pay some attention to the interval of existence of solutions of our models.

The solution curve of (2.1) starting from x_0 at $t = 0$ is called the *system trajectory* starting from x_0 and is denoted by $\phi(\cdot, x_0)$. The system trajectory starting from x_0 is a function of time; given a specified time, the system trajectory function maps the specified time into a point in the state space. The parametrization $t \rightarrow \phi(t, x_0)$ generates a curve in R^n, which is called the **orbit** or **trajectory** of (2.1) passing through x_0. The entire orbit passing through x_0 will be denoted by $\phi_t(x_0)$ and is defined as $\phi_t(x_0) = \{\phi(t, x_0) \in R^n : t \in R\}$. In some cases, we will be interested not only in the trajectory passing through a single point but also in the collection of trajectories passing through a set of initial conditions. In this case, if A is a subset of R^n, then $\phi(t, A)$ denotes the set $\{\phi(t, x) \in R^n : x \in A\}$.

The point $\bar{x} \in R^n$ is said to be an **equilibrium point** of (2.1) if $f(\bar{x}) = 0$, i.e. the equilibrium point is a particular type of solution that does not change in time. Hence, equilibrium points are degenerated trajectories that do not move. The set of all equilibrium points of the nonlinear dynamical system (2.1) will be denoted by E, where $E = \{x \in R^n : f(x) = 0\}$.

Another important type of trajectory is the closed orbit. By a **closed orbit** of a dynamical system (2.1) we mean the image of a nonconstant periodic solution of (2.1), i.e. a trajectory γ is a closed orbit if γ is not an equilibrium point and, for any $x \in \gamma$, there exists a time $T > 0$ such that $\phi(T, x) = x$. Equilibrium points and closed orbits can be stable or unstable. The formal definition of stability properties of equilibrium points and closed orbits will be presented in Section 2.3.

The concept of invariance is important to dynamical system theory. A set $M \in R^n$ is called an **invariant set** of (2.1) if every trajectory of (2.1) starting in M remains in M for all t. An orbit is an example of an invariant set, which has the equilibrium points and closed orbits as particular cases. The union and intersection of invariant sets are naturally invariant and the closure of an invariant set is also invariant.

A set $M \in R^n$ is called a *positively (negatively) invariant set* of (2.1) if every trajectory of (2.1) starting in M remains in M for all $t \geq 0$ ($t \leq 0$). Solutions that enter a positively invariant set cannot leave them at future times. Positively invariant sets play an important role in the theory of stability and in estimating stability regions of nonlinear dynamical systems.

We next discuss the asymptotic behavior of solutions of system (2.1). More precisely, we will be interested in the behavior of the system when time tends to infinity. The concepts of nonwandering points and limit sets are relevant for the theory of nonlinear dynamical systems, capturing the asymptotic behaviors of their trajectories.

A point p is said to be *nonwandering* for the flow $\phi(t, \cdot)$ if, for any neighborhood U of p, there exists an arbitrarily large time T such that $\phi(T, U) \cap U \neq \emptyset$. The **nonwandering set** Ω is the collection of all nonwandering points. Equilibrium points and closed orbits are clearly contained in Ω. A point that is not nonwandering is called a **wandering point** for the flow $\phi(t, \cdot)$. The set of wandering points is open while Ω is an invariant closed set.

A point p is said to be in the **ω-limit set** of x if corresponding to each $\varepsilon > 0$ and $T > 0$ there is a $t > T$ with the property that $|\phi(t, x) - p| < \varepsilon$. This is equivalent to saying that there is a sequence $\{t_i\}$ in R, $t_i \rightarrow \infty$, with the property that $p = \lim_{i \rightarrow \infty} \phi(t_i, x)$. A point p is said to be in the **α-limit set** of x if corresponding to each $\varepsilon > 0$ and $T < 0$, there is a $t < T$ with the property that $|\phi(t, x) - p| < \varepsilon$. This is equivalent to saying that there is a sequence $\{t_i\}$ in R, $t_i \to -\infty$, with the property that $p = \lim_{i \to \infty} \phi(t_i, x)$. Hence, the ω-limit set captures the asymptotic behavior of trajectories in positive time while the α-limit set captures the asymptotic behavior of trajectories in negative time. The ω-limit set and the α-limit set of x will be respectively denoted by $\omega(x)$ and $\alpha(x)$.

One fundamental property of the limit sets is described in the next theorem.

THEOREM 2-2 **(Properties of limit sets)**
The ω-limit set and the α-limit set of a trajectory $\phi(t, x)$ of system (2.1) are closed and invariant. In addition, if a trajectory $\phi(t, x)$ of system (2.1) is bounded for $t \geq 0$ (or $t \leq 0$), then its ω-limit set (or α-limit set) is non-empty, compact and connected, moreover $d(\phi(t, x), \omega(x)) \to 0$ as $t \to \infty$.

Asymptotically stable equilibrium points are the simplest type of limit set. However, limit sets can be very complex; they can be equilibrium points, limit cycles (closed orbits), quasi-periodic solutions and chaos. These complex limit sets will be the subject of discussion in Chapter 5.

2.3 Stability

Stability is a very broad subject, and the concept of stability can be formulated in a variety of different ways, depending on the intended use of stability analysis and design. The stability property of an equilibrium point will be discussed next. We first review the concepts of (Lyapunov) stability and asymptotic stability.

DEFINITION **(Lyapunov stability)**
*An equilibrium point $\bar{x} \in R^n$ of (2.1) is said to be **(Lyapunov) stable** if for each open neighborhood U of $\bar{x} \in R^n$, there exists an open neighborhood V of $\bar{x} \in R^n$ such that $\phi(t, x) \in U$ for all $x \in V$ and for all $t > 0$. Otherwise, $\bar{x}$ is unstable.*

The concept of (Lyapunov) stability is illustrated in Figure 2.1. Intuitively speaking, an equilibrium point is stable if nearby trajectories stay nearby. In many applications, the requirement of "nearby trajectories stay nearby" is not sufficient; instead, the requirement becomes "nearby trajectories stay nearby and all converge to the equilibrium

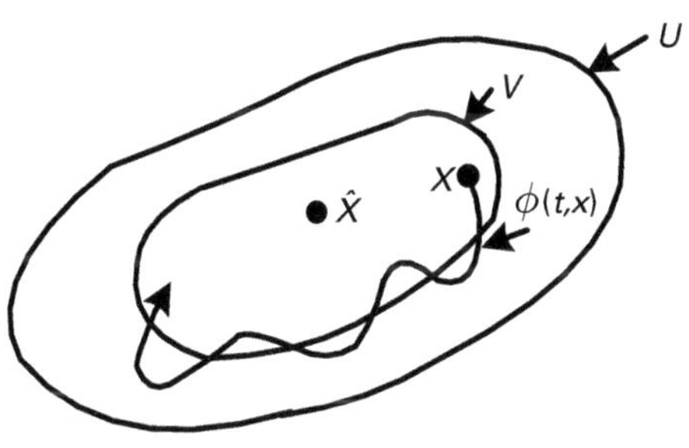

Figure 2.1 An illustration of the definition of (Lyapunov) stability.

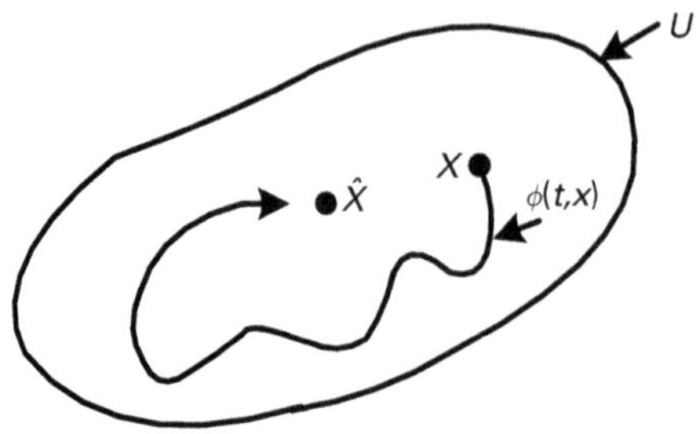

Figure 2.2 An illustration of the definition of asymptotic stability.

point." In this situation, the concept of (Lyapunov) stability can be sharpened into the concept of asymptotic stability as defined in the following.

DEFINITION **(Asymptotic stability)**
An equilibrium point $\bar{x} \in R^n$ of (2.1) is said to be ***asymptotically stable*** *if it is stable and there exists an open neighborhood U of $\bar{x} \in R^n$, such that every trajectory $\phi(t, x)$ starting from this neighborhood U converges to the equilibrium point $\bar{x}$ as $t \to \infty$ or, equivalently, $\lim_{t\to\infty} \|\phi(t,x) - \bar{x}\| = 0$ for every $x \in U$.*

The concept of asymptotic stability is illustrated in Figure 2.2. Intuitively speaking, an equilibrium point is asymptotically stable if it is the sink of nearby trajectories. An asymptotically stable equilibrium point is also termed a sink in the classical literature of nonlinear dynamical systems. Sometimes, it is desired to make the neighborhood U of the asymptotic stability the entire state space so that every trajectory converges to the equilibrium point as time tends to infinity. In these cases, the concept of global asymptotic stability is relevant. Global stability is rare in nonlinear dynamical systems. Actually this common feature of nonlinear systems makes the concept of stability region relevant.

DEFINITION **(Global asymptotic stability)**
An equilibrium point $\bar{x} \in R^n$ of (2.1) is said to be ***globally asymptotically stable*** *if it is stable and for all $x_0 \in R^n$, $\phi(t,x) \to \bar{x}$ as $t \to \infty$.*

The concepts of stability of equilibrium points can be easily extended to general closed invariant sets, with the closed orbits as a particular case. Stable invariant sets are very important in experimental and numerical settings because they are the only kind of limit sets that can be observed naturally! The concept of stable invariant set is similar to that of stable equilibrium point.

DEFINITION **(Stable invariant set)**
A closed and invariant set γ is said to be ***Lyapunov stable*** *if for each open neighborhood U of γ, there exists an open neighborhood V of γ such that $\phi(t, x) \in U$ for all $x \in V$ and for all $t > 0$. Otherwise, γ is unstable.*

DEFINITION **(Asymptotically stable invariant set)**
A closed and invariant set γ is ***asymptotically stable*** *if it is stable and there exists an open neighborhood V of γ such that the ω-limit set of every point in V is contained in γ.*

In this book we will also explore the concept of an **attracting set**, which is weaker than the concept of asymptotically stable invariant set.

DEFINITION **(Attracting set)**
A closed invariant set $\gamma \subset R^n$ *is called an* ***attracting set*** *if there exists some open neighborhood* U *of* γ *such that, for all* $x_0 \in U$, $\phi(t, x_0) \in U$ *for all* $t \geq 0$ *and* $\phi(t, x_0) \to \gamma$ *as* $t \to \infty$.

Actually, a stable and attracting invariant set is an asymptotically stable invariant set. Roughly speaking, a set γ is an attracting set if orbits nearby stay nearby and converge to γ as $t \to \infty$. A more detailed discussion of attracting sets and attractors will be given in Chapter 5.

2.4 Hyperbolicity and invariant manifolds

In order to determine the stability of an equilibrium point $\bar{x}$ we need to understand the nature of trajectories near $\bar{x}$. We need tools to describe, at least qualitatively, the dynamical behavior of orbits near the equilibrium point. For linear systems, in the form $\dot{x} = Ax$, the dynamical behavior of orbits can be completely determined by calculating the eigenvalues and eigenvectors of matrix A. In this section we will see that locally, under some conditions, the nonlinear system dynamics are very similar, from the qualitative point of view, to the dynamics of an associated linear system. Consequently, the local dynamic behavior of the nonlinear system can be studied by means of an eigenanalysis of the associated linear system.

With that in mind, let $\bar{x}$ be an equilibrium point of the nonlinear dynamical system (2.1) and consider the following change of variables:

$$x(t) = \bar{x} + y(t). \tag{2.2}$$

Substituting Eq. (2.2) into Eq. (2.1) and Taylor expanding about $\bar{x}$ gives

$$\dot{x}(t) = \dot{y}(t) = f(\bar{x}) + Df(\bar{x})y + O(\|y\|^2) \tag{2.3}$$

where Df is the derivative of f and $\|.\|$ denotes a norm on R^n. Using the fact that $0 = f(\bar{x})$, Eq. (2.3) becomes

$$\dot{y} = Df(\bar{x})y + O(\|y\|^2). \tag{2.4}$$

Equation (2.4) describes the evolution of trajectories near the equilibrium $\bar{x}$. For stability analysis, we are concerned with the behavior of trajectories arbitrarily close to $\bar{x}$, so it is reasonable that the stability analysis can be performed by studying the associated linear system

$$\dot{y}(t) = Df(\bar{x})y. \tag{2.5}$$

An equilibrium point $\bar{x}$ of the nonlinear system (2.1) is **hyperbolic** if the corresponding Jacobian matrix $Df(\bar{x})$ has no eigenvalues with zero real part; otherwise, it is a nonhyperbolic equilibrium point.

Hyperbolic equilibrium points have some nice properties. Some of them will be studied in this section. As a direct consequence of the inverse function theorem, hyperbolic equilibrium points are isolated, i.e. there exists a neighborhood (an open set) of the equilibrium that does not contain any other equilibrium point. We define the type of hyperbolic equilibrium point according to the number of eigenvalues of the corresponding Jacobian matrix with a positive real part. More precisely, a hyperbolic equilibrium point $\bar{x}$ of system (2.1) is of **type-*k*** if k eigenvalues of the Jacobian matrix $Df(\bar{x})$ have positive real part and $n-k$ have negative real part. If $Df(\bar{x})$ has exactly one eigenvalue with a positive real part, we call *it* a *type-one equilibrium point.* It will be shown in later chapters that type-one equilibrium points play a key role in the characterization of stability boundaries and quasi-stability boundaries. Hence, type-one equilibrium points and stable equilibrium points are important equilibrium points in our study of stability regions.

Let λ be an eigenvalue of $Df(\bar{x})$. We denote E_λ the generalized eigenspace associated with the eigenvalue λ. Remember that E_λ is a vector subspace of R^n, which is invariant with respect to the linear system (2.5). If the origin of the linear system is a hyperbolic equilibrium point, then we can decompose the space R^n as a direct sum of two invariant subspaces $R^n = E^s \oplus E^u$, respectively called stable and unstable eigenspaces of $Df(\bar{x})$, where $E^s = \oplus_{Re(\lambda) > 0} E_\lambda$ and $E^s = \oplus_{Re(\lambda) < 0} E_\lambda$.

If the hyperbolic equilibrium is of type k, then the eigenspaces E^s and E^u respectively possess dimensions $n-k$ and k. Trajectories of the linear system (2.5) in E^s approach the origin exponentially when $t \to \infty$ while those in E^u approach the origin exponentially when $t \to -\infty$.

It can be shown that if the eigenvalues of the associated linear system have nonzero real part, then the orbit (i.e. trajectory) structure *near an equilibrium point* of the nonlinear vector field (2.1) is essentially the same as that of the linear vector field (2.5). This is addressed by the *Hartman–Grobman theorem.*

THEOREM 2-3 **(Hartman–Grobman theorem)**
Consider the system (2.1) with an equilibrium point $\bar{x}$. If $Df(\bar{x})$ has no zero and no purely imaginary eigenvalues, then there is a homeomorphism h, defined on a neighborhood U of $\bar{x}$, taking orbits of the flow $\phi(t,.)$ of the nonlinear system (2.1) to those of the linear flow $e^{tDf(\bar{x})}$ of (2.5). The homeomorphism preserves the sense of the orbits and is chosen to preserve parameterization by time.

The Hartman and Grobman theorem establishes a continuous one-to-one correspondence between orbits of the nonlinear system and those of the associated linear system in the neighborhood of a hyperbolic equilibrium point. The proof of this theorem is beyond the scope of this book. An elegant proof is given in [209]. Note that the Hartman–Grobman theorem can be generalized to arbitrary invariant manifolds (rather than merely equilibrium points), including closed orbits.

The stability of the linear system (2.5) can be easily checked by computing the eigenvalues of the Jacobian matrix $Df(\overline{x})$. A corollary of Theorem 2-3 and fundamental linear system theory lead to the following sufficient condition for an equilibrium point of the nonlinear system to be asymptotically stable.

THEOREM 2-4 **(Asymptotic stability)**
Suppose all of the eigenvalues of $Df(\overline{x})$ of the associated linear system (2.5) have negative real parts. Then the equilibrium solution $x = \overline{x}$ of the nonlinear system (2.1) is asymptotically stable.

This is a well-known stability evaluation method that checks the eigenvalues of the Jacobian matrix at an equilibrium point. The computational effort required to compute these eigenvalues can be immense for high-dimensional systems. Moreover, there are non-hyperbolic equilibrium points whose stability property cannot be derived from Theorem 2-4. An alternative yet popular approach to checking the stability property of an equilibrium point is based on the Lyapunov function theory, which is discussed in the next section.

The Hartman and Grobman theorem also suggests that the decomposition of the space of the linear system into a stable and an unstable eigenspace must be true somehow for the nonlinear system in the neighborhood of the hyperbolic equilibrium point. We will see that the stable and the unstable manifolds will play the role of these invariant eigenspaces for the nonlinear case. This is exactly what establishes the stable manifold theorem.

Let $\hat{x}$ be a hyperbolic equilibrium point and $U \subset R^n$ be a neighborhood of $\hat{x}$. We define the local stable manifold of $\hat{x}$ as follows:

$$W^s_{loc}(\hat{x}) := \{x \in U : \phi(t,x) \rightarrow \hat{x} \quad \text{as} \quad t \rightarrow \infty\}.$$

The local unstable manifold of $\overline{x}$ is defined as

$$W^u_{loc}(\hat{x}) := \{x \in U : \phi(t,x) \rightarrow \hat{x} \quad \text{as} \quad t \rightarrow -\infty\}.$$

Note that $W^s_{loc}(\hat{x})$ is a positive-invariant set while $W^u_{loc}(\hat{x})$ is a negative-invariant set. They may not be manifolds when $\hat{x}$ is non-hyperbolic. Figure 2.3 illustrates these local manifolds.

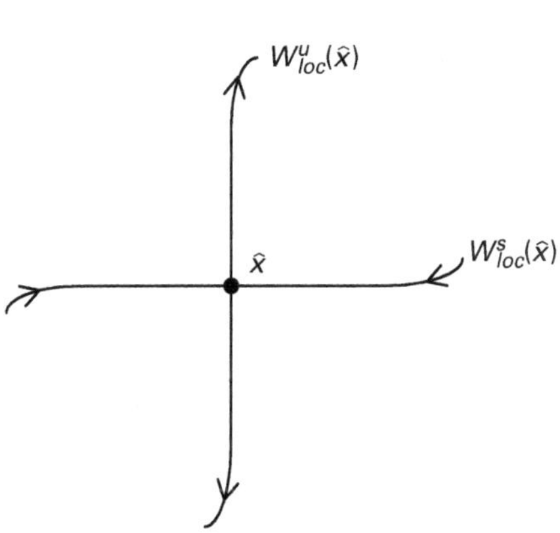

Figure 2.3 The local stable and unstable manifolds of an equilibrium point.

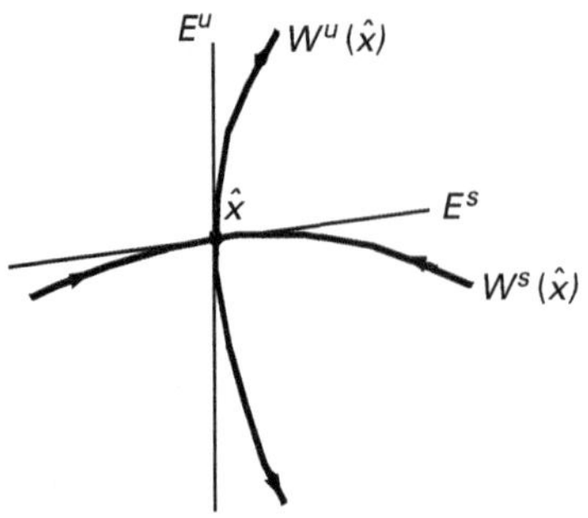

Figure 2.4 The relationship between a stable eigen-splicing, an unstable eigen-splicing, and stable and unstable manifolds at a hyperbolic equilibrium point.

THEOREM 2-5 **(Unstable–stable manifold theorem)**
Suppose that the nonlinear system (2.1) has a hyperbolic equilibrium point $\bar{x}$. Then the sets $W^s_{loc}(\bar{x})$ and $W^u_{loc}(\bar{x})$, often referred to as the local stable and unstable manifolds, are manifolds of the same dimensions n_s, n_u as those of the stable and unstable eigenspaces E^s, E^u of the linearized system (2.5). These manifolds are also tangent to E^s, E^u at $\bar{x}$.$W^s_{loc}(\bar{x})$ and $W^u_{loc}(\bar{x})$ are as smooth as the vector field f(x) of (2.1). They are invariant with respect to the nonlinear system (2.1) and trajectories of the nonlinear system in these manifolds have the same asymptotic properties as solutions of the linear system in their corresponding eigenspaces.

Figure 2.4 illustrates the manifold theorem.

The stable manifold $W^s(\bar{x})$ and the unstable manifold $W^u(\bar{x})$ are obtained by letting points in $W^s_{loc}(\bar{x})$ flow backwards in time and points in $W^u_{loc}(\bar{x})$ flow forwards in time:

$$W^s(\bar{x}) := \bigcup_{t\leq 0} \phi(W^s_{loc}(\bar{x}), t) \tag{2.6}$$

$$W^u(\bar{x}) := \bigcup_{t\geq 0} \phi(W^u_{loc}(\bar{x}), t). \tag{2.7}$$

Remarks

[1] Clearly, $\hat{x}$ is the ω-limit set of every point in $W^s(\hat{x})$, as well as the α-limit set of every point in $W^u(\hat{x})$. For a hyperbolic equilibrium point, the dimension of the stable manifold $W^s(\hat{x})$ equals the number of eigenvalues of the Jacobian $Df(\hat{x})$ with negative real part.

[2] For the flow $\phi(t, \cdot)$, since the stable manifold of a critical element coincides with the unstable manifold of the critical element for the flow $\phi(-t, \cdot)$, this dual property enables us to translate each property of stable manifolds into that of unstable manifolds.

[3] The existence and uniqueness of the solutions ensure that neither $W^s(\bar{x})$ nor $W^u(\bar{x})$ can intersect themselves. However, $W^s(\bar{x})$ and $W^u(\bar{x})$ can intersect with each other. The intersection is in fact a source of complex behaviors occurring in dynamical systems.

[4] Stable and unstable manifolds are invariant sets. Every trajectory in the stable manifold $W^s(\hat{x})$ converges to $\hat{x}$ as time goes to positive infinity, while every trajectory in the unstable manifold $W^u(\hat{x})$ converges to $\hat{x}$ as time goes to negative infinity.

[5] It is generally very difficult, if not impossible, to derive closed form expressions for the stable and unstable manifolds of equilibrium points.

[6] The concepts and definitions of stable and unstable manifolds are applicable to other classes of hyperbolic limit sets such as periodic solutions, quasi-periodic solutions and chaos.

Example 2-1 To illustrate the stable and unstable manifolds numerically, we consider the following unforced Duffing oscillator:

$$\begin{cases} \dot{x} = y \\ \dot{y} = x - x^3 - \varepsilon y, \quad \varepsilon > 0. \end{cases}$$

It is noted that there are two stable equilibrium points (1,0) (–1,0). There is one type-one equilibrium point, (0,0), that has a one-dimensional stable manifold and a one-dimensional unstable manifold (see Figure 2.5).

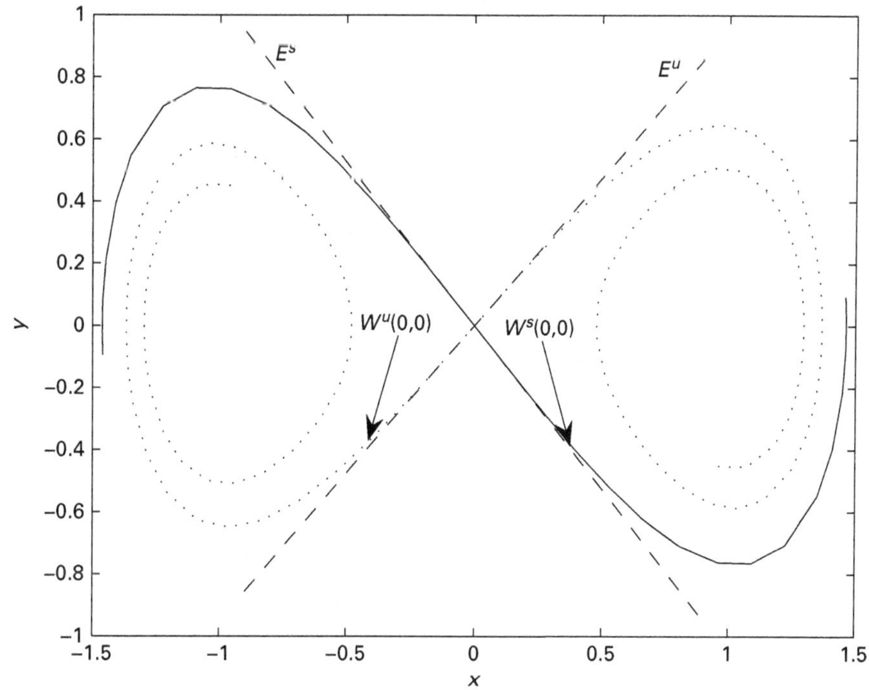

Figure 2.5 The relationship between stable and unstable manifolds and eigenspaces.

The associated linearized system is given by

$$\begin{cases} \dot{x} = y \\ \dot{y} = x - \varepsilon y \end{cases}$$

with the following stable and unstable eigenspaces

$$E^s = \left\{ (x, y) : y = \left(-\frac{\varepsilon}{2} + \frac{\sqrt{\varepsilon^2 + 4}}{2} \right) x \right\} \quad \text{and} \quad E^u = \left\{ (x, y) : y = \left(-\frac{\varepsilon}{2} + \frac{\sqrt{\varepsilon^2 + 4}}{2} \right) x \right\}.$$

The local stable and unstable manifolds are tangent to the stable and unstable eigenspaces E^s, E^u at the type-one equilibrium point (0,0).

Example 2-2 We consider the following system:

$$\begin{cases} \dot{x} = x \\ \dot{y} = -y + x^2 \end{cases}$$

having one hyperbolic equilibrium point at $(x,y) = (0,0)$. The associated linearized system is

$$\begin{cases} \dot{x} = x \\ \dot{y} = -y \end{cases}$$

having eigenvalues of -1 and $+1$ with stable and unstable eigenspaces given by (see Figure 2.6)

$$E^s = \{((x, y) \in R^2) : x = 0\}$$
$$E^u = \{((x, y) \in R^2) : y = 0\}.$$

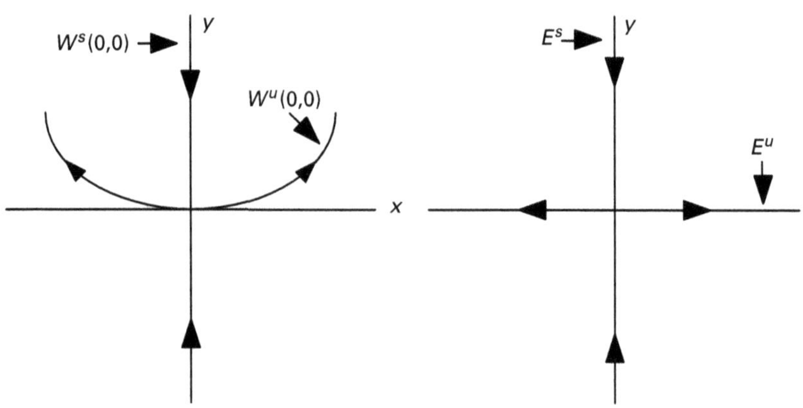

Figure 2.6 (a) The stable and unstable manifolds of (0.0,0.0) and (b) the eigenspace.

To derive $W^u_{loc}(0,0)$, we notice that the solution can be obtained explicitly as follows:

$$\frac{\dot{y}}{\dot{x}} = \frac{dy}{dx} = \frac{-y}{x} + x \quad \text{or} \quad y(x) = \frac{x^2}{3} + \frac{c}{x}.$$

Since $W^u_{loc}(0,0)$ can be represented by a graph over the x variables, i.e. $y = h(x)$ with $h(0) = h'(0) = 0$, it follows that

$$W^u_{loc}(0,0) = \left\{(x,y) \in R^2 : y = \frac{x^2}{3}\right\}.$$

We also note that all initial conditions (0,y), for all $y \in R$, make the solution stay on the y-axis and approach (0,0) at $t \to \infty$; hence, it follows that the unstable eigenspace E^u is tangent to W^u(0,0) at (0.0, 0.0) (see Figure 2.6).

2.5 Transversality

The idea of transversality is basic in the study of dynamical systems. If A and B are injectively immersed manifolds in M, we say they satisfy the **transversality condition** if either (i) at every point of intersection $x \in A \cap B$, the tangent spaces of A and B span the tangent space of M at x, i.e.

$$T_x(A) + T_x(B) = T_x(M) \quad \text{for} \quad x \subset A \cap B$$

or (ii) they do not intersect at all.

One of the most important features of a hyperbolic equilibrium point $\hat{x}$ is that its stable and unstable manifolds intersect transversely at $\hat{x}$. This transverse intersection is important because it persists under perturbation of the vector field.

2.6 Lyapunov function theory

One of the most important developments in stability theory is the Lyapunov function theory. Lyapunov, a Russian mathematician and engineer, laid down the foundations for Lyapunov theory. Lyapunov stability theorems give sufficient conditions for Lyapunov stability and asymptotic stability. The appealing feature of the Lyapunov function theory is that it derives the stability properties of the equilibrium point without solving the underlying ordinary differential (difference) equations. This is the "spirit of Lyapunov."

There are several versions of the proof of the Lyapunov function theorem. In this section, we present a fundamental Lyapunov function theorem and provide a rigorous proof. We denote the following notation as the time derivative of a function $V(x)$, taken along the system trajectory:

$$\dot{V}(x(t)) = \frac{\partial V(x(t))^T}{\partial x} \bullet \dot{x}(t)$$
$$= \frac{\partial V(x)^T}{\partial_x} \bullet f(x). \tag{2.8}$$

Since the vector field $f(x)$ and the gradient of the function $V(x)$ are available without explicit knowledge of the system trajectory, the time derivative of $V(x(t))$ can be determined without knowledge of the system trajectory.

THEOREM 2-6 **(Lyapunov stability)**

Let $\hat{x}$ be an equilibrium point of $\dot{x} = f(x)$, where $f{:}R^n \to R^n$. Let $V{:}\ U \to R$ be a continuous function defined on a neighborhood U of $\hat{x}$, differentiable on U, such that

(a) $V(\hat{x}) = 0$ *and* $V(x) > 0$ *if* $x \neq \hat{x}$, *and* $x \in U$,
(b) $\dot{V}(x) \leq 0$ *in* $U - \hat{x}$.
Then $\hat{x}$ *is stable. Furthermore, if also*
(c) $\dot{V}(x) < 0$ *in* $U - \hat{x}$, *then* $\hat{x}$ *is asymptotically stable.*

Proof Let $\delta > 0$ be small enough that the ball $B_\delta(\hat{x}) = \{x \in R^n : \|x - \hat{x}\| < \delta\}$ lies entirely in U (see Figure 2.2). Let $\partial B_\delta(\hat{x}) = \{x \in R^n : \|x - \hat{x}\| = \delta\}$. This set denotes the boundary of the ball $B_\delta(\hat{x})$. Let $\alpha = \min_{x \in \partial B_\delta(\hat{x})} V(x)$. Note that the minimum value α exists due to condition (a) and that the scalar $\alpha > 0$.

Next, we select another neighborhood (open set) to show that $\hat{x}$ is stable. Let $U_1 = \{x \in B_\delta(\hat{x}) : V(x) < \alpha\}$. Since the function $V(x(t))$ is non-increasing along any system trajectory $x(t)$ (by condition (b)), no trajectory starting in U_1 can leave the ball $B_\delta(\hat{x})$ and intersect with the set $\partial B_\delta(\hat{x})$. This is due to the fact that the function value $V(x)$ at any point of the set $\partial B_\delta(\hat{x})$ is greater than or equal to the scalar α. Hence, we have shown that every trajectory starting in U_1 never leaves $B_\delta(\hat{x})$. This proves that $\hat{x}$ is stable. Note that we have also shown that U_1 is a positive-invariant set.

Now, assume that condition (c) holds as well. Since the function $V(x(t))$ does not increase ($\dot{V} \leq 0$) along the system trajectory $x(t)$, and $x(t)$ stays inside the compact set $B_\delta(\hat{x})$ for $t \geq 0$, we conclude, from the continuity of V, that $V(x(t))$ is bounded from below for $t \geq 0$. It follows that $\lim_{t \to \infty} V(x(t))$ exists and equals a constant, say $\lim_{t \to \infty} V(x(t)) = b$.

We claim that $b = 0$ and thus $x(t) \to \hat{x}$ as $t \to \infty$.

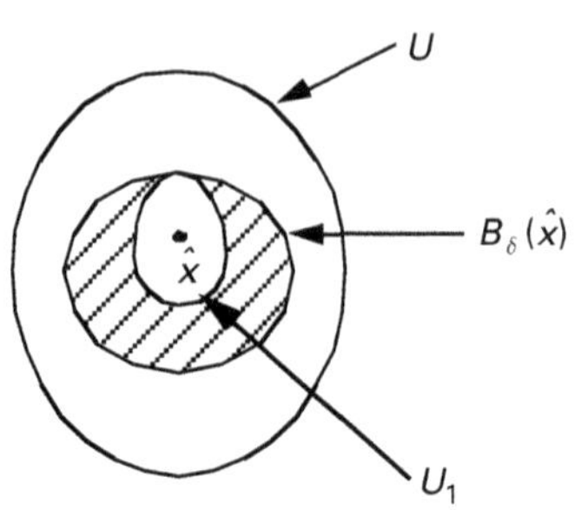

Figure 2.7 An illustration of the relationship between the open set and the closed ball.

Proof of the claim: suppose by contradicition that $b > 0$. Then, from condition (a), there exists a time $T > 0$ and a neighborhood $U \subset B_\delta(\hat{x})$ of $\hat{x}$ such that $x(t) \in B_\delta(\hat{x}) \setminus U$ for all $t > T$. By condition (c) we know that $\dot{V}(z) < -\varepsilon,\ \varepsilon > 0$, for all $z \in B_\delta(\hat{x}) \setminus U$. Then a contradiction is reached because

$$\int_0^\infty \dot{V}(x(t))dt < \int_T^\infty \dot{V}(x(t))dt < -\int_T^\infty \varepsilon dt$$

or

$$\int_0^\infty \dot{V}(x(t))dt = \lim_{t\to\infty} V(x(t)) - V(x_0) < -\varepsilon(\infty - T)$$

or

$$b - V(x_0) < -\infty.$$

The above equation is a contradiction because the left hand side of the equation is bounded while the right hand side of the equation is unbounded. Thus every trajectory in $B_\delta(\hat{x})$ converges to the equilibrium point and this concludes the proof of this theorem.

Remarks

[1] The Lyapunov function theory asserts not only the stability property of the equilibrium point (a local result) but also that there does not exist any limit cycle (oscillation behavior) or bounded complicated behavior such as an almost periodic trajectory, chaotic motion, etc. in the subset of the state space where a Lyapunov function exists.

[2] It should be pointed out that the Lyapunov function theory only furnishes sufficient conditions for stability. If for a particular Lyapunov function candidate V, the required conditions on the derivative of V, i.e. $\dot{V}$, are not met, then conclusions regarding the stability or instability of the equilibrium point cannot be drawn.

[3] Many Lyapunov functions may exist for the same nonlinear system. For instance, if V is a Lyapunov function for a system, so is $V_1 = pV^a$, where $p > 0$, $\alpha > 1$. Moreover, specific choices of Lyapunov functions may yield more precise results than others.

There is no systematic way of constructing Lyapunov functions for general nonlinear systems. This is a fundamental drawback of the Lyapunov direct method. Therefore, faced with specific nonlinear dynamical systems, one often has to use experience, intuition, trial and error, and physical insights (e.g. the energy function for electrical and mechanical systems) to search for an appropriate Lyapunov function. In the literature, a number of methods and techniques facilitating the search of Lyapunov functions have been proposed.

2.7 Stability regions

For an asymptotically stable equilibrium point $\hat{x}$, there exists a number $\delta > 0$ such that $\|x_0 - \hat{x}\| < \delta$ implies $\phi(t, x_0) \to \hat{x}$ as $(t \to \infty)$. If δ is arbitrarily large, then $\hat{x}$ is called a

global stable equilibrium point. There are many physical systems containing stable equilibrium points but not globally stable equilibrium points. In the rest of this book, asymptotically stable equilibrium point and stable equilibrium point will be used interchangeably.

The **stability region** of a stable equilibrium point x_s is the set of all points x such that

$$\lim_{t\to\infty}\phi(t,x)=x_s. \tag{2.9}$$

We will denote the stability region of x_s by $A(x_s)$, and its closure by $\overline{A}(x_s)$, respectively; hence

$$A(x_s):=\{x\in R^n:\lim_{t\to\infty}\phi(t,x)=x_s\}. \tag{2.10}$$

When it is clear from the context, we write A for $A(x_s)$, etc. Alternatively, the stability region can be expressed as

$$A(x_s)=\{x\in R^n:\omega(x)=x_s\} \tag{2.11}$$

where $\omega(x)$ denotes the ω-limit set of x.

The (topological) boundary of the stability region $A(x_s)$ is called the **stability boundary** (also called the **separatrix**) of x_s and will be denoted by $\partial A(x_s)$. Since the stability region of a stable equilibrium point is in fact the stable manifold, the topological properties of stability regions are such that the stability region $A(x_s)$ is an open, invariant set diffeomorphic to R^n. Hence, every trajectory in a stability region lies entirely in the stability region, and the dimension of the stability region is n. Since the boundary of an invariant set is also invariant and the boundary of an open set is a closed set, the stability boundary $\partial A(x_s)$ is a closed invariant set of dimension less than n. If $A(x_s)$ is not dense in R^n, then $\partial A(x_s)$ is of dimension $n-1$. As time increases, every trajectory in the stability region $A(x_s)$ converges to the stable equilibrium point x_s and every trajectory on the stability boundary remains on the boundary.

Both the stability region $A(x_s)$ and the stability boundary $\partial A(x_s)$ can be very complex. The degree of complexity can range from the simple situation that every trajectory on the stability boundary converges to one of the equilibrium points to complex situations. One example of a complex situation is that every trajectory either diverges or converges to one of the limit sets such as equilibrium points, limit cycles, quasi-periodic solutions and chaos.

2.8 Remarks

The determination of stability regions continues to play an important role in many emerging research areas of engineering and the sciences. In this book, a comprehensive theory for stability regions will be derived in Chapter 4 through Chapter 9 for the following classes of nonlinear systems:

- nonlinear continuous systems,
- nonlinear discrete-time systems,

- nonlinear constrained systems,
- nonlinear two-time-scale systems and
- nonlinear non-hyperbolic continuous systems.

Methods for estimating stability regions of the above classes of dynamical systems will be developed. In addition, these methods will be shown to be effective and yet scalable, i.e. they will be applied to effectively estimate the stability region of very large nonlinear systems. The topic of how to estimate stability regions for various high-dimensional nonlinear dynamical systems will be discussed in Chapter 10 through Chapter 18.

3 Energy function theory

Energy functions play an important role in characterizing stability regions and offer a practical and effective way to estimate stability regions. Energy functions are useful, for example, for global analysis of system trajectories and predicting the structure of the limit sets of trajectories. A comprehensive energy function theory for general nonlinear autonomous dynamical systems will be presented in this chapter. This energy function theory can be applied to a variety of nonlinear dynamical systems. Applications of energy functions to optimally estimate stability regions of large-scale nonlinear systems and their theoretical basis will be developed in later chapters.

3.1 Energy functions

We consider a general nonlinear autonomous dynamical system described by the following equation:

$$\dot{x}(t) = f(x(t)). \tag{3.1}$$

We say a C^r-function $V{:}R^n \rightarrow R$, with $r \geq 1$, is an **energy function** for the system (3.1) if the following three conditions are satisfied.

(E1) The derivative of the energy function $V(x)$ along any system trajectory $x(t)$ is non-positive, i.e.

$$\dot{V}(x(t)) \leq 0.$$

(E2) If $x(t)$ is a nontrivial trajectory (i.e. $x(t)$ is not an equilibrium point), then along the nontrivial trajectory $x(t)$, the set $\{t \in R : \dot{V}(x(t)) = 0\}$ has measure zero in R.

(E3) That a trajectory $x(t)$ has a bounded value of $V(x(t))$ for $t \in R^+$ implies that the trajectory $x(t)$ is also bounded for $t \in R^+$. Stating this in brief: if $V(x(t))$ is bounded then $x(t)$ itself is also bounded.

Property (E1) indicates that the energy function is non-increasing along its trajectory, but it alone does not imply that the energy function is strictly decreasing along its trajectory. There may exist a time interval $[t_1, t_2]$ such that $\dot{V}(x(t)) = 0$ for $t \in [t_1, t_2]$. However, properties (E1) and (E2) together imply that the energy function is strictly decreasing along any system trajectory. Property (E3) states that the energy function is a proper map along any system trajectory, but it need not be a proper map for the entire state space.

Recall that a proper map is a function $f{:}X \rightarrow Y$ such that for each compact set $D \in Y$, the set $f^{-1}(D)$ is compact in X. Property (E3), which can be viewed as a "dynamic" proper map, is useful in the characterization of a stability boundary. From the above definition of energy functions, it is obvious that an energy function may not be a Lyapunov function and a Lyapunov function may not be an energy function.

Example 3-1 Consider the class of gradient systems

$$\dot{x} = -\nabla V(x) \tag{3.2}$$

where $V{:}R^n \rightarrow R$ is a scalar, proper C^1-function. We will show that V is an energy function for system (3.2). Differentiating $V(x(t))$, one obtains:

$$\dot{V}(x) = \langle \nabla V(x), f(x) \rangle = -\|\nabla V(x)\|^2 \leq 0$$

and therefore condition (E1) holds. Moreover $\dot{V}(x) = 0$ if and only if x is an equilibrium point. Thus (E2) is also satisfied. The condition that $V(x)$ is proper is a sufficient condition for the satisfaction of (E3). Consequently, $V(x)$ is an energy function for system (3.2).

Example 3-2 We consider the following classical model for transient stability analyses in power systems and derive an energy function for it. Consider a power system consisting of n generators. Let the loads be modeled as constant impedances. Under the assumption that the transfer conductance of the reduced network, after eliminating all load buses, is zero, the dynamics of the ith generator can be represented by the equations

$$\begin{aligned} \dot{\delta}_i &= \omega_i \\ M_i\dot{\omega}_i &= P_i - D_i\omega_i - \sum_{j=1}^{n+1} V_iV_jB_{ij}\sin(\delta_i - \delta_j) \end{aligned} \tag{3.3}$$

where the voltage angle at node $i+1$ is served as the reference, i.e. $\delta_{i+1}{:=}0$. This is a version of the so-called *classical model* of the power system, where M_i and D_i are positive constants for $i = 1, \ldots, n$ and B_{ij} are the elements of the susceptance matrix. We next show that there exists an energy function $V(\delta, \omega)$ for the classical model (3.3).

Consider the following function:

$$V(\delta,\omega) = \frac{1}{2}\sum_{i=1}^{n} M_i\omega_i^2 - \sum_{i=1}^{n} P_i(\delta_i - \delta_i^s) - \sum_{i=1}^{n}\sum_{j=i+1}^{n+1} V_iV_jB_{ij}\{\cos(\delta_i - \delta_j) - \cos(\delta_i^s - \delta_j^s)\} \tag{3.4}$$

where $x^s = (\delta^s, 0)$ is the stable equilibrium point of (3.3) under consideration.

Differentiating V along the trajectory $(\delta(t), \omega(t))$ of (3.3) gives

$$\dot{V}(\delta(t),\omega(t)) = \sum_{i=1}^{n}\left(\frac{\partial V}{\partial \delta_i}\dot{\delta}_i + \frac{\partial V}{\partial \omega_i}\dot{\omega}_i\right) = -\sum_{i=1}^{n}(D_i\omega_i^2)\leq 0 \tag{3.5}$$

and therefore condition (E1) of an energy function holds.

Suppose that there is an interval $t \in [t_1, t_2]$ such that

$$\dot{V}(\delta(t),\omega(t)) = 0, \;\; t\in[t_1,t_2]. \tag{3.6}$$

Hence

$$\omega(t) = 0, \;\; t\in[t_1,t_2]. \tag{3.7}$$

But this implies $\omega(t) = 0$ and $\delta(t)$ = constant for $t \in [t_1, t_2]$. It then follows from (3.3) that

$$P_i - \sum_{j=1}^{n+1} V_iV_jB_{ij}\sin(\delta_i - \delta_j) = 0 \tag{3.8}$$

which are precisely the equations for the equilibrium point of (3.3). Therefore, ($\delta(t)$, $\omega(t)$), $t \in [t_1, t_2]$, must be on an equilibrium point. Consequently, condition (E2) holds.

In order to prove (E3), let us first integrate (3.3) for ω_i

$$\begin{aligned}\omega_i(t) = \;& e^{-D_i/M_it}\omega_i(0)\\ &+\int_0^t e^{-D_i/M_it(t-s)}\left\{P_i - \sum_{j=1}^{n+1}V_iV_jB_{ij}\sin(\delta_i(s) - \delta_j(s))\right\}ds\end{aligned}\cdot \tag{3.9}$$

The term in the bracket is uniformly bounded, say, by a_i, i.e.

$$\left|P_i - \sum_{j=1}^{n+1} V_iV_jB_{ij}\sin(\delta_i(s) - \delta_j(s))\right|\leq a_i. \tag{3.10}$$

Since D_i and M_i are positive numbers, we have from (3.9):

$$|\omega_i(t)|\leq|\omega_i(0)| + a_i\frac{M_i}{D_i}. \tag{3.11}$$

That is, $\omega_i(t)$ is bounded by $b_i := |\omega_i(0)| + a_iM_i/D_i$. We next show that condition (E3) of the energy function is satisfied. Suppose $V(\delta(t), \omega(t))$ is bounded below and above, say, by c_1 and c_2 respectively. Then we have

$$\begin{aligned}c_1 - \frac{1}{2}\sum_{i=1}^{n}M_ib_i^2 < -\sum_{i=1}^{n}P_i(\delta i \; - \; \delta_i^s)\\ -\sum_{i=1}^{n}\sum_{j=i-1}^{n+1}V_iV_jB_{ij}\left\{\cos(\delta_i - \delta_j) - \cos(\delta_i^s - \delta_j^s)\right\} < c_2\end{aligned} \tag{3.12}$$

But the second term on the right hand side is uniformly bounded, say, by c, i.e.

$$\left| \sum_{i=1}^{n} \sum_{j=i-1}^{n+1} V_i V_j B_{ij} \{ \cos(\delta_i - \delta_j) - \cos(\delta_i^s - \delta_j^s) \} \right| < c. \tag{3.13}$$

Substituting (3.13) into (3.12), we get

$$c_1 - \frac{1}{2} \sum_{i=1}^{n} M_i b_i^2 - c < -\sum_{i=1}^{n} P_i(\delta_i - \delta_i^s) < c_2 + c. \tag{3.14}$$

Hence, the term $P^T\delta$ is bounded, which implies $\delta(t)$ is bounded. This result, together with Eq. (3.11), asserts that the trajectory ($\delta(t)$, $\omega(t)$) is bounded. Hence, condition (E3) of energy function is satisfied.

3.2 Energy function theory

The dynamic behaviors of general nonlinear systems can be very complicated. The asymptotic behavior (i.e. the ω-limit set) of trajectories can be manifested as equilibrium points, or closed orbits, or quasi-periodic trajectories or chaotic trajectories. However, as shown in Theorem 3-1, if the underlying dynamical system admits an energy function, then the system only allows simple trajectories. For instance, every trajectory of system (3.1) admitting an energy function has only two modes of behavior: its trajectory either converges to an equilibrium point or goes to infinity (becomes unbounded) as time increases or decreases. This result is explained in the following theorem.

THEOREM 3-1 **(Global behavior of trajectories)**
If there exists a function satisfying condition (E1) and condition (E2) of the energy function for system (3.1) and all equilibrium points are isolated, then every bounded trajectory of system (3.1) converges to one of the equilibrium points.

Proof Let S be the ω-limit set of the bounded trajectory $x(t)$. This set is non-empty according to Theorem 2.2. In order to prove that S consists of only equilibrium points, we prove that (a) S is contained in the set at which the derivative of function V is zero, and (b) $\hat{x} \notin S$ if $\hat{x} \notin E$.

Suppose that $\hat{x} \in S$. Then, there exists a sequence of increasing times $\{t_n\}$ such that $x(t_n) \rightarrow \hat{x}$ as $n \rightarrow \infty$. Condition (E1) and the boundedness of $x(t)$ ensures that $V(x(t))$ is a non-increasing function bounded from below. Thus there is a real number α such that $V(x(t)) \rightarrow \alpha$ as $t \rightarrow \infty$. In particular, $V(x(t_n)) \rightarrow \alpha$ as $n \rightarrow \infty$. Therefore, by the continuity of V, $V(\hat{x}) = \alpha$ for every $\hat{x} \in S$. The invariance of S implies that $\dot{V}(\hat{x}) = 0$ for every $\hat{x} \in S$. Thus S is contained in the set at which the derivative of V is zero and (a) is true.

Suppose now that $\hat{x} \in S$ and $\hat{x} \notin E$. Since S is an invariant set and S is contained in the set at which the derivative of V is zero, there is an interval I at which the derivative of the solution passing through $\hat{x} \in S$ is zero. This contradicts condition (E2). Thus $x \in E$ and (b) is true. The connectedness of S and the fact that all equilibrium points are isolated guarantee that every bounded trajectory of (3.1) must converge to one of the equilibrium points. This completes the proof.

Theorem 3-1 asserts that there does not exist any limit cycle (oscillation behavior) or bounded complicated behavior (such as an almost periodic trajectory, chaotic motion, etc.) in the system. In Theorem 3-1 we have shown that the trajectory of system (3.1) either converges to one of the equilibrium points or goes to infinity. We next show a sharper result, asserting that every trajectory on the stability boundary must converge to one of the equilibrium points on the stability boundary.

THEOREM 3-2 **(Trajectories on the stability boundary)**
If there exists an energy function for system (3.1), then every trajectory on the stability boundary $\partial A(x_s)$ converges to one of the equilibrium points on the stability boundary $\partial A(x_s)$.

The significance of this theorem is that it offers an effective way to characterize the stability boundary. In fact, Theorem 3-2 asserts that the stability boundary $\partial A(x_s)$ is contained in the union of stable manifolds of the unstable equilibrium points (UEPs) on the stability boundary. One corollary of Theorem 3-2 shown below provides a characterization of stability boundaries.

THEOREM 3-3 **(Energy function and stability boundary)**
If there exists an energy function for system (3.1) which has an asymptotically stable equilibrium point x_s (but not globally asymptotically stable), then the stability boundary $\partial A(x_s)$ is contained in the set which is the union of the stable manifolds of the UEPs on the stability boundary $\partial A(x_s)$, i.e.

$$\partial A(x_s) \subseteq \bigcup_{x_i \in \{E \cap \partial A(x_s)\}} W^s(x_i).$$

The following two theorems give interesting results on the structure of the equilibrium points on the stability boundary, and present necessary condition for the existence of certain types of equilibrium points on a *bounded* stability boundary.

THEOREM 3-4 **(Structure of equilibrium points on the stability boundary)**
If there exists an energy function for the system (3.1) which has an asymptotically stable equilibrium point x_s (but not globally asymptotically stable), then the stability boundary $\partial A(x_s)$ must contain at least one type-one equilibrium point. If, furthermore, the stability region is bounded, then the stability boundary $\partial A(x_s)$ must contain at least one type-one equilibrium point and one source.

The contra-positive of Theorem 3-4 leads to the following corollary, which is useful in predicting the unboundedness of a stability region.

THEOREM 3-5 **(Sufficient condition for an unbounded stability region)**
If there exists an energy function for the system (3.1) which has an asymptotically stable equilibrium point x_s (but not globally asymptotically stable) and if $\partial A(x_s)$ contains no source, then the stability region $A(x_s)$ is unbounded.

3.3 Generalized energy functions

In this section, generalized energy functions, which are generalizations of energy functions and Lyapunov functions, are presented. One distinguishing feature of a generalized energy function is that its derivative along system trajectories can be positive in some bounded sets, while the derivative along a system trajectory of Lyapunov functions and of energy functions must be negative semi-definite. Generalized energy functions are useful in providing global information about the limit sets of general nonlinear systems, including those systems exhibiting complex behaviors in their limit sets. In addition, generalized energy functions can be explored to estimate stability regions of complex attractors.

It was shown in Section 3.2 that the existence of an energy function implies that (i) the limit set of the underlying dynamical system is composed only of equilibrium points, (ii) every bounded trajectory converges to an equilibrium point, and therefore (iii) there is no limit cycle or other complex behavior in these systems. As a consequence, nonlinear dynamical systems that exhibit complex behavior, such as closed, quasi-periodic orbits and chaos, cannot admit energy functions.

Many nonlinear system models exhibit complex behavior, such as closed orbits and chaos in their limit sets, see for example [19,70,165]. For instance, extended power system transient stability models can exhibit equilibrium points, closed orbits, quasi-periodic solutions and chaos in their limit sets [57,240,263]. Consequently, energy function theory is not applicable to this class of nonlinear model. To this end, by following the same "spirit" as energy functions, we generalize an energy function such that it has a zero derivative at every limit point. In [24] for instance, the existence of Lyapunov-like functions with derivatives equal to zero in the attracting set was studied. However, the complex structure of limit sets makes the task of finding an energy function difficult even if one can prove its existence.

A generalized energy function is a practical alternative for analyzing the dynamics of systems that exhibit complex behaviors in their limit sets. The feature of generalization is achieved by allowing the derivative of an energy function to have positive values in some bounded sets. This feature will be further explored in Chapter 11 to obtain estimates of the stability region of an attractor for nonlinear dynamical systems exhibiting complex behaviors such as closed orbits, quasi-periodic orbits and chaos.

Let $V: R^n \rightarrow R$ be a C^r-function, $r \geq 1$, and define the following set:

$$C : \{x \in R^n : \dot{V}(x) \geq 0\} \tag{3.15}$$

composed of the points where the derivative of $V(x)$ is positive. Set C is generally composed of several connected components, denoted by C_i, the ith connected component of C. These components are generally isolated (i.e. there exists a collection of disjoint open sets D_{is} satisfying $C_i \subset D_i$ for every i). If zero is a regular value of $\dot{V}$, then the boundary of set C is an $(n-1)$-dimensional C^{r-1} submanifold of R^n.

A C^r function $V:R^n \rightarrow R$, with $r \geq 1$, is a generalized energy function for system (3.1) if it satisfies the following three conditions:

(G1) the number of connected components C_i of C is finite;

(G2) every component C_i is bounded and

(G3) that a trajectory $x(t)$ has a bounded value of $V(x(t))$ for $t \in R^+$ implies that the trajectory $x(t)$ is also bounded.

Condition (G3) is identical to condition (E3) for an energy function. Conditions (G1) and (G2) imply that the derivative of generalized energy functions along trajectories can be positive in some bounded sets C_{is}. In contrast to energy functions, the definition of a generalized energy function does not rely on any type of limit set. It allows a variety of limit sets with complex behaviors including chaotic, quasi-periodic and closed trajectories. In the next section, the implications of the existence of a generalized energy function on global dynamics of nonlinear dynamical systems will be presented.

3.4 Generalized energy function theory

It was shown in Section 3.2 that the limit sets of nonlinear dynamical systems admitting an energy function are strictly located on the set $M = \{x \in R^n : \dot{V}(x) = 0\}$, where the derivative of the energy function is zero. The existence of generalized energy functions can also provide useful information about the location of limit sets. We next show that the existence of a generalized energy function ensures that the limit sets of the underlying system have to intersect the bounded sets C_{is} where the derivative of the generalized energy function is non-negative. Nevertheless, the existence of a generalized energy function does not preclude the possibility of the limit set being entirely located on the set $M = \{x \in R^n : \dot{V}(x) = 0\}$, where the derivative of the generalized energy function is zero.

THEOREM 3-6 **(Location of the ω-limit set)**
Let $V(x)$ be a generalized energy function for system (3.1). Suppose the trajectory $\phi(t, x_0)$ of the dynamical system (3.1) is bounded for $t \geq 0$. Then there exists at least one component C_j of C such that $\omega(x_0) \cap C_j \neq \varnothing$.

Proof Since $\phi(t, x_0)$ is bounded, the ω-limit set $\omega(x_0)$ is a non-empty, closed, invariant connected set. Suppose $\phi(t, x_0) \notin C$ for all $t \geq 0$. Then $V(t) = V(\phi(t, x_0))$ is a non-increasing function of t bounded from below. Hence, there exists a real number l such that $V(t) \to l$ as $t \to \infty$. If $p \in \omega(x_0)$, there exists a sequence of times $\{t_n\} \to \infty$ such that $\phi(t_n, x_0) \to p$ as $n \to \infty$. Therefore $V(\phi(t_n, x_0)) \to l$ as $n \to \infty$ and due to the continuity of V, $V(p) = l$. Since this is true for any point in $\omega(x_0)$, $\omega(x_0) \subset \{x \in R^n : x \in V^{-1}(L)\}$. Using the invariance of $\omega(x_0)$, we conclude that $\dot{V}(p) = 0$ for any $p \in \omega(x_0)$, and so $\omega(x_0) \subset M \subset C$. The connectedness of the limit set $\omega(x_0)$ guarantees the existence of a component C_j such that $\omega(x_0) \subset C_j$.

Suppose now the trajectory $\phi_t(x_0) := \{\phi(t, x_0) \in R^n : t \geq 0\}$ has a non empty intersection with set C. Then there exists a connected component C_{j_1} such that either $x_o \in C_{j_1}$ or there exists a pair of times t_1 and t_1^* such that $\phi(t, x_0) \notin C$ for $0 \leq t < t_1$ and $\phi(t, x_o) \in C_{j_1}$ for

$t_1 \le t \le t_1^*$. If $\phi(t, x_0)$ stays inside C_{j_1} for all $t \ge t_1$, that is $t_1^* = +\infty$, then $V(t)$ is a non-decreasing function of t bounded from above for $t \ge t_1$. Using arguments similar to those used in the first part of the proof, we conclude that $\omega(x_o) \subset C_{j_1}$.

If $t_1^* < \infty$, two dynamical behaviors can occur. Either $\phi(t, x_0) \notin C$ for $t \ge t_1^*$ or there exists a connected component C_{j_2} and a pair of times t_2 and t_2^* such that $\phi(t, x_0) \notin C$ for $t_1^* < t < t_2$ and $\phi(t, x_o) \in C_{j_2}$ for $t_2 \le t \le t_2^*$. Beyond this point, the analysis is repeated. If the number of times this analysis is repeated is finite, then $\omega(x_0) \subset M$ and $\omega(x_0) \subset C_j$ for some j. Otherwise, there exists a sequence of times $\{t_n\} \to \infty$ and a sequence of connected components C_{j_n} such that $\phi(t_n, x_o) \in C_{j_n}$. Since the number of connected components C_i of C is finite, there exists at least one component C_{j_k} that is visited by the trajectory an infinite number times. In other words, there exists a subsequence of times t_{n_i} of $\{t_n\}$ such that $x_i = \phi(t_{n_i}, x_o) \in C_{j_k}$. Since C_{j_k} is a compact set, there exists a convergent subsequence $\{x_{i_v}\}$ of $\{x_i\}$ converging to some point $\tilde{x} \in C_{j_k}$. By definition, $\tilde{x}$ is an ω-limit point of x_0 and so $\omega(x_o) \cap C_{j_k} \ne \emptyset$. This completes the proof.

Theorem 3-6 provides great insight into the location of the limit sets of bounded trajectories for the class of nonlinear dynamical systems (3.1) that admit generalized energy functions. It asserts that the ω-limit set of bounded solutions must intersect at least one connected component C_i of C. Figure 3.1 illustrates the two possible limit sets of bounded trajectories. It is important to emphasize that the ω-limit set of complex nonlinear systems can intersect more than one connected component C_j of C as shown in Figure 3.2.

Of all the possible limit sets, the equilibrium points must lie on the following set

$$M = \{x \subset R^n : \dot{V}(x) = 0\} \tag{3.16}$$

for all nonlinear dynamical systems admitting either an energy function or a generalized energy function. The generalized energy function does not preclude the possibility of other types of limit sets (such as closed orbits, quasi-periodic orbits, and chaotic orbits) lying on M, even though this can be rare given the complex nature of these sets.

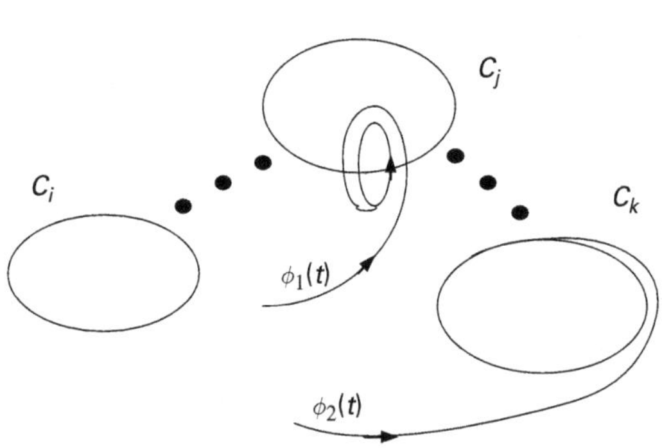

Figure 3.1 Illustration of Theorem 3-6. Two cases can occur: (i) the bounded trajectory $\phi_1(t)$ has a non-empty intersection with the interior of C_j or (ii) $\phi_2(t)$ approaches the set M, where the derivative of the generalized energy function equals zero on the boundary of component C_k.

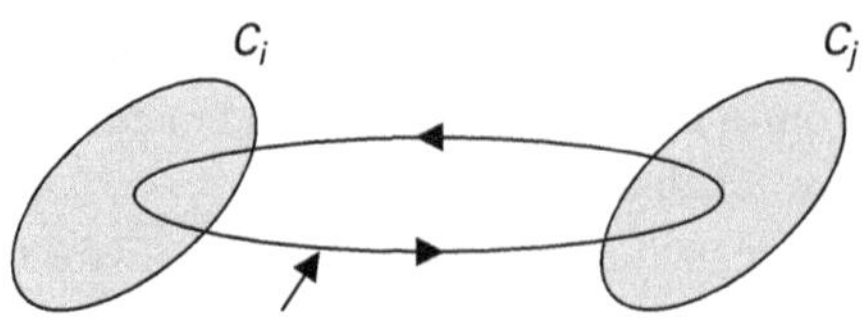

Figure 3.2 Illustration of Theorem 3-6. The ω-limit set can intersect with more than one connected component C_j of C. For instance, a limit cycle can have a non-empty intersection with two connected components C_i and C_j of C.

3.5 Energy functions for second-order dynamical systems

Energy functions can give sharp information regarding the global behavior of nonlinear dynamical systems. While the task of finding energy functions is not trivial, there is no systematic procedure to search for energy functions and for generalized energy functions. We next show how to derive an energy function for an important class of nonlinear dynamical systems.

Many physical systems are modeled by a second-order nonlinear dynamical system of the form:

$$M\ddot{x} + D\dot{x} + f(x) = 0$$

whose state space representation is:

$$\begin{aligned} \dot{x} &= y \\ M\dot{y} &= -Dy - f(x) \end{aligned} \tag{3.17}$$

where M is a diagonal matrix with positive elements, D is a symmetric, diagonally dominant matrix with positive diagonal elements and $f{:}R^n \to R^n$ is a C^1-class function. Let E denote the set of equilibrium points of (3.17) and suppose the number of equilibrium points on any stability boundary is finite. We use the notation $d(M,D)$ to denote system (3.17). We next present a sufficient condition for the existence of an energy function and also show how to derive an energy function for system (3.17).

THEOREM 3-7 **(Sufficient condition)**
If $f(x)$ is a conservative vector field, i.e. there exists a scalar C^1-function $V_p{:}R^n \to R$ such that $f = \nabla V_p$, then there exists a C^1 function $V{:}R^{2n} \to R$ for $d(M,D)$ such that

(a) $\dot{V}(x(t), y(t)) \leq 0$,
(b) Let $(x(0), y(0)) \notin E$, *then the set* $\{t \subset R : \dot{V}(x(t), y(t)) = 0\}$ *has measure zero in R.*

Proof We define $V{:}R^n \times R^n \to R$ by

$$V(x,y) = \frac{1}{2}\langle y, My\rangle + V_p(x). \tag{3.18}$$

The derivative of $V(\cdot)$ along the trajectory of $d(M,D)$ is

$$\begin{aligned}\dot{V}(x,y) &= \frac{\partial V}{\partial x}\dot{x} + \frac{\partial V}{\partial y}\dot{y} \\ &= -\langle y, Dy\rangle \leq 0.\end{aligned} \tag{3.19}$$

So, part (a) is true. Suppose that part (b) is not true, then there exists an interval $T = (t_1, t_2)$ with $t_2 > t_1 \geq 0$ such that $\dot{V}(x(t), y(t)) = 0$ for $t \in T$. From (3.19), we have $y(t) = 0$ for $t \in T$. This implies that $y(t) = 0$ and $x(t) =$ constant, for $t \in (t_1, t_2)$. From (3.17), this implies that $f(x(t)) = 0$. So, we have $(x(t), y(t)) \in E$ for $t \in (t_1, t_2)$. Since (3.17) is an autonomous dynamical system, it follows that $(x(t), y(t)) \in E$ for $t \in R$. This contradicts the fact that $(x(0), y(0)) \notin E$. Therefore, part (b) is also true. The proof is complete.

Theorem 3-7 shows that function (3.18) satisfies conditions (E1) and (E2) of an energy function. Thus, as a direct consequence of Theorem 3-1, the ω-limit set of every bounded trajectory of system (3.17) consists of only equilibrium points. In other words, the class of systems in the form of (3.17) does not admit complex behavior such as limit cycles or chaos. We hence conclude that every bounded trajectory of system (3.17) converges to one of the equilibrium points.

THEOREM 3-8 **(Existence of an energy function)**
Consider the nonlinear system d(M,D). If function f is bounded and $V_p(x)$ is a proper function, then there exists a C^1-energy function $V: R^{2n} \to R$ for d(M,D).

Proof According to Theorem 3-7, function (3.18) satisfies conditions (E1) and (E2) of an energy function. We conclude the proof of this theorem by showing that function (3.18) also satisfies condition (E3) of an energy function under the condition that f is bounded and $V_p(x)$ is a proper function. From system (3.17), one obtains:

$$\dot{y} = -M^{-1}Dy - M^{-1}f(x). \tag{3.20}$$

As a direct application of the variation of constants formula, one has:

$$\|y(t)\| \leq \|e^{At}\|\|y_0\| + \int_0^t \|e^{A(t-s)}\|\|M^{-1}\|\|f(x(s))\|ds \tag{3.21}$$

with $A = -M^{-1}D$. Since both M and D are diagonal matrices with positive elements, then $A = -M^{-1}D$ has all eigenvalues on the left side of the complex plane. Therefore, there exist real constants $C > 0$ and $\alpha > 0$ such that $\|e^{At}\| \leq Ce^{-\alpha t}$ for all $t \geq 0$. If b is a bound for the norm of f, one has:

$$\|y(t)\| \leq Ce^{-\alpha t}\|y_0\| + \int_0^t Ce^{-\alpha(t-s)}\|M^{-1}\|bds. \tag{3.22}$$

Then we conclude, after some calculation, that:

$$\|y(t)\| \leq C\|y_0\| + \frac{C\|M^{-1}\|b}{\alpha}. \tag{3.23}$$

In other words, $y(t)$ is bounded. Suppose now that the trajectory $(x(t),y(t))$ is unbounded for $t \geq 0$ while the value of $V(x,y)$ calculated along this trajectory is bounded.

Since $y(t)$ is always bounded, we conclude: (i) that $x(t)$ must be unbounded and (ii) the term $V_p(x)$ must be bounded. This contradicts the hypothesis that $V_p(x)$ is a proper function. This completes the proof.

Consider now the following class of second-order nonlinear dynamical systems:

$$\begin{aligned} \dot{x} &= y \\ M\dot{y} &= -Dy = \frac{\partial W(x)}{\partial x} + \varepsilon g(x) \end{aligned} \tag{3.24}$$

where M and D are diagonal matrices with positive entries and ε is a small real number. Function $W{:}R^n \to R$ is C^2 and g is a uniformly bounded C^1 function. System (3.24) is a perturbed version of system (3.17) in which the perturbation $g{:}R^n \to R^n$ is a non-gradient vector field. This class of systems appears, for example, on transient stability analysis of power systems in the presence of transfer conductances [29,47].

Nonlinear dynamical systems in the form of (3.24) do not admit energy functions. For instance, the following simple power system model,

$$\begin{aligned} \dot{x}_1 &= y_1 \\ \dot{x}_2 &= y_2 \\ 0.053\dot{y}_1 &= 1.78 - 3.16\sin x_1 - 0.28\cos x_1 - 0.9\sin(x_1 - x_2) - \varepsilon\cos(x_1 - x_2) - 0.1y_1 \\ 0.079\dot{y}_1 &= 3.83 - 7.85\sin x_2 - 0.255\cos x_2 - 0.9\sin(x_2 - x_1) - \varepsilon\cos(x_2 - x_1) - 0.1y_2, \end{aligned} \tag{3.25}$$

can be put into the form (3.24) by choosing the following function:

$$\begin{aligned} W(x) = &-1.78x_1 - 3.83x_2 - 3.16\cos x_1 + 0.28\sin x_1 - 7.85\cos x_2 \\ &+0.255\sin x_2 - 0.9\cos(x_1 - x_2). \end{aligned}$$

It can be shown that this simple model admits a limit cycle for $\varepsilon > 3.3$. It is therefore inferred that system (3.25) cannot admit an energy function, according to Theorem 3-1. Figure 3.3 shows this limit cycle for $\varepsilon = 3.5$ and Figure 3.4 depicts a bifurcation diagram of this system. For $\varepsilon > 3.3$, this system exhibits complex behavior and cannot admit an energy function.

Although nonlinear dynamical systems in the form of (3.24) do not admit energy functions, it is possible to show the existence of a general generalized energy function for this class for sufficiently small ε. For this purpose, we assume system (3.24) has a finite number of isolated equilibrium points. We next show the following function with $\beta > 0$ is a generalized energy function of system (3.24) for sufficiently small ε:

$$V(x, y) = \frac{1}{2}y^T My + W(x) + \beta\left[\frac{\partial W(x)}{\partial x} - \varepsilon g(x)\right]^T y. \tag{3.26}$$

PROPOSITION 3-9 **(Conditions (G1) and (G2))**

Consider the nonlinear dynamical system (3.24) where $W(x)$: $R^n \to R$ is a C^2 function and $g{:}R^n \to R^n$ is a C^1 function. If g is a bounded function, then for a sufficiently small ε, function (3.26) satisfies conditions (G1) and (G2) of a generalized energy function.

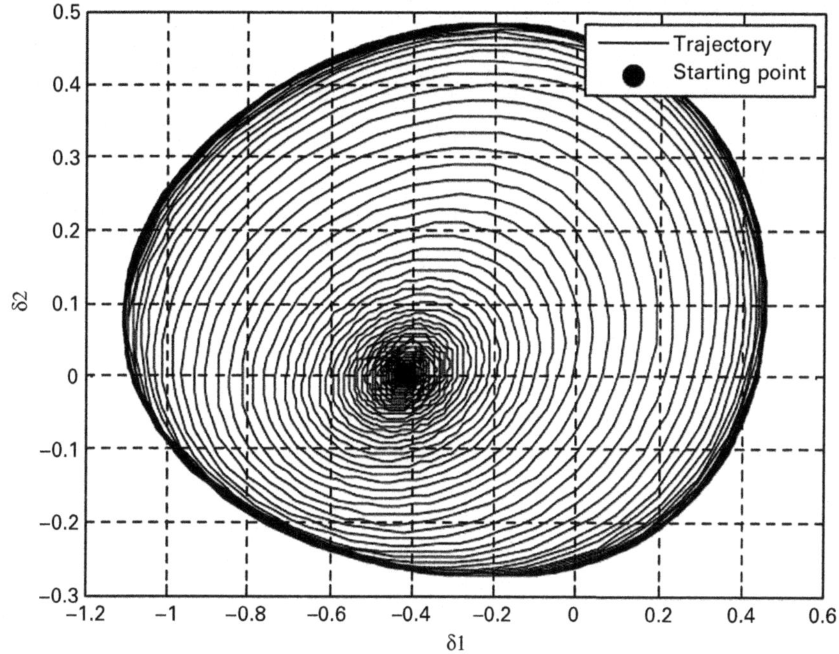

Figure 3.3 Limit cycle of system (3.25) for $\varepsilon = 3.5$ projected to the space $\omega_1 = \omega_2 = 0$.

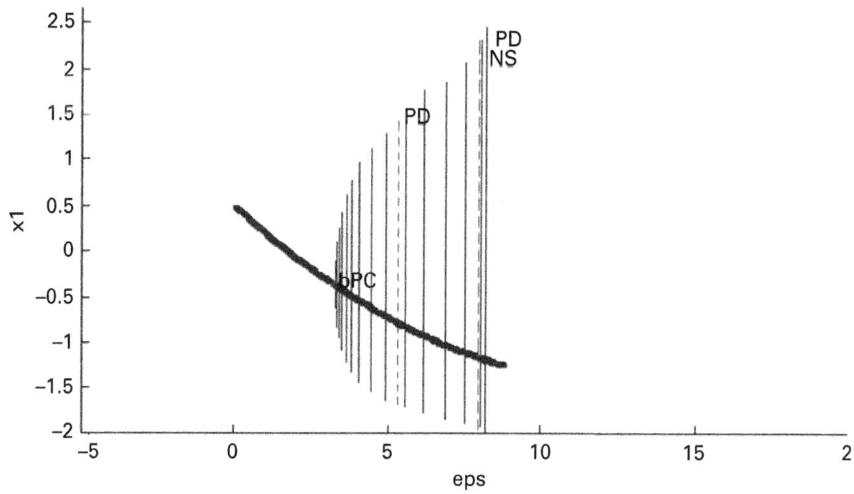

Figure 3.4 Bifurcation diagram of system (3.25). At $\varepsilon = 3.3$ a Hopf bifurcation occurs. The asymptotic stable equilibrium point loses stability and an asymptotic stable limit cycle is created.

Proof The derivative of V along the trajectory of (3.24) is given by:

$$\dot{V}(x,y) = -\left[y^T \left[\frac{\partial W(x)}{\partial x} - \varepsilon g(x) \right]^T \right] Q \left[\begin{matrix} y \\ \dfrac{\partial W(x)}{\partial (x)} - \varepsilon g(x) \end{matrix} \right] + \varepsilon g^T(x) y$$

where

$$Q = \begin{bmatrix} \left(D - \beta\dfrac{\partial^2 W(x)}{\partial x^2} + \beta\varepsilon\dfrac{\partial g}{\partial x}\right) & \left(\dfrac{1}{2}\beta D M^{-1}\right) \\ \dfrac{1}{2}\beta D M^{-1} & \beta M^{-1} \end{bmatrix}.$$

First of all, we show that matrix Q is positive definite for a small enough $\beta > 0$. By the continuity of the determinants with respect to the matrix entries, the following relationship holds for sufficiently small β,

$$A : = D - \beta\frac{\partial^2 W(x)}{\partial x^2} + \beta\varepsilon\frac{\partial g}{\partial x} > 0.$$

Let A_{ij} denote the minor of A obtained by deleting the ith row and jth column. Consider now the following minor of matrix Q:

$$B_1 := \begin{bmatrix} & \frac{1}{2}\beta D_1 M_1^{-1} \\ A & 0 \\ & 0 \\ & \cdots \\ & 0 \\ \frac{1}{2}\beta D_1 M_1^{-1} 0 0 \ldots 0 & \beta M_1^{-1} \end{bmatrix}.$$

Using the Laplacian expansion of determinants by minors, one obtains

$$\det B_1 = \beta M_1^{-1}\left[\det A - \frac{1}{4}\beta D_1^2 M_1^{-1}\det A_{11}\right].$$

Since $A > 0$ and symmetric, then $\det A > 0$ and $\det A_{11} > 0$. As a consequence, $\det B_1 > 0$ if $0 > \beta < \dfrac{4M_1 \det A}{D_1^2 \det A_{11}}$. We next consider

$$B_2 : \begin{bmatrix} & \frac{1}{2}\beta D_2 M_2^{-1} \\ B_1 & 0 \\ & 0 \\ & \cdots \\ & 0 \\ 0\frac{1}{2}\beta D_2 M_2^{-1} 0 0 \ldots 0 & \beta M_2^{-1} \end{bmatrix}.$$

It follows that

$$\det B_2 = \beta M_2^{-1}\left[\det B_1 - \frac{1}{4}\beta D^2 M_2^{-1}\det B_{1_{22}}\right].$$

Since $\det B_1 = O(\beta) > 0$ and $\det B_{1_{22}} = O(\beta) > 0$, $\det B_2 > 0$ for sufficiently small β. Repeating this procedure n times we prove that the determinant of all main minors of Q are positive for sufficiently small $\beta > 0$. Then, as an implication of Sylvester's criterion, $Q > 0$.

The quadratic term of V equals zero only at the set of equilibrium points E. Since g is uniformly bounded, for ε sufficiently small, the regions where the derivative of V is positive are contained in small bounded connected sets C_{i_s} close to the equilibrium points. Since the equilibrium points are isolated, for sufficiently small ε, the sets C_{i_s} are isolated; every two sets C_i and C_j have an empty intersection and the distance between them is greater than zero. Therefore, condition (G2) is satisfied. A finite number of equilibrium points on the stability boundary proves condition (G1). This completes the proof.

To examine condition (G3), the vector field for this class of dynamical systems is explored. The next theorem provides some sufficient conditions to ensure that condition (G3) is satisfied.

PROPOSITION 3-10 **(Condition (G3))**
If W is a proper function and both $\partial W/\partial x$ and g are bounded, then the function $V(x)$ of (3.26) satisfies condition (G3).

Proof Since both $\partial W/\partial x$ and g are bounded, as a direct application of the variation of constants formula, we prove that $y(t)$ is always bounded. Suppose by contradiction that $\sup_{t\geq 0}\|V\| < \infty$ and $\phi(t, x_0) = (x(t), y(t))$ is unbounded for $t \geq 0$. Since $y(t)$ is bounded, then $x(t)$ must be unbounded for $t \geq 0$. This implies that $\sup_{t\geq 0}\|W(x)(t)\| = \infty$. Since $\|y(t)\|$ is bounded for $t \geq 0$, $\sup_{t\geq 0}\|W(x(t))\| = \infty$ and both $\partial W/\partial x$ and g are uniformly bounded, we conclude that $\sup_{t\geq 0}\|V\| = \infty$. This is a contradiction, and so $\sup_{t\geq 0}\|V\| < \infty$ implies that $\phi(t, x_0) = (x(t), y(t))$ is bounded for $t \geq 0$. This concludes the proof.

3.6 Numerical studies

We illustrate the analytical results of generalized energy function developed in this chapter on several simple numerical examples taken from power system models, non-linear control models and Lorenz systems.

Example 3-3 (Power system stability model) The following system of equations was obtained from the power system literature and was proposed to model the dynamical behavior of a two-generator system versus an infinite bus with transfer conductance [29]:

$$\begin{cases} \dot{x}_1 = y_1 \\ \dot{x}_2 = y_2 \\ M_1\dot{y}_1 = P_1 - G_1 \sin x_1 - B_1 \cos x_1 \\ \qquad\qquad - G_{12} \sin(x_1 - x_2) - \varepsilon \cos(x_1 - x_2) - D_1 y_1 \\ M_2\dot{y}_2 = P_2 - G_2 \sin x_2 - B_2 \cos x_2 \\ \qquad\qquad - G_{12} \sin(x_2 - x_1) - \varepsilon \cos(x_2 - x_1) - D_2 y_2. \end{cases} \tag{3.27}$$

Parameter ε represents the transfer conductance. For $\varepsilon = 0$, this system admits a general energy function:

$$W(x_1,x_2) := -P_1x_1 - G_1\cos x_1 + B_1\sin x_1 - P_2x_2 \\ -G_2\cos x_2 + B_2\sin x_2 - G_{12}\cos(x_1-x_2) + \alpha$$

where α is an arbitrary constant. We leave it for the reader to show that W is an energy function for system (3.27) if $\varepsilon = 0$. However, it has been shown that a general energy function does not exist for this system when $\varepsilon \neq 0$ [47]. With the development of generalized energy functions, a generalized energy function that satisfies conditions (G1)–(G3) can be derived.

The previous set of differential equations can be put in the general form (3.24) by choosing $g(x_1, x_2)$: $\cos(x_1 - x_2)$. It is straightforward to check that (i) both $\partial W/\partial x$ and g are uniformly bounded, and (ii) although function W is not proper, condition (G3) is generically satisfied. Hence, the following function

$$V(x_1,x_2,y_1,y_2) = M_1\frac{y_1^2}{2} + M_2\frac{y_2^2}{2} + W(x_1,x_2) \\ -\beta y_1[P_1 - G_1\sin x_1 - B_1\cos x_1 - G_{12}\sin(x_1-x_2), \\ -\varepsilon\cos(x_1-x_2)] \\ -\beta y_2[P_2 - G_2\sin x_2 - B_2\cos x_2 - G_{12}\sin(x_2-x_1) \\ -\varepsilon\cos(x_2-x_1)]$$

is a generalized energy function provided both $\beta > 0$ and ε are small enough.

An estimate for β can be obtained by computing the derivative of this function along the system orbits:

$$-\dot{V} = \begin{bmatrix} P_{l_1}(x_1,x_2) \\ y_1 \\ P_{l_2}(x_1,x_2) \\ y_2 \end{bmatrix}^T A \begin{bmatrix} P_{l_1}(x_1,x_2) \\ y_1 \\ P_{l_2}(x_1,x_2) \\ y_2 \end{bmatrix} + \\ \begin{bmatrix} y_1 \\ y_2 \end{bmatrix}^T B \begin{bmatrix} y_1 \\ y_2 \end{bmatrix} + \varepsilon[\cos(x_1-x_2) - 1(y_1+y_2)]$$

where

$$A = \begin{bmatrix} A_{11} & 0 \\ 0 & A_{22} \end{bmatrix}, \quad B = \begin{bmatrix} \dfrac{D_1}{2} & \beta G_{12}\cos(x_1-x_2) \\ \beta G_{12}\cos(x_1-x_2) & \dfrac{D_2}{2} \end{bmatrix},$$

$$A_{11} = \begin{bmatrix} \dfrac{\beta}{M_1} & -\dfrac{\beta D_1}{2M_1} \\ -\dfrac{\beta D_1}{2M_1} & \begin{matrix} \dfrac{D_1}{2} + \beta[-G_1\cos x_1 + B_1\sin x_1 \\ -G_{12}\cos(x_1-x_2) + \varepsilon\sin(x_1-x_2)] \end{matrix} \end{bmatrix}$$

and

$$A_{22} = \begin{bmatrix} \dfrac{\beta}{M_2} & -\dfrac{\beta D_2}{2M_2} \\ -\dfrac{\beta D_2}{2M_2} & \begin{matrix} \dfrac{D_2}{2} + \beta[-G_2\cos x_2 + B_2\sin x_2 \\ -G_{12}\cos(x_2-x_1) + \varepsilon\sin(x_2-x_1)] \end{matrix} \end{bmatrix}$$

The parameter β can be chosen to make the quadratic term positive definite. By applying Silvester's criterion one can easily find that the positive definiteness is guaranteed if

$$\beta^2 < \frac{D_1 D_2}{4G_{12}^2}, \quad 0 < \beta < \frac{D_1}{2\left(\frac{D_1^2}{4M} + G_1 + B_1 + G_{12} + \varepsilon\right)}$$

and

$$0 < \beta \frac{D_2}{2\left(\frac{D_2^2}{4M} + G_2 + B_2 + G_{12} + \varepsilon\right)}.$$

The set with a positive derivative of $V(x_1, x_2, y_1, y_2)$ is composed of two small bounded sets C_1 and C_2. These sets intersect the stability boundary, and they are close to the unstable equilibrium points (0.91,0,2.72,0) and (2.71,0,0.6,0). Their intersections with the subset $\{(x_1, x_2, y_1, y_2) : x_1 \in R, x_2 \in R, y_1 = y_2 = 0.8\}$ are depicted in Figure 3.5 with the following parameters: $P_1 = 1.78$, $P_2 = 3.83$, $G_1 = 3.16$, $G_2 = 7.85$, $B_1 = 0.28$, $B_2 = 0.255$, $G_{12} = 0.9$, $\varepsilon = 0.1$, $D_1 = D_2 = 0.1$, $M_1 = 0.053$, $M_2 = 0.079$, $\alpha = 13.017$ and $\beta = 0.005$.

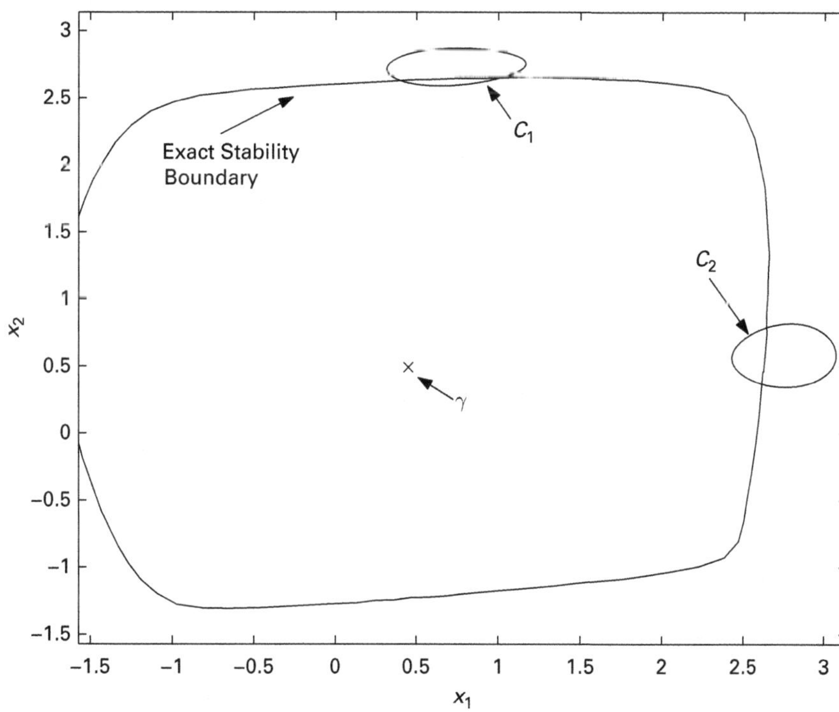

Figure 3.5 The set with a positive derivative of $V(x_1, x_2, y_1, y_2)$ is composed of two small bounded sets C_1 and C_2. These sets intersect the stability boundary, and they are close to the unstable equilibrium points (0.91,0,2.72,0) and (2.71,0,0.6,0).

Example 3-4 (Lorenz systems) Consider the Lorenz system:

$$\begin{cases} \dot{x} = -\sigma x + \sigma y \\ \dot{y} = -y - xz + rx \\ \dot{z} = -bz + xy \end{cases}$$

where, $\sigma = 10$, $r = 28$ and $b = 8/3$. The Lorenz system possesses a globally stable chaotic attractor and as a consequence it does not admit an energy function. Let us show that the following candidate:

$$V(x, y, z) = rx^2 + 4\sigma y^2 + 4\sigma(z - 5/4r)^2$$

is a generalized energy function for the Lorenz systems. It is obvious that function $V(x, y, z)$ is radially unbounded; hence, condition (G3) is satisfied. The derivative of function $V(x, y, z)$ along the system trajectory is given by:

$$\dot{V}(x, y, z) = -2\sigma(rx^2 + 4y^2 + 4bz^2 - 5rbz).$$

Set C is given by $C : \{x \in R^3 : rx^2 + 4y^2 + 4bz^2 - 5rbz < 0\}$. It is straightforward to verify that the boundary of set C is an ellipsoid centered at ($x = 0$, $y = 0$, $z = 5/8r$). Consequently, set C is composed of a single bounded component and conditions (G1) and (G2) are also satisfied. Thus function $V(x, y, z)$ is a generalized energy function for the Lorenz system. Every trajectory of the Lorenz system is bounded and must intersect set C. Figure 3.6 illustrates this intersection.

Example 3-5 (Nonlinear control of DC motor) Consider the nonlinear system:

$$\begin{aligned} \dot{x}_1 &= x_2 \\ \dot{x}_2 &= -k_d x_2 - x_1 - g x_1^2\left(\frac{x_2}{k_d} + x_1 + 1\right) + u \end{aligned} \tag{3.28}$$

with $k_d = 1$ and $g = 6$. Function u is the control input. This model comes from the problem of speed control in a direct current machine using the field voltage as a control input and taking into account some nonlinearities related to the back electromotive force [92].

It is easily verified that the origin is an asymptotically stable equilibrium point of the open loop system for every $k_d > 0$. However, for $g > 4$ the system has three equilibrium points, indicating that the origin is not a globally stable equilibrium point. In this example, the concept of a generalized energy function and some peculiarities of this system will be explored to design a feedback control law $u = h(x_1, x_2)$ so that global stability of the origin is achieved. To this end, we consider the following candidate for a generalized energy function:

$$V(x, u) = k_1 x_1^2 + k_2 x_2^2 + k_3 x_1 x_2 + k_4 g \frac{x_1^4}{4} + k_5 g \frac{x_1^3}{3} \tag{3.29}$$

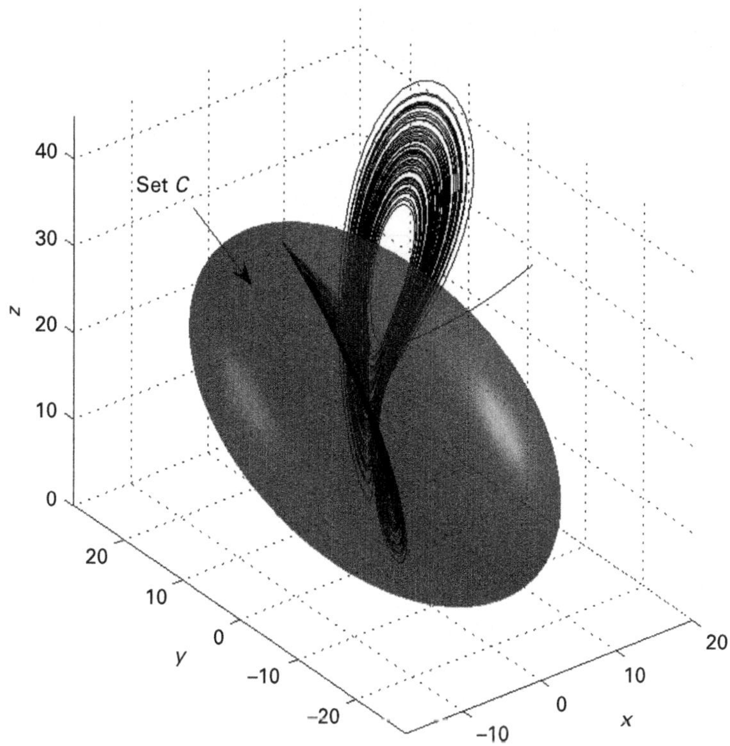

Figure 3.6 The Lorenz attractor intersects set C where the derivative of the generalized energy function is positive.

Choosing $k_1 = (k_3 k_d/_2) + k_2$, $k_4 = 2k_2 + k_3/k_d$ and $k_5 = 2k_2$ and, after calculating the derivative of function $V(x, u)$ along trajectories of (3.28), it follows that

$$\dot{V}(x,u) = -k_3 g x_1^4 - k_3 g x_1^3 - k_3 x_1^2 + (k_3 - 2k_2 k_d) x_2^2 - 2g\frac{k_2}{k_d} x_1^2 x_2^2 + 2k_2 x_2 u + k_3 x_1 u. \tag{3.30}$$

The constants can be chosen such that $2k_2 k_d > k_3$. In this example, we choose $k_3 = k_d$ and $k_2 = 1$. With this choice, the quadratic term for variable x_2 becomes negatively definite, and it is not difficult to check that $V(x, u)$ is an energy function of the open loop system (3.28) if $g < 4$. However, for $g > 4$ condition (E1) of the energy function is not satisfied.

Next, we design a feedback control law and show that $V(x, u)$ is a generalized energy function of the closed loop system. Let $u = h(x_1, x_2) = gx_1^2$ and observe that $h(0, 0) = 0$. Thus, the origin is an equilibrium point of the closed loop system and the origin is the unique equilibrium point of the closed loop system. In order to show that $V(x, u)$ is a generalized energy function of the closed loop system, we substitute the feedback law u into the expression of $\dot{V}$ and obtain the following:

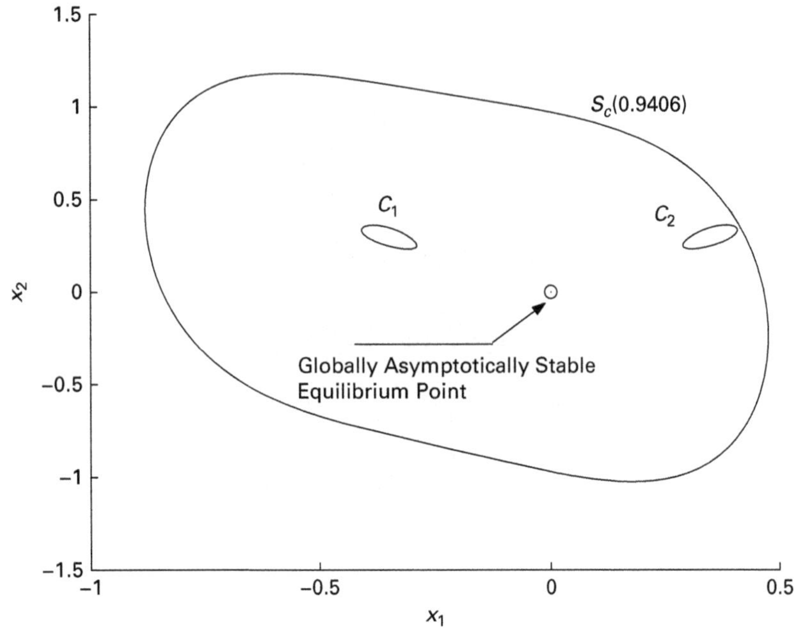

Figure 3.7 Level set $S_c(0.9406)$ of the generalized energy function $V(x, u)$ associated with the closed loop system for $k_3 = k_d$ and $k_2 = 1$. Using the generalized energy function and exploring some properties of system (3.28), it is shown that the origin is a globally asymptotically stable equilibrium point.

$$\dot{V}(x) = -k_3 g x_1^4 - k3x_1^2 - (2k_2 k_d - k_3)x_2^2 - 2g\frac{k_2}{k_d}x_1^2 x_2^2 + 2k_2\, g x_1^2\, x_2. \tag{3.31}$$

We then choose the constants such that $2k_2k_d - k_3 > 0$ and all the terms of equation (3.31) except the term $2k_2 g x_1^2 x_2$ are non-positive. The regions where the derivative of $V(x, u)$ is positive are shown in Figure 3.7. Set C is composed of three connected and bounded components: they are the sets C_1, C_2 and the origin. Therefore, conditions (G1) and (G2) of generalized energy functions are satisfied. If the parameters are chosen such that $2k_2k_d > k_3$, then the function $V(x, u)$ becomes radially unbounded and, consequently, condition (G3) of generalized energy functions is also satisfied. Therefore, $V(x, u)$ is a generalized energy function of the closed loop system.

By choosing $l = \max_{x \in C} V = 0.9406$ and using the fact that $V(x, u)$ is a radially unbounded function, we conclude that all the trajectories of the closed loop system enter the positively invariant bounded set $S_c(0.9406) = \{x \in R^2 \colon (V(x) < 0.9406)\}$ for some positive time. Hence, every trajectory of the closed loop system enters the set $S_c(0.9406)$ and approaches its limit set, which has a non-empty intersection with set C.

Now, some distinguishing features of the closed loop system will be explored to show that every trajectory approaches the origin, which is an asymptotically stable equilibrium point of the closed loop system. According to the Poincaré–Bendixson theorem

[107], if the ω-limit set does not contain equilibrium points, then it is a closed orbit. One can verify that

$$\frac{\partial f_1}{\partial x_1} + \frac{\partial f_2}{\partial x2} = -k_d - \frac{g}{k_d} x_1^2 < 0 \tag{3.32}$$

for $k_d > 0$. Therefore, according to Bendixson's criterion [107], there are no closed orbits inside the set $S_c(0.9406)$. Hence, the limit sets contain the unique equilibrium point of the closed loop system that is the origin. Therefore, the origin is a globally asymptotic stable equilibrium point of the closed loop system.

3.7 Concluding remarks

In this chapter, we have presented a comprehensive energy function theory and generalized energy function theory for general nonlinear autonomous dynamical systems. Analytical results on the structure of ω-limit sets and on the global behavior of trajectories using energy functions and generalized energy functions have been presented.

The dynamic behaviors of general nonlinear systems can be very complicated. It has been shown that, if the underlying dynamical system admits an energy function, then the system only allows simple trajectories: every trajectory either converges to an equilibrium point or goes to infinity (becomes unbounded). Moreover, the stability boundary is contained in the set which is the union of the stable manifolds of the UEPs on the stability boundary, which gives a characterization of the stability boundary.

It is well recognized that there is no systematic procedure for deriving Lyapunov functions for general nonlinear systems. Likewise, there is no systematic procedure for deriving energy functions or generalized energy functions for general nonlinear dynamical systems. In order to avoid searching for an energy function for each individual nonlinear dynamical system, it is beneficial to develop energy functions for a class of nonlinear dynamical systems. We have derived an energy function for an important class of nonlinear dynamical systems: second-order nonlinear dynamical system.

Both energy function theory and generalized energy function theory will be further explored in later chapters to develop optimal schemes for estimating stability regions of general nonlinear systems. It will be shown that the stability regions estimated via energy functions or generalized energy functions using the schemes to be developed are optimal in a certain sense.

4 Stability regions of continuous dynamical systems

The determination of stability regions continues to play an important role in many emerging research areas of engineering and the sciences. A comprehensive theory of stability regions for continuous dynamical systems will be presented in this chapter and Chapter 5. The topic of how to estimate stability regions for high-dimensional nonlinear continuous, dynamical systems will be discussed in Chapter 10 and Chapter 11.

We consider the following nonlinear dynamical system

$$\dot{x} = f(x). \tag{4.1}$$

It is natural to assume that the function (i.e. the vector field) f: $R^n \to R^n$ satisfies a sufficient condition for the existence and uniqueness of the solution. The stability region of a stable equilibrium point x_s, denoted by $A(x_s)$ is the set of all points x such that

$$A(x_s) := \{x \in R^n : \lim_{t \to \infty} \phi(t, x) = x_s\}. \tag{4.2}$$

The (topological) boundary of the stability region $A(x_s)$ is called the *stability boundary* (also called the *separatrix*) of x_s and will be denoted by $\partial A(x_s)$. Since the stability region of a stable equilibrium point is in fact its stable manifold, the topological properties of stability regions are such that the stability region $A(x_s)$ is an open and invariant set diffeomorphic to R^n.

If there are at least two stable equilibrium points, then the dimension of each stability boundary is $n-1$; in this case stability boundaries are non-empty. Characterization of stability regions can be achieved via characterization of the stability boundary. We will develop a comprehensive theory for stability regions of nonlinear dynamical systems (4.1) in the next section.

4.1 Equilibrium points on the stability boundary

In this section, several dynamical and topological properties of a stability boundary will be derived. A complete characterization for two fundamental limit sets of general nonlinear dynamical systems (i.e. equilibrium points and limit cycles) lying on the stability boundary will also be derived. Our approach starts from a local characterization of the stability boundary and then extends toward a global characterization of the stability boundary.

We first derive a complete characterization of an equilibrium point, or limit cycle, lying on the stability boundary, which is a key step in the characterization of the stability region $A(x_s)$. We do this in two steps. First we impose only one assumption on the nonlinear dynamical system (4.1), namely, that equilibrium points are hyperbolic, and derive necessary and sufficient conditions for an equilibrium point to lie on the stability boundary in terms of both its stable and unstable manifolds. Additional conditions are then imposed on the dynamical system and the results are further sharpened. We also derive the characterizations of closed orbits on the stability boundary. We use the term *critical element* to denote equilibrium points and limit cycles.

Let x be a hyperbolic critical element (which is either an equilibrium point or a limit cycle). Let U be a neighborhood of x in $W^s(x)$ whose boundary ∂U is transversal to the vector field f. We call ∂U a *fundamental domain* of $W^s(x)$. A cross section $V \subset R^n$ of a vector field f is the manifold V of dimension $n - 1$, which need not be a hyperplane but must be in a manner such that the flow of f is everywhere transversal to it. Any neighborhood of ∂U with an empty intersection with $W^u(x)$ is a *fundamental neighborhood* $G(x)$ associated with $W^s(x)$. It follows that

(1) $W^s(x) = \cup_{t \in R}\phi(t, \partial U) \cup \{x\}$, and
(2) $\cup_{t \geq 0}\phi(t, G(x)) \cup W^u(x)$ contains a neighborhood of x.

We next derive a complete characterization for an equilibrium point lying on the stability boundary of a general nonlinear continuous dynamical system (4.1).

THEOREM 4-1 **(Characterization of equilibrium point on the stability boundary)** *Consider a general nonlinear continuous dynamical system (4.1). Let $A(x_s)$ be the stability region of an asymptotically stable equilibrium point x_s. Let $\hat{x} \neq x_s$ be a hyperbolic equilibrium point. Then:*

(a) if $\{W^u(\hat{x}) - \hat{x}\} \cap A(x_s) \neq \emptyset$, then $\hat{x} \in \partial A(x_s)$; conversely, if $\hat{x} \in \partial A(x_s)$, then $\{W^u(\hat{x}) - \hat{x}\} \cap \overline{A(x_s)} \neq \emptyset$;
(b) suppose $\hat{x}$ is not a source (i.e. $\{W^s(\hat{x}) - \hat{x}\} \neq \emptyset$), then $\hat{x} \in \partial A(x_s)$ if and only if $\{W^s(\hat{x}) - \hat{x}\} \cap \partial A(x_s) \neq \emptyset$.

Proof

(a) If $y \in W^u(x) \cap \overline{A(x_s)}$, then $\lim_{t \to \infty} \phi(-t, y) = \hat{x}$.

But since $\overline{A(x_s)}$ is invariant, we have

$$\phi(-t, y) \in \overline{A(x_s)} \text{ for all } t \in R.$$

It follows that

$$\hat{x} \in \overline{A(x_s)}.$$

Since $\hat{x}$ cannot be in the stability region, $\hat{x}$ is on the stability boundary.

Suppose conversely that $\hat{x} \in \partial A(x_s)$. Let $G \subset \{W^u(\hat{x}) - \hat{x}\}$ be a fundamental domain of $W^u(\hat{x})$; this means that G is a compact set such that

$$\cup_{t \in R} \phi(t, G) = \{W^u(\hat{x}) - \hat{x}\}.$$

Let G_ε be the ε-neighborhood of G in R^n. Set G_ε a fundamental neighborhood associated with $W^u(\hat{x})$. Then $\cup_{t<0}\phi(t, G_\varepsilon)$ contains a set of the form $\{U - W^s(\hat{x})\}$, where U is a neighborhood of $\hat{x}$. Since $\hat{x} \in \partial A(x_s)$, it follows that $U \cap A(x_s) \neq \emptyset$. Indeed $W^s(\hat{x}) \cap A(x_s) = \emptyset$. Therefore we have

$$\{U - W^s(\hat{x})\} \cap A(x_s) \neq \emptyset$$

or

$$\cup_{t<0} \phi_t(G_\varepsilon) \cap A(x_s) \neq \emptyset.$$

This implies that $\phi(t, G_\varepsilon) \cap A(x_s) \neq 0$ for some t. Since a stability region is invariant under the flow it follows that

$$G_\varepsilon \cap A(x_s) \neq \emptyset.$$

Since $\varepsilon > 0$ is arbitrary and G is a compact set, we conclude that G contains at least a point of $\overline{A(x_s)}$. The proof of (b) is similar to the proof of (a), thus completing the proof.

The above characterization of an equilibrium point lying on the stability boundary can be extended to another critical element, i.e. closed orbit (limit cycle). Let $\Phi_t(p) = R^n \to R^n$ be the map $\Phi_t(p) = \phi(t, p)$. A closed orbit γ is said to be *hyperbolic* if for any $p \in \gamma$, $n - 1$ of the eigenvalues of the Jacobian of $\Phi_t(\gamma)$ at p have a modulus not equal to 1 (one eigenvalue must always be 1). A *critical element* of the vector field f is either a closed orbit or an equilibrium point.

Recall that the stable and unstable manifolds of a hyperbolic closed orbit γ are defined by following:

$$W^s(\gamma) = \{x \in R^n : \lim_{t\to\infty} \|\phi(t, x) - \gamma\| = 0\}$$
$$W^u(\gamma) = \{x \in R^n : \lim_{t\to-\infty} \|\phi(t, x) - \gamma\| = 0\}.$$

A complete characterization for a closed orbit (limit cycle) to lie on the stability boundary is presented in Theorem 4-2. The proof is similar to that of Theorem 4-1, and hence is omitted.

THEOREM 4-2 **(Characterization of closed orbit on the stability boundary)** *Consider a general nonlinear continuous dynamical system (4.1). Let $A(x_s)$ be the stability region of an asymptotically stable equilibrium point x_s. Let γ be a hyperbolic closed orbit. Then*

(a) $\gamma \subseteq \partial A$ if and only if $\{W^u(\gamma) - \gamma\} \cap \overline{A(x_s)} \neq \emptyset$;
(b) suppose $\{W^s(\gamma) - \gamma\} \neq \emptyset$, then $\gamma \subseteq \partial A(x_s)$ if and only if $\{W^s(\gamma) - \gamma\} \cap \partial A(x_s) \neq \emptyset$.

We next seek to develop alternative characterizations that can be effectively checked by numerical simulation for Theorem 4-1 and Theorem 4-2. As a corollary to Theorem 4-1, if $\{W^u(\hat{x}) - \hat{x}\} \cap A(x_s) \neq \emptyset$, then $\hat{x}$ must be on the stability boundary. Since any trajectory in $A(x_s)$ approaches x_s, we see that a sufficient condition for $\hat{x}$ to be on the stability boundary is the existence of a trajectory in $W^u(\hat{x})$ which approaches x_s. One nice thing about this condition is that it can be checked numerically. From a practical point of view one would like to see when this condition is necessary. We will show that this condition becomes necessary under two additional assumptions.

So far we have made only one assumption that the critical elements are hyperbolic. This assumption in fact is a generic property of nonlinear dynamical systems. Roughly speaking we say a property is generic for a class of systems if that property is true for *almost all* systems in the class. A formal definition is given in [234]. It was shown in [234] that among C^r ($r \geq 1$) vector fields, the following properties are generic: (i) all equilibrium points and closed orbits are hyperbolic and (ii) the intersections of the stable and unstable manifolds of critical elements satisfy the transversality condition. Theorem 4-1 can be sharpened under two conditions, one of which is generic for a nonlinear dynamical system (4.1). That is the transversality condition. The other condition requires that every trajectory on the stability boundary approaches one of the critical elements.

Now, we present a key theorem which characterizes an equilibrium point which is on the stability boundary in terms of both its stable and unstable manifolds. From the practical point of view, this result is useful for checking numerically whether or not an equilibrium point lies on a stability boundary.

THEOREM 4-3 **(Further characterization of an equilibrium point on the stability boundary)**
Let $A(x_s)$ be the stability region of an asymptotically stable equilibrium point x_s of the nonlinear dynamical system (4.1). Let $\hat{x}$ be an equilibrium point. Assume the following:

(A1) all the equilibrium points on $\partial A(x_s)$ are hyperbolic;
(A2) the stable and unstable manifolds of equilibrium points on $\partial A(x_s)$ satisfy the transversality condition;
(A3) every trajectory on $\partial A(x_s)$ approaches one of the equilibrium points as $t \to \infty$.

Then

(1) $\hat{x} \in \partial A(x_s)$ if and only if $W^u(\hat{x}) \cap A(x_s) \neq \emptyset$,
(2) $\hat{x} \in \partial A(x_s)$ if and only if $W^s(\hat{x}) \subseteq \partial A(x_s)$.

Proof (1) Because of Theorem 4-1 we only need to prove that, under these assumptions, $W^u(\hat{x}) \cap \overline{A(x_s)} \neq \emptyset$ implies $W^u(\hat{x}) \cap A(x_s) \neq \emptyset$. Let $n_u(x)$ denote the type of equilibrium point x, i.e. the dimension of its unstable manifold. It follows from assumption (A1) that $n_u(x) \geq 1$ for all equilibrium points $x \in \partial A(x_s)$. Let $\hat{x} \in \partial A(x_s)$ and $n_u(\hat{x}) = h$. By Theorem 4-1 there exists a point $y \in \{W^u(\hat{x}) - \hat{x}\} \cap \overline{A(x_s)}$. If $y \in A(x_s)$, the proof is complete. If $y \in \partial A(x_s)$, by assumption (A3) there exists an equilibrium point $\hat{z} \in \partial A(x_s)$ and $y \in \{W^s(\hat{z}) - \hat{z}\}$. Let $n_u(\hat{z}) = m$. By assumption (A2) $W^u(\hat{x})$ and $W^s(\hat{z})$ meet transversely at y, and thus by the following lemma, $h > m$.

LEMMA 4-4 *Let x_i and x_j be hyperbolic critical elements of the nonlinear dynamical system (4.1). Suppose that the intersection of the stable and unstable manifolds of x_i, x_j satisfies the transversality condition and $\{W^u(x_i) - x_i\} \cap \{W^s(x_j) - x_j\} \neq \emptyset$. Then*

dim $W^u(x_i) \geq \dim W^u(x_j)$, *where the equality sign is true only when x_i is an equilibrium point and x_j is a closed orbit.*
Now, consider two cases.

(a) $h = 1$: *then m must be zero (i.e. $\hat{z}$ must be a stable equilibrium point), which is a contradiction to the fact that no stable equilibrium point exists on the stability boundary. Consequently,* $W^u(\hat{x}) \cap A(x_s) \neq \emptyset$.
(b) $h > 1$: *without loss of generality, we assume inductively that* $W^u(\hat{z}) \cap A(x_s) \neq \emptyset$. *Since $W^u(\hat{x})$ and $W^s(\hat{z})$ intersect transversely at y, $W^u(\hat{x})$ contains an m-disk N centered at y, transverse to $W^s(\hat{z})$. Applying the following lemma with $\upsilon = \hat{z}$, we have $\phi(t, N) \cap A(x_s) \neq \emptyset$ for some $t > 0$.*

LEMMA 4-5 *Let υ be a hyperbolic critical element of the nonlinear dynamical system (4.1) with dimension $W^u(\hat{\upsilon}) = m$. If υ is an equilibrium point, let D be an m-disk in $W^u(\hat{\upsilon})$. If υ is a closed orbit, let D be an $(m-1)$-disk in $W^u(\hat{\upsilon}) \cap S$, where S is a cross section at $p \in \upsilon$. Let N be an m-disk (if υ is an equilibrium point) or $(m-1)$-disk (if υ is a closed orbit) having a point of transversal intersection with $W^s(\hat{\upsilon})$. Then D is contained in the closure of the set $\cap_{t \geq 0} \phi(t, N)$.*

Since $A(x_s)$ is invariant, this implies that $N \cap A(x_s) \neq \emptyset$, hence, $W^u(\hat{x}) \cap A(x_s) \neq \emptyset$. This completes part (1).

(2) If $W^s(\hat{x}) \subset \partial A(x_s)$, then $\hat{x} \in \partial A$ since $\hat{x} \in W^s(\hat{x})$. Conversely, suppose $\hat{x} \in \partial A(x_s)$. By part (1), $W^u(\hat{x}) \cap A(x_s) \neq \emptyset$. Let $D \subset W^u(\hat{x}) \cap A(x_s)$ be an m-disk, $m = \dim W^u(\hat{x})$. Let $y \in W^s(\hat{x})$ be arbitrary. For any $\varepsilon > 0$, let N be an m-disk transversal to $W^s(\hat{x})$ at y, contained in the ε-neighborhood of y. By Lemma 4-5 with $\upsilon = \hat{x}$, there exists a $t > 0$ such that $\phi(t, N)$ is so close to D that $\phi(t, N)$ contains a point $p \in A(x_s)$. Thus, $\phi(-t, p) \in N$. Since the stability region is invariant, this shows that $N \cap A(x_s) \neq \emptyset$. Letting $\varepsilon \to 0$ proves $y \in \overline{A(x_s)}$. Thus $W^s(\hat{x}) \subset \overline{A(x_s)}$. Since $W^s(\hat{x})$ is disjoint from A, it follows that $W^s(\hat{x}) \subset \partial A(x_s)$. This completes the proof.

To show that the transversality condition is needed in Theorem 4-3, let us consider the example taken from [254]. In Figure 4.1 the transversality condition is not satisfied because the intersection of the unstable manifold of x_1 and the stable manifold of x_2 is a portion of the manifold whose tangent space has a dimension of one. Note that the unstable manifold of x_1 intersects with the stability boundary (see Theorem 4-1), but not the stability region (see Theorem 4-3). A part of the stable manifold of x_1 (the upper part in Figure 4.1) is not in the stability boundary (see Theorem 4-3).

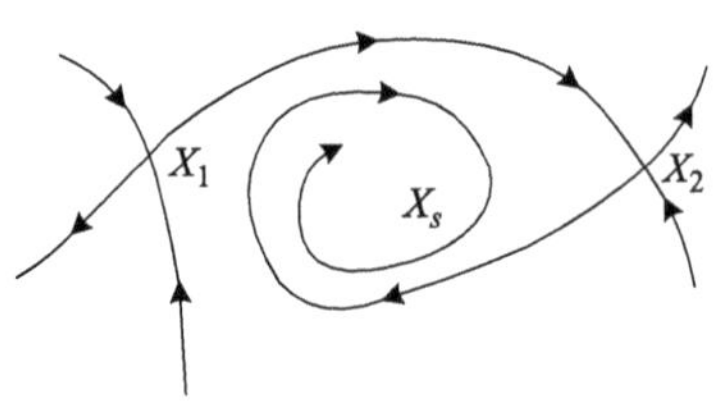

Figure 4.1 The intersection between the unstable manifold of x_1 and the stable manifold of x_2 does not satisfy the transversality condition.

Theorem 4-6 below extends the result of Theorem 4-3 to accommodate closed orbits on the stability boundary.

THEOREM 4-6 **(Characterization of critical element on the stability boundary)** *Let $A(x_s)$ be the stability region of a stable equilibrium point x_s of the nonlinear dynamical system (4.1). Let r be a critical element. Assume the following.*

(B1) All the critical elements on $\partial A(x_s)$ are hyperbolic.
(B2) The stable and unstable manifolds of critical elements on $\partial A(x_s)$ satisfy the transversality condition.
(B3) Every trajectory on $\partial A(x_s)$ approaches one of the critical elements as $t \rightarrow \infty$.

Then

(1) $\hat{r}$ is on the stability boundary $\partial A(x_s)$ if and only if $W^u(\hat{r}) \cap A(x_s) \neq \emptyset$,
(2) $\hat{r}$ is on the stability boundary $\partial A(x_s)$ if and only if $W^s(\hat{r}) \subseteq \partial A(x_s)$.

The next result concerns the number of equilibrium points on the stability boundary. We say that $S \subset R^n$ is a *smooth* manifold of dimension s if, for each point $p \in S$, there exist a neighborhood $U \subset S$ of p and a homeomorphism $h{:}U \rightarrow V$, where V is an open subset of R^s, such that the inverse homeomorphism $h^{-1}{:}V \rightarrow U \subset R^n$ is an immersion of class C^1.

THEOREM 4-7 **(Number of equilibrium points on the stability boundary)** *Let $A(x_s)$ be the stability region of a stable equilibrium point x_s of the nonlinear dynamical system (4.1). If the stability boundary $\partial A(x_s)$ of a stable equilibrium point is a smooth compact manifold and all the equilibrium points on $\partial A(x_s)$ are hyperbolic, then the number of equilibrium points on ∂A is even.*

Proof The proof is based on the following fact [122, Exercise 7, p.139]: the Euler characteristic of the boundary of a compact manifold is even. From the Poincaré Hopf index theorem [109, p.134], it follows that the sum of the indices of equilibrium points that lie on the smooth, compact stability boundary $\partial A(x_s)$ are even. But the index of f at a hyperbolic equilibrium point is either +1 or −1 [186, p.37]. Consequently, this theorem follows.

Theorem 4-7 is also true if instead of hyperbolicity, we assume only that every equilibrium point is *non-degenerate* in the sense that the corresponding Jacobian matrix of the vector field is invertible. The proof is the same as shown previously.

4.2 Complete characterization of the stability boundary

In this section we present a complete characterization of the stability boundary for a fairly large class of the nonlinear dynamical systems (4.1) whose stability boundary is non-empty. This global characterization is built on the local characterizations developed in previous section.

We start the analysis with the class of nonlinear dynamical systems (4.1) admitting an energy function. If there exists an energy function for system (4.1), then every trajectory on the stability boundary $\partial A(x_s)$ converges to one of the equilibrium points on the stability boundary $\partial A(x_s)$. In fact, the significance of Theorem 3-3 is that it offers a way to characterize the stability boundary. It asserts that the stability boundary $\partial A(x_s)$ is contained in the union of stable manifolds of the UEPs on the stability boundary.

We next present sharper results, compared to Theorem 3-3, on the complete characterization of stability boundary for system (4.1).

We make the following assumptions concerning the vector field.

(A1) All the equilibrium points on the stability boundary are hyperbolic.
(A2) The stable and unstable manifolds of equilibrium points on the stability boundary satisfy the transversality condition.
(A3) Every trajectory on the stability boundary approaches one of the equilibrium points as $t \to \infty$.

Theorem 4-8 asserts that if assumptions (A1) to (A3) are satisfied, then the stability boundary is the union of the stable manifolds of the equilibrium points on the stability boundary.

THEOREM 4-8 **(Characterization of stability boundary)**
For a nonlinear autonomous dynamical system (4.1) which satisfies assumptions (A1) to (A3), let x_i, $i = 1, 2, \ldots$ be the equilibrium points on the stability boundary $\partial A(x_s)$ of the asymptotically stable equilibrium point x_s. Then

(a) $x_i \in \partial A(x_s)$ *if and only if* $W^u(x_i) \cap A(x_s) \neq \emptyset$,
(b) $\partial A(x_s) = \cup W^s(x_i)$.

Proof Part (a) is shown in Theorem 4-3. We prove part (b). Let x_i, $i = 1, 2, \ldots$ be the equilibrium points on the stability boundary $\partial A(x_s)$. Theorem 4-3 implies the following

$$\partial A(x_s) \supseteq \bigcup_i W^s(x_i). \tag{4.3}$$

Assumption (A3) implies

$$\partial A(x_s) \subseteq \bigcup_i W^s(x_i). \tag{4.4}$$

Combining Eqs. (4.3) and (4.4) we complete the proof for part (b).

Remarks

[1] Assumption (A1) is a generic property of C^1 dynamical systems, and can be checked for a particular system by direct computation of the eigenvalues of the corresponding Jacobian matrix of the vector field.
[2] Assumption (A2) is also a generic property; however, it is not easy to check.
[3] Assumption (A3) is not a generic property, thus it needs to be verified. Several methods of verifying this assumption will be described in a later part of this chapter. The existence of an energy function provides a sufficient condition for assumption (A3) to hold.

Theorem 4-8 can be generalized to allow closed orbits to exist on the stability boundary.

THEOREM 4-9 **(Characterization of a stability boundary)**
A nonlinear dynamical system (4.1) satisfies the following assumptions.

(B1) All the critical elements on the stability boundary are hyperbolic.
(B2) The stable and unstable manifolds of critical elements on the stability boundary satisfy the transversality condition.
(B3) Every trajectory on the stability boundary approaches one of the critical elements as $t \to \infty$.

Let x_i, $i = 1, 2, \ldots$, be the equilibrium points and γ_j, $j = 1, 2, \ldots$, be the closed orbits on the stability boundary $\partial A(x_s)$ of the asymptotically stable equilibrium point x_s. Then

$$\partial A(x_s) = \bigcup_i W^s(x_i) \bigcup_j W^s(\gamma_j).$$

Proof By Theorem 4-6, the stable manifolds of critical elements which are on the stability boundary $\partial A(x_s)$ of an asymptotically stable equilibrium point lie in the stability boundary $\partial A(x_s)$; hence

$$\partial A(x_s) \supseteq \bigcup_i W^s(x_i) \bigcup_j W^s(\gamma_j). \tag{4.5}$$

By assumption (B3), every point on the stability boundary $\partial A(x_s)$ is on the stable manifold of one of the critical elements on stability boundary $\partial A(x_s)$; hence

$$\partial A(x_s) \subseteq \bigcup_i W^s(x_i) \bigcup_j W^s(\gamma_j). \tag{4.6}$$

Combining Equations (4.5) and (4.6) we complete the proof.

Remarks

[1] Assumption (B1) is a generic property of C^1 dynamical systems; however, because closed orbits are hard to determine it is difficult to check assumption (B1) for a given system except perhaps for planar systems.

[2] Assumption (B2) is also a generic property, but it is even harder to check. Assumption (B3) is a generic property only for planar systems. For higher dimensional systems no general methods for verifying this assumption are known.

The following theorem develops analytical results on the structure of the equilibrium points on the stability boundary. Moreover, it presents a necessary condition for the existence of certain types of equilibrium points on a *bounded* stability boundary.

THEOREM 4-10 **(Structure of equilibrium points on the stability boundary)**
For the nonlinear autonomous dynamical system (4.1) containing two or more stable equilibrium points, if the system satisfies assumptions (A1) to (A3), then the stability boundary $\partial A(x_s)$ of the stable equilibrium point x_s must contain at least one type-one

equilibrium point. If, furthermore, the stability region $A(x_s)$ is bounded, then $\partial A(x_s)$ must contain at least one type-one equilibrium point and one source.

Proof Since there are at least two stable equilibrium points, including, say x_s, it follows that the dimension of $\partial A(x_s)$ is $(n-1)$. Since $\partial A(x_s) = \cup W^s(x_j)$, where $x_j \in \partial A(x_s)$, at least one of the x_j must be a type-one equilibrium point, say x_1, so that the dimension of $\cup W^s(x_j)$ is $(n-1)$. Repeating the same argument, if $\partial W^s(x_1)$ is non-empty, then the dimension of $\partial W^s(x_1)$ is less than $(n-2)$, say $(n-k)$. Application of Theorem 4-8 yields $\partial W^s(x_1) = \cup W^s(x_j)$, $x_j \in \partial W^s(x_1)$. In order for $\cup W^s(x_j)$ to have dimension $(n-k)$, at least one of the x_j must be a type-k equilibrium point. If the stability region is bounded, the same argument can be repeated until we reach a type-n equilibrium point (a source). This completes the proof.

The contra-positive of Theorem 4-10 leads to the following corollary, which is useful in predicting the unboundedness of the stability region.

THEOREM 4-11 **(Sufficient condition for an unbounded stability region)**
Consider the nonlinear autonomous dynamical systems (4.1) with a stable equilibrium point x_s whose stability boundary is non-empty. If assumptions (A1) to (A3) are satisfied and if $\partial A(x_s)$ contains no source, then the stability region $A(x_s)$ is unbounded.

4.3 Trajectories on the stability boundary

The characterization of stability boundaries in the previous section is valid for dynamical systems satisfying assumptions (A1) to (A3). Since assumptions (A1) and (A2) are generic properties, assumption (A3) is the crucial one in the application of the complete characterization. In this section, we will show that many dynamical systems arising from physical system models satisfy assumption (A3). We first present two theorems that give sufficient conditions for this assumption.

It should be stressed that the main results in this section are independent of the existence of Lyapunov functions. For a convenient sufficient condition guaranteeing assumption (A3), we will introduce a function in the following theorems resembling a Lyapunov function. Recall that E denotes the set of equilibrium points of (4.1). If V is a scalar function on R^n, then $\dot{V}(x) = d/dt|_{t=0} V(\phi(t,x)) = \nabla V(x) \cdot f(x)$.

THEOREM 4-12 **(Sufficient condition for (A3))**
Let the set E be the collection of all the equilibrium points that lie on the stability boundary ∂A of the nonlinear autonomous dynamical systems (4.1). Suppose there exists a C^1 function $V: R^n \to R$ f such that

(1) $\dot{V}(x) < 0$ if $x \notin E$.

Suppose also that there exists a $\delta > 0$ such that for any $\hat{x} \in E$, the open ball $B_\delta(\hat{x}) := \{x : |x - \hat{x}| < \delta\}$ contains no other point in E and that the distance between *any two such balls is at least δ. Furthermore, suppose that there exist a positive continuous function $\alpha:R^n \to R^+$ and two constants, $c_1 > 0$ and $c_2 > 0$, such that*

(2) $\alpha(x)|f(x)| < c_1$ for all $x \in \partial A$;
(3) $\alpha(x)\dot{V}(x) < -c_2$ unless $x \in B_\delta(\hat{x})$ for some $\hat{x} \in E \cap \partial A$.

Under these conditions, the assumption (A3) holds for the nonlinear autonomous dynamical systems (4.1): every trajectory on ∂A converges to an equilibrium point as $t \to \infty$.

Proof Let $x(t) := \phi(t, x)$ be a trajectory on the stability boundary. Suppose $x(t)$ does not approach one of the equilibrium points. We will show that this leads to a contradiction. We consider two cases.

Case 1: There exists a $T > 0$ such that for all $t > T$, $\phi(t, x)$ is not in $B_\delta(\hat{x})$, for any $\hat{x} \in E$. Therefore, by condition (3) we have

$$\dot{V}(x(t)) < -\frac{c_2}{\alpha(x(t))} \quad \text{for all} \quad t > T.$$

We estimate that for $t > T$,

$$\begin{aligned} V(x(t)) - V(x(T)) &= \int_T^t \dot{V}(x(\tau))d\tau \\ &< -c_2 \int_T^t \frac{1}{\alpha(x(\tau))} d\tau < -\frac{c_2}{c_1}\int_T^t |f(x(\tau))| d\tau. \end{aligned}$$

This shows that $\lim_{t\to\infty} V(x(t)) = -\infty$. But this contradicts the fact that $V(.)$ is bounded below (by $V(x_s)$) along any trajectory on the stability boundary, which follows from condition (1) and the continuity property of the function $V(.)$.

Case 2: There is an infinite sequence $\{\hat{p}_i\}$ of equilibrium points and an increasing sequence $r_i \to \infty$ such that $x(r_j) \in B_\delta(\hat{p}_i)$.

Let us define two increasing sequences $\{t_i\}$ and $\{s_i\}$ where the sequence $\{t_i\}$ is the first time $x(t)$ enters the δ-ball $B_\delta(\hat{p}_i)$ and s_i is the first time $t > t_i$ that $x(t)$ leaves the 2δ-ball $B_{2\delta}(\hat{p}_i)$.

Fix an integer $m > 0$; then for $t \geq t_{m+1}$ we have

$$\begin{aligned} V(x(t)) - V(x(0)) &= \int_0^t \dot{V}(x(\tau))d\tau \\ &< \sum_{i=1}^m \int_{t_i}^{s_i} \dot{V}(x(\tau))d\tau < -\frac{c_2}{c_1}\sum_{i=1}^m \left| \int_{t_i}^{s_i} \dot{x}(\tau)d\tau \right| \\ &< -\frac{c_2}{c_1} m\delta. \end{aligned}$$

Letting $m \to \infty$, we contradict the fact that $V(.)$ is bounded below on the stability boundary. Therefore every trajectory on the stability boundary must approach one of the equilibrium points. This completes the proof.

Theorem 4-12 provides a large class of sufficient conditions for assumption (A3) to hold. One example of these sufficient conditions is stated below.

COROLLARY 4-13 **(Sufficient condition for (A3))**
If the nonlinear autonomous dynamical systems (4.1) satisfy the following conditions,

(1) it has a finite number of equilibrium points on its stability boundary;
(2) there exists a C^1 function $V{:}R^n \to R$ and two positive numbers ε and δ such that (i) $\dot{V}(x) < 0$ if $x \notin E$ and $\dot{V}(x) < -\delta$ if $x \notin B_\varepsilon(\hat{x}), \hat{x} \in E$, and (ii) $|f(x)|$ is bounded for $x \in R^n$,

then assumption (A3) holds for the nonlinear autonomous dynamical systems (4.1).

THEOREM 4-14 **(Sufficient condition for (A3))**
If there exists a C^1 function $V{:}R^n \to R$ for the nonlinear autonomous dynamical systems (4.1) such that

(1) $\dot{V}(x) \leq 0$ at every point $x \notin E$,
(2) if $x \notin E$, then the set $\{t \in R : \dot{V}(\phi(t,x)) = 0\}_{t \geq 0}$ has measure 0 in R, and either
(3) the map $V: R^n \to R$ is proper or
(3') for each $x \in R^n$, the condition of $\{V(\phi(t, x))\}_{t \geq 0}$ being bounded ensures that $\{\phi(t, x)\}_{t \geq 0}$ is bounded,

then assumption (A3) holds for the nonlinear autonomous dynamical systems (4.1).

Many second-order dynamical systems frequently encountered in physical system models are systems of the form

$$M\ddot{x} + D\dot{x} + f(x) = 0$$

whose state space representation is

$$\begin{aligned} \dot{x} &= y \\ M\dot{y} &= -Dy - f(x) \end{aligned} \tag{4.7}$$

where M is a diagonal matrix with positive elements, D is a symmetric, diagonally dominant matrix with positive diagonal elements, $f{:}R^n \to R^n$ is a bounded gradient vector field with bounded Jacobian, and the number of equilibrium points on any stability boundary is finite. Moreover, $f(.)$ is assumed to satisfy the following condition. There exists an $\varepsilon > 0$ and $\delta > 0$ such that (i) the distance between the two balls $B_\varepsilon(x_i)$ and $B_\varepsilon(x_j)$ is greater than ε, for all $x_i\, x_j \in E$ and (ii) $|f(x)| > \delta$, for $x \notin \cup_{\bar{x} \in E} B_\varepsilon(\bar{x})$, where $E{:} = \{x{:}f(x) = 0\}$, $B_r(\hat{x}) := \{x : |x - \hat{x}| \leq r\}$. We consider the function $V(x,y) = \frac{1}{2}\langle y, My\rangle + \int_0^x \langle f(u), du\rangle$. By applying Theorem 4-14 to the second-order system (4.7), it follows that assumption (A3) is satisfied for the system. Hence, the stability boundary of the second-order system (4.7) can be completely characterized. In addition, its stability region is unbounded as stated in the following theorem.

THEOREM 4-15 **(Unbounded stability region)**
The stability boundary $\partial A(x_s)$ of an (asymptotically) stable equilibrium point of the second-order system (4.7) is unbounded.

Proof We will show that the equilibrium points of (4.7) do not contain a source. Specifically we will show that the Jacobian $J(\hat{x})$ always has eigenvalues with a negative real part. This is done by applying the inertia theorem [278] which states that if H is a nonsingular Hermitian matrix and A has no eigenvalues on the imaginary axis (i.e. $n_c(A) = n_c(H) = 0$), then $AH + HA^* \geq 0$ implies

$$In(A) = In(H). \tag{4.8}$$

Now we choose H to be the symmetric matrix

$$H = \begin{bmatrix} -F(\hat{x})^{-1} & 0 \\ 0 & -M^{-1} \end{bmatrix}. \tag{4.9}$$

We have

$$J(\hat{x})H + HJ(\hat{x})^T = 2\begin{bmatrix} 0 & 0 \\ 0 & M^{-1}DM^{-1} \end{bmatrix}. \tag{4.10}$$

Hence

$$In(J(\hat{x})) = In\begin{bmatrix} -F(\hat{x})^{-1} & 0 \\ 0 & -M^{-1} \end{bmatrix}. \tag{4.11}$$

Since M is a diagonal matrix with positive elements, the relationship stated in (4.11) implies that, at any equilibrium point $\hat{x}$, the Jacobian $J(\hat{x})$ has at least n eigenvalues with negative real parts. This completes the proof.

4.4 Algorithm to determine a complete stability boundary

Theorem 4-8 and Theorem 4-9 lead to the following conceptual algorithm for determining the stability boundary of an asymptotically stable equilibrium point of the nonlinear autonomous dynamical systems (4.1) satisfying assumptions (A1) to (A3).

Algorithm (To determine the stability boundary $\partial A(x_s)$)

Step 1: Find all the equilibrium points.

Step 2: Identify those equilibrium points whose unstable manifolds contain trajectories approaching the stable equilibrium point x_s.

Step 3: The stability boundary of x_s is the union of the stable manifolds of the equilibrium points identified in Step 2.

Step 1 in the algorithm involves finding all the solutions to $f(x) = 0$. Step 2 can be accomplished numerically. The following procedure is suggested.

(i) Find the Jacobian at the equilibrium point (say, $\hat{x}$).

(ii) Find many of the generalized unstable eigenvectors of the Jacobian having unit length.

(iii) Find the intersection of each of these normalized, generalized, and unstable eigenvectors (say, y_i) with the boundary of an ε-ball of the equilibrium point (the intersection points are $\hat{x} + \varepsilon y_i$ and $\hat{x} - \varepsilon y_i$).

(iv) Integrate backwards the vector field (i.e. in reverse time) from each of these intersection points up to some specified time (say, 20 time-steps). If the trajectory remains inside this ε-ball, then go to the next step. Otherwise, we replace the value ε by $\alpha\varepsilon$ and also the intersection points $\hat{x} \pm \varepsilon y_i$ by $\hat{x} \pm \alpha\varepsilon y_i$, where $0 < \alpha < 1$. Repeat this step.
(v) Numerically integrate the vector field starting from these intersection points.
(vi) Repeat steps (iii) through (v). If any of these trajectories approaches x_s, then the equilibrium point is on the stability boundary.

For a planar system, the equilibrium point on the stability boundary is either a type-one equilibrium point or a type-two equilibrium point, which is a source. The stable manifold of a type-one equilibrium point in this case has dimension one, which can easily be determined numerically as follows.

(a) Find a normalized stable eigenvector y of the Jacobian at the equilibrium point $\hat{x}$.
(b) Find the intersection of this stable eigenvector with the boundary of an ε-ball of the equilibrium point $\hat{x}$ (where the intersection points are $\hat{x} + \varepsilon y$ and $\hat{x} - \varepsilon y$).
(c) Integrate the vector field from each of these intersection points after some specified time. If the trajectory remains inside this ε-ball, then go to the next step. Otherwise, we replace the value ε by $\alpha\varepsilon$ and also the intersection points $\hat{x} \pm \varepsilon y_i$ by $\hat{x} \pm \alpha\varepsilon y_i$, where $0 < \alpha < 1$. Repeat this step.
(d) Numerically integrate the vector field backward (reverse time) starting from these intersection points.
(e) The resulting trajectories are the stable manifold of the equilibrium point.

For higher dimensional systems, a numerical procedure similar to the one above can only provide a set of trajectories on the stable manifold. Finding the stable manifold and unstable manifold of an equilibrium point is a nontrivial problem, and advanced numerical methods for computing stable and unstable manifolds are needed [33,150].

4.5 Numerical examples

The method for a complete determination of a stability region presented in the previous section will be illustrated by two simple examples. In each example, two figures will be derived; one compares the stability region estimated using the previous methods and using the present method, while the other gives the phase portrait of the system to verify the results of this method. Throughout these examples we *assume* the transversality condition (A2) is satisfied.

Example 4-1 This is an example studied in [98,182].

$$\begin{aligned} \dot{x}_1 &= -2x_1 + x_1x_2 \\ \dot{x}_2 &= -x_2 + x_1x_2. \end{aligned} \tag{4.12}$$

There are two equilibrium points: (0, 0) is a stable equilibrium point and (1, 2) is a type-one equilibrium point. Assumption (A1) is satisfied. The trajectory on the unstable manifold of (1, 2) converges to the stable equilibrium point (0, 0); hence, (1, 2) is on the stability boundary $\partial A(0,\ 0)$ (see Theorem 4-1). Next, we check assumption (A3). Consider the following function:

$$V(x_1,x_2) = x_1^2 - 2x_1x_2 + x_2^2.$$

The derivative of $V(x_1, x_2)$ along the trajectory of (4.12) is

$$\begin{aligned}\dot{V}(x_1,x_2) &= \frac{\partial V}{\partial x_1}\dot{x}_1 + \frac{\partial V}{\partial x_2}\dot{x}_2 \\ &= -2(2x_1 - x_2)(x_1 - x_2).\end{aligned}$$

Hence,

$$\dot{V}(x_1,x_2) < 0 \text{ for } (x_1,x_2) \in R^2 - B$$

where $B{:} = \{(x_1, x_2){:}2x_1 - x_2 \geq 0 \text{ and } x_1 - x_2 \leq 0\}$.

Define the following sets:

$$\widetilde{B} := B_1 \cup B_2 \cup B_3$$

where $B_1 = \{(x_1, x_2){:}x_1 < 1, x_2 < 2\}$, $B_2 = \{(x_1,x_2) : x_1 \geq 1, x_2 \leq 2\} \cap B$, and $B_3 = \{(x_1, x_2){:}x_1 > 1,\ x_2 > 2\}$. Since, in the set B_1, both $|x_1(t_i)|$ and $|x_2(t_i)|$ are strictly decreasing sequences, we conclude that B_1 is inside the stability region (0, 0). In other words, the stability boundary $\partial A(0,0)$ cannot lie in B_1. On the other hand, every trajectory of (4.12) in the set B_3 is unbounded as $t \to \infty$; therefore, the stability boundary $\partial A(0,\ 0)$ cannot lie in B_3 either. However, by checking the vector field of (4.12) in B_2 we find that every trajectory in B_2 will either enter into B_1 or B_3 or converge to the point (1, 2).

We have shown that the stability boundary $\partial A(0,\ 0)$ cannot be in B_1 or B_3 and the part of the stability boundary $\partial A(0,\ 0)$ in B_2 must converge to (1, 2). Next, we will show that the part of the stability boundary $\partial A(0,\ 0)$ in $R^2 - \widetilde{B}$ also converges to (1, 2). Then, we may claim that assumption (A3) is satisfied. Note that $\dot{V}(x_1,x_2) < 0$ for $(x_1,x_2) \in R^2 - \widetilde{B} \subseteq R^2 - B$ and the function $V(x_1,\ x_2)$ is a proper map in $R^2 – \widetilde{B}$. Thus, every trajectory of $\partial A(0,\ 0)$ in $R^2 – \widetilde{B}$ is bounded and, if it converges in $R^2 – \widetilde{B}$, it must converge to an equilibrium point in $R^2 – \widetilde{B}$. However, there is no equilibrium point in $R^2 – \widetilde{B}$. So, the part of the stability boundary $\partial A(0,\ 0)$ in $R^2 – \widetilde{B}$ must enter the set $\widetilde{B}$. But, we have shown that the stability boundary $\partial A(0,\ 0)$ in $\widetilde{B}$ converges to (1, 2). Therefore, the trajectories on the stability boundary $\partial A(0,\ 0)$ converge to (1, 2), and assumption (A3) is shown to be satisfied.

It follows that the stability boundary is the stable manifold of (1, 2) (see Theorem 4-8), which is the curve C in Figure 4.2. Because there is no source, the stability region is unbounded (Theorem 4-11). Curves A and B in Figure 4.2 are obtained by the methods in [182] and [98], respectively. Figure 4.3 plots the phase portrait of this system, which confirms that curve C represents the exact stability boundary, which confirms the theoretical characterization of Theorem 4-8.

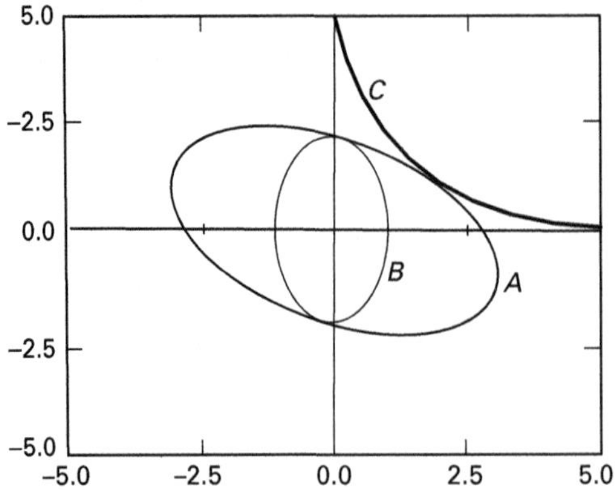

Figure 4.2 Predictions of the stability region of Example 4-1 by different methods. Curves A and B are obtained using the methods in [182] and [98]. Curve C is obtained using the present method.

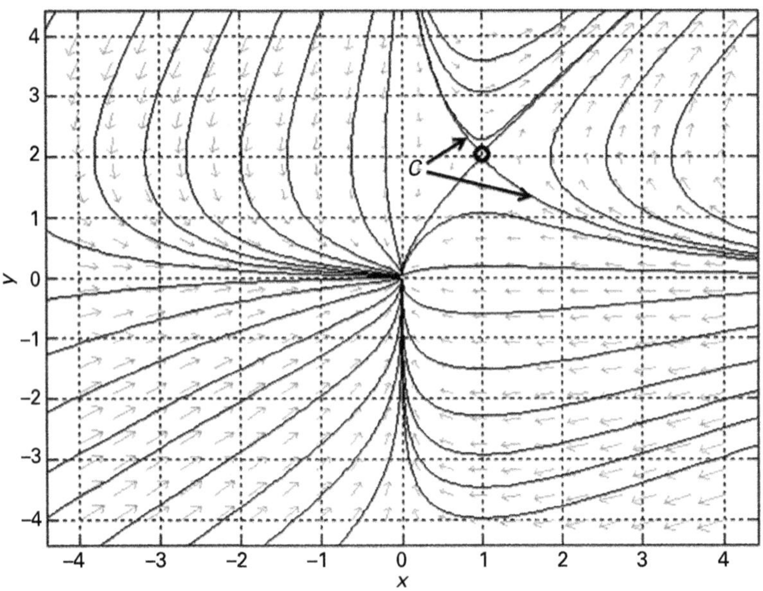

Figure 4.3 The phase portrait of the system of Example 4-1. Note that all the points inside the curve C converge to the stable equilibrium point, which verifies that curve C is the exact stability boundary.

Example 4-2 The following system is close to a power system transient stability model and is considered in [182]:

$$\dot{x}_1 = x_2$$

$$\dot{x}_2 = 0.301 - \sin(x_1 + 0.4136) + 0.138 \sin 2(x_1 + 0.4136) \quad 0.279 x_2. \tag{4.13}$$

Since this system is of the same form as (4.7), it follows that assumption (A3) is satisfied. The equilibrium points of (4.13) are periodic on the subspace $\{(x_1, x_2)\, x_2 = 0\}$ and the Jacobian matrix of (4.13) at (x_1, x_2) is

$$J(x) = \begin{bmatrix} 0 & 1 \\ a & -0.279 \end{bmatrix} \tag{4.14}$$

where $a = -\cos(x_1 + 0.4136) + 0.276\cos 2(x_1 + 0.4136)$. Let λ_1, λ_2 be the eigenvalues of $J(x)$:

$$\lambda_1 + \lambda_2 = -0.279$$
$$\lambda_1 \times \lambda_2 = -a\ .$$

The following observations are immediate.

(1) Assumption (A1) is satisfied.
(2) At least one of the eigenvalues must be negative, which implies that there is no source in the system (4.13). According to Theorem 4-11 we conclude that the stability region of a stable equilibrium point is unbounded.
(3) The stable equilibrium points and the type-one equilibrium points are located alternately on the x_1-axis.

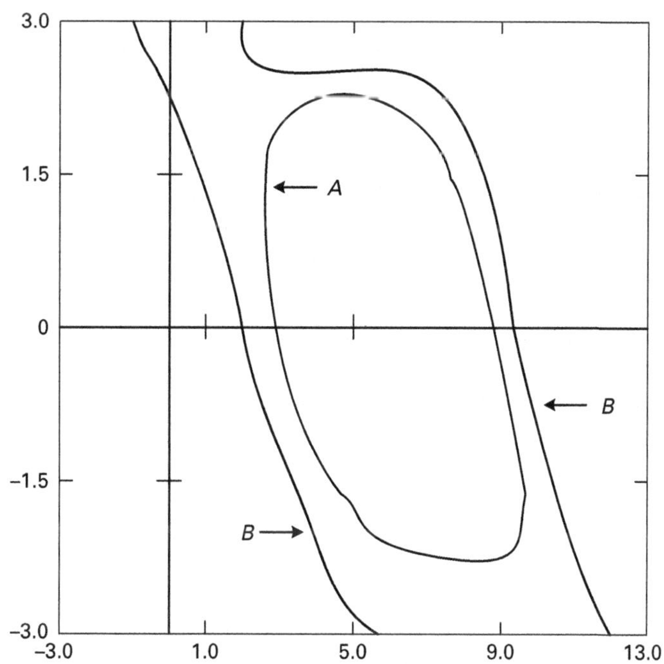

Figure 4.4 Estimations of the stability region of Example 4-2 by different methods. Curve *A* is the stability boundary estimated using the methods in [182] (after a shift in coordinates). Curve *B* is the stability boundary estimated using the present method.

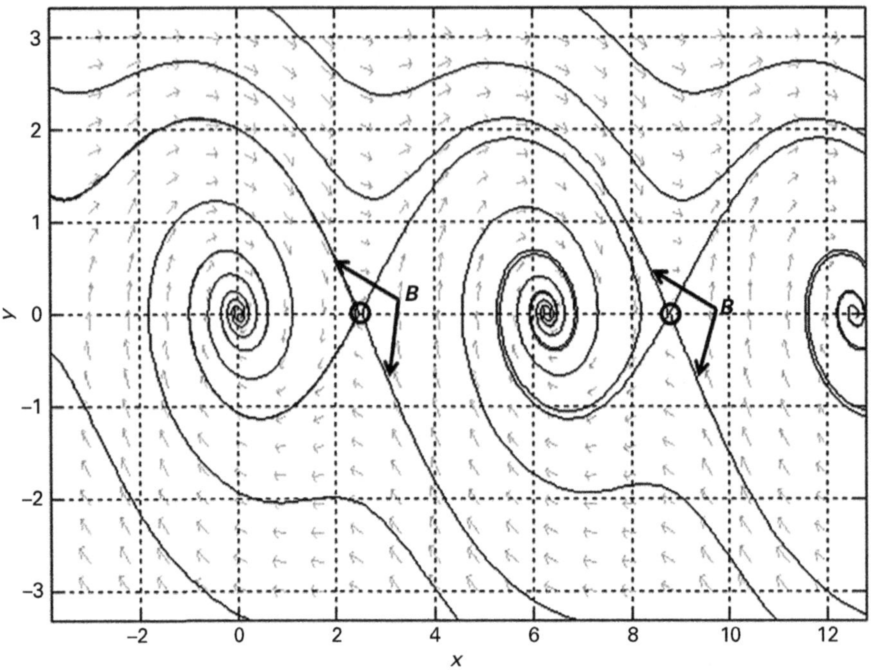

Figure 4.5 The phase portrait of the system of Example 4-2, which confirms that curve B in Figure 4.4 is the exact stability boundary.

Note that (6.284098, 0.0) is a stable equilibrium point of (4.13). Let us consider its stability region. The application of Theorem 4-1 asserts that the type-one equilibrium points (2.488345, 0.0) and (8.772443, 0.0) are on the stability boundary of (6.284098, 0.0). The stability region is again unbounded due to the absence of a source. The stability boundary obtained by the algorithm of Section 4.5 is the curve B shown in Figure 4.4 which is the union of the stable manifolds of the equilibrium points (2.488345, 0.0) and (8.772443, 0.0). Curve A is the estimated stability boundary obtained using the method described in [182] (after a shift in coordinates). It is clear from the phase portrait in Figure 4.5 that the trajectories of the points inside curve B converge to the stable equilibrium point, which verifies that curve B is the exact stability boundary.

4.6 Conclusion

In this chapter, a comprehensive theory of the stability regions of stable equilibrium points for nonlinear autonomous dynamical systems has been presented. A complete dynamical characterization of the stability boundary of a fairly large class of nonlinear autonomous dynamical systems has been derived. Several topological and dynamical characterizations of equilibrium points and limit cycles that lie on the stability boundary have been derived.

A conceptual method based on complete characterization for the complete determination of a stability boundary is proposed but it requires the determination of stable manifolds of equilibrium points. For low-dimensional systems this method can be implemented numerically. For high-dimensional systems, efficient computational methods to derive stable manifolds are needed.

The analytical results on the characterizations of critical elements on the stability boundary and on the complete characterization of the stability boundary are very useful in the development of practical methods for estimating stability regions of large-scale nonlinear systems. Details of these practical methods will be presented in later chapters.

5 Stability regions of attracting sets of complex nonlinear dynamical systems

A comprehensive theory of the stability region was developed for a fairly large class of nonlinear continuous dynamical systems in Chapter 4. In particular, the stability boundary of this class of nonlinear dynamical systems was shown to be equal to the union of the stable manifolds of the critical elements lying on the stability boundary. This complete characterization was derived under the key assumption that critical elements (limit sets) on the stability boundary are either hyperbolic equilibrium points or hyperbolic closed orbits.

However, many nonlinear dynamical systems arising from physical models exhibit limit sets with complex structure. Chua's electrical circuit, for example [70], exhibits chaos. Complex behavior has also been observed in certain electric power system models; for instance, the existence of limit cycles on the stability boundary of a power system with a DC transmission line has been reported in [240], and chaos and strange attractors have been observed in both 3-bus and 9-bus power system models [52,57,263]. It has been found that the bifurcation phenomena observed in the 3-bus and 9-bus power systems are similar. These bifurcation phenomena have been observed in the 39-bus power system as well, including Hopf, period-doubling, and cyclic fold bifurcations. Furthermore, some bifurcation phenomena not appearing in the 3-bus and 9-bus systems have surfaced in the 39-bus system [52]. These numerical studies favor the claim that various types of bifurcations, and thus various types of limit sets, can occur in real power systems. Complex behavior is common in biological models. The coexistence of multi-limit cycles, for example, has been observed in models for species competition [19]. The coexistence of multiple attractors has also been observed in economic models.

5.1 Complex limit sets

For a fairly large class of nonlinear dynamical systems, the structure of their limit sets can be very complex. Limit sets of general dynamical systems may contain equilibrium points, closed orbits, quasi-periodic solutions, and chaos. We use the term *complex* (or *general*) *dynamical systems* to denote a class of nonlinear dynamical systems that admit complex structure on their limit sets. A precise definition of this class will be given in Section 5.5. The body of theory developed in Chapter 4 is not applicable to this class of *complex nonlinear dynamical systems* whose stability boundaries admit a complex structure such as quasi-periodic solutions and chaos.

This chapter generalizes the comprehensive theory of the stability region developed in Chapter 4 in the following way:

- it develops a characterization of the stability boundary of an *attracting set* (which can be an equilibrium, a stable closed orbit, a stable quasi-periodic orbit, or a stable chaotic attractor) of a continuous dynamical system,
- it develops a characterization of the stability boundary admitting limit sets with complex structure, such as equilibrium points, closed orbits, quasi-periodic solutions and chaotic trajectories.

It will be shown that the stability boundary of this class of dynamical systems is composed of the union of stable manifolds of critical elements on the stability boundary. The critical elements in this case will be *indecomposable maximal invariant sets* on the stability boundary, which include, as particular cases, hyperbolic equilibrium points and hyperbolic closed orbits. As a consequence, the analytical results developed in this chapter extend the results of Chapter 4 into a larger class of nonlinear systems by relaxing the constraints on the structure of critical elements on the stability boundary.

The extension of the theory of the stability boundary, however, comes with a price in the sense that the results developed in this chapter are not as sharp as those derived in Chapter 4. In spite of that, the analytical developments made in this chapter, among others, will serve as the basis for the developments of solution methodologies for estimating stability regions of complex nonlinear dynamical systems, as presented in Chapter 11.

5.2 Stability region of attracting sets

We consider the following general class of autonomous nonlinear dynamical system

$$\dot{x} = f(x), \quad x \subset R^n \tag{5.1}$$

with the assumption that solutions exist and are unique for all $t \geq 0$. We will be interested in studying the stability region of *attracting sets*, a particular type of invariant set.

DEFINITION **(Attracting sets)**
A closed invariant set $\gamma \subset R^n$ *is called an* attracting set *if there exists some open neighborhood* V *of* γ *such that, for all* $x_0 \in V$, $\phi(t, x_0) \in V$ *for all* $t \leq 0$ *and* $\phi(t, x_0) \to \gamma$ *as* $t \to \infty$.

Generally speaking, a closed and invariant set γ is an attracting set if every trajectory starting in a neighborhood of γ stays close to γ for $t \geq 0$ and approaches γ as $t \to \infty$. The definition of an attracting set resembles the definition of asymptotic stability of an equilibrium point. The neighborhood V is called an attracting neighborhood of γ. An asymptotically stable equilibrium point is the simplest example of an attracting set. Asymptotically stable limit cycles are another example of attracting sets. Figure 5.1 illustrates the concept of an attracting set.

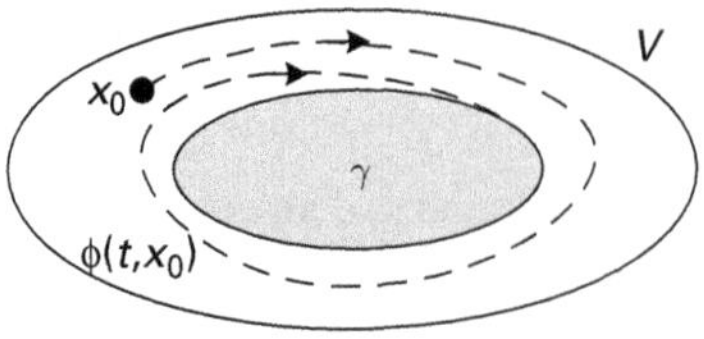

Figure 5.1 Illustration of an attracting set. Solutions starting in a neighborhood V of an attracting invariant set γ stay in V for all positive times and tend to set γ as t tends to infinity.

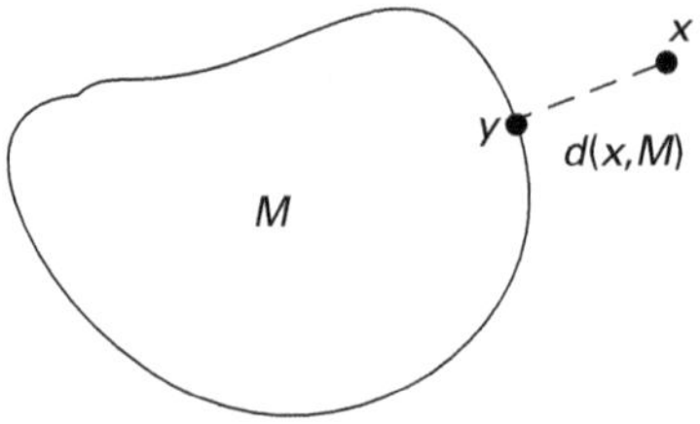

Figure 5.2 The distance from a point x to a set M. The distance of a point $x \in R^n$ to a set $M \subset R^n$ is given by $d(x, M) = \inf_{y \in M} d(x, y)$, the distance from x to the closest point y in M.

In order to make the definition of an attracting set precise, the meaning of the terminology "$\phi(t, x_0) \to \gamma$ as $t \to \infty$" needs to be explained. Let $d(.,.)$ be a distance in R^n. The distance of a point $x \in R^n$ to a set $\gamma \subset R^n$ is given by $d(x, \gamma) = \inf_{y \in \gamma} d(x, y)$, the distance from x to the closest point y in γ. Figure 5.2 illustrates the notion of distance from a point to a set. Then, the terminology "$\phi(t, x_0) \to \gamma$ as $t \to \infty$" implies $d(\phi(t, x), \gamma) \to 0$ as $t \to \infty$.

There are several definitions of attracting sets and attractors in literature. A comprehensive discussion on this subject can be found in [185]. The following definition of an attractor is more restrictive than the previous definition of attracting sets.

DEFINITION **(Attractor)**
An attractor is an attracting set γ that contains a dense orbit in γ.

It is difficult to identify dense orbits, therefore, the concept of attracting set is more practical, albeit less restrictive. We note that attracting sets are not necessarily *recurrent*. Consider the dynamical system:

$$\begin{aligned} \dot{x} &= x - x^3 \\ \dot{y} &= -y. \end{aligned} \tag{5.2}$$

The line segment $\gamma = \{(x, 0) \in R^2 | x \in [-1, 1]\}$ is an attracting set of system (5.2) and the set U in Figure 5.3 is an attracting neighborhood. This attracting set contains points that are not ω-limit points of any orbit. More precisely, every point in the set $\{(x, 0) \in R^2 | x \in (-1, 0) \cup (0, 1)\}$ is not an ω-limit point of any trajectory. In fact, every point in this set is a wandering point; those to the left of the origin have the equilibrium $(-1,0)$ as ω-limit point, and those to the right have the equilibrium $(1,0)$ as ω-limit point. Hence, the attracting set γ does not contain a dense orbit and, therefore, is not an attractor. It is also interesting to note that this attracting set contains an unstable

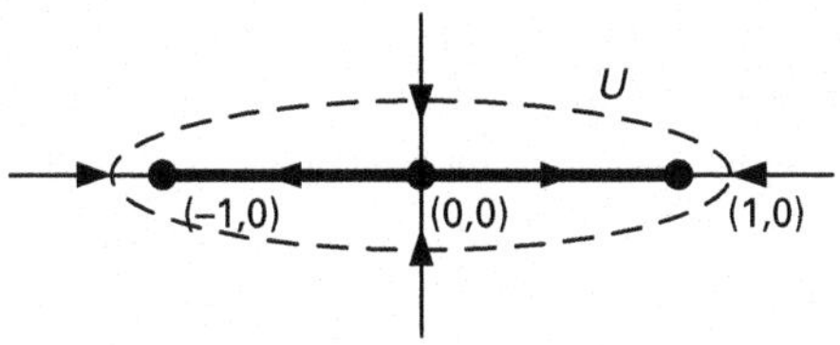

Figure 5.3 The attracting set of system (5.2).

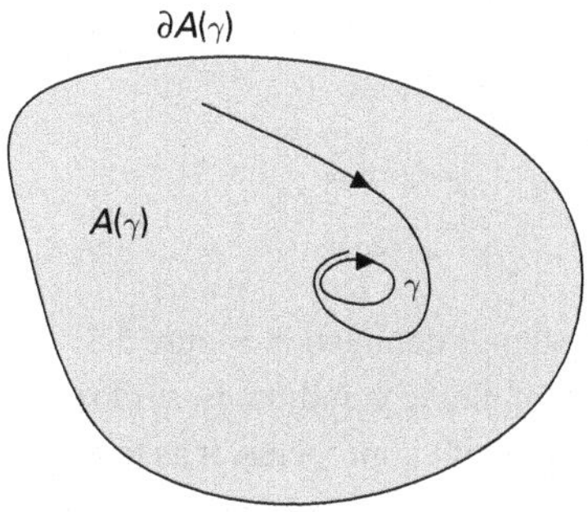

Figure 5.4 Stability region and stability boundary of an attracting set γ.

hyperbolic equilibrium point at the origin. Attracting sets can be very complex. An attracting set can contain a countable collection of equilibrium points, including a countable collection of unstable equilibrium points [107].

For general nonlinear dynamical systems, attracting sets, such as asymptotically stable equilibrium points and asymptotically stable closed orbits, are usually not globally stable. In fact, there is a region of the underlying state space, called the *stability region*, in which trajectories starting from this region converge to the attracting set as time goes to infinity. The stability region of an attracting set γ is defined as

$$A(\gamma) := \{x_o \in R^n : \phi(t, x_o) \to \gamma \text{ as } t \to \infty\}. \tag{5.3}$$

We denote the topological boundary of $A(\gamma)$ by $\partial A(\gamma)$, termed the stability boundary (of the attracting set γ). Figure 5.4 illustrates the stability region and stability boundary of an attracting set γ.

5.3 Topological properties

In this section we discuss some topological properties of the stability region and stability boundary of attracting sets. These properties are quite similar to the topological properties of the stability region and stability boundary of asymptotically stable equilibrium points.

THEOREM 5-1 **(Topological property)**
The stability region $A(\gamma)$ of an attracting set γ is a non-empty, open and invariant set.

Clearly $\gamma \subset A(\gamma)$ and the attracting neighborhood V of γ are also contained in $A(\gamma)$; therefore, $A(\gamma)$ is non-empty. The above result asserts that every trajectory in a stability region lies entirely in the stability region, and that the dimension of the stability region is n. Unlike the stability region of asymptotically stable equilibrium points, the stability regions of general attracting sets are not diffeomorphic to R^n. The next example illustrates the stability region of an asymptotically stable limit cycle that is not homeomorphic to R^2.

Example 5-1 Consider the nonlinear dynamical system:

$$\begin{aligned} \dot{x} &= (1 - \sqrt{x^2 + y^2})x - y \\ \dot{y} &= x + (1 - \sqrt{x^2 + y^2})y. \end{aligned} \tag{5.4}$$

The phase portrait of this dynamical system is depicted in Figure 5.5. The origin is an unstable focus while the circumference of unitary radius is a closed orbit. Every trajectory on the neighborhood of this closed orbit approaches it as time goes to infinity. As a result, the closed orbit is both an attracting set and an attractor. The stability region of this attracting set, which is not homeomorphic to R^2, is the entire space (R^2) excluding the origin.

Attracting invariant sets may not be connected. Furthermore, the union of two isolated attracting sets is also an attracting set. Hence, the stability region of general attracting

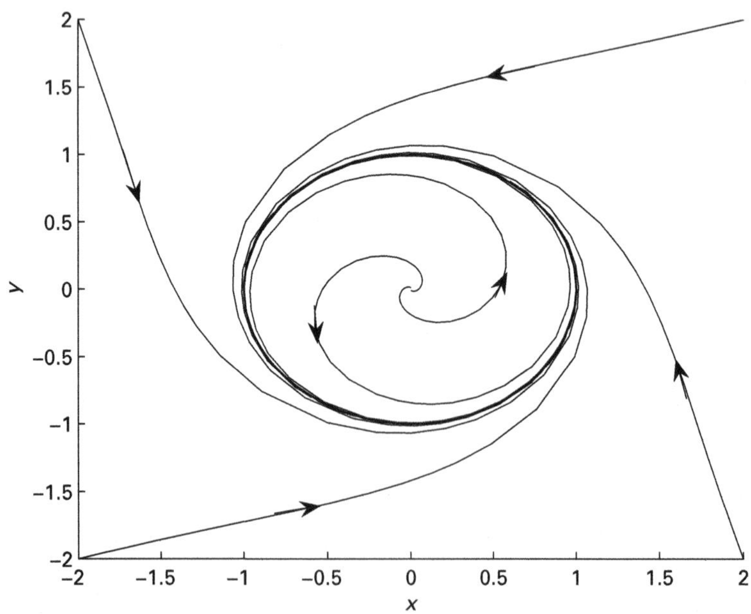

Figure 5.5 The closed orbit is both an attracting set and an attractor. The stability region of this attracting set is the entire space (R^2) excluding the origin, which is not homeomorphic to R^2.

sets can be disconnected. However, if the attracting set is connected, then the stability region is also connected. Since the boundary of an invariant set is also invariant and the boundary of an open set is a closed set, the set of the stability boundary has the following properties.

THEOREM 5-2 **(Topological property)**
The stability boundary $\partial A(\gamma)$ is a closed invariant set of dimension less than n. If $A(\gamma)$ is not dense in R^n, then $\partial A(\gamma)$ is of dimension $n-1$.

If system (5.1) possesses at least two isolated attracting sets, then the dimension of each stability boundary is $n-1$. The complete characterization of a stability region can be achieved via a complete characterization of its stability boundary. We will develop a comprehensive theory for characterizing stability boundaries of nonlinear dynamical systems (5.1) in the next sections.

5.4 Invariant sets on the stability boundary

Our aim is to present a comprehensive theory of stability boundaries for the general class of dynamical systems (5.1). In this section, several dynamical and topological properties of stability boundaries will be derived. A complete characterization of the stability boundary will then be given in terms of the *maximal indecomposable invariant sets* lying on the stability boundary.

We will take an approach which is similar to that of Chapter 4, starting from a local characterization and then generalizing toward a global characterization of the stability boundary. We first derive conditions for a closed and invariant set to lie on the stability boundary, which is a key step in the characterization of the stability region $A(\gamma)$. We then show that under additional conditions imposed on the dynamical system (5.1), sharper characterization results can be derived.

It will be shown that closed invariant sets play a fundamental role in the characterization of stability boundaries of general nonlinear dynamical systems (5.1). Similar to hyperbolic equilibrium points and hyperbolic closed orbits, invariant sets also have their stable and unstable manifolds as defined below.

DEFINITION **(Stable and unstable manifolds of invariant sets)**
Let Λ be an invariant set of the dynamical system (5.1). The stable manifold *of Λ, termed $W^s(\Lambda)$, is the set of all points q such that $\omega(q) \subset \Lambda$. Similarly, the* unstable manifold *of Λ, $W^u(\Lambda)$, is the set of all points q such that $\alpha(q) \subset \Lambda$.*

Although the terms stable and unstable *manifolds* are used for a general invariant set, we notice that depending on the structure of Λ, the stable and unstable manifolds *may not be* manifolds. If the invariant set Λ has a *hyperbolic structure*, then the sets $W^s(\Lambda)$ and $W^u(\Lambda)$ are manifolds as smooth as the vector field [216, p.243]. Hyperbolic equilibrium points and hyperbolic closed orbits are particular examples of compact invariant sets with a hyperbolic structure.

The next proposition provides sufficient conditions to ensure a compact invariant set intersects with the stability boundary in terms of its stable and unstable manifolds of general nonlinear dynamical systems (5.1).

PROPOSITION 5-3 **(Invariant sets on the stability boundary)**
Let $A(\gamma)$ be the stability region of an attracting set γ of the dynamical system (5.1) and let Λ be a closed invariant set, which has an empty intersection with γ. Then:

(a) if $\{W^u(\Lambda) - \Lambda\} \cap A(\gamma) \neq \varnothing$, then $\Lambda \cap \partial A(\gamma) \neq \varnothing$;
(b) if $\{W^s(\Lambda) - \Lambda\} \cap \partial A(\gamma) \neq \varnothing$, then $\Lambda \cap \partial A(\gamma) \neq \varnothing$.

Proof (a) Suppose $y \in \{W^u(\Lambda) \cap A(\gamma)\}$, then $\alpha(y) \subset \Lambda$. Since $A(\gamma)$ is invariant, there exists at least one point x of Λ that belongs to $\overline{A}(\gamma)$. We claim that x belongs to $\partial A(\gamma)$. Suppose on the contrary that $x \in A(\gamma)$. This implies the existence of a trajectory on Λ that converges to γ. Since Λ is closed, then there exists at least one point in γ that also belongs to Λ. This fact contradicts the assumption that $\Lambda \cap \gamma \neq \varnothing$. As a result, x belongs to $\partial A(\gamma)$ and the proof of result (a) is completed. (b) Suppose $y \in \{W^s(\Lambda) - \Lambda\} \cap \partial A(\gamma)$, then $\omega(y) \subset \Lambda$. Since $\partial A(\gamma)$ is a closed and invariant set, $\omega(y) \subset \partial A(\gamma)$. Consequently, $\Lambda \cap \partial A(\gamma) \neq \varnothing$ and the proof of result (b) is completed.

Generally speaking, an invariant set intersecting the stability boundary may not be entirely contained in the stability boundary. The next proposition nevertheless shows that every compact invariant set intersecting the stability boundary has a compact invariant subset lying entirely in the stability boundary.

PROPOSITION 5-4 **(Compact invariant sets on the stability boundary)**
If a compact invariant set Λ intersects the stability boundary $\partial A(\gamma)$ of an attracting set γ of the general nonlinear dynamical system (5.1), then $\Lambda' = \Lambda \cap \partial A(\gamma)$ is a compact invariant set that is entirely contained in the stability boundary $\partial A(\gamma)$ of γ.

Proof The proof is based on the facts that (i) the intersection of a compact set with a closed set is also compact, and the intersection between invariant sets is also invariant, and (ii) the stability boundary is an invariant set. This completes the proof.

Compact invariant sets on the stability boundary might not be maximal, however Proposition 5-4 shows that every compact invariant set that intersects with the stability boundary has a maximal compact invariant subset that is entirely contained in the stability boundary. We note that closed and invariant sets can be decomposed into maximal indecomposable components.

DEFINITION **(Indecomposable invariant sets)**
The closed and invariant set γ is indecomposable if for every pair of points $x, y \in \gamma$ and $\varepsilon > 0$, there exist points $x = x_0, x_1, \ldots, x_n = y$ and times $t_1, t_2, \ldots, t_n \geq 1$ such that the distance from $\phi(t_i, x_{i-1})$ to x_i is smaller than ε.

Indecomposable invariant sets are recurrent sets, which means that neighborhoods of points on the set will be recurrently visited by the forward trajectory. If two indecomposable sets have non-empty intersection, then the union is also indecomposable. Consequently, it is natural to define a maximal indecomposable invariant set.

DEFINITION **(Maximal indecomposable set)**
A compact and invariant set γ is a maximal indecomposable set if there exists an open neighborhood of γ that does not intersect any other indecomposable invariant set.

We note that the concept of indecomposability is relative. A set $X \subset Y \subset Z$ which is indecomposable in the space Y might not be indecomposable in a larger space Z. The stability boundary itself is a closed and invariant set, but the stability boundary is usually not indecomposable. There are usually many wandering points on the stability boundary. The indecomposable sets on the stability boundary will serve as critical elements on that boundary. Hyperbolic equilibrium points and hyperbolic closed orbits are particular examples of indecomposable sets. We note that maximal indecomposable sets are disjoint. If the number of indecomposable compact invariant sets is finite, then each indecomposable component stays at a positive distance from the other components.

5.5 Characterization of the stability boundary

In this section, we will derive a complete characterization of the stability boundary for a very general class of dynamical systems whose stability boundary is non-empty. It will be shown that limit points and limit sets on the stability boundary of system (5.1) will play a key role in a complete characterization of the stability boundary.

The asymptotic behavior of a *specific trajectory* can be studied in terms of its ω-limit and α-limit sets, the *positive* and *negative limit sets* of a *vector field* capture the asymptotic behaviors of trajectories in a global sense.

DEFINITION **(Positive limit set)**
The positive limit set $L_+(f)$ *of the vector field f is the closure of the union of all ω-limit sets, $L_+(f) = \overline{\cup\,\omega(x)}$ and the* negative limit set $L_-(f)$ *is the closure of all α-limit sets, $x \in R^n$.*

Every bounded trajectory of a dynamical system converges to $L_+(f)$ as $t \to \infty$. In some sense, $L_+(f)$ captures all the possible "steady state" behavior of a nonlinear dynamical system, which can be an equilibrium point, a closed orbit, a quasi-periodic orbit, or even chaos. We next characterize the stability boundary of a very large class of nonlinear autonomous dynamical systems (5.1). The stability boundaries of this class of nonlinear systems admit a finite number of disjoint compact invariant components of the limit set on it.

We make the following general assumptions concerning dynamical systems (5.1).

(C1) The positive limit set $L_+(f|_{\partial A(\gamma)})$ lying on the stability boundary is composed of a finite number of disjoint indecomposable compact invariant sets Λ_i, $i = 1, \ldots, m$.
(C2) Every trajectory on the stability boundary is bounded (for $t > 0$).

THEOREM 5-5 **(Characterization of the stability boundary)**
Consider a nonlinear system (5.1) satisfying assumptions (C1) and (C2) and let $A(\gamma)$ be the stability region of an attracting set γ. If Λ_i, i=1,2, ...,m are the disjoint indecomposable compact invariant components of $L_+(f|_{\partial A(\gamma)})$ lying on the stability boundary, then

the stability boundary $\partial A(\gamma)$ is contained in the union of the stable manifolds of all compact invariant sets Λ_i on the stability boundary, i.e.:

$$\partial A(\gamma) \subset \bigcup_i W^s(\Lambda_i). \tag{5.5}$$

Proof Let x be a point on $\partial A(\gamma)$. Assumption (C2) implies the solution $\phi(t, x)$ is bounded. Then $\omega(x) \subset \partial A(\gamma)$ is non-empty, compact, connected and invariant. Moreover, $\phi(t, x) \rightarrow \omega(x)$ as $t \rightarrow \infty$. Since $\omega(x) \subset L_+(f)$, then $\omega(x)$ belongs to Λ_i, for some i. Therefore $x \in \cup_i W^s(\Lambda_i)$ and, as a result, $\partial A(\gamma) \subset \cup_i W^s(\Lambda_i)$, and the proof is completed.

Theorem 5-5 is a generalization of Theorems 4-8 and 4-9 of Chapter 4. Theorems 4-8 and 4-9 are particular versions of Theorem 5-5 which were obtained under additional conditions on the structure of the compact invariant sets on the stability boundary.

5.6 Sufficient conditions for (C2)

The characterization of stability boundaries in the previous section is valid for a class of dynamical systems that satisfy assumptions (C1) and (C2). Assumption (C1) holds for many nonlinear dynamical systems. For instance, it is a generic property of certain classes of vector fields on compact manifolds [216]. Assumptions (A1) and (A2) of Chapter 4 imply assumption (C1), therefore assumption (C1) is more general than assumptions (A1) and (A2) and includes a larger class of dynamical systems. However, assumption (C2), like assumption (A3) of Chapter 4, is not generic. In this section, we will show that many dynamical systems arising from physical system models satisfy assumption (C2). More precisely, we will show that the existence of a generalized energy function is a sufficient condition for ensuring assumption (C2).

The following theorem shows that the existence of a generalized energy function is a sufficient condition for the satisfaction of assumption (C2), that is, all the trajectories on the stability boundary of a dynamical system (5.1) that admits a generalized energy function are bounded. The boundedness of trajectories on the stability boundary is useful for deriving important information about limit sets on the stability boundary, and crucial for the complete characterization of the stability boundary.

THEOREM 5-6 **(Sufficient condition for (C2))**
If the nonlinear dynamical system (5.1) admits a generalized energy function, and γ is an attracting set, then every trajectory on the stability boundary $\partial A(\gamma)$ is bounded for $t \geq 0$.

Theorem 5-6 guarantees assumption (C2) is held for every nonlinear dynamical system that admits a generalized energy function. Before proving Theorem 5-6, we derive the following lemma that guarantees the lower boundedness of the generalized energy function over a stability boundary, even if the stability boundary is unbounded.

LEMMA 5-7 **(Generalized energy function lower bound)**
Let γ be an attracting set of the dynamical system (5.1). If there exists a C^1-function $V{:}R^n \to R$ that satisfies assumptions (G1) and (G2) of a generalized energy function, then $V(x)$ has a lower bound on the stability boundary $\partial A(\gamma)$.

Proof Let B be the union of all connected components C_i of C with a non-empty intersection with $\overline{A(\gamma)}$. Let $\hat{x} \in \partial A(\gamma)$. If $\hat{x} \in B$ then, using the continuity of V, assumption (G1), and the compactness of every connected component C_j of C, one concludes that $V(\hat{x}) \geq \min_{y \in B} V(y)$. If $\hat{x} \notin B$, then, since $A(\gamma)$ is an open set arbitrarily close to $\hat{x}$, there exists an $x_0 \in A(\gamma)$ such that $x_0 \notin B$. Now, we analyze the following two cases.

(i) Suppose $\phi(t, x_0) \notin B$ for all $t \geq 0$, then $V(\phi(t, x_0)) \leq V(x_0)$ for $t \geq 0$. Using the continuity of V and the fact that $\phi(t, x_0)$ converges to γ as $t \to \infty$ and x_0 is arbitrarily close to $\hat{x}$, we conclude that $V(\hat{x}) \geq \min_{y \in \gamma} V(y)$.
(ii) If (i) is not true, then there exists a connected component C_{j_1} and a pair of times t_1 and t_1^*, with $0 < t_1 < t_1^*$, such that $\phi(t, x_0) \notin B$ for $0 < t < t_1$ and $\phi(t, x_o) \notin C_{j_1}$ for $t_1 \leq t \leq t_1^*$. Then, $V(\phi(t_1, x_0)) \leq V(\phi(t, x_0)) \leq V(x_0)$ for $0 \leq t \leq t_1$ and $V(\phi(t_1, x_0)) \leq V(\phi(t, x_0))$ for $t_1 \leq t \leq t_1^*$. As a result, a local minimum of V along the trajectory $\phi(t, x_0)$ is attained at time t_1 when the trajectory reaches ∂C_{j_1}. Using compactness of C_{j_1} and continuity of V, one concludes that $\min_{y \in C_{j_1}} V(y)$ is a lower bound for $V(\hat{x})$. From assumption (G1) (there exists a finite number of C_i intersecting with $A(\gamma)$) and the continuity of V, we conclude that $V(x)$ has a lower bound on $\partial A(\gamma)$ given by the following number $\min\{\min_{y \in \gamma} V(y), \min_{y \in B} V(y)\}$,This concludes the proof.

We are now in a position to prove Theorem 5-6.

Proof of Theorem 5-6 Let V be a generalized energy function for system (5.1) and B be as in the proof of the previous lemma. According to Lemma 5-7, function $V(x)$ has a lower bound, say ρ, on $\partial A(\gamma)$. Let $x_0 \in \partial A(\gamma)$ and let ω_+ be the maximal time of existence for $\phi(t, x_0)$. Because $\partial A(\gamma)$ is an invariant set, it follows that $\phi(t, x_0) \in \partial A(\gamma)$ for $t \in [0, \omega_+)$. If $\phi(t, x_0) \notin C$ for all $t \in [0, \omega_+)$ then $V(x_0) \geq V(\phi(t, x_0)) \geq \rho$ for all $t \in [0, \omega_+)$. Hence, $V(\phi(t, x_0))$ is bounded for $t \geq 0$ and assumption (G3) of a generalized energy function implies that $\phi(t, x_0)$ is bounded for $t \geq 0$. As a direct consequence of Theorem 2.1 of Chapter 2, we conclude that $\omega_+ = \infty$, i.e. the trajectory on the stability boundary exists for $t \geq 0$ and is bounded.

If the intersection of $\phi(t, x_0)$ with C is non-empty, then there exists a connected component C_{j1} of C and a pair of times t_1 and t_a, with $0 \leq t_1 < t_a$, such that $\phi(t, x_0) \notin C$ for $0 < t < t_1$ and $\phi(t, x_0) \in C_{j1}$ for $t_1 < t < t_a$. If $t_a = \infty$, that is $\phi(t, x_0)$ does not leave C_{j1} for all $t \in [t_1, \omega_+)$, then $\omega_+ = \infty$ and $\phi(t, x_0)$ is bounded for $t \geq 0$. If this is not the case, that is if $t_a < \infty$, then two situations can occur:

(i) either $\phi(t, x_0) \notin C$ for all $t > t_a$ or
(ii) there exists a connected component C_{j2} and a pair of times t_2 and t_b satisfying $t_a < t_2 < t_b$ such that $\phi(t, x_0) \notin C$ for $t_1 < t < t_2$ and $\phi(t, x_0) \in C_{j2}$ for $t_2 < t < t_b$.

In case (i), $\max\{V(x_0), V(\phi(t_a, x_0))\} \geq V(\phi(t, x_0)) \geq \rho$ for all $t \in [0, \omega_+)$. Assumption (G3) and the boundedness of $V(\phi(t, x_0))$ imply that $\phi(t, x_0)$ is bounded for $t \in [0, \omega_+)$, and we conclude that $\omega_+ = \infty$. In case (ii), if $t_b = \infty$, then $\phi(t, x_0)$ is bounded for $t \geq 0$. Otherwise, if $t_b < \infty$ then the previous analysis is repeated. From assumptions (G1) and (G2), one concludes that $\rho \leq V(\phi(t, x_o)) \leq \max\{V(x_o), \max_{y \in \overline{B}} V(y)\}$ for any $x_0 \in \partial A(\gamma)$ and $t \geq 0$. Then, from assumption (G3), we conclude that every trajectory on the stability boundary is bounded. This completes the proof.

From the proof of Theorem 5-6, we obtain the following analytical result regarding the characterization of stability boundary of nonlinear dynamical systems that admit generalized energy functions.

THEOREM 5-8 **(Complete characterization of the stability boundary)**
Consider an attracting set γ of the dynamical system (5.1). Let Λ_i, i=1,2, . . .,m be the disjoint indecomposable compact invariant components of $L_+(f|_{\partial A(\gamma)})$ lying on the stability boundary $\partial A(\gamma)$. If the system (5.1) admits a generalized energy function, then the stability boundary $\partial A(\gamma)$ is contained in the union of the stable manifolds of all compact invariant sets Λ_i on the stability boundary, i.e.:

$$\partial A(\gamma) \subset \bigcup_i W^s(\Lambda_i). \tag{5.6}$$

For a proof of Theorem 5-8, see [8].

5.7 Numerical example

We illustrate some of the theoretical developments presented in this chapter in a simple numerical example. This example is a three-dimensional nonlinear dynamical system that exhibits complex behavior on the stability boundary of an asymptotically stable equilibrium point. We will show that it admits a generalized energy function and that, therefore, the stability boundary characterization developed in Theorem 5-8 holds. Consider the following nonlinear autonomous system:

$$\begin{cases} \dot{x} = \dfrac{(\sigma - \alpha c^2)x - \sigma y}{1 + \alpha c^2} \\ \dot{y} = \dfrac{(1 - \alpha c^2)y + xz - rx}{1 + \alpha c^2} \\ \dot{z} = \dfrac{(b - \alpha c^2)z - xy}{1 + \alpha c^2} \end{cases} \tag{5.7}$$

where $c^2 = x^2 + y^2 + z^2$, $\sigma = 10$, $r = 28$, $b = 8/3$ and $\alpha = 0.0001$.

This system has an asymptotically stable equilibrium point at approximately (315, −0.65, 28). The stability boundary of this asymptotically stable equilibrium point is quite complex. Figure 5.6 shows a cross section of the stability boundary. The white area belongs to the stability region of the asymptotically stable equilibrium point x_s, while the black area represents its complement. The stability boundary is shown between white and black areas. In this example, numerical simulations suggest the presence of chaos on

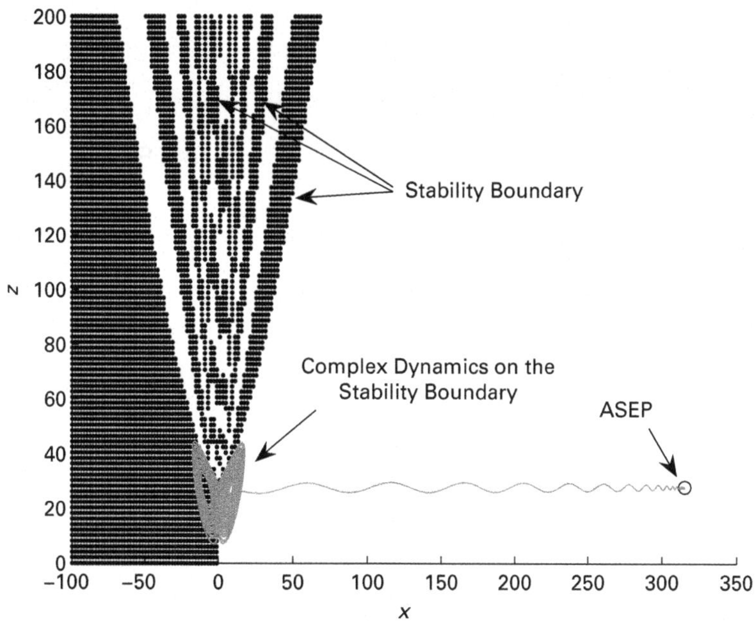

Figure 5.6 The cross section of the exact stability boundary in which a chaotic trajectory exists. The white area belongs to the stability region of the asymptotically stable equilibrium point x_s while the black area represents its complement. The stability boundary lies between white and black areas.

the stability boundary. Figure 5.6 depicts the projection of an orbit, which approaches the stability boundary for large negative times, on the plane $\{(x, y, z) \in R^3: y = -0.65\}$, suggesting chaotic behavior on the stability boundary.

We next construct a generalized energy function for this system. We consider the following candidate as a generalized energy function

$$V(x, y, z) = ax^2 + c(y - \xi)^2 + d(z - \beta)^2 \tag{5.8}$$

where $a = 2, c = d = 8, \xi = -0.65$ and $\beta = 28$. We will numerically show that $V(x,y,z)$ is a generalized energy function for system (5.7). Figure 5.7 shows the set C where the derivative of function $V(x,y,z)$ is positive. Set C is composed of a single bounded component and consequently, assumptions (G1) and (G2) are satisfied. Moreover, function $V(x,y,z)$ is a radially unbounded function, and so assumption (G3) is also satisfied. As a consequence, $V(x,y,z)$ is a generalized energy function for system (5.7).

Since $V(x,y,z)$ is a generalized energy function, according to Theorem 5-6, every trajectory of this complex system lying on the stability boundary must be bounded and converge to a limit set that has a non-empty intersection with the set C. Numerical simulation suggests that almost all trajectories lying on the stability boundary approach an invariant set in a positive time. This invariant set exhibits complex behavior on the stability boundary, as shown in Figure 5.7.

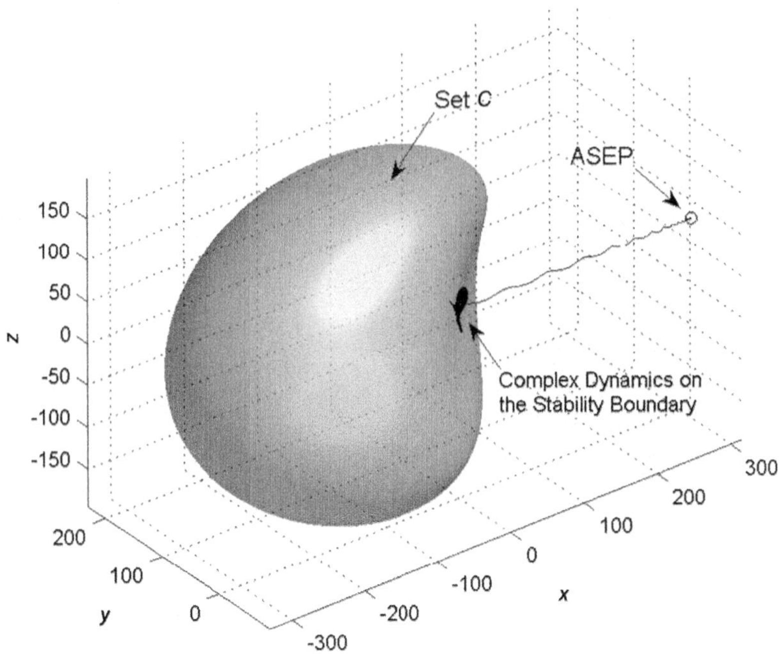

Figure 5.7 A complex invariant set lying on the stability boundary of an asymptotically stable equilibrium point. Set C, where the derivative of function $V(x,y,z)$ is positive, is composed of a single bounded component and intersects the invariant set on the stability boundary.

5.8 Concluding remarks

In this chapter, we have presented a comprehensive theory of stability boundaries for a class of complex nonlinear autonomous dynamical systems. A dynamical characterization of the stability boundary has also been derived. The characterization is a generalization of those derived in Chapter 4 for stability boundaries of asymptotically stable equilibrium points.

The characterization of the stability boundary developed in this chapter requires the determination of stable manifolds of invariant sets. The structure of these manifolds is usually very complex and poses enormous difficulty for the development of practical computational methods for estimating stability regions, even for low-dimensional systems. The analytical results derived for the characterization of invariant sets lying on the stability boundary and for the characterization of the stability boundary can lead to several applications. One application is that they provide deep insights into the development of practical computational methods to estimate stability regions of complex nonlinear systems via generalized energy functions. Computational methods for estimating the stability regions of complex nonlinear dynamical systems exploring the theory of generalized energy functions will be presented in Chapter 11.

6 Quasi-stability regions of continuous dynamical systems

The structure of the stability boundary of general nonlinear dynamical systems can be very complex. A simple three-dimensional example shows that the closure of stability regions may contain subsets of the interior of the closure of stability regions [285]. The stability boundary of a simple power system stability model may have a truncated fractal structure [258]. There are several factors that contribute to the complexity of the stability boundary. One of them is the presence of critical elements (i.e. equilibrium points and limit cycles) in the interior of the closure of the stability region.

This motivates study of the notions of the quasi-stability region and quasi-stability boundary. Indeed, from an engineering viewpoint, the quasi-stability region is a "practical" stability region while the quasi-stability boundary is less complex than the stability boundary.

In this chapter we present the concept of quasi-stability region and develop a comprehensive theory for the quasi-stability boundary. It will be shown that the quasi-stability boundary is the union of the stable manifolds of all the critical elements on the boundary. More sharply, it will be shown that the quasi-stability boundary is the union of the stable manifolds of all the type-one critical elements on the boundary. A characterization of critical elements lying inside the quasi-stability regions will be derived. In addition, the class of nonlinear dynamical systems whose stability regions equal their quasi-stability regions will be characterized. Moreover, it will be shown that the set of critical elements lying inside the quasi-stability region is robust relative to small perturbations of its vector field. This robust property provides another justification for the concept of quasi-stability region.

6.1 Quasi-stability region

To illustrate the concept of the quasi-stability region and quasi-stability boundary, we first consider the following two-dimensional system, which is a reduced version of the system presented in [285]

$$\begin{aligned} \dot{x} &= \{(\sqrt{x^2+y^2}-3)[x^2+y^2+(y-2)\sqrt{x^2+y^2}-2y+1.5]+y\}x \\ \dot{y} &= \{(\sqrt{x^2+y^2}-3)[x^2+y^2+(y-2)\sqrt{x^2+y^2}-2y+1.5]y\}-x^2. \end{aligned} \tag{6.1}$$

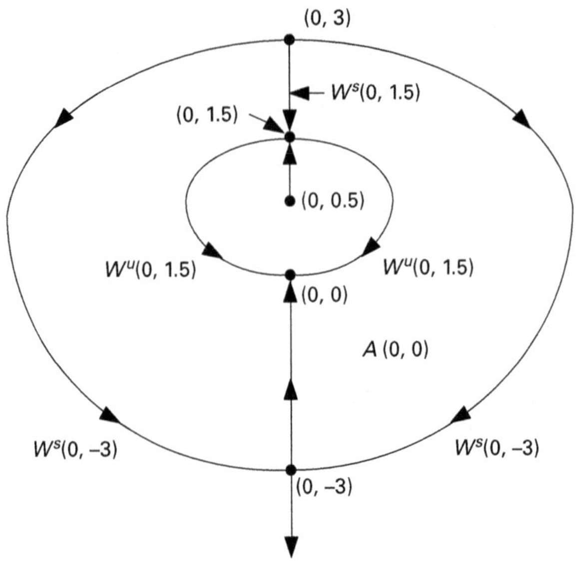

Figure 6.1 The stability boundary of the stable equilibrium point (0,0) of system (6.1) is composed of the stable manifold of (0, −3), the stable manifold of (0, 1.5), the stable manifold of (0,0.5) and the stable manifold of (0, 3). On the other hand, the quasi-stability boundary of the stable equilibrium point (0, 0) of system (6.1) is composed only of the stable manifold of (0, −3) and the stable manifold (0, 3).

It can be easily verified that (0, 0) is an asymptotic stable equilibrium point of the system. The stability region $A(0, 0)$ and stability boundary $\partial A(0, 0)$ are shown in Figure 6.1. There are four equilibrium points on the stability boundary, namely (0, 0.5), (0, 3), (0, 1.5) and (0, −3). Of them, two equilibrium points, (0, 0.5), (0, 3) are of type-two equilibriums (i.e. sources) while (0, 1.5) and (0, −3) are type-one equilibriums. The stability boundary of the stable equilibrium point (0,0) of system (6.1) is composed of the stable manifold of (0, −3), the stable manifold of (0, 1.5), the stable manifold of (0, 0.5) and the stable manifold of (0, 3). On the other hand, the quasi-stability boundary of the stable equilibrium point (0, 0) of system (6.1) is composed only of the stable manifold of (0, −3) and the stable manifold (0, 3). In other words, the quasi-stability boundary of the stable equilibrium point (0, 0) is the union of the stable manifold of (0, −3) and the unstable equilibrium point (0, 3) (which is also the stable manifold of (0, 3)). The following observations can be made.

- The unstable manifold of each equilibrium point on the stability boundary intersects the stability region.
- The phase portrait of the system shows that the stability boundary is composed of two parts. One part consists of the stable manifold of (0,−3) and the stable manifold of (0, 1.5) while the other part contains only two equilibrium points (0, 0.5), (0, 3). These two equilibrium points are the stable manifolds of themselves (i.e. the degenerate stable manifold).

- The unstable manifold of (0, 1.5) and the unstable manifold of (0, 0.5) intersect the stability region $A(0,0)$ while they do not intersect the complement of the closure of the stability region.
- The unstable manifold of (0, 3.0) and the unstable manifold of (0, −3.0) all intersect the stability region $A(0,0)$ as well as the complement of the closure of the stability region.

We note that the stable manifold of (0, 1.5) lies in the interior of the closure of the stability region $A(0,0)$. A slight perturbation of any trajectory lying on the stable manifold of (0, 1.5) can make this trajectory converge to the SEP (0,0). Thus, this part of the stability boundary effectively behaves as if it were part of the stability region. This motivates us to view this part of the stability boundary practically as a subset of the stability region. The second part of the stability boundary, which consists of the stable manifold of (0, −3) and the type-two UEP (0, 3) (i.e. the stable manifold of (0, 3)), divides the state space into two regions. One region consists of all the points whose trajectories converge to the SEP (0,0) or practically converge to (0,0). We will term this region the *quasi-stability region*, denoted A_q. The second region consists of points whose trajectories move away from the closure of the stability region. The boundary (i.e. the curve) that separates these two regions will be termed the *quasi-stability boundary*, denoted ∂A_q.

The above two-dimensional system suggests an intuitive definition of a quasi-stability boundary by observing the phase portrait of the system. For the two-dimensional example, we note that this system satisfies assumptions (A1) to (A3). Thus, for an equilibrium point to be on the stability boundary, its unstable manifold must intersect the stability region. This property is obvious for the two sources on the boundary because their unstable manifolds are two dimensional. Furthermore, the two type-one equilibrium points on the stability boundary also have this property. There is, however, a substantial difference between these two type-one equilibrium points. The unstable manifold of (0, 1.5) is contained in the closure of the stability region A, i.e. $W^u(0,1.5)\cap(\overline{A})^c = \emptyset$. In contrast, the unstable manifold of (0, −3) is composed of two parts: one part is contained in the closure of A, while the other part is contained in the complement of $(\overline{A})$, $(\overline{A})^c$ (i.e. $(R^n - \overline{A})$).

The above example suggests an intuitive definition of a critical element on ∂A_q as follows.

DEFINITION

A critical element σ is on the quasi-stability boundary ∂A_q if and only if the following two conditions are met:

(a) $\sigma \in \partial A$,

(b) $W^u(\sigma)\cap(\overline{A})^c \neq \emptyset$.

In order to show that the above definition is consistent with the definition of stability region, one needs to show that the (topological) boundary of the quasi-stability region equals the quasi-stability boundary ∂A_q. This will be shown in the following section.

6.2 Quasi-stability boundary

We consider the following (autonomous) nonlinear dynamical system

$$\dot{x} = f(x),\ x \in R^n. \tag{6.2}$$

It is natural to assume the function (i.e. the vector field) f: $R^n \rightarrow R^n$ satisfies a sufficient condition for the existence and uniqueness of the solution.

The next analytical result gives a necessary condition for a critical element to be on the quasi-stability boundary.

THEOREM 6-1 **(A necessary condition)**
Let A be the stability region of a stable equilibrium point of system (6.2). If σ is a hyperbolic critical element lying on the quasi-stability boundary, then the stable manifold of the element is contained in the complement of the stability region, i.e. $\sigma \in \partial A_q \Rightarrow W^s(\sigma) \subseteq (A)^c$.

Proof This proof requires the following simplified version of the λ-lemma.

Lemma: let σ be a hyperbolic critical element of system (6.2) with $\dim W^u(\sigma) = m$. If σ is a hyperbolic equilibrium point, let D be an m-disk in $W^u(\sigma)$. If σ is a closed orbit, let D be an $(m-1)$-disk in $W^u(\sigma) \cap S$, where S is a cross section at $p \in \sigma$. Let N be an m-disk (if σ is an equilibrium point) or let N be an $(m-1)$-disk (if σ is a closed orbit) with a point of transversal intersection with $W^s(\sigma)$. Then, the disk D is contained in the closure of the set $\cap_{t\geq 0}\phi(t, N)$.

Suppose $\sigma \in \partial A_q$. Let $D \subset W^u(\sigma) \cap (\overline{A})^c$ be an m-disk, $\dim W^u(\sigma) = m$. Let $y \in W^s(\sigma)$ be arbitrary. For any $\varepsilon > 0$, let N be an m-disk transversal to $W^s(\sigma)$ at y, contained in the ε-neighborhood of y. By the above lemma, there exists a $t > 0$ such that $\phi(t, N)$ is so close to D that $\phi(t, N)$ contains a point $p \in (\overline{A})^c$. Thus $\phi(-t, p) \in N$. Since $(A)^c$ is an invariant set and $N \cap (\overline{A})^c \neq \emptyset$, it follows that $y \in (A)^c$ as $\varepsilon \rightarrow 0$; in other words, $W^s(\sigma) \subseteq (A)^c$. This completes the proof.

We next establish a relationship between the critical elements belonging to the sets ∂A_q and $\partial\overline{A}$.

THEOREM 6-2 **(A relation between ∂A_q and $\partial\overline{A}$)**
Let A be the stability region of an asymptotically stable equilibrium point of system (6.2). If σ is a hyperbolic critical element lying on ∂A_q, then σ must lie on $\partial\overline{A}$.

Proof From the definition, we have $\sigma \in \partial A_q$ if and only if (a) $\sigma \in \partial A$ and (b) $W^u(\sigma) \cap (\overline{A})^c \neq \emptyset$. If we let $y \in \{W^u(\sigma) - \sigma\} \cap (\overline{A})^c$, then $\lim_{t\rightarrow -\infty}\phi(t, y) = \sigma$. Since $(\overline{A})^c$ is invariant, we have $\sigma \in \overline{(\overline{A})^c}$. Also, since $\sigma \in \partial A$, $\sigma \in \overline{A}$. Thus $\sigma \in \overline{(\overline{A})^c} \cap \overline{A}$, which is the definition of $\partial\overline{A}$. This completes the proof.

In the following we present two analytical results which characterize a critical element on $\partial\overline{A}$. These results provide insight into the definition of the quasi-stability boundary. In particular, it will be shown that $\partial A_q = \partial\overline{A}$.

THEOREM 6-3 **(Characterization of critical elements on $\partial\overline{A}$)**
Let A be the stability region of a stable equilibrium point of system (6.2). If σ is a hyperbolic critical element, then $\sigma \in \partial\overline{A}$ implies $\{W^u(\sigma) - \sigma\} \cap (\overline{A})^c \neq \emptyset$. Conversely, if $\{W^u(\sigma) - \sigma\} \cap (\overline{A})^c \neq \emptyset$ and $\{W^u(\sigma) - \sigma\} \cap \overline{A} \neq \emptyset$, then $\sigma \in \partial\overline{A}$.

Proof Let $G \subset \{W^u(\sigma) - \sigma\}$ be a fundamental domain of $W^u(\sigma)$, i.e. G is compact and $\cup\phi(t, G) = \{W^u(\sigma) - \sigma\}$. Let G_ε be an ε-neighborhood of G in R^n. Then $U_{t<0}\,\phi(t_\varepsilon, G)$ contains a set of the form $\{U - W^s(\sigma)\}$, where U is a neighborhood of σ. Since $\sigma \in \partial\overline{A}$ is equivalent to $\sigma \in \partial(\overline{A})^c$, $U \cap (\overline{A})^c \neq \emptyset$. But by the topological inclusion $\partial\overline{A} \subset \partial A$ we get $W^s(\sigma) \subset \partial A$; hence, $W^s(\sigma) \cap (\overline{A})^c = \emptyset$. Therefore, we have $\{U - W^s(\sigma)\} \cap (\overline{A})^c \neq \emptyset$. Equivalently $U_{t<0}\,\phi(t_\varepsilon, G) \cap (\overline{A})^c \neq \emptyset$. This implies that $G_\varepsilon \cap \phi(t, (\overline{A})^c) \neq \emptyset$ for some t.

Since $(\overline{A})^c$ is an invariant open set, it follows that $G_\varepsilon \cap (\overline{A})^c \neq \emptyset$. Since $\varepsilon > 0$ is arbitrarily chosen and G is compact, G contains a point in $\overline{(\overline{A})^c}$; thus $\{W^u(\sigma) - \sigma\} \cap \overline{(\overline{A})^c} \neq \emptyset$. For the second part, let $y \in W^u(\sigma) \cap \overline{(\overline{A})^c}$, it follows that $\lim_{t \to -\infty} \phi(t, y) = \sigma$; hence, $\sigma \in \overline{(\overline{A})^c}$. Also from the assumptions $\sigma \in \overline{A}$, we have $\sigma \in \partial\overline{A}$. This completes the proof.

We make the following assumptions concerning the vector field (6.2).

(B1) All the equilibrium points and closed orbits on the stability boundary are hyperbolic.
(B2) The stable and unstable manifolds of equilibrium points and closed orbits on the stability boundary satisfy the transversality condition.
(B3) Every trajectory on the stability boundary approaches one of the equilibrium points or the closed orbits as $t \to \infty$.

THEOREM 6-4 **(Characterization of critical elements on $\partial\overline{A}$)**
Let A be the stability region of a stable equilibrium point of system (6.2) satisfying assumptions (B1) and (B2). If σ is a hyperbolic critical element, then $\sigma \in \partial\overline{A}$ implies that $\{W^u(\sigma) - \sigma\} \cap (\overline{A})^c \neq \emptyset$.

Proof We shall prove this theorem for an equilibrium point $\hat{x}$. The proof is similar for the case of a closed periodic orbit. According to Theorem 6-3, we have $\{W^u(\hat{x}) - \hat{x}\} \cap \overline{(\overline{A})^c} \neq \emptyset$. Let $n_u(\hat{x})$ denote the type of equilibrium point $\hat{x}$. Let $\hat{x} \in \partial A$, then one has $n_u(\hat{x}) \geq 1$. Call y a point in $\{W^u(\hat{x}) - \hat{x}\} \cap \overline{(\overline{A})^c}$. If $y \in (\overline{A})^c$, then the proof is complete; otherwise, there exists an equilibrium point $\hat{z} \in \partial\overline{A}$ and $y \in \{W^s(\hat{z}) - \hat{z}\}$. Let $h = n_u(\hat{x})$ and $m = n_u(\hat{z})$, where h and m are positive integers. Assumption (B2) implies that $W^u(\hat{x})$ intersects $W^s(\hat{z})$ transversally at y. Thus, $h > m$. Two cases may arise.

(a) If $h = 1$, then $m = 0$, which is a contradiction. Hence, $\{W^u(\hat{x}) - \hat{x}\} \cap (\overline{A})^c \neq \emptyset$.
(b) If $h > 1$, then $m \leq h - 1$. Assume inductively that $\{W^u(\hat{x}) - \hat{x}\} \cap (\overline{A})^c \neq \emptyset$. $W^u(\hat{x})$ contains an m-disk N at y and is transverse to $W^s(\hat{z})$. Hence, we have $\phi(t, N) \cap (\overline{A})^c \neq \emptyset$ for some $t > 0$. Since $(\overline{A})^c$ is an invariant set, we have $N \cap (\overline{A})^c \neq \emptyset$. This completes the proof.

We are now in a position to derive the following main result which provides greater insight into the concept of quasi-stability regions.

THEOREM 6-5 **(Relation between ∂A_q and $\partial \overline{A}$)**
Consider a nonlinear dynamical system which satisfies assumptions (B1) and (B2). If σ is a critical element, then $\sigma \in \partial A_q$ if and only if $\sigma \in \partial \overline{A}$.

Proof From Theorem 6-2, we have $\sigma \in \partial A_q$, which implies $\sigma \in \partial \overline{A}$. Now, if $\sigma \in \partial \overline{A}$, then $\sigma \in \partial A$ because $\partial \overline{A} \subset \partial A$. Moreover, Theorem 6-4 implies that $\{W^u(\sigma) - \sigma\} \cap (\overline{A})^c \neq \emptyset$. In other words, $\sigma \in \partial A_q$. This completes the proof.

We have so far developed the concept of a quasi-stability boundary via critical elements. To define the quasi-stability boundary, we need to develop a formal definition that includes points other than equilibrium points and periodic solutions. Theorem 6-5 in conjunction with the following topological fact

$$\partial \overline{\text{int}\overline{A}} \subset \partial \text{int}\overline{A} \subset \partial \overline{A} \subset \partial A \tag{6.3}$$

suggests that either $\partial \text{int}\overline{A}$ or $\partial \overline{\text{int}\overline{A}}$ or $\partial \overline{A}$ can be chosen as ∂A_q. Theorem 6-5 asserts that the last relationship is the most appropriate choice and thus the following formal definition of a quasi-stability region is presented.

DEFINITION (QUASI-STABILITY BOUNDARY)
Let A be the stability region of a stable equilibrium point of the nonlinear dynamical system (6.2). The quasi-stability boundary *∂A_q is $\partial \overline{A}$ and the* quasi-stability region *A_q is the open set* $\text{int}\overline{A}$.

In order to have a consistent definition, one needs to show that the boundary of A_q equals ∂A_q. The next analytical result, which is based on a known topological property that for any open set A, it is true that $\partial \text{int}\overline{A} = \partial \overline{A}$, shows that the quasi-stability region and the quasi-stability boundary are well defined.

THEOREM 6-6 **(Topological property)**
Let A be the stability region of a stable equilibrium point of system (6.2) which satisfies assumptions (B1) and (B2). Let the quasi-stability boundary be denoted as ∂A_q and the quasi-stability region be denoted as A_q, then $\partial \text{int}\overline{A} = \partial A_q$.

A graphical illustration of the difference between the stability region and quasi-stability region is shown in Figure 6.2. Based on the definition of quasi-stability regions, we have the following result.

PROPOSITION 6-7 **(Topological property)**
The quasi-stability region A_q is an open invariant set which is diffeomorphic to R^n and the quasi-stability boundary ∂A_q is a closed invariant set. Moreover, if $\overline{A} \neq R^n$, then ∂A_q has dimension $n-1$.

Proof The first part follows from the facts that $A \subset A_q$, A is an invariant set diffeomorphic to R^n, and $\partial \overline{A}$ is an invariant set. To prove the second part, observe that if $A = R^n$, then $\partial A_q = \partial \overline{\text{int}\overline{A}} = \partial R^n = \emptyset$. And if $\overline{A} \neq R^n$, then $\partial A_q = \partial \overline{\text{int}\overline{A}}$. Since $\text{int}\overline{A}$ is an open set and $\overline{\text{int}\overline{A}} \neq R^n$, ∂A_q has dimension $n - 1$. This completes the proof.

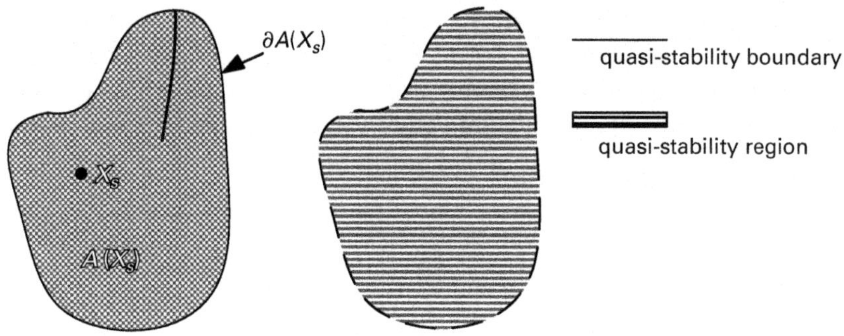

Figure 6.2 A graphical illustration of the difference between the stability region and quasi-stability region. The left figure illustrates the stability region and its boundary while the right figure illustrates the quasi-stability region and its boundary.

6.3 Characterization of quasi-stability boundaries

In this section we will derive a complete characterization of the quasi-stability boundary for a fairly large class of nonlinear dynamical systems. To this end, we will first derive conditions which an equilibrium point (or closed orbit) must possess to be on the quasi-stability boundary. We will then show that the quasi-stability boundary is the union of the stable manifolds of all the critical elements on the quasi-stability boundary. Moreover, the quasi-stability boundary is the union of the closure of the stable manifolds of all the type-one critical elements on the quasi-stability boundary.

We first present conditions for an equilibrium point (or limit cycle) to lie on the quasi-stability boundary $\partial A_q(x_s)$. This is a key step for developing a complete characterization of $\partial A_q(x_s)$.

THEOREM 6-8 **(Critical elements on the quasi-stability boundary)**
Consider the general nonlinear dynamical system (6.2). Let A_q be the quasi-stability region of x_s and A be the stability region of x_s. Let $\sigma \neq x_s$ be a hyperbolic critical element. If assumptions (B1), (B2) and (B3) are satisfied, then the following results hold:
(a) $\sigma \in \partial A_q$ *if and only if* $W^u(\sigma) \cap A \neq \emptyset$ *and* $W^u(\sigma) \cap (\bar{A})^c \neq \emptyset$,
(b) $\sigma \in \partial A_q$ *if and only if* $W^s(\sigma) \subseteq \partial A_q$.

Proof The proof of part (a) follows from the definition and Theorem 6-6. We prove part (b). It suffices to show that $\sigma \in \partial\bar{A} \Leftrightarrow W^s(\sigma) \subset \partial\bar{A}$.

Sufficiency: if $W^s(\sigma) \subseteq \partial\bar{A}$, then $\sigma \in \partial Aq$ since $\sigma \in W^s(\sigma)$.

Necessity: suppose $\sigma \in \partial Aq$, then it follows that $W^s(\sigma) \subseteq \overline{(\bar{A})^c}$. But Theorem 6-1 implies that $W^s(\sigma) \subseteq \partial A$, therefore $W^s(\sigma) \not\subset (\bar{A})^c$. Now $\overline{(\bar{A})^c} = (\bar{A})^c \cup \partial(\bar{A})^c = (\bar{A})^c \cup \partial\bar{A}$; thus $W^s(\sigma) \subseteq \partial\bar{A}$. This completes the proof.

We next develop a complete characterization of the quasi-stability boundary for a fairly large class of nonlinear dynamical systems whose quasi-stability boundary is non-empty.

THEOREM 6-9 **(Characterization of the quasi-stability boundary)**
Consider a nonlinear dynamical system described by (6.2) which satisfies assumptions (B1) to (B3). Let σ_i, $i = 1, 2, \ldots$, be the critical elements on the quasi-stability boundary $\partial A_q(x_s)$ of the stable equilibrium point x_s. Then, $\partial A_q(x_s) = \underset{\sigma_i \in \partial A_q(x_s)}{\cup} W^s(\sigma_i)$

Proof We deduce from Theorem 6-8 that

$$\underset{\sigma_i \in \partial A_q(x_s)}{\cup} W^s(\sigma_i) \subseteq \partial A_q(x_s). \tag{6.4}$$

Since $\partial A_q(x_s)$ is an invariant set, assumption (B3) implies that every point of $\partial A_q(x_s)$ must lie on $W^s(\sigma)$ for some critical element $\sigma \in \partial A_q(x_s)$; in other words, assumption (B3) implies

$$\partial A_q(x_s) \subseteq \underset{\sigma_i \in \partial A_q(x_s)}{\cup} W^s(\sigma_i). \tag{6.5}$$

By combining Eqs. (6.4) and (6.5), the proof is completed.

We next study a topological relationship between critical elements on a quasi-stability boundary. In particular, type-one critical elements and their stable manifolds play a dominant role in the characterization of the quasi-stability boundary.

THEOREM 6-10 **(A topological relation)**
Consider a nonlinear dynamical system described by (6.2) which satisfies assumptions (B1) to (B3). Let A_q be the quasi-stability region of x_s and $\sigma \in \partial A_q$ be a hyperbolic critical element whose type is greater than one. Then there exists a type-one critical element $\sigma_1 \in \partial A_q$ such that $\sigma \in \overline{W^s(\sigma_1)}$.

Proof Let Σ_2 denote the set of critical elements s on ∂A_q with type greater than or equal to two, i.e. $\dim W^s(\sigma) < n-1$, and let $M = \underset{\sigma_i \in \Sigma_2}{\cup} W^s(\sigma_i)$. Let S be the space of all smooth one-dimensional (1-D) manifolds such that any element in S can be parameterized as $s(t)$, $t \in R$. Let S be endowed with the C^∞-topology. Now for any $s(t) \in S$ and any $\sigma_i \in \Sigma_2$, the co-dimension of S equals $n-1$, which is greater than dim $W^s(\sigma_i)$. We now invoke the following corollary, which is a simplified version of Thom's transversality theorem.

Corollary: let M and S be two submanifolds of R^n. Then the property that M intersects S transversally is generic, and, if co-dim $S >$ dim M, then $M \cap S = \emptyset$ generically (i.e. the manifolds S that do not intersect M form a dense countable intersection of open sets everywhere).

This corollary asserts that $M \cap S = \emptyset$. Let N be a neighborhood of the critical element σ. Since $\sigma \in \partial A_q$, N must contain a point $x_0 \in A_q$ and a point $x_1 \in \overline{(A)}^c$. According to Thom's theorem, there is a smooth path $s(t)$ in N, $t \in [0,1]$ such that $s(0) = x_0$, $s(1) = x_1$ and for all $t \in [0,1]$, $s(t) \cap M = \emptyset$. If we let

$$\tilde{t} = \sup\{(t' \in [0,1]) : s(t) \in \text{int}\overline{A},\ t \in (0,t')\},$$

then $\tilde{t} \neq 0$ or 1 and $s(\tilde{t}) \in \partial A$. From the construction of set M, $s(\tilde{t})$ cannot be on the stable manifold of a critical element of type greater than or equal to two. Hence, by Theorem 6-9, it follows that $s(\tilde{t}) = x_{\tilde{t}}$ must lie on the stable manifold of a type-one critical element

$\sigma_1 \in \partial A_q$. Now the neighborhood N was arbitrary and for each neighborhood there exists a point in $W^s(\sigma_1)$ for some σ_1. Since the number of type-one critical elements is finite, it follows that for a fixed σ_1 we can construct a countable sequence $x_{\bar{i}}(k)$ such that $\lim_{k\to\infty} x_{\bar{i}}(k) = \sigma$. This leads to $\sigma \in \overline{W^s(\sigma_1)}$. Hence, the proof is completed.

In fact, it can be further shown using arguments similar to those of Theorem 6-1 that $W^u(\sigma) \cap W^s(\sigma_1) \neq \emptyset$. Let x be a point at which $W^u(\sigma)$ and $W^s(\sigma_1)$ meet transversally and consider a closed disk $D \subset W^s(\sigma_1)$ containing x. Since the intersection is transversal, the following holds

$$W^s(\sigma) \subset \overline{\bigcup_{t\le 0} \phi(t, D)} \subset \overline{W^s(\sigma_1)}$$

By Theorem 6-9, it follows that

$$\partial A_q \subset \bigcup_{\sigma_i \in \Sigma_1 \cap \partial A_q} \overline{W^s(\sigma_i)} \tag{6.6}$$

where Σ_1 is the set of type-one critical elements of the system. Since ∂A_q is a closed and invariant set, assumption (B3) implies the following relationship

$$\bigcup_{\sigma_i \in \Sigma_1 \cap \partial A_q} \overline{W^s(\sigma_i)} \subset \partial A_q.$$

Thus, we have the following result that is a refined version of Theorem 6-9.

THEOREM 6-11 **(Characterization of the quasi-stability boundary)**
Consider a nonlinear dynamical system described by (6.2) which satisfies assumptions (B1) to (B3). Let σ_i^1, $i = 1, 2, \ldots$, be the type-one critical elements on the quasi-stability boundary $\partial A_q(x_s)$ of stable equilibrium point x_s. Then,

$$\partial A_q(x_s) = \bigcup_{\sigma_i^1 \in \partial A_q(x_s)} \overline{W^s(\sigma_i^1)}.$$

Theorem 6-11 asserts that the quasi-stability boundary equals the union of the closure of the stable manifolds of all type-one critical elements on the boundary. In addition, the unstable manifolds of type-one UEPs and of type-one closed orbits all converge to the stable equilibrium point; according to Theorem 6-8.

6.4 Robustness and characterizations

In this section we will show that the set of critical elements lying inside the quasi-stability region is robust relative to small perturbations of its vector field. We will then derive a characterization for critical elements lying inside the quasi-stability regions. A characterization for the class of nonlinear dynamical systems whose stability regions equal their quasi-stability regions will be developed. None of the results derived in this section require the transversality condition as stated in assumption (B2), which is hard to check numerically.

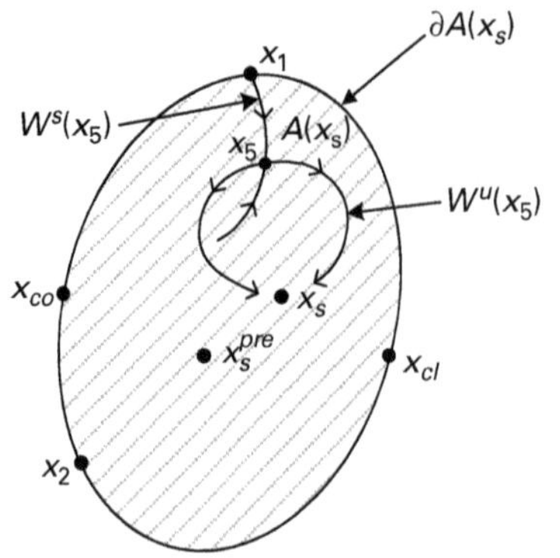

Figure 6.3 The unstable manifold of a critical element lying in the quasi-stability region is contained in the quasi-stability region.

We start this section with a characterization of the critical elements which lie inside a quasi-stability region in terms of its unstable manifold (see Figure 6.3). In this figure, the UEPs x_{c0}, x_{c1}, x_1, x_2 all lie on the quasi-stability boundary while the UEP x_5 does not; instead it lies inside the quasi-stability region. The unstable manifold of the UEP x_5 does not intersect the complement of the closure of the quasi-stability region while it is contained in the quasi-stability region. On the other hand, the unstable manifolds of the UEPs x_{c0}, x_{c1}, x_1, x_2 intersect the quasi-stability region as well as the complement of its closure.

THEOREM 6-12 **(Critical element in a quasi-stability region)**
Let A and A_q be the stability region and the quasi-stability region of x_s of system (6.2), respectively. Let $\sigma \neq x_s$ be a hyperbolic critical element of system (6.2). Then, $\sigma \in A_q$ if and only if $W^s(\sigma) \subset \partial A \cap A_q$.

Proof Suppose $\sigma \in \partial A \cap \text{int}\overline{A}$. From Theorem 6-8, the following results hold:

(a) $W^u(\sigma) \cap A \neq \emptyset$ or $W^u(\sigma) \cap W^s(x_s) \neq \emptyset$, and
(b) the intersection is transversal because $W^s(x_s)$ has dimension n.

Now let y be a point of the set $W^u(\sigma) \cap W^s(x_s)$ and consider a closed disk $D \subset W^s(x_s)$ containing y. Since the intersection is transversal, the simplified version of the λ-lemma in the proof of Theorem 6-1 asserts the following:

$$W^s(\sigma) \subset \overline{\bigcup_{t \le 0} \phi(t, D)} \subset \overline{W^s(x_s)} = \overline{A}.$$

Since $W^s(\sigma)$ is disjoint from A, it follows that

$$W^s(\sigma) \subset \partial A.$$

On the other hand, $\partial A \cap \text{int}\overline{A}$ is an invariant set and $\sigma \in \text{int}\overline{A}$, so it follows that $W^s(\sigma) \subset \partial A \cap \text{int}\overline{A}$. This completes the proof.

The set $\partial A \cap \text{int}\overline{A}$ is essentially the "difference" between the quasi-stability region and the stability region. We next present a characterization of the set $\partial A \cap \text{int}\overline{A}$ for system (6.2) under assumptions (B1) and (B3) without assumption (B2).

THEOREM 6-13 **(Characterization of $\partial A \cap$ int $\overline{A}$)**
Consider the nonlinear dynamical system (6.2) which satisfies assumptions (B1), (B2) and (B3). Let A and A_q be the stability region and the quasi-stability region of x_s of system (6.2), respectively. Let $\sigma_i \neq x_s$ be a critical element of system (6.2), $i = 1, 2, \ldots,$. Then

(a) the set $\partial A \cap \mathrm{int}\overline{A}$ is characterized by

$$\partial A \cap A_q = \bigcup_{\sigma_i \in A_q} W^s(\sigma_i)$$

(b) $\partial A \cap A_q$ is a repellent set (relative to A) of measure 0 in R_n.

Proof Since $\partial\overline{A} = \overline{A} - \mathrm{int}\overline{A}$, the following equality follows:

$$\partial A \cap A_q = \partial A - \partial\overline{A}.$$

According to Theorem 6-12,

$$\bigcup_{\sigma_i \in A_q} W^s(\sigma_i) \subset \partial A \cap A_q.$$

The invariance property of the set is $\partial A \cap A_q$ and assumption (B3) implies that

$$\partial A \cap A_q \subset \bigcup_{\sigma_i \in A_q} W^s(\sigma_i).$$

Combining the above two equations leads to the following:

$$\partial A \cap A_q = \bigcup_{\sigma_i \in A_q} W^s(\sigma_i).$$

Hence, the proof of part (a) is completed. Since the stability boundary is a repellent set, the set $\partial A \cap A_q$ is also repellent and is of dimension less than $(n-1)$. Hence, part (b) holds.

We are now in a position to present a characterization of a class of nonlinear dynamical systems whose stability regions equals their quasi-stability regions.

THEOREM 6-14 **(A characterization)**
Consider the nonlinear dynamical system (6.2) which satisfies assumptions (B1), (B2) and (B3). Let A and A_q be the stability region and the quasi-stability region of x_s of system (6.2), respectively. If there do not exist critical elements, say $\sigma_i \neq x_s$, whose unstable manifold is contained in A, i.e. $\{W^u(\sigma) - \sigma\} \subset A$, then $A = A_q$.

Proof This theorem follows from Theorem 6-12 and Theorem 6-13.

The difference between the stability region and the quasi-stability region is the set $\partial A \cap A_q$. If a small perturbation on the dynamical system (6.2) can destroy the set $\partial A \cap A_q$ completely, in other words if $\partial A \cap A_q$ is not robust under small perturbations, then the concept of quasi-stability regions will be practically and theoretically irrelevant.

The robustness of quasi-stability regions can be rephrased as follows: whether the set $(\partial A \cap A_q)$ persists under small perturbations of the vector field (6.2). The next theorem shows that the set $(\partial A \cap A_q)$ is topologically persistent under small C^1 perturbations of the vector field (6.2). Hence, quasi-stability regions are robust under small perturbations of their vector fields.

Let $X(R^n)$ be the space of C^1 vector fields on (the closed and smooth n-dimensional Euclidean space) R^n with the C^1-topology. Two vector fields will be close if the vector fields and their first derivatives are close at all points of R^n. A perturbation of the vector field f is a vector field in $X(R^n)$ which is close to f.

THEOREM 6-15 **(Robustness of quasi-stability regions)**
Consider the nonlinear dynamical system (6.2) which satisfies assumptions (B1), (B2) and (B3). Let A and A_q be the stability region and the quasi-stability region of x_s of system (6.2), respectively. If the set $\partial A \cap A_q$ contains a hyperbolic critical element, then the set $\partial A \cap A_q$ is robust under small C^1 perturbations of the vector field f.

Proof Let σ be a critical element on the set $\partial A \cap A_q$ which is non-empty. Note that if σ is hyperbolic, then there exists a neighborhood $N \subset R^n$ of σ that contains a hyperbolic critical element near σ for any vector field, say g, which is a small C^1 perturbation of the vector field f of system (6.2). In addition, this perturbation keeps all the topological properties of σ invariant. Thus, $(x_s^{'})$ is a hyperbolic stable equilibrium point of g with a well-defined stability region $A(x_s^{'})$. If σ' denotes the perturbed critical element of σ, then $W^u(\sigma')$ has the same dimension as $W^u(\sigma)$ and they are close in a neighborhood U. We need to show that σ' lies in $A_q(x_s^{'})$. Since $\sigma \in A_q(x_s)$, Theorem 6-11 asserts that $\{W^u(\sigma) - \sigma\} \subset A$. This in turn implies that

$$W^u(\sigma) \cap (\overline{A})^c = \varnothing.$$

Because empty intersections are transversal and transversal intersections persist under small perturbations, it follows that

$$W^u(\sigma^{'}) \cap N \cap (\overline{A(x_s{'})})^c = \varnothing.$$

Hence, $W^u(\sigma') \cap N \subset A_q(x_s')$. Since $A_q(x_s')$ is an invariant set, $W^u(\sigma') \subset A_q(x_s')$. Now applying Theorem 6-11, we have $\sigma^{'} \in A_q(x_s')$ or $\sigma' \in \partial A(x_s')$. Thus, $\partial A(x_s') \cap A_q(x_s') \neq \varnothing$. This completes the proof.

The robustness property of quasi-stability regions stated in Theorem 6-15 ensures that there exists a neighborhood $U \in X(R^n)$ of the vector field f of system (6.2) such that if $g \in U$ and (x_s') is the corresponding nearby stable equilibrium point of g, then $\partial A^{'} \cap \operatorname{int}\overline{A'} \neq \varnothing$, where A' is the stability region of (x_s').

6.5 Conclusions

The concept of quasi-stability regions of general nonlinear dynamical systems has been introduced. From an engineering viewpoint, the quasi-stability region is a "practical" stability region while the quasi-stability boundary is less complex than the stability boundary. A comprehensive theory for the quasi-stability boundary has been presented.

The quasi-stability boundary is shown to be the union of stable manifolds of equilibrium points and limit cycles on the boundary. In addition, the quasi-stability boundary is shown to be the union of the closure of stable manifolds of type-one equilibrium points and limit cycles on the boundary. Dynamic and topological properties of quasi-stability regions have also been derived. Whereas a stability region is always a subset of a quasi-stability region, the class of nonlinear dynamical systems whose stability regions equal their quasi-stability regions has also been characterized.

The quasi-stability regions persist under small perturbations of their vector fields. Furthermore, the set of critical elements and their stable manifolds which lie inside the quasi-stability regions are robust relative to small perturbations of their vector fields. The robust property makes the concept of quasi-stability region practically and theoretically relevant.

7 Stability regions of constrained dynamical systems

Constrained nonlinear dynamical systems have been used to model a variety of practical nonlinear systems. Constraints usually have their origin in equations of balance such as the balance of energy or power in physical systems. These balance equations are a consequence of the physical laws of conservation, such as the energy conservation principle and the Kirchoff law of currents. Usually, these constraints appear in mathematical models in the form of algebraic equations and, consequently, constrained dynamical systems are often represented by a set of differential and algebraic equations. For instance, power system transient stability models are described by a set of differential-algebraic equations. The set of algebraic equations defines a constraint set or constraint manifold that constrains the system dynamics. Hence, the system trajectories of constrained nonlinear dynamical systems are confined to the constraint manifold, making the analysis of constrained dynamical systems challenging. Depending on the nature of the constraint set, bounded trajectories might not be defined for all times. This again makes the analysis of the asymptotic behavior of solutions of constrained dynamical systems a challenge.

Constrained nonlinear dynamical systems can exhibit trajectories whose sources and sinks all lie in singularities of the constraint set. These singular sets can exist on the stability boundary, making the task of characterizing the stability boundary of these systems even more complicated than the characterization of stability boundaries of ordinary differential equations developed in Chapter 4.

A comprehensive theory of stability regions and of stability boundaries for unconstrained continuous dynamical systems was presented in Chapters 4 and 5. In this chapter, a comprehensive theory of the stability boundary and of the stability regions of constrained nonlinear dynamical systems will be developed. Two approaches for studying the stability region of constrained dynamical systems will be presented in this chapter. The first relies on an approximation of the stability region via the singular perturbation theory and the second relies on a regularization of the vector field on the singular surface. For the latter approach, we follow the developments presented in [262,264].

7.1 Constrained nonlinear dynamical systems

Consider the following class of autonomous constrained nonlinear dynamical systems:

$$(\Sigma)\quad \begin{cases} \dot{x} = f(x,y) \\ 0 = g(x,y) \end{cases} \tag{7.1}$$

where $x \in R^n$ is the vector of dynamical state variables, $y \in R^m$ is the vector of algebraic or instantaneous state variables and $f: R^n \times R^m \to R^n$ and $g: R^n \times R^m \to R^m$ are smooth functions. This system is composed of a set of differential and algebraic equations and will be called, for short, a DAE system.

The state variables of this system belong to the Euclidean space R^{n+m}, however, trajectories of this system are constrained to a subset of R^{n+m}, the *constraint set*:

$$\Gamma = \{(x, y) \in R^{n+m} | 0 = g(x, y)\}. \tag{7.2}$$

A continuous and differentiable function $\phi(t) = (x(t), y(t))$: $I \to R^{n+m}$ defined on an interval $I \subset R$ is a solution of the DAE system (7.1) if (i) $(x(t), y(t)) \in \Gamma$ for all $t \in I$ and (ii) $(x(t), y(t))$ satisfies equation (7.1) for all $t \in I$. The solution of the DAE system (7.1) starting in (x, y) at time $t=0$ is denoted $\phi(t, (x, y))$. The maximal interval of definition of a solution will be denoted (ω_-, ω_+).

A point $(\bar{x}, \bar{y}) \in R^{n+m}$ is an equilibrium point of the DAE system (7.1) if $f(\bar{x}, \bar{y}) = 0$ and $g(\bar{x}, \bar{y}) = 0$. The definitions of stability and asymptotic stability of equilibrium points of DAE systems are analogous to those of equilibrium points for ODE systems, see Section 2.3 of Chapter 2. The only difference is that the open neighborhoods of the equilibrium points in these definitions have to be taken on the constraint set Γ and not in R^{n+m}.

The associated linearized system, in the neighborhood of the equilibrium point $(\bar{x}, \bar{y})$, is given by:

$$\begin{cases} \dot{\zeta} = D_x f(\bar{x}, \bar{y})\zeta + D_y f(\bar{x}, \bar{y})\xi \\ 0 = D_x g(\bar{x}, \bar{y})\zeta + D_y g(\bar{x}, \bar{y})\xi. \end{cases}$$

If $D_y g$, calculated at the equilibrium point $(\bar{x}, \bar{y})$, is invertible, then we can solve the algebraic equation and derive the following reduced linearized system:

$$\dot{\zeta} = [D_x f - D_y f (D_y g)^{-1} D_x g]\zeta.$$

The equilibrium point $(\bar{x}, \bar{y})$ is hyperbolic if all eigenvalues of the reduced Jacobian matrix $J_{red} = D_x f - D_y f (D_y g)^{-1} D_x g$ have real part different from zero. Moreover, a hyperbolic equilibrium point is said to be a type-*k* equilibrium point if J_{red} possesses exactly *k* eigenvalues with real part greater than zero.

If the algebraic equation $0 = g(x, y)$ can be solved, i.e. if there exists a smooth function $h: R^n \to R^m$ such that $0 = g(x, h(x))$ for all x, then the instantaneous state variables can be eliminated from (7.1), reducing the problem to the analysis of the following ordinary differential equation:

$$\dot{x} = f(x, h(x)). \tag{7.3}$$

The theory developed in Chapters 4 and 5 is applicable and sufficient to establish a complete characterization of the stability region and stability boundary of (7.3) and consequently of (7.1). However, the condition that instantaneous state variables can be completely eliminated may not hold for many dynamical systems in the form of (7.1). In spite of that, we can locally solve the algebraic equation and use the theory of ordinary differential equations to guarantee the local existence and uniqueness of solutions of DAE systems in *regular points* of the constraint set Γ.

A point $(x_0, y_0) \in \Gamma$ is regular if the derivative $D_y g(x_0,y_0)$ of g with respect to y, calculated at $(x_0,y_0) \in \Gamma$, is invertible. If $(x_0,y_0) \in \Gamma$ is a regular point, then the implicit function theorem guarantees the existence of neighborhoods U of x_0 and W of y_0, and a smooth function $h: U \to W$, with $h(x_0) = y_0$, that solves the algebraic equation $0 = g(x, h(x))$ in the neighborhood $U \times W$ of the point $(x_0,y_0) \in \Gamma$. Then the smoothness of f and g and the theory of existence of solutions of ordinary differential equations guarantee the existence and uniqueness of the solution of (7.1) passing through (x_0,y_0) in a neighborhood of $(x_0,y_0) \in \Gamma$. More precisely, there is an interval of time I containing the origin and a function $\alpha(t)$: $I \to R^n$ such that $\alpha(t) \in U$ and

$$\dot{\alpha}(t) = f(\alpha(t),\, h(\alpha(t))) \quad \text{for all } t \in I; \tag{7.4}$$

consequently, $(x(t) = \alpha(t), y(t) = h(\alpha(t)))$ is a solution of the constrained system (7.1) for all $t \in I$. In other words, at regular points of Γ, the local existence and uniqueness of trajectories of the DAE system (7.1) are inherited from the local existence and uniqueness of trajectories of ordinary differential equations.

Unlike ordinary differential equations (see Theorem 2.1 of Chapter 2), bounded trajectories of DAE systems cannot always be extended infinitely because of the presence of singular points in Γ. Trajectories can be extended until they reach these singularities and, therefore, the singular surface plays a role in the characterization of stability boundaries. Next, the constraint set and its singular points will be examined.

7.1.1 Constraint set and singular points

The constraint set Γ may have several connected components and, typically, each connected component of Γ is an n-dimensional submanifold of R^{n+m}. A real number α is said to be a regular value of a scalar function $h(x)$ if the derivative Dh is surjective at every point of $h^{-1}(\alpha)$. In addition, zero is said to be a regular value for the vector function g if rank$[D_x g\ D_y g] = m$ at every point of Γ. The pre-image theorem [109] ensures that each connected component of Γ is usually an n-dimensional submanifold of R^{n+m}. Define the following subset of Γ:

$$M = \{(x,y) \in \Gamma \mid \text{rank}\ [D_x g(x,y)\, D_y g(x,y)] = m\}.$$

The derivative of g is surjective in every point of M and therefore M is an n-dimensional submanifold in R^{n+m}. Typically M is a dense set in Γ. Actually, we will show that $\Gamma - M$, the set of points of Γ at which rank$[D_x g\ D_y g] < m$, is a thin set in Γ. While the points $(x_0, y_0) \in \Gamma$ where the derivative $D_y g(x_0, y_0)$ is invertible are called *regular points* of Γ, the points where the derivative $D_y g(x_0, y_0)$ is non-invertible are called *singular points*. The collection of singular points is called the *singular surface* and will be denoted by S:

$$S = \{(x,y) \in \Gamma \mid \Delta(x,y) = 0\}, \tag{7.5}$$

where $\Delta(x, y) = \det D_y g(x, y)$.

The following rank condition is usually satisfied

$$\operatorname{rank}\begin{bmatrix} D_x g(x,y)\, D_y g(x,y) \\ D_x \Delta(x,y)\, D_y \Delta(x,y) \end{bmatrix} = m+1$$

for every $(x, y) \in S$, thus, from the pre-image theorem [109], it is concluded that S is usually a submanifold of dimension $n-1$ embedded in Γ.

Typically, trajectories do not exist at singular points. Physically speaking, these singular points are usually due to un-modeled or neglected dynamics, such as parasitic dynamics, and therefore these singular points signal the onset of unpredictable system behavior. The following example, which appears in [246], illustrates a singular point on a two-dimensional system:

$$\begin{aligned} \dot{x} &= y \\ 0 &= (y-1)^2 + x. \end{aligned} \tag{7.6}$$

The constraint set $\Gamma = \{(x, y) \in R^2 | 0 = (y-1)^2 + x\}$ is a parabola passing through the point (0,1). This parabola is shown in Figure 7.1. Indeed $\operatorname{rank}[D_x g\, D_y g] = \operatorname{rank}[2(y-1)\ 1] = 1$ for all $(x, y) \in \Gamma$ and therefore Γ is a manifold of dimension one in R^2. Calculating the derivative of the algebraic equation with respect to y, one obtains $D_y g = 2(y-1)$. In this example, matrix $D_y g$ is a real number and therefore $\det D_y g = 2(y-1)$. This determinant is equal to zero if and only if $y = 1$, which implies, from the algebraic equation (7.6), that x–0. It follows that the point (0,1) is the unique singular point in Γ, see Figure 7.1. At every other point of Γ, the trajectories exist locally and are unique.

Consider the trajectory that passes through the point ($x_0 = -1$, $y_0 = 2$). This trajectory moves along the constraint set Γ and approaches the singular point in finite time as illustrated in Figure 7.1. However, at the singular point (0,1), there is no function that

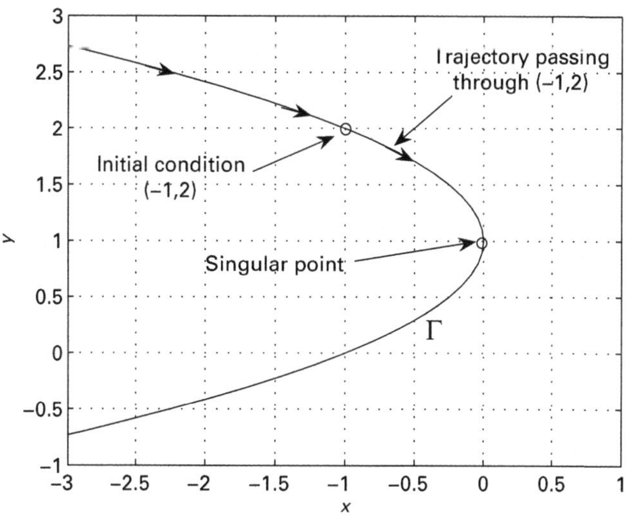

Figure 7.1 Phase portrait of system (7.6). The trajectory passing through the regular point (−1,2) reaches the singularity point (0,1) in finite time.

satisfies the conditions for a solution of (7.1). Therefore, the trajectory passing through the point $(-1, 2)$ can only be extended up to the time when the trajectory reaches the singular point.

The constraint manifold Γ typically possesses a number of isolated connected components. Within each connected component, the singular surface S decomposes Γ into several smaller components Γ_i, such that $\Gamma - S = \bigcup_i \Gamma_i$. Since trajectories of the differential-algebraic system (7.1) cannot cross singular surfaces, they are confined to a single component Γ_i of Γ. If $\text{rank}[D_x g\ D_y g] < m$ at a point $(x_0, y_0) \in \Gamma$, then in particular rank $D_y g < m$ and thus $(x_0, y_0) \in S$. Therefore $\Gamma - S$ is always an n-dimensional submanifold of R^{n+m}. In particular, every component Γ_i of Γ is an n-dimensional submanifold of R^{n+m}.

7.2 Stability region of DAE systems

Let Γ_s be a component of Γ and $(x_s, y_s) \in \Gamma_s$ an asymptotically stable equilibrium point of a DAE system (7.1). The stability region of (x_s, y_s) is defined as

$$A(x_s, y_s) = \{(x,y) \in \Gamma_s : \lim_{t \to \infty} \phi(t, (x,y)) = (x_s, y_s)\}. \tag{7.7}$$

The continuity of solutions with respect to the initial conditions at regular points of the constraint set Γ ensures that the stability region $A(x_s, y_s)$ is an open set relative to $\Gamma_s \subset M$. The topological boundary of the stability region $A(x_s, y_s)$ will be denoted $\partial A(x_s, y_s)$. The stability boundary is a closed set. The stability boundary has maximal dimension $n-1$ at every point. However, the stability boundary may have parts with lower dimension.

Characterizations of the stability boundary $\partial A(x_s, y_s)$ of DAE systems have recently been developed. It has been shown that under certain conditions, the stability boundary $\partial A(x_s, y_s)$ consists of two parts: the first part is the stable manifolds of the equilibrium points on the stability boundary, while the second part contains points whose trajectories reach singular surfaces [264]. The second part can be further delineated as a union of the stable manifolds of pseudo-equilibrium points and semi-singular points on the stability boundary and parts of singular surfaces [262,264]. These characterizations will be presented in the later part of this chapter.

The next chapter studies the stability region of constrained dynamical systems by establishing a relationship with the stability region of an associated family of unconstrained dynamical systems, the so-called singularly perturbed systems. The advantages of this approach are (i) we avoid the difficulties of dealing with DAE models and (ii) the theory of stability regions for unconstrained systems has already been developed.

7.3 Singular perturbation approach

The singular perturbation approach treats the set of algebraic equations describing a DAE system as a limit of the fast dynamics $\varepsilon\dot{y} = g(x, y)$. In other words, for ε sufficiently small, the dynamics will quickly approach the algebraic manifold Γ and the solution of this approximated system will quickly converge to a solution of the DAE system.

With a possible change of sign of function g, the component of interest of the constraint manifold Γ will be an attractor for these pseudo fast dynamics. If the points on a component Γ_i of Γ are such that the corresponding Jacobian matrix $(\partial g/\partial y)\,(x,y)$ has all the eigenvalues with negative real parts, then the component is stable; otherwise, it is an unstable component.

Therefore, for the DAE system (7.1), we can define an associated singularly perturbed system:

$$\begin{aligned} \dot{x} &= f(x,y) \\ \varepsilon\dot{y} &= g(x,y) \end{aligned} \tag{7.8}$$

where ε is a sufficiently small positive number. The state variables of system (7.8) have very different rates of dynamics and they can be separated into two distinct time scales: slow variable x and fast variable y.

The trajectory of the singularly perturbed system (7.8) starting in (x,y) at time t=0 will be denoted by $\varphi_\varepsilon(t, x, y)$. If (x_s, y_s) is an asymptotically stable equilibrium point of the singularly perturbed system (7.8), then its stability region is defined as:

$$A_\varepsilon(x_s,y_s) = \{(x,y) \in R^{n+m} : \varphi_\varepsilon(t,x,y) \rightarrow (x_s,y_s) \text{ as } t \rightarrow \infty\}.$$

A relationship exists between the stability region of the DAE system and that of the singularly perturbed system. In particular, we will see that for sufficiently small ε, their stability boundaries are close, at least in compact sets of these boundaries.

Note that trajectories of the singularly perturbed system (7.8) are not confined to the algebraic manifold Γ and are not exactly the same as those of the original DAE system (7.1). However, trajectories generated by the singularly perturbed system are still valid approximations to those of the DAE system. Indeed, a theoretical justification ensuring that the difference of trajectories between the original DAE (7.1) and the singularly perturbed system (7.8) is uniformly bounded by the order of $O(\varepsilon)$ is provided by Tikhonov's theorem. Tikhonov's theorem is formally stated in Chapter 16. Figure 7.2

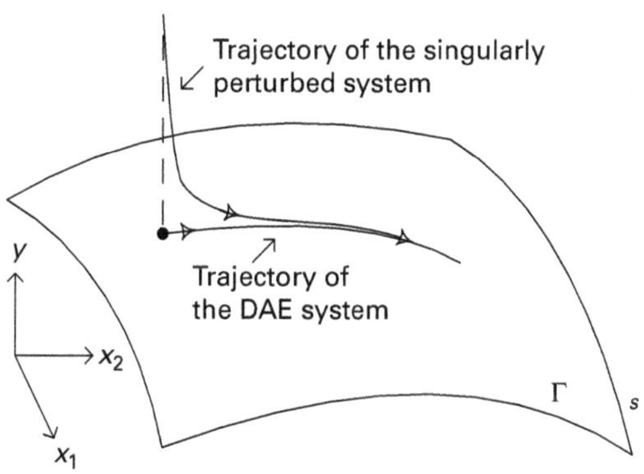

Figure 7.2 Relationship between the trajectory of the singularly perturbed system and that of the DAE system.

illustrates the relationship between the trajectory of the singularly perturbed system and that of the DAE system. In the beginning, the trajectory of the singularly perturbed system quickly approaches the constraint manifold Γ and then follows very close to a solution of the DAE system in the neighborhood of Γ.

A DAE system and its corresponding singularly perturbed system share several similar dynamical properties, including a close relationship between their stability regions. One obvious property is that they have the same set of equilibrium points, i.e. a point $(\bar{x},\bar{y})$ is an equilibrium point of the DAE system (7.1) if and only if it is an equilibrium point of the singularly perturbed system (7.8) for all ε. In a stable component Γ_s of the constraint manifold Γ, the equilibrium points of the DAE system are the same type as those of the singularly perturbed system. The following results show this invariant topological relationship between the equilibrium points of a DAE system and those of its associated singularly perturbed system.

THEOREM 7-1 **(Invariant topological relationship)**
If an equilibrium point, say $(\bar{x},\bar{y})$ of system (7.1) lies on one stable component Γ_s of the constraint manifold Γ, then there exists an ε > 0 such that for all ε ∈ (0, ε), it follows that

(a) if $(\bar{x},\bar{y})$ is a hyperbolic equilibrium point of the DAE system (7.1), then $(\bar{x},\bar{y})$ is a hyperbolic equilibrium point of the singularly perturbed system (7.8), moreover
(b) if $(\bar{x},\bar{y})$is a type-k equilibrium point of the DAE system (7.1), then $(\bar{x},\bar{y})$ is a type-k equilibrium point of the singularly perturbed system (7.8).

Proof Since $(\bar{x},\bar{y})$ lies on a stable component Γ_s of the constraint manifold Γ, $D_zg(\bar{x},\bar{y})$ possesses no eigenvalues on the right hand side of the complex plane and m eigenvalues on the left hand side. Therefore, there exists a real number $\alpha > 0$ such that $|\mathrm{Re}\lambda\}| > \alpha$ for every eigenvalue λ of $D_zg(\bar{x},\bar{y})$.

Consider the following time scale change $t = \varepsilon\tau$. In this new time scale, the singularly perturbed system (7.8) assumes the form:

$$\begin{aligned} \frac{dx}{d\tau} &= \varepsilon f(x,y) \\ \frac{dy}{d\tau} &= g(x,y). \end{aligned} \tag{7.9}$$

Consider the linearization of system (7.9) in the neighborhood of the equilibrium $(\bar{x},\bar{y})$:

$$\begin{bmatrix} \frac{d\Delta x}{d\tau} \\ \frac{d\Delta y}{d\tau} \end{bmatrix} = J_\varepsilon^{fast} \begin{bmatrix} \Delta x \\ \Delta y \end{bmatrix}$$

where

$$J_\varepsilon^{fast} = \begin{bmatrix} \varepsilon D_x f & \varepsilon D_y f \\ D_x g & D_y g \end{bmatrix}.$$

The complex number μ is an eigenvalue of matrix J_ε^{fast} if there exists a vector $(\Delta x, \Delta y) \neq 0$ satisfying:

$$\begin{bmatrix} \varepsilon D_x f - \mu I_n & \varepsilon D_y f \\ D_x g & D_y g - \mu I_m \end{bmatrix} \begin{bmatrix} \Delta x \\ \Delta y \end{bmatrix} = 0.$$

For $\mu \neq 0$ and sufficiently small ε, the submatrix $\varepsilon D_x f - \mu I_n$ is invertible. Then, one can solve Δx as a function of Δz and write:

$$[D_y g - \varepsilon D_x g(\varepsilon D_x f - \mu I_n)^{-1} D_y f - \mu I_m]\Delta z = 0.$$

Then μ is an eigenvalue of a matrix that can be seen as a perturbation of the matrix $D_y g$. Define the function $p_\varepsilon(\mu) = \det[D_y g - \varepsilon C(\varepsilon, \mu) - \mu I_m]$ where $C(\varepsilon, \mu) = D_x g(\varepsilon D_x f - \mu I_n)^{-1}$ $D_y f$. For $\mu \neq 0$ and sufficiently small ε, C is a continuous function of μ and ε.

Consider a simple closed curve γ in the complex plane such that all the eigenvalues of $D_y g$ are contained in the area that is delimited by this curve. Since all the m eigenvalues of $D_y g$ are located on the left hand side of the complex plane, the curve γ can be chosen such that $\gamma \subset \{\mu\colon \mathrm{Re}\{\mu\} < \alpha < 0\}$. Therefore $p_0(\mu) \neq 0$ for all $\mu \in \gamma$ and, as a consequence, $\inf_{\mu \in \gamma} |p_0(\mu)| = m > 0$. Using the continuity of $p_\varepsilon(\mu)$ with respect to ε, one concludes that $\inf_{\mu \in \gamma} |p_\varepsilon(\mu)| > 0$ for sufficiently small ε. Then

$$v(\varepsilon) = \frac{1}{2\pi i} \oint_\gamma \frac{p_\varepsilon'(\mu)}{p_\varepsilon(\mu)}.$$

is well defined and represents, according to the theory of complex variables, the number of zeros of $p_\varepsilon(\mu)$ inside γ. Since $v(\varepsilon)$ must be an integer number, we conclude, from the continuity of $v(\varepsilon)$, that $m = v(0) = v(\varepsilon)$ for sufficiently small ε. In other words, the existence of m eigenvalues of $D_y g$ with real part less than zero implies the existence of m eigenvalues of J_ε^{fast} with real part less than zero.

Suppose that $(\overline{x}, \overline{y})$ is a type-k equilibrium point of the DAE system (7.1). Then the matrix $J_{red} - D_x f - D_y f(D_y g)^{-1} D_x g$ has k eigenvalues with real part grater than zero and $n-k$ eigenvalues with real part less than zero. Beyond that, there exists a number $M>0$ such that every eigenvalue λ satisfies $|\lambda| < M$.

Consider the linearization of the singularly perturbed system (7.8) in the neighborhood of the equilibrium point $(\overline{x}, \overline{y})$:

$$\begin{bmatrix} \frac{d\Delta x}{d\tau} \\ \frac{d\Delta y}{d\tau} \end{bmatrix} = J_\varepsilon \begin{bmatrix} \Delta x \\ \Delta y \end{bmatrix}$$

where

$$J_\varepsilon = \begin{bmatrix} D_x f & D_y f \\ \frac{1}{\varepsilon} D_x g & \frac{1}{\varepsilon} D_y g \end{bmatrix}.$$

The complex number μ is an eigenvalue of matrix J_ε if there exists a vector $(\Delta x, \Delta y) \neq 0$ satisfying

$$\begin{bmatrix} D_x f - \mu I_n & D_y f \\ \frac{1}{\varepsilon} D_x g & \frac{1}{\varepsilon} D_y g - \mu I_m \end{bmatrix} \begin{bmatrix} \Delta x \\ \Delta y \end{bmatrix} = 0$$

which is equivalent to

$$\begin{bmatrix} D_x f - \mu I_n & D_y f \\ D_x g & D_y g - \varepsilon\mu I_m \end{bmatrix} \begin{bmatrix} \Delta x \\ \Delta y \end{bmatrix} = 0.$$

If μ is bounded and ε is sufficiently small, then $D_y g - \varepsilon\mu I_m$ is invertible. Hence, one can solve Δy as a function of Δx and write:

$$[D_x f - D_y f (D_y g - \varepsilon\mu I_m)^{-1} D_x g - \mu I_n]\Delta x = 0.$$

Using the identity $B^{-1} = A^{-1} - B^{-1}(B - A)A^{-1}$ with $B = D_y g - \varepsilon\mu I_m$ and $A = D_y g$, one obtains:

$$[J_{red} - \varepsilon\mu C(\mu, \varepsilon) - \mu I_n]\Delta x = 0,$$

where, in this case, $C(\mu, \varepsilon) = D_y f(D_y g - \varepsilon\mu I_m)^{-1}(D_y g)^{-1} D_x g$. Hence, the eigenvalue μ of J_ε can be seen as an eigenvalue of a matrix that is a perturbation of J_{red}.

Define function $q_\varepsilon(\mu) = \det[J_{red} - \varepsilon\mu C(\mu, \varepsilon) - \mu I_n]$. For bounded μ and sufficiently small ε, C is a continuous function of μ and ε. Consider again a simple closed curve γ in the complex plane such that all the eigenvalues of J_{red} with real part greater than zero are contained in the area delimited by this curve. This curve can be chosen such that $\gamma \subset \{\mu: \text{Re}\{\mu\} > 0 \text{ and } |\mu| < M\}$.

Therefore $q_0(\mu) \neq 0$ for all $\mu \in \gamma$ and, as a consequence, $\inf_{\mu\in\gamma}|q_0(\mu)| = b > 0$. Using the continuity of $q_\varepsilon(\mu)$ with respect to ε, one concludes that $\inf_{\mu\in\gamma}|q_\varepsilon(\mu)| > 0$ for sufficiently small ε. Then

$$v(\varepsilon) = \frac{1}{2\pi i}\oint_\gamma \frac{q_\varepsilon{}'(\mu)}{q_\varepsilon(\mu)}$$

is well defined and represents, according to the theory of complex variables, the number of zeros of $q_\varepsilon(\mu)$ inside γ. Since $v(\varepsilon)$ must be an integer number, we conclude, from the continuity of $v(\varepsilon)$, that $k = v(0) = v(\varepsilon)$ for sufficiently small ε. In other words, the existence of k eigenvalues of J_{red} with real part greater than zero implies the existence of k eigenvalues of J_ε with real part greater than zero for sufficiently small ε.

Similar arguments can be used to show that the existence of $n{-}k$ eigenvalues of J_{red} with real part less than zero implies the existence of $n{-}k$ eigenvalues of J_ε with real part less than zero for sufficiently small ε.

Using the fact that $J_\varepsilon^{fast} = \varepsilon J_\varepsilon$, one concludes that λ is an eigenvalue of J_ε^{fast} if and only if λ/ε is an eigenvalue of J_ε. Then, for sufficiently small ε, the m eigenvalues of J_ε obtained in the fast time scale analysis via $D_y g$ have modulus sufficiently large to be different from the n eigenvalues obtained in the analysis in the slow time scale via J_{red}.

Thus, J_ε possesses k eigenvalues with real part greater than zero and $m+n-k$ eigenvalues with real part less than zero. Consequently, $(\bar{x},\bar{y})$ is a type-k hyperbolic equilibrium point of the singularly perturbed system (7.8) for sufficiently small ε. This completes the proof.

The above result shows that the type of equilibrium point of the singularly perturbed system is the same as the type of the corresponding equilibrium point of the DAE system, provided ε is sufficiently small. The converse of this theorem is not true. The following example illustrates an equilibrium point of a DAE system that does not have the same type of equilibrium points as the singularly perturbed system.

Example 7-1 Consider the following DAE system:

$$\begin{aligned}\dot{x} &= -x - z\\ \dot{y} &= y + z\\ 0 &= -x - y - z\end{aligned} \tag{7.10}$$

and the associated singularly perturbed linear system

$$\begin{aligned}\dot{x} &= -x - z\\ \dot{y} &= y + z\\ \varepsilon\dot{z} &= -x - y - z.\end{aligned} \tag{7.11}$$

The origin is an equilibrium point of these systems. In this case the constraint manifold is the plane $\Gamma = \{(x, y, z) \in R^3 : z = -x - y\}$. In this simple DAE system, the algebraic equation can easily be solved and the variable z can be eliminated. The reduced system is the harmonic oscillator, and the eigenvalues of the DAE system calculated at the origin are purely imaginary. On the other hand, the origin is a type-two equilibrium point of the singularly perturbed system for all $\varepsilon > 0$. Figure 7.3 presents the eigenvalues of the full system for the range $0.01 < \varepsilon < 0.5$. Observe that the eigenvalues of the singularly perturbed system approach the imaginary axis as ε approaches zero.

Nevertheless, under the condition that eigenvalues of the singularly perturbed system do not approach the imaginary axis for sufficiently small ε, the converse implication of Theorem 7.1 can be proven. We show this result in the next theorem.

THEOREM 7-2 **(Invariant topological relationship)**
Let $(\bar{x},\bar{y})$ be a type-k hyperbolic equilibrium point of the singularly perturbed system (7.8) on a stable component Γ_s of the constraint manifold Γ for every sufficiently small ε. If for every sufficient small $\varepsilon > 0$ there exists a real number $\alpha > 0$ such that $|\mathrm{Re}\{\lambda\}| > \alpha$ for every eigenvalue λ of the Jacobian matrix of the singularly perturbed system (7.8) calculated at $(\bar{x},\bar{y})$, then $(\bar{x},\bar{y})$ is a type-k hyperbolic equilibrium point of the DAE system (7.1).

Proof Suppose, by contradiction, that $(\bar{x},\bar{y})$ is not a hyperbolic equilibrium point of the DAE system (7.1). Then there are eigenvalues of $J_{red} = D_x f - D_y f (D_y g)^{-1} D_x g$ lying on

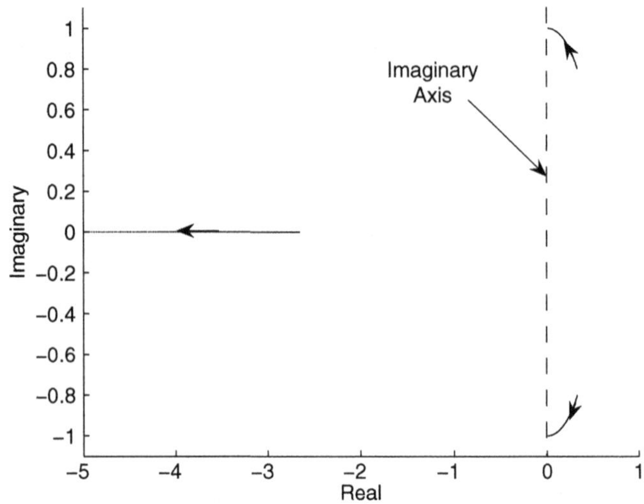

Figure 7.3 Root locus of system (7.11) for $0.01 < \varepsilon < 0.5$. Arrows indicate the movement of eigenvalues in the complex plane in the direction of decreasing ε. A pair of complex conjugate eigenvalues approaches the imaginary axis as $\varepsilon \to 0$.

the imaginary axis. Consider a simple closed curve γ in the complex plane such that all the eigenvalues of J_{red} on the imaginary axis belong to the area delimited by this curve. The curve γ can be chosen such that $\gamma \subset \{\mu\text{:Re}\{\mu\} < \alpha \text{ and } |\mu| < M\}$ for some sufficiently large $M>0$. Using arguments similar to those employed in the proof of Theorem 7.1, one proves the existence of eigenvalues of J_ε in the set $\{\mu\text{: Re}\{\mu\} < \alpha \text{ and } |\mu| < M\}$ for sufficiently small ε. This leads us to a contradiction and, therefore, $(\bar{x}, \bar{y})$ is a hyperbolic equilibrium point of the DAE system (7.1) for sufficiently small ε. Suppose now that the type of hyperbolic equilibrium $(\bar{x}, \bar{y})$ of the DAE system (7.1) is $r \neq k$, then Theorem 7.1 implies that $(\bar{x}, \bar{y})$ is a type-r hyperbolic equilibrium of the singularly perturbed system (7.8). Thus we reach a contradiction. Consequently, $(\bar{x}, \bar{y})$ is a type-k hyperbolic equilibrium of the DAE system (7.1) for sufficiently small ε. This completes the proof.

Theorem 7-1 and Theorem 7-2 assert, under certain conditions, that a point $(\bar{x}, \bar{y})$ is a hyperbolic type-k equilibrium point of the DAE system if and only if $(\bar{x}, \bar{y})$ is a hyperbolic type-k equilibrium point of the singularly perturbed system for all small $\varepsilon > 0$.

These local results have been extended into global results to establish the relationship between the stability boundaries of the DAE system and the singularly perturbed system. It will be shown that the stability boundaries of these two systems contain the same set of equilibrium points on the stable components of the constraint manifold.

THEOREM 7-3 **(Dynamic relationship)**
Let (x_s, y_s) and (x_u, y_u) be a hyperbolic asymptotically stable and an unstable equilibrium point of the DAE system (7.1) on the stable component Γ_s of Γ respectively. Suppose that for each $\varepsilon > 0$, the associated singularly perturbed system (7.8) has an energy function. Then there exists an $\varepsilon > 0$ such that if (x_u, y_u) lies on the stability boundary $\partial A(x_s, y_s)$ of

the DAE system (7.1), then (x_u, y_u) *lies on the stability boundary* $\partial A(x_s, y_s)$ *of the singularly perturbed system (7.8) for all* $\varepsilon \in (0, \varepsilon)$.

Proof Let N be a neighborhood in R^{n+m} of the asymptotically stable equilibrium point (x_s, y_s) of the singularly perturbed system (7.8) and M a neighborhood in R^{n+m} of (x_u, y_u). The asymptotically stability property of (x_s, y_s) ensures that the neighborhood N can be chosen sufficiently small such that every trajectory of the DAE system starting in $N \cap \Gamma_s$ stays bounded and approaches (x_s, y_s) as $t \rightarrow \infty$. Let B be an open ball strictly contained in N. If (x_u, y_u) belongs to the stability boundary of the DAE system (7.1), then there exists a point (x^*, y^*) in the induced neighborhood $M \cap \Gamma_s$ such that the trajectory $\varphi_{DAE}(t, x^*, y^*)$ of the DAE system starting at (x^*, y^*) enters B in finite time. More precisely, there is a time $T > 0$ such that $\varphi_{DAE}(T, x^*, y^*) \in B$. Using Tikonov's theorem for finite interval time, one can prove that trajectories of the singularly perturbed system will enter the neighborhood N in finite time for sufficiently small ε. In other words, $\varphi_\varepsilon(T, x^*, y^*) \in N$ for sufficiently small ε. Inside N, we can apply Tikonov's theorem for infinite time intervals to conclude that every trajectory of the singularly perturbed system starting in N is bounded and stays close to the asymptotically stable equilibrium point (z_s, y_s) for sufficiently small ε. The existence of an energy function implies that the trajectory approaches the asymptotically stable equilibrium point (x_s, y_s) as $t \rightarrow \infty$. As a result, we have proven that (x^*, y^*) is a point in the neighborhood M that belongs to the stability region of the singularly perturbed system for every sufficiently small ε. This completes the proof.

In addition, the converse of this theorem is not true. However, by imposing the condition that the eigenvalues of the equilibriums of the singularly perturbed system do not approach the imaginary axis as ε approaches zero, the converse holds as well. The next theorem formally states this result.

THEOREM 7-4 **(Invariant property)**
Let (x_s, y_s) *and* (x_u, y_u) *be a hyperbolic asymptotically stable equilibrium point and a hyperbolic unstable equilibrium point, respectively, of the singularly perturbed system (7.8) for all sufficiently small* ε. *Suppose the existence of a constant* $\alpha > 0$ *such that every eigenvalue* λ *of the Jacobian matrix of the singularly perturbed system (7.8) satisfies* $|\mathrm{Re}\{\lambda\}| > \alpha$ *for all* $\varepsilon \leq \bar{\varepsilon}$. *Then, if* (x_u, y_u) *lies on the stability boundary of the singularly perturbed system (7.8) for* $\varepsilon \leq \bar{\varepsilon}$, *then* (x_u, y_u) *lies on the stability boundary of the DAE system (7.1).*

Proof According to Theorem 7-3, (x_s, y_s) is a hyperbolic asymptotically stable equilibrium point and (x_u, y_u) is a hyperbolic unstable equilibrium point of the DAE system. For a given neighborhood $M \subset R^{n+m}$ of (x_u, y_u) and every sufficiently small $\varepsilon > 0$, there exists a point $(x_\varepsilon, y_\varepsilon) \in W_\varepsilon^u(x_u, y_u) \cap M$ such that $(x_\varepsilon, y_\varepsilon) \in A_\varepsilon(x_s, y_s)$. Consider a monotonically decreasing sequence $\{\varepsilon_j\}$ with $\varepsilon_j \rightarrow 0$ as $j \rightarrow \infty$. Since $\overline{M}$ is compact, the sequence $\{(x_{\varepsilon_j}, y_{\varepsilon_j})\}$ possesses a convergent subsequence, i.e. there exists (x_0, y_0) such that $(x_{\varepsilon_{j_k}}, y_{\varepsilon_{j_k}}) \rightarrow (x_0, y_0)$ as $\varepsilon_{j_k} \rightarrow 0$. We can also chose this sequence such that $(x_0, y_0) \neq (x_u, y_u)$. We know that $W_\varepsilon^u(x_u, y_u) \cap \overline{M}$ is ε-close to $W^u(x_u, y_u) \cap \overline{M}$. Then $d\left((x_{\varepsilon_{j_k}}, y_{\varepsilon_{j_k}}), W^u(x_u, y_u)\right) \rightarrow 0$ as $\varepsilon_{j_k} \rightarrow 0$. Then $d((x_0, y_0), W^u(x_u, y_u)) \leq d((x_0, y_0), (x_{\varepsilon_{j_k}}, y_{\varepsilon_{j_k}})) + d\left((x_{\varepsilon_{j_k}}, y_{\varepsilon_{j_k}}), W^u(x_u, y_u)\right) \rightarrow 0$ as $\varepsilon_{j_k} \rightarrow 0$ and $(x_0, y_0) \in \overline{W^u(x_u, y_u)} \subset \Gamma$.

Suppose, by contradiction, that $(x_0, y_0) \notin A(x_s, y_s)$. Then, there exists a number $\rho > 0$ such that the solution of the DAE system $\phi(t, x_0, y_0) \notin B_\rho(x_s, y_s)$ for all $t > 0$, where $B_\rho(x_s, y_s)$ is an open ball of radius ρ centered at (x_s, y_s). According to Tikhonov's theorem for finite time intervals, for every $T>0$, there exists an $\varepsilon(T)$ such that $\phi_\varepsilon(t, x_\varepsilon, y_\varepsilon) \notin B_{\rho/2}(x_s, y_s)$ for all $t \in [0, T]$ and $\varepsilon \in (0, \varepsilon)$. But this contradicts the fact that $\phi_\varepsilon(t, x_\varepsilon, y_\varepsilon) \rightarrow (x_s, y_s)$ as $t \rightarrow \infty$. Hence, $(x_0, y_0) \in A(x_s, y_s) \cap W^u(x_u, y_u)$ and $(x_u, y_u) \in \partial A(x_s, y_s)$. This completes the proof.

The above theorems provide a theoretical basis for characterizing an unstable equilibrium point on the stability boundary of the DAE system (7.1) by characterizing the same equilibrium in the stability boundary of the singularly perturbed system (7.8). In other words, we can check whether an equilibrium lies on the stability boundary of a DAE system, checking whether it lies on the stability of the singularly perturbed system for small ε, which is a unconstrained dynamical system.

Although the stability boundaries of the singular perturbed system and DAE system share the same unstable equilibrium points, the stability region and the stability boundary lie on completely different spaces. While the stability region of the singularly perturbed system is a set of dimension $n+m$ in the space R^{n+m}, the stability region of the DAE system is a set of dimension n on the set Γ. However, it can be proved that the intersection of the stability boundary of the singularly perturbed system with the constraint set Γ is a good approximation of the stability region of the DAE system for small ε. The next theorem establishes this property.

THEOREM 7-5 **(Stability boundary approximation)**
Let (x_s, y_s) be an asymptotically stable equilibrium point of the DAE system (7.1) in a stable component Γ_s of Γ. Suppose that for each $\varepsilon>0$, the associated singularly perturbed system (7.8) has an energy function. If (x, y) belongs to the stability region $A(x_s, y_s)$ of the DAE system (7.1) then there exists $\bar{\varepsilon} > 0$ such that (x, y) belongs to the stability region $A_\varepsilon(x_s, y_s)$ of the singularly perturbed system for all $\varepsilon \in (0, \bar{\varepsilon}]$.

The proof of this theorem is a direct consequence of Tikhonov's results. It is also very similar to the proof of Theorem 7-3 and for this reason will be omitted. Theorem 7-5 asserts that $A_\varepsilon(x_s, y_s) \cap \Gamma_s$ is a good candidate for approximating the stability region of the DAE system.

The following example, illustrates the results of this section and how the stability region of the singularly perturbed system can approximate the stability region of the DAE system.

Example 7-2 (Simple DAE system) The following dynamical system models a power system composed of a single generator and one load bus:

$$\begin{aligned} \dot{\omega} &= -\frac{1}{M_g} D_g \omega - \frac{1}{M_g} f(\alpha, V) \\ \dot{\alpha} &= -\frac{1}{D_l} f(\alpha, V) + \omega \\ 0 &= -g(\alpha, V) \end{aligned} \tag{7.12}$$

where $f(\alpha, V) = B_{12} V \sin\alpha - P_l$ and $g(\alpha, V) = \frac{1}{V}(Q_l - B_{12} V \cos\alpha - B_{22} V^2)$.

For the following set of parameters, $M_g = 20$, $D_g = 9$, $D_l = 50$, $P_l = 4$, $Q_l = -0.5$, $B_{12} = 10$ and $B_{22} = -10$, system (7.12) possesses the stable equilibrium point (0, 0.4291, 0.9613) and the following two unstable equilibrium points (0, 1.2660, 0.4193) and (0, −5.0172, 0.4193). These equilibrium points lie on the constraint manifold $\Gamma = \{(\omega, \alpha\ V): g(\alpha, V) = 0\}$. Figure 7.4 illustrates a stable component of this constraint manifold, which contains these three equilibrium points.

Both unstable equilibrium points lie on the stability boundary of the stable equilibrium point of the DAE system and as a consequence of Theorem 7-3, they also lie on the stability boundary of the singularly perturbed system for sufficiently small ε. The stability boundary of the DAE system is composed of the union of the stable manifolds of these two UEPs, as indicated in Figure 7.4. We compare the stability region of the DAE system with the stability region of the corresponding singularly perturbed system with different values of ε. The comparison is made on the subspace of (ω, α); i.e. the projection of the stability region in the subspace of (ω, α). As can be seen from Figure 7.5–Figure 7.7, the intersection of the stability region of the corresponding singularly perturbed system with the constraint manifold approaches the stability region of the DAE system as the values of ε approach zero, and this observation is in agreement with our theoretical development. When ε becomes smaller at the value of 0.1, the stability region of the corresponding singularly perturbed system captures that of the DAE system more accurately as we can see in Figure 7.7.

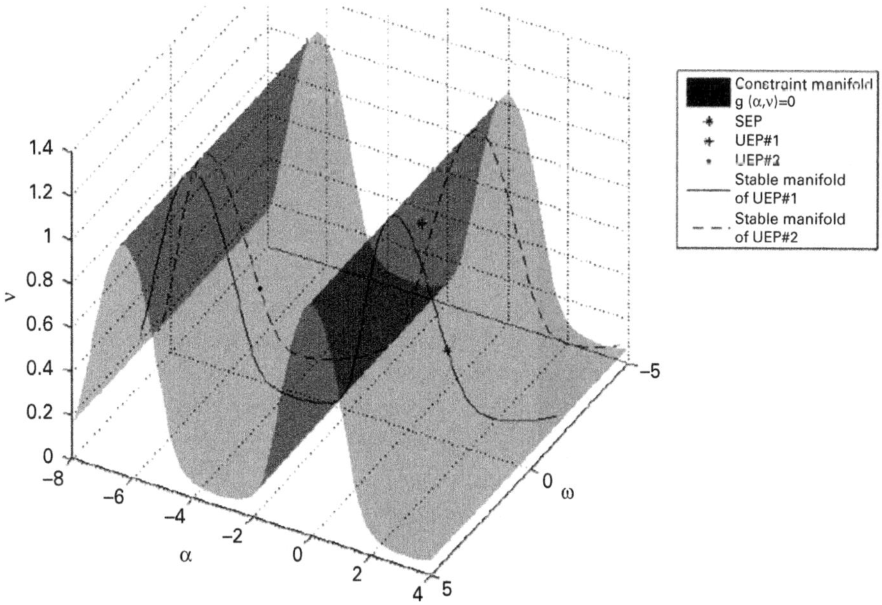

Figure 7.4 The constraint manifold of the DAE system (7.12). Both unstable equilibrium points (0, 1.2660, 0.4193) and (0, −5.0172, 0.4193), marked with the symbols * and °, lie on the stability boundary of the stable equilibrium point (0, 0.4291, 0.9613), marked with the symbols *.

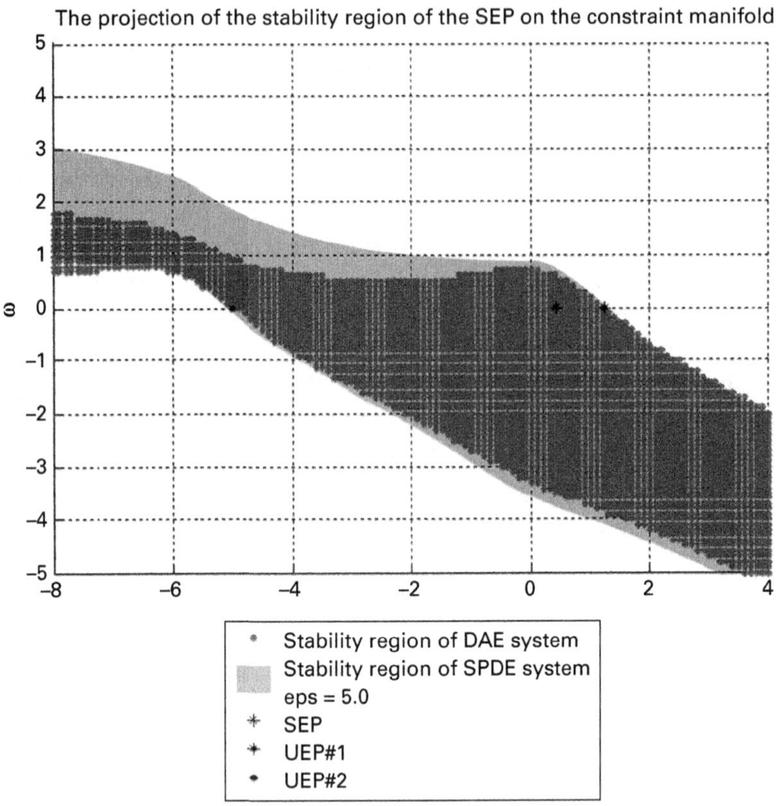

Figure 7.5 The stability region (on $\omega = 0$ and $g(x,y) = 0$) of the singularly perturbed system with $\varepsilon = 5.0$ is close to that of the corresponding DAE system.

7.4 Regularization of the DAE vector field

Another approach to studying the properties of the stability region and stability boundary of the DAE system (7.1) is to construct a transformed vector field that is well defined in the singular surface and that is equivalent to the DAE system (7.1) in $\Gamma - S$. A close relationship between these two systems, and in particular between their stability regions, exists. Exploring this relationship, the regular tools of ordinary differential equations can be employed to study the stability region of the transformed system and consequently the stability region of the DAE system (7.1).

To avoid the problem of singularities, we will extend the vector field of the DAE system (7.1) defined on the constraint set $\Gamma - S$ to a vector field that is globally defined, smooth and equivalent to system (7.1) when restricted to $\Gamma - S$. This approach allows the application of the regular theory of ODEs to study dynamical behaviors of DAE systems. We will achieve this extension in two steps. First we will extend the vector field of the DAE system to the set R^{n+m} with the exception of singular points, and then we will regularize the vector field at singular points by a convenient change of time scale.

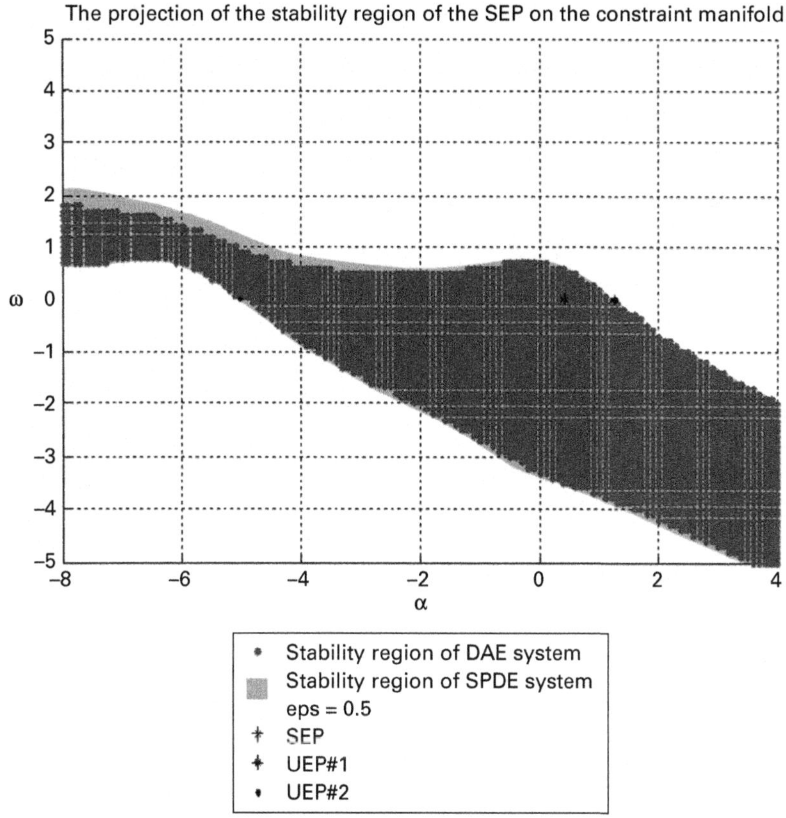

Figure 7.6 The stability region (on $\omega = 0$ and $g(x,y) = 0$) of the singularly perturbed system with $\varepsilon = 0.5$ is close to that of the corresponding DAE system.

The system of differential-algebraic equations (7.1) can be interpreted as a dynamical system on the manifold Γ. The vector field of this dynamical system is a function that take values from the manifold Γ to vectors in the tangent space $T\Gamma$ of Γ. This vector field can be calculated by implicit differentiation of the algebraic equation of (7.1):

$$D_x g(x,y) f(x,y) + D_y g(x,y)\dot{y} = 0. \tag{7.13}$$

If (x,y) is a regular point of Γ, then $D_y g(x,y)$ is invertible and one obtains:

$$\dot{y} = -(D_y g(x,y))^{-1} D_x g(x,y) f(x,y). \tag{7.14}$$

Thus the vector field on $\Gamma - S$ assumes the form:

$$(\Sigma') \begin{cases} \dot{x} = f(x,y) \\ \dot{y} = -(D_y g(x,y))^{-1} D_x g(x,y) f(x,y). \end{cases} \tag{7.15}$$

Actually, this vector field is defined for all points in R^{n+m} excluding those where $D_y g(x,y)$ is not invertible. The manifold Γ is invariant to (Σ') and its restriction to $\Gamma - S$ is

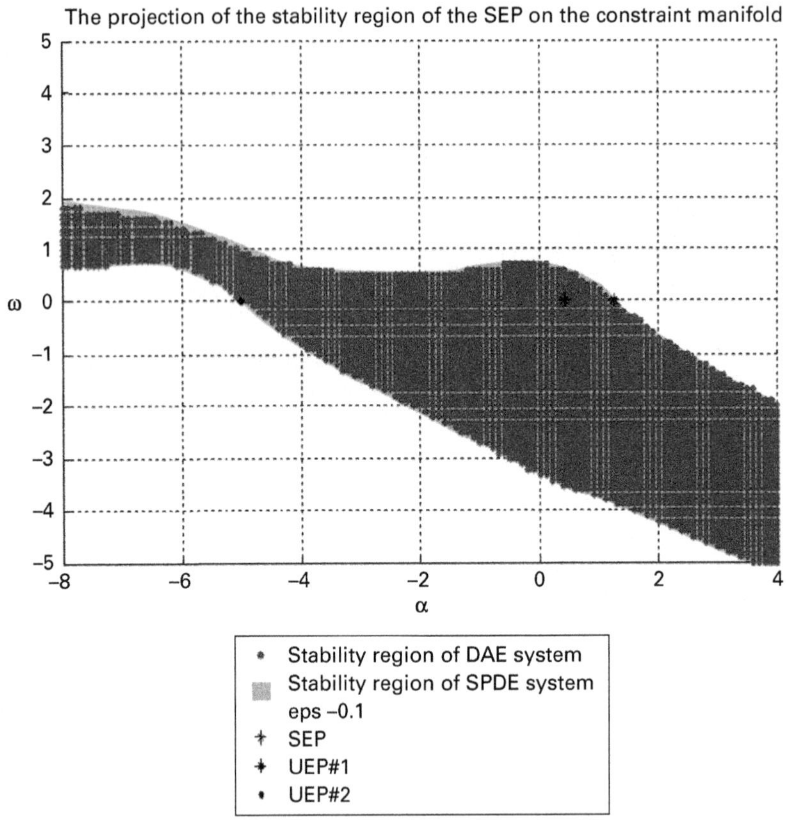

Figure 7.7 The stability region (on $\omega=0$ and $g(x,y)=0$) of the singularly perturbed system with $\varepsilon = 0.1$ is close to that of the corresponding DAE system.

completely equivalent to the dynamical system (7.1), in the sense that every trajectory of (7.1) is also a trajectory of (7.15).

In order to analyze the solutions of the DAE system (7.1) near singular points, we employ a singular transformation that was suggested by Takens [246] and extensively explored in [262,264]. More precisely, we multiply the vector field (7.15) with $\Delta(x, y) = \det(D_y g(x, y))$, the determinant of $D_y g(x,y)$, and exploit the following property of the adjoint matrix

$$\mathrm{adj}(D_y g) D_y g = D_y g\, \mathrm{adj}(D_y g) = \det(D_y g) I_n \tag{7.16}$$

to obtain the following transformed vector field:

$$(\Sigma'') \begin{cases} \dot{x} = f(x,y)\Delta(x,y) \\ \dot{y} = -\mathrm{adj}(D_y g(x,y)) D_x g(x,y) f(x,y). \end{cases} \tag{7.17}$$

The transformed vector field is globally defined and smooth and leaves the constraint manifold Γ invariant. The advantage of using this transformed vector field is that it allows study of the differential-algebraic dynamical system (7.1) using the standard tools for analysis of the set of ordinary differential equations (7.17). The relationship between system (7.17) and the original system (7.1) will be now investigated.

7.4.1 Relationship between (Σ), (Σ') and (Σ'')

The continuity of the determinant of a matrix with respect to its entries ensures that the sign of $\Delta(x, y)=\det D_y g(x,y)$ is constant in each component Γ_i of Γ. Exploring these properties, we show that the vector field (Σ'), in each component Γ_i, is equivalent either to (Σ'') or ($-\Sigma''$), where ($-\Sigma''$) is the same vector field (Σ'') but with opposite sign (i.e. its trajectories are equal to the trajectories of (Σ'') but they flow in the reverse direction).

THEOREM 7-6 **(Equivalency between (Σ') and (Σ''))**
We consider the system of differential-algebraic equations (7.1) and its vector field on $\Gamma - S$ expressed in (7.15). We consider the transformed vector field of the differential-algebraic equations (7.1) expressed in (7.17), which is globally defined and smooth. Define:

$$(\Gamma - S)_+ = \{(x,y) \in \Gamma - S \mid \Delta(x,y) > 0\} \tag{7.18}$$

$$(\Gamma - S)_- = \{(x,y) \in \Gamma - S \mid \Delta(x,y) < 0\}. \tag{7.19}$$

Then the following results hold.

(a) The vector fields (Σ') and (Σ'') are equivalent in $(\Gamma - S)_+$.
(b) The vector fields (Σ') and ($-\Sigma''$) are equivalent in $(\Gamma - S)_-$.

Proof The vector field (Σ'') is obtained from (Σ') by a pointwise scaling of time. If t denotes the time parametrizing the orbits of system (Σ'), it is straightforward to see that the transformed vector field (Σ'') can be obtained by the following change of time scale:

$$\frac{d\tau}{dt} = \frac{1}{\Delta(x(t), y(t))} \tag{7.20}$$

where τ denotes the time parametrizing orbits of system (Σ''). Consequently, (Σ') and (Σ'') are equivalent when $\Delta(x, y) > 0$, and (Σ') and ($-\Sigma'$)) are equivalent when $\Delta(x, y) < 0$. This completes the proof.

Theorem 7-6 shows that systems (Σ') and (Σ'') possess, besides orientation and time scale, the same set of orbits in $\Gamma - S$. In particular, the vector fields (Σ') and (Σ'') have the same set of equilibrium points in the set $\Gamma - S$. If x_s is an asymptotically stable equilibrium point of (Σ) and $x_s \in \Gamma_s \subset (\Gamma - S)_+$, then x_s is also an asymptotically stable equilibrium point of (Σ'') restricted to Γ. Consequently, to study the stability region of the DAE system (Σ), it is sufficient to study the stability region of the transformed system (Σ'') restricted to the component Γ_s of the constraint set Γ. More precisely, we can define the stability region of the DAE system (Σ) in terms of trajectories of the transformed system (Σ'') as:

$$A(x_s) = \{x \in \Gamma_s | \varphi_{(\Sigma'')}(t,x) \in \Gamma_s \quad \text{for all} \quad t \geq 0, \varphi_{(\Sigma'')}(t,x) \to x_s \quad \text{as} \quad t \to \infty\}. \tag{7.21}$$

The vector field (Σ') is not defined in S while (Σ'') is a smooth vector field in all Γ, consequently, the stability region of the transformed system (Σ'') may contain points in Γ_s whose trajectories leave the set Γ_s, by crossing the singular surface S, to return later on to Γ_s, by crossing again the singular surface S. These points on the stability region of the transformed system do not belong to the stability region of the DAE system. Hence, the stability region of the DAE system $A(x_s)$ is not the intersection of the stability region of the transformed system $A_{(\Sigma'')}(x_s)$ with the component Γ_s of Γ. This explains why we have to restrict the trajectories of system (Σ'') to the set Γ_s in the definition (7.21).

The previous observation regarding orbits of (Σ'') also indicates the possibility of the existence of singular points on the boundary of $A(x_s)$. Thus, understanding the structure of the singular surface is important for the characterization of the stability boundary of constrained dynamical systems.

7.4.2 Structure of the singular surface

The dynamics of the transformed system (Σ'') in the neighborhood of the singular surface suggests a decomposition of the singular surface into three disjoint sets: the pseudo-equilibrium surface, the semi-singular surface and the remaining points of S. Every equilibrium of (Σ) is also an equilibrium of (Σ''); however, in S, the vector field (Σ'') possesses an additional set of equilibrium points. Define the function

$$k(x,y) = \operatorname{adj}(D_y g(x,y)) D_x g(x,y) f(x,y) \tag{7.22}$$

and consider the set:

$$\Psi = \{(x,y) \in S | k(x,y) = 0\}. \tag{7.23}$$

We note that $\Delta(x, y)=0$ for every point in S and hence every point in Ψ is an equilibrium point of the transformed system (7.17). These points are called *pseudo-equilibriums* and the set Ψ is called the *pseudo-equilibrium surface*. Typically, set Ψ is an $(n-2)$-dimensional submanifold embedded in S [262]. In the pseudo-equilibrium surface, the transformed vector field (7.17) is null. Moreover, we can prove that $\Gamma - M \subset \Psi$. In other words, the points of Γ where Γ cannot be locally described as a submanifold of R^{n+m} is a subset of the pseudo-equilibrium surface, which is usually a very thin subset of Γ.

Another important subset of S is the one composed of points at which the vector field is not null but tangent to S. Let us first investigate the tangent spaces of manifolds. Consider a point (x_0,y_0) in the manifold Γ, i.e. $g(x_0,y_0)=0$. Now consider a small variation $(\Delta x, \Delta y)$ such that the perturbed point $(x_0 + \Delta x, y_0 + \Delta y)$ belongs to Γ, i.e. $g(x_0 + \Delta x, y_0 + \Delta y) = 0$. Expanding this equation in a power series and truncating at the first order we obtain

$$D_x g(x_0,y_0)\Delta x + D_y g(x_0,y_0)\Delta y = 0, \tag{7.24}$$

which is equivalent to

$$\begin{bmatrix} D_xg(x_0,y_0) & D_yg(x_0,y_0) \end{bmatrix} \begin{bmatrix} \Delta x \\ \Delta y \end{bmatrix} = 0. \tag{7.25}$$

Therefore, the vector $v = [\Delta x^T \ \Delta y^T]^T$ belongs to the tangent space of Γ at (x_0,y_0) if and only if v belongs to the kernel of the linear operator $[D_xg(x_0, y_0), D_yg(x_0, y_0)]$. In addition, the transformed vector field (7.17) belongs to the kernel of the linear operator $[D_xg(x_0, y_0), D_yg(x_0, y_0)]$ and therefore is tangent to Γ and leaves Γ invariant.

Now consider a point (x_0,y_0) in the submanifold $S \subset \Gamma$, i.e. $g(x_0,y_0) = 0$ and $\Delta(x_0, y_0) = 0$. It is straightforward to see that the tangent space of S at (x_0,y_0) is a vector subspace of the tangent space of Γ at (x_0,y_0). Thus, a vector v belongs to the tangent space of S at (x_0,y_0) if v belongs to the kernel of the following linear operators $[D_xg(x_0, y_0), D_yg(x_0, y_0)]$ and $[D_x\Delta(x_0, y_0), D_y\Delta(x_0, y_0)]$. Now, the points in the set S where the transformed vector field is tangent to S satisfy:

$$\begin{bmatrix} D_x\Delta(x_0,y_0) & D_y\Delta(x_0,y_0) \end{bmatrix} \begin{bmatrix} f(x_0,y_0)\Delta(x_0,y_0) \\ -k(x_0,y_0) \end{bmatrix} = 0. \tag{7.26}$$

Since $\Delta(x, y)$=0 in every point of S, the condition (7.26) is reduced to:

$$D_y\Delta(x_0,y_0)k(x_0,y_0) = 0. \tag{7.27}$$

Consequently, the following set:

$$\Xi = \{(x,y) \in S - \Psi | D_y\Delta(x_0, y_0)k(x_0, y_0) = 0\} \tag{7.28}$$

is the set of points in S that are not pseudo-equilibrium points but at which the vector field is tangent to S. These points will be called *semi-singular points* and the set Ξ the *semi-singular surface*. The reason for this name is that solutions of the DAE system at semi-singular points are somewhat well defined. At these points, the solutions of the system (Σ'') intersect the singular surface tangentially and, although they are not real solutions of the DAE system (Σ), they can be continuously extended to the singular surface. The semi-singular surface is typically an $(n-2)$-dimensional manifold embedded in S.

Consider now the following set,

$$T = S - (\Psi \cup \Xi)$$

which is composed of points of the singular surface that are neither pseudo-equilibrium points nor semi-singular points. Then, the singular surface is composed of the union of three disjoint subsets:

$$S = \Psi \cup \Xi \cup T.$$

In T, the vector field (Σ'') is not null and is not tangent to S. Consequently, the vector field is necessarily transversal to T at every point (x,y) in T, which is an $(n-1)$-dimensional submanifold embedded in S [262].

Since both Ψ and Ξ are typically $(n-2)$-dimensional submanifolds embedded in S, T is usually dense in S. Hence, S is typically an $(n-1)$-dimensional submanifold embedded in Γ. Usually, set T possesses several connected $(n-1)$-dimensional components in S that are separated by $(n-2)$-dimensional components of either the pseudo-surface or the semi-singular surface.

The following examples illustrate the concepts of singular surface, pseudo-equilibrium surface and semi-singular surface.

Example 7-3 Consider the following system of differential-algebraic equations:

$$\begin{cases} \dot{x}_1 = -2x_1 + 1 + y^2 \\ \dot{x}_2 = -x_2 \\ 0 = y^2 - x_1. \end{cases} \tag{7.29}$$

The constraint set Γ is given by:

$$\Gamma = \{(x_1, x_2, y) \in R^3 | 0 = y^2 - x_1\}.$$

Calculating the derivative of the algebraic equation, one obtains $[D_x g\ D_y g] = [-1\ 0\ 2y]$, which has complete rank independently of y. Thus $\Gamma = M$ is a two-dimensional submanifold embedded in R^3. Figure 7.8 illustrates the constraint manifold Γ. The derivative of the algebraic equation with respect to the instantaneous variable y is given by $D_y g = 2y$. Therefore $\Delta(x_1, x_2, y) = 2y$ equals zero if and only if $y = 0$. Consequently, the singular surface is given by:

$$S = \{(x_1, x_2, y) \in R^3 | x_1 = 0, y = 0\}.$$

The singular surface is depicted in Figure 7.8. It splits the constraint set Γ into two connected components Γ_s and Γ_u, where

$$\Gamma_s = \{(x_1, x_2, y) \in R^3 | 0 = y^2 - x_1, y > 0\}$$
$$\Gamma_u = \{(x_1, x_2, y) \in R^3\} | 0 = y^2 - x_1, y < 0,$$

such that $\Gamma - S = \Gamma_s \cup \Gamma_u$. The set Γ_s contains an asymptotically equilibrium point $x_s = (1,0,1)$ of (7.29).

By implicit differentiation, we calculate:

$$\dot{y} = \frac{-2x_1 + 1 + y^2}{2y}.$$

Then, the system (Σ') assumes the form:

$$(\Sigma') \begin{cases} \dot{x}_1 = -2x_1 + 1 + y^2 \\ \dot{x}_2 = -x_2 \\ \dot{y} = \dfrac{-2x_1 + 1 + y^2}{2y}. \end{cases}$$

Multiplying the vector field by $\Delta(x_1, x_2, y) = 2y$, one obtains the transformed system:

$$(\Sigma'') \begin{cases} \dot{x}_1 = (-2x_1 + 1 + y^2)2y \\ \dot{x}_2 = (-x_2)2y \\ \dot{y} = -2x_1 + 1 + y^2. \end{cases}$$

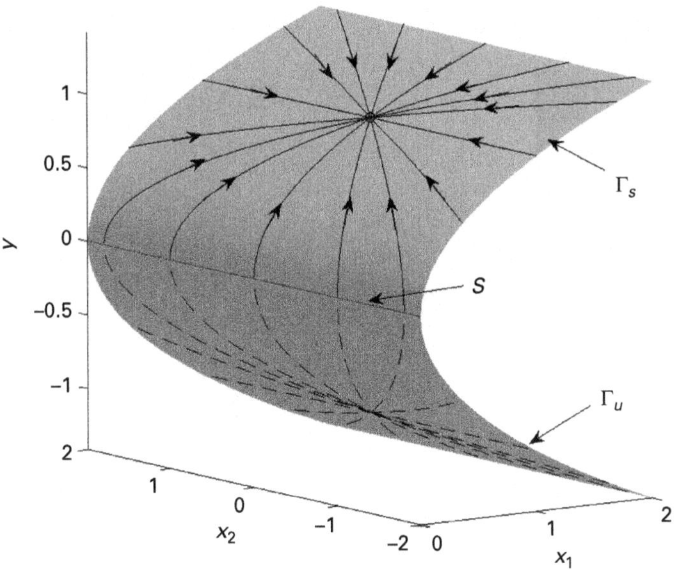

Figure 7.8 The phase portrait of system (7.29). The singular surface is composed of points of the set *T*. Trajectories of the transformed system (Σ'') are transversal to *T*.

In the pseudo-equilibrium surface, the following conditions must be satisfied: $x_1 = 0$, $y = 0$ and $-2x_1+1+y^2 = 0$, therefore, the pseudo-equilibrium surface is an empty set. The semi-singular surface is also an empty set because the following conditions must be satisfied simultaneously in this surface: $x_1 = 0$, $y = 0$ and $D_y\Delta(x, y)k(x, y) = 2(-2x_1 + 1+y^2) = 0$. As a consequence, $S = T$ is a one-dimensional submanifold embedded in Γ. Figure 7.8 illustrates the vector field (Σ'') transversal to T in Γ. Since $\Delta(x_1, x_2, y) > 0$ in Γ_s, systems (Σ) and (Σ'') are topologically equivalent in Γ_s. Figure 7.8 illustrates the orbits of (Σ'') in Γ.

Example 7-4 Consider now the following system of differential-algebraic equations, which is a slight variation of the previous example:

$$\begin{cases} \dot{x}_1 = -2x_1 + x_2 + 1 + y^2 \\ \dot{x}_2 = -x_2 \\ 0 \ = y^2 - x_1. \end{cases} \tag{7.30}$$

The algebraic equation is equal to that of the previous example, therefore the constraint set Γ and the singular surface are equal to those of the previous example and are given by:

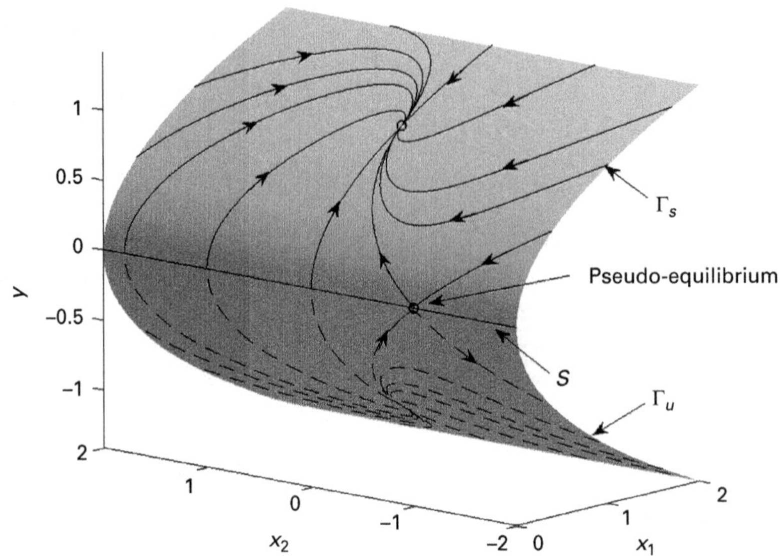

Figure 7.9 The phase portrait of system (7.30). A pseudo-saddle equilibrium point appears in the singular surface, which is composed by two connected components of set T separated by the pseudo-saddle.

$$\Gamma = \{(x_1, x_2, y) \in R^3\}|0 = y^2 - x_1\}$$
$$S = \{(x_1, x_2, y) \in R^3|x_1 = 0, y = 0\}.$$

Figure 7.9 illustrates the constraint manifold Γ and the singular surface S. Again, the singular surface splits the constraint set Γ into two connected components Γ_s and Γ_u, where

$$\Gamma_s = \{(x_1, x_2, y) \in R^3|0 = y^2 - x_1, y > 0\}$$
$$\Gamma_u = \{(x_1, x_2, y) \in R^3|0 = y^2 - x_1, y < 0\},$$

such that $\Gamma - S = \Gamma_s \cup \Gamma_u$. The set Γ_s contains an asymptotically equilibrium point x_s=(1, 0, 1) of (7.30).

By implicit differentiation, we calculate:

$$\dot{y} = \frac{-2x_1 + x_2 + 1 + y^2}{2y}.$$

Then, the system (Σ′) assumes the form:

$$(\Sigma')\begin{cases} \dot{x}_1 = -2x_1 + 1 + y^2 \\ \dot{x}_2 = -x_2 \\ \dot{y} \ = \dfrac{-2x_1 + x_2 + 1 + y_2}{2y}. \end{cases}$$

Multiplying the vector field by $\Delta(x_1, x_2, y) = 2y$, one obtains the transformed system:

$$(\Sigma'')\begin{cases} \dot{x}_1 = (-2x_1 + 1 + y^2)2y \\ \dot{x}_2 = (-x_2)2y \\ \dot{y} \ = -2x_1 + x_2 + 1 + y^2. \end{cases}$$

In the pseudo-equilibrium surface, the following conditions must be satisfied: $x_1 = 0$, $y = 0$ and $-2x_1+x_2+1+y^2 = 0$, therefore, the pseudo-equilibrium surface is the set

$$\Psi = \{(0, -1, 0)\}.$$

The pseudo-equilibrium point (0,−1,0) is a type-one hyperbolic equilibrium point of system (Σ'') on the constraint manifold Γ. This pseudo-equilibrium point and the dynamics of system (Σ'') on Γ are depicted in Figure 7.9.

The semi-singular surface is empty because the following conditions must be satisfied on this surface: $x_1 = 0$, $y = 0$ and $D_y\Delta(x, y)k(x, y) = 2(-2x_1 + x_2 + 1 + y_2) = 0$. The only point that satisfies these conditions is the pseudo-equilibrium point. Consequently, there is no point in the set $S - \Psi$ that satisfies these conditions and the set of semi-singular points is empty. Set $T=S-\Psi$ is a one-dimensional submanifold embedded in Γ. The pseudo-equilibrium splits T into two connected components. Figure 7.9 illustrates the vector field (Σ'') transversal to T in Γ. Since $\Delta(x_1, x_2, y) > 0$ in Γ_s, systems (Σ) and (Σ'') are topologically equivalent in Γ_s. Figure 7.9 illustrates the orbits of (Σ'') in Γ.

Example 7-5 Consider the following system of differential-algebraic equations:

$$\begin{cases} \dot{x}_1 & = 1 - x_1 \\ \dot{x}_2 & = 2 - x_2 \\ 0 & = y^3 + x_1 y - x_2. \end{cases} \quad (7.31)$$

The constraint set Γ is given by:

$$\Gamma - \{(x_1, x_2, y) \subset R^3 | 0 = y^3 + x_1 y - x_2\}.$$

Calculating the derivative of the algebraic equation, one obtains $[D_x g \; D_y g] = [y - 1 \;\; 3y^2 + x_1]$, which has complete rank independently of (x_1, x_2, y). Thus $\Gamma = M$ is a two-dimensional submanifold embedded in R^3. Figure 7.10 illustrates the constraint manifold Γ. The derivative of the algebraic equation with respect to the instantaneous variable y is given by $D_y g - 3y^2 + x_1$. Therefore $\Delta(x_1, x_2, y) = 3y^2 + x_1$. On the singular surface, $\Delta(x_1, x_2, y) = 0$, which is equivalent to $x_1 = -3y^2$. Substituting this condition into the algebraic constraint, we conclude that the singular surface is given by:

$$S = \{(x_1, x_2, y) \in R^3 | x_1 = -3y^2, x_2 = -2y^3, y \in R\}.$$

The singular surface is the thick line depicted in Figure 7.10. It splits the constraint set Γ into two connected components Γ_s and Γ_u. The set Γ_s contains an asymptotically equilibrium point x_s=(1, 2, 1) of (7.31).

By implicit differentiation, we calculate:

$$\dot{y} = \frac{-x_2 + 2 - y + x_1 y}{(x_1 + 3y^2)}.$$

Then, the system (Σ') assumes the form:

$$(\Sigma') \begin{cases} \dot{x}_1 = 1 - x_1 \\ \dot{x}_2 = 2 - x_2 \\ \dot{y} \; = \dfrac{-x_2 + 2 - y + x_1 y}{(x_1 + 3y^2)}. \end{cases}$$

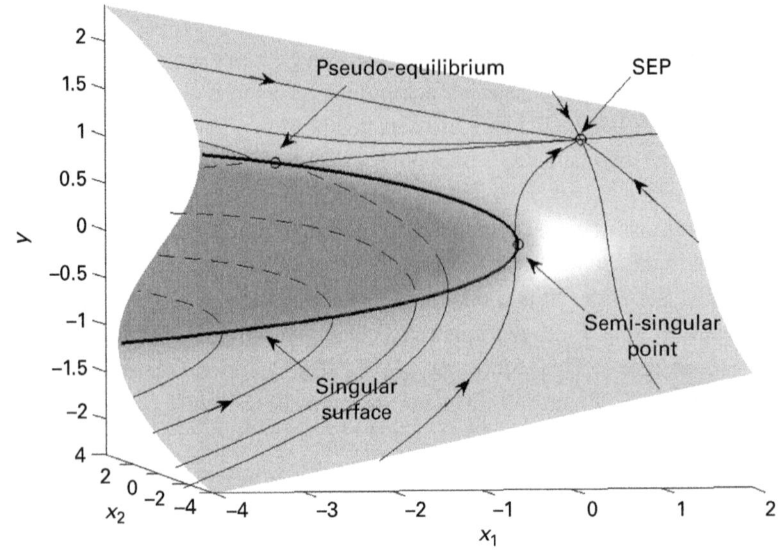

Figure 7.10 Phase portrait of system (7.31).

Multiplying the vector field by $\Delta(x_1, x_2, y) = (x_1 + 3y^2)$, one obtains the transformed system:

$$(\Sigma'') \begin{cases} \dot{x}_1 = (1 - x_1)(x_1 + 3y^2) \\ \dot{x}_2 = (2 - x_2)(x_1 + 3y^2) \\ \dot{y} = -x_2 + 2 - y + x_1 y. \end{cases}$$

On the pseudo-equilibrium surface, the following conditions must be satisfied: $x_1{=}{-}3y^2$, $x_2{=}{-}2y^3$ and $-x_2{+}2{-}y{+}x_1y{=}0$. Substituting the first two equations in the third equation, variables x_1 and x_2 are eliminated and we obtain the following equation: $-y^3{-}y{+}2{=}0$, which has a unique solution $y{=}1$. Therefore, the point $(-3, -2, 1)$ is a pseudo-equilibrium point and the pseudo-equilibrium surface is the set $\Psi = \{(-3, -2, 1)\}$. In the semi-singular surface, the following conditions must be satisfied: $x_1{=}{-}3y^2$, $x_2{=}{-}2y^3$ and $D_y\Delta(x, y)k(x, y) = -6y(-x_2 + 2 - y + x_1y) = 0$. Thus the origin is a semi-singular point. Figure 7.10 illustrates the phase portrait of the vector field (Σ'') in Γ. Since $\Delta(x_1, x_2, y) > 0$ in Γ_s, systems (Σ) and (Σ'') are topologically equivalent in Γ_s. A trajectory of (Σ'') touches the semi-singular point tangent to the singular surface S.

7.4.3 Dynamics near the singular surface

The behavior of the DAE system in the neighborhood of the semi-singular surface is illustrated by a three-dimensional example in Figure 7.10 and also in the diagram of Figure 7.11.

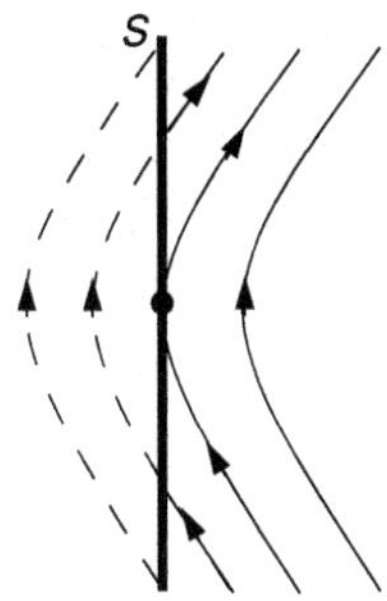

Figure 7.11 The dynamical behavior of the constrained system (Σ) in the neighborhood of the semi-singular surface. Trajectories tangentially touch the singular surphace at the semi-singular point. On the right side of S, the dynamical behavior in the neighborhood of the semi-singular point resembles the dynamics of a saddle-node equilibrium point. On the left side, the dynamical behavior resembles the dynamics of a focus.

Let us call Γ_s the component of interest (usually the component that contains the asymptotically stable equilibrium point of interest) and suppose, without loss of generality, that $\Delta(x, y) > 0$ in Γ_s. The singular surface S and in particular the semi-singular surface Ξ are on the boundary of Γ_s. In each side of the singular surface, the dynamical behavior is different. On one side, the dynamics resembles the dynamics of a saddle-node, while on the other side, the dynamics resembles the dynamics of a focus. Depending on the case, the behavior of the dynamics in Γ_s will be equivalent to one of these. This observation suggests the following subdivision of the semi-singular surface [262]:

$$\Xi'_{sa} = \{(x,y) \in \Xi | D_y\{(D_y\Delta)k\}k > 0\} \tag{7.32}$$

$$\Xi'_{fo} = \{(x,y) \in \Xi' | D_y\{(Dy\Delta)k\}k > 0\}. \tag{7.33}$$

In the neighborhood of the semi-focus surface Ξ'_{fo}, orbits which are born in the singular surface S circle around the semi-singular set dying at the singular surface in finite time, see Figure 7.11. Consequently, the semi-focus surface Ξ'_{fo} cannot lie on the stability boundary of any asymptotically stable equilibrium point in Γ_s. Due to the lack of importance of the semi-focus surface in the stability boundary characterization, this surface will not be analyzed in detail.

If $(x,y) \in \Xi'_{sa}$, then there exists a solution of the DAE system (7.1) in the component Γ_s that reaches the point $(x,y) \in \Xi'_{sa}$ in a finite time and immediately leaves the point, returning to the same component Γ_s as illustrated in Figure 7.11. These points in the semi-singular surface will be called semi-saddles.

7.4.4 Dynamics near pseudo-saddles

Pseudo-equilibriums are real equilibrium points of the transformed system (Σ''). Every point in Ψ is actually a non-hyperbolic equilibrium point of the transformed system (Σ'')

[262] with, at maximum, two eigenvalues different from zero. The center manifold of these non-hyperbolic equilibrium points is the set Ψ itself, which possesses dimension $n-2$ [262]. Depending on the signs of the nonzero eigenvalues, we have different types of behavior in the neighborhood of Ψ. We will divide set Ψ into three disjoint subsets according to the stability type of the pseudo-equilibrium points. To this end, define:

$$\Psi'_{so} = \{(x,y) \in \Psi' \,|\, \text{the Jacobian of } (\Sigma'') \text{ has two eigenvalues in } C^+\}, \tag{7.34}$$

$$\Psi'_{sa} = \{(x,y) \in \Psi' \,|\, \text{the Jacobian of } (\Sigma'') \text{ has one eigenvalue in } C^+ \text{ and one in } C^-\}, \tag{7.35}$$

$$\Psi'_{si} = \{(x,y) \in \Psi' \,|\, \text{the Jacobian of } (\Sigma'') \text{ has two eigenvalues in } C^-\}. \tag{7.36}$$

The points in Ψ'_{so}, Ψ'_{sa} and Ψ'_{si} will be called respectively pseudo-sources, pseudo-saddles and pseudo-sinks.

The dynamical behavior of trajectories in the neighborhood of pseudo-sinks indicates that they cannot lie on the stability boundary of any asymptotically stable equilibrium point. Even though pseudo-sources may exist on the stability boundary, since $W^s(\Psi'_{so}) - \Psi'_{so}$ is an empty set, they are not important in the characterization of stability boundaries. Pseudo-saddles are relevant to the characterization of stability boundaries and will be studied in some detail.

The behavior of the DAE system in the neighborhood of a pseudo-saddle can be represented in a convenient coordinate system as sketched in Figure 7.12.

A pseudo-saddle is called *transverse* if neither the stable nor the unstable manifold is tangent to the singular surface S. The set of transverse pseudo-saddles will be denoted Ψ'_{trsa}. Transverse saddles are generic in the set Ψ'_{sa}. The stable and unstable manifolds of transverse pseudo-saddles are one dimensional and $W^s(\Psi'_{trsa})$ and $W^u(\Psi'_{trsa})$ are $(n-1)$-dimensional manifolds [262]. (Figure 7.12 illustrates these manifolds.)

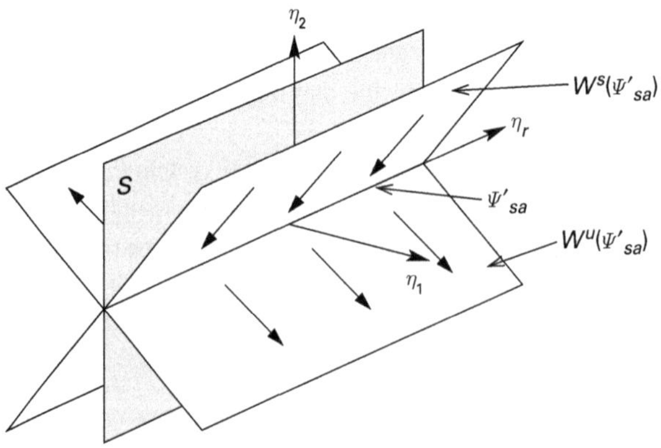

Figure 7.12 A sketch of the dynamical behavior of a system (Σ) in the neighborhood of a pseudo-saddle surface.

7.5 Stable and unstable manifolds

In this section, we will study the properties of stable and unstable manifolds of hyperbolic equilibrium points, semi-singular points and pseudo-equilibrium points. It will be shown that the stable and unstable manifolds of these points can be written in terms of trajectories of both systems: the DAE system (Σ) and the transformed system (Σ''). As a consequence, the characterization of the stability boundary of the DAE system (Σ) can be represented in terms of the stable and unstable manifolds of certain critical points of the transformed system (Σ'').

We will consider the following generic assumption for the DAE system (Σ).

(A0) Equilibrium points do not lie on the singular set S.

We will also assume, without loss of generality, that $\Delta(x, y) > 0$ in the component Γ_s of Γ that contains the asymptotically stable equilibrium point of interest.

For a hyperbolic equilibrium point $z_0 = (x_0, y_0) \in \Gamma_s$ of the DAE system (Σ), the stable and unstable manifolds are respectively defined as:

$$W^s(z_0) = \{z \in \Gamma_s | \varphi(t, z) \in \Gamma_s \text{ for all } t \geq 0 \text{ and } \varphi(t, z) \to z_0 \text{ as } t \to +\infty\} \tag{7.37}$$

$$W^u(z_0) = \{z \in \Gamma_s | \varphi(t, z) \in \Gamma_s \text{ for all } t \leq 0 \text{ and } \varphi(t, z) \to z_0 \text{ as } t \to -\infty\}. \tag{7.38}$$

The stable and unstable manifolds are subsets of the component Γ_s of the constraint set Γ. Points on the singular set will not be considered to lie on the stable or unstable manifolds of any hyperbolic equilibrium point in Γ_s. If $z_0 = (x_0, y_0) \in \Gamma_s$ is a type-k hyperbolic equilibrium point of the DAE system (Σ), then $W^u(z_0)$ is a k-dimensional manifold while $W^s(z_0)$ is an $(n-k)$ dimensional manifold. These manifolds intersect transversally at the equilibrium z_0.

We can also define the stable and unstable manifolds of pseudo-equilibrium points and of semi-singular points. However, these definitions are different from the definition of invariant manifolds of hyperbolic equilibrium points because solutions of the DAE system (Σ) reach these points in finite time. More precisely, if $z_0 = (x_0, y_0) \in S$ is either a pseudo-equilibrium point or a semi-singular point, then the stable and unstable manifolds are respectively defined as:

$$W^s(z_0) = \{z \in \Gamma_s | \exists t_0 > 0 \text{ such that } \varphi(t, z) \in \Gamma_s \text{ for } 0 \leq t \leq t_0 \text{ and } \varphi(t, z) \to z_0 \text{ as } t \to t_0\} \tag{7.39}$$

$$W^u(z_0) = \{z \in \Gamma_s | \exists t_0 < 0 \text{ such that } \varphi(t, z) \in \Gamma_s \text{ for } 0 \geq t \geq t_0 \text{ and } \varphi(t, z) \to z_0 \text{ as } t \to t_0\}. \tag{7.40}$$

Since the DAE system (Σ) and the transformed system (Σ'') have the same set of trajectories in Γ_s, we can define the stable and unstable manifolds of hyperbolic equilibrium points, pseudo-equilibrium points and semi-singular points of the DAE system (Σ) in terms of trajectories of the transformed system (Σ'').

If $z_0 = (x_0, y_0) \in \Gamma_s$ is a hyperbolic equilibrium point of the DAE system (Σ), then it is also a hyperbolic equilibrium point of the transformed system (Σ'') and the stable and unstable manifolds of z_0 of the DAE system (7.1) are respectively defined as:

$$W^s(z_0) = \{z \in \Gamma_s \big| \varphi_{(\Sigma'')}(t, z) \in \Gamma_s \text{ for all } t \geq 0 \text{ and } \varphi_{(\Sigma'')}(t, z) \to z_0 \text{ as } t \to +\infty\} \quad (7.41)$$

$$W^u(z_0) = \{z \in \Gamma_s \big| \varphi_{(\Sigma'')}(t, z) \in \Gamma_s \text{ for all } t \leq 0 \text{ and } \varphi_{(\Sigma'')}(t, z) \to z_0 \text{ as } t \to -\infty\}. \quad (7.42)$$

Observe that (7.41) and (7.37) are two representations of the same set. The same is true for (7.42) and (7.38). Note that the stable and unstable manifolds of the transformed system (Σ''), respectively denoted $W^s_{(\Sigma'')}(z_0)$ and $W^u_{(\Sigma'')}(z_0)$, are not necessarily restricted to Γ_s. Moreover, $W^s_{(\Sigma'')}(z_0) \cap \Gamma_s$ is not necessarily equal to $W^s(z_0)$. Trajectories of the transformed system (Σ'') might leave the component Γ_s, by crossing the singular surface, and return to this component later on.

Since pseudo-equilibrium points are equilibrium points of the transformed system (Σ''), the stable and unstable manifolds of a pseudo-equilibrium z_0 are well defined for the transformed system (Σ''). Restricting the trajectories to the component Γ_s, we can define the stable and unstable manifolds of a pseudo-equilibrium z_0 for the DAE system (7.1) in terms of trajectories of the transformed system (Σ''). More precisely, the stable and unstable manifolds of a pseudo-equilibrium z_0 are also respectively defined by (7.41) and (7.42). Observe that trajectories of the DAE system (Σ) reach pseudo-equilibrium points in finite time while the trajectories of the transformed system (Σ'') approach them asymptotically as time tends to infinity. Observe that for a pseudo-equilibrium point z_0, (7.41) and (7.39) represent the same set. It is also important to note that just a subset of the stable and unstable manifolds of the transformed system contains the stable and unstable manifolds of the DAE system (7.1). Figure 7.13 illustrates these stable manifolds of the pseudo-saddle of system (7.30).

Semi-singular points are regular points of the transformed system (Σ''), therefore trajectories of the transformed system also reach these points in finite time. If $z_0 = (x_0, y_0) \in S$ is a semi-singular point, then the stable and unstable manifolds of the DAE system (7.1) are respectively defined as:

$$\begin{aligned} W^s(z_0) = \{z \in \Gamma_s \big| \exists t_0 > 0 \text{ such that } & \varphi_{(\Sigma'')}(t, z) \in \Gamma_s \text{ for } 0 \leq t \leq t_0 \\ & \text{and } \varphi_{(\Sigma'')}(t, z) \to z_0 \text{ as } t \to t_0\} \end{aligned} \quad (7.43)$$

$$\begin{aligned} W^u(z_0) = \{z \in \Gamma_s \big| \exists t_0 > 0 \text{ such that } & \varphi_{(\Sigma'')}(t, z) \in \Gamma_s \text{ for } 0 \leq t \leq t_0 \\ & \text{and } \varphi_{(\Sigma'')}(t, z) \to z_0 \text{ as } t \to t_0\}. \end{aligned} \quad (7.44)$$

Observe that, for a semi-singular point, (7.43) and (7.39) represent the same set. The same is true for (7.44) and (7.40). Now consider the set of semi-saddle-points Ξ'_{sa}, which is an $(n-2)$-dimensional manifold. We define the stable manifold

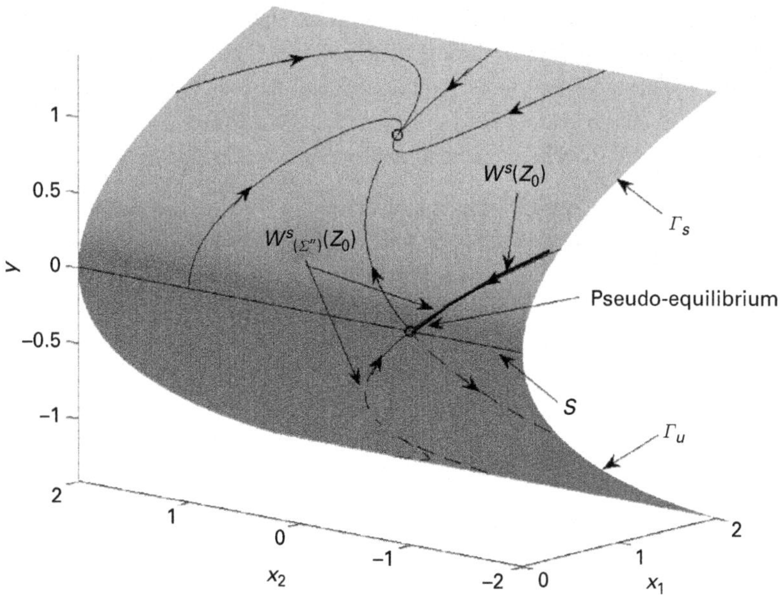

Figure 7.13 The stable manifold of a transverse pseudo-saddle.

$$W^s(\Xi'_{sa}) = \bigcup_{z_0 \in \Xi'_{sa}} W^s(z_0)$$

and the unstable manifold

$$W^u(\Xi'_{sa}) = \bigcup_{z_0 \in \Xi'_{sa}} W^u(z_0)$$

of the semi-saddle surface as the union of the manifolds of the semi-saddle-points. It is clear from the analysis of dynamics in the neighborhood of a semi-saddle that both $W^s(\Xi'_{sa})$ and $W^u(\Xi'_{sa})$ are $(n-1)$-dimensional manifolds embedded in the component Γ_s. Figure 7.12 illustrates these manifolds.

7.6 Generalized critical points on the stability boundary

The approach to developing a characterization of the stability boundary of DAE systems is similar to the approach that was employed to develop a characterization of the stability boundary of ordinary differential equations in Chapter 4. More precisely, we start from a local characterization of the stability boundary and then we extend it to a global one. The local characterization of the stability boundary is developed in this section while the global characterization is developed in the following section.

We will call equilibrium points, pseudo-equilibrium points and semi-singular points on the stability boundary generalized critical points because they play a key role in the stability boundary characterization. Without imposing any condition on the vector field, we will investigate these critical points on the stability boundary. In particular, we will derive characterizations of these critical points in the stability boundary in terms of their stable and unstable manifolds. Additional conditions on the vector field are then imposed in order to obtain sharper results on the characterizations of these critical points on the stability boundary. These characterizations will be derived in the next section.

THEOREM 7-7 **(Characterization of critical points on the stability boundary)**
Let $A(x_s)$ be the stability region of an asymptotically stable equilibrium point of the DAE system (7.1) in the component Γ_s of the constraint set Γ.
For a hyperbolic equilibrium point $z \in \Gamma_s$

$$z \in \partial A(x_s) \Leftrightarrow \{W^u(z) - \{z\}\} \cap \overline{A(x_s)} \neq \emptyset \tag{7.45}$$

and if z is not a source

$$z \in \partial A(x_s) \Leftrightarrow \{W^s(z) - \{z\}\} \cap \partial A(x_s) \neq \emptyset. \tag{7.46}$$

For a transverse pseudo-saddle or a semi-saddle v:

$$v \in \partial A(x_s) \Leftrightarrow W^u(v) \cap \overline{A(x_s)} \neq \emptyset. \tag{7.47}$$

Theorem 7-7 offers conditions to check whether a critical point lies on the stability boundary in terms of the properties of their unstable manifolds. More precisely, if the unstable manifold of a critical point intersects with the stability region, then the critical point lies on the stability boundary. Points in the surface T can also lie on the stability boundary. The following theorem offers a practical condition to check whether a point of T lies on the stability boundary.

THEOREM 7-8 **(Points of surface on the stability boundary)**
Let $A(x_s)$ be the stability region of an asymptotically stable equilibrium point of the DAE system (7.1) in the component Γ_s of the constraint set Γ. A point $v \in T$ lies on $\partial A(x_s)$ if and only if there exists a time $t_0>0$ such that $\varphi_{(\Sigma'')}(t, v) \cap A(x_s) \neq \emptyset$ for all $t \in (0, t_0]$.

Theorem 7-8 offers a means of checking whether a point of the surface T lies on the stability boundary. It asserts that it is sufficient to check whether the trajectory of the transformed system (Σ'') starting from this point enters the stability region for a sufficiently small time.

7.7 Characterization of the stability boundary of DAE systems

Characterizations of the stability boundary $\partial A(x_s, y_s)$ of DAE systems have recently been developed. It has been shown that under certain conditions, the stability boundary $\partial A(x_s, y_s)$ consists of two parts: the first part is the stable manifolds of the equilibrium points on the stability boundary while the second part contains points whose trajectories reach singular surfaces [262]. The second part can be further delineated as a union of the stable manifolds of pseudo-equilibrium points and semi-singular points on the stability boundary and parts of the singular surface [262,264].

A complete characterization of the stability boundary $\partial A(x_s,)$ of the DAE system (Σ) will be developed in this section. It will be shown that the stability boundaries of constrained dynamical systems are composed of stable manifolds of generalized critical points on the stability boundary and pieces of the singular surface. To this end, consider the following assumptions.

(A1) Every equilibrium point on the stability boundary $\partial A(x_s)$ is hyperbolic and, except for a set of dimension n−3, every pseudo-saddle in $\partial A(x_s,)$ is transverse.

(A2) Stable manifolds of equilibrium points and of connected components of Ψ, Ξ, intersect transversally with unstable manifolds of the same elements.

(A3) Every trajectory in $\bar{A}$ converges to an equilibrium point, a pseudo-equilibrium point or a semi-singular point.

Assumptions (A1) and (A2) are generic while assumption (A3) is not generic. The existence of an energy function for the DAE system (Σ) is a sufficient condition for the satisfaction of assumption (A3).

Assumption (A1) guarantees that every equilibrium point on the stability boundary is isolated. Thus there is a countable collection of equilibrium points z_i, i=1,2, . . ., on the stability boundary. Pseudo-equilibrium points and semi-singular points are not isolated. The connected components of pseudo-equilibrium points will be denoted Ψ'_j, j=1,2, ... The connected components of semi-singular points will be denoted Ξ'_l, l=1,2, ...

The stability boundary has maximal dimension n−1 and the quasi-stability boundary possesses a dense (n−1)-dimensional set. Connected components of the set T have dimension n−1 and therefore they are important pieces in the characterization of the stability boundary. Generalized critical elements that possess (n−1)-dimensional stable manifolds also have an important contribution to the characterization of the stability boundary. They are type-one hyperbolic equilibrium points, (n−2)-dimensional connected components of transverse pseudo-saddles (a component of the set Ψ'_{trsa}) and (n−2) dimensional connected components of semi-saddles (a component of the set Ξ'_{sa}).

Next we will study further local characterization of these critical points on the quasi-stability boundary.

THEOREM 7-9 **(Equilibrium points on the quasi-stability boundary)**
Let $A(x_s)$ be the stability region of an asymptotically stable equilibrium point of the DAE system (7.1) in the component Γ_s of the constraint set Γ and $z \in \Gamma_s$, with $z \neq x_s$, be a hyperbolic type-one equilibrium point. If assumptions (A0)–(A3) are satisfied, then

$$z \in \partial\overline{A(x_s)} \Leftrightarrow W^u(z) \cap A(x_s) \neq \emptyset \;\; and \;\; \mathrm{W}^u(z) \cap \overline{A(x_s)}^C \neq \emptyset \tag{7.48}$$

$$z \in \partial\overline{A(x_s)} \Leftrightarrow W^s(z) \subset \partial\overline{A(x_s)}. \tag{7.49}$$

Now let us consider transverse pseudo-saddles on the stability boundary. Isolated transverse pseudo-saddles do not have an important contribution to the characterization of the stability boundary for systems with n>2. However, an (n−2)-dimensional component of Ψ'_{trsa}, composed of continuous transverse pseudo-saddles, has an important contribution to the characterization of the stability boundary. Consider an $(n-2)$-dimensional connected component Ψ'_j of the set of pseudo-equilibrium points such that $\Psi'_j \subset \Psi'_{trsa}$. This component might intersect with the stability boundary but it might not be entirely contained on the stability boundary. Thus, we consider the subset

$$N\Psi'_j = \{v \in \Psi'_j \cap \partial\overline{A(x_s)} \,|\, \exists\, a\ \delta\text{-neighborhood}\ \delta(v)\ \text{in}\ \Psi'_j\ \text{such that}\ \delta(v) \subset \partial\overline{A(x_s)}\} \tag{7.50}$$

of points of Ψ'_j that are entirely contained on the stability boundary. The following theorem offers necessary and sufficient conditions to guarantee that a pseudo-transverse saddle belongs to $N\Psi'_j$.

THEOREM 7-10 **(Transverse pseudo-saddles on the quasi-stability boundary)**
Let $A(x_s)$ be the stability region of an asymptotically stable equilibrium point of the DAE system (7.1) in the component Γ_s of the constraint set Γ and $v \in \Psi'_j$, with $\Psi'_j \subset \Psi'_{trsa}$ an $(n-2)$-dimensional component of pseudo-equilibrium points composed exclusively of transverse pseudo-saddles. If assumptions (A0) – (A3) are satisfied, then $v \in N\Psi'_j$ if and only if there exists δ_0 such that $W^u(\delta(v)) \cap A(x_s)$ is dense in $W^u(v)$ for every δ-neighborhood $\delta(v)$ of v in Ψ'_j with $\delta < \delta_0$. Moreover, $v \in N\Psi'_j$ if and only if $W^s(v) \subset \partial\overline{A(x_s)}$.

A similar result can be stated for semi-singular points. We have already observed that semi-foci cannot belong to the stability boundary of any asymptotically stable equilibrium point. However, sets of semi-saddles have an important contribution to the characterization of stability boundaries. Again, consider an $(n-2)$-dimensional connected component Ξ'_l of the semi-singular surface such that $\Xi'_l \subset \Xi'_{sa}$. This component might intersect with the stability boundary but it might not be entirely contained on the stability boundary. Thus, we consider the subset

$$N\Xi'_l = \{v \in \Xi'_l \cap \partial\overline{A(xs)} \exists \,|\, a\ \delta\text{-neighborhood}\ \delta(v)\ in\ \Xi'_l\ \text{such that}\ \delta(v) \subset \partial\overline{A(x_s)}\} \tag{7.51}$$

of points of Ξ'_l that are contained on the stability boundary. The following theorem offers necessary and sufficient conditions to guarantee that a semi-saddle belongs to $N\Xi'_l$.

THEOREM 7-11 **(Semi-saddles on the quasi-stability boundary)**
Let $A(x_s)$ be the stability region of an asymptotically stable equilibrium point of the DAE system (7.1) in the component Γ_s of the constraint set Γ and $v \in \Xi'_l$, with $\Xi'_l \subset \Xi'_{sa}$ an $(n-2)$-dimensional component of the semi-singular surface composed exclusively of semi-saddles. If assumptions (A0)–(A3) are satisfied, then $v \in N\Xi'_l$ if and only if there exists δ_0 such that $W^u(\delta(v)) \cap A(x_s)$ is dense in $W^u(v)$ for every δ-neighborhood $\delta(v)$ of v in Ξ'_l with $\delta < \delta_0$. Moreover, $v \in N\Xi'_l$ if and only if $W^s(v) \subset \partial A(x_s)$.

Now we are in a position to establish a complete characterization of the quasi-stability boundary of constrained dynamical systems.

THEOREM 7-12 **(Characterization of the quasi-stability boundary)**
Let $A(x_s)$ be the stability region of an asymptotically stable equilibrium point x_s in the component Γ_s of Γ. Under assumptions (A0)–(A3), the boundary of the closure of the stability region $\partial\overline{A(x_s)}$ is composed of:

(a) the stable manifolds of type-one hyperbolic equilibrium points;
(b) stable manifolds of transverse pseudo-saddles;
(c) stable manifolds of semi-saddles;
(d) pieces of singular points.

More precisely, if z_i, i=1,2, . . ., are the type-one hyperbolic equilibrium points on the stability boundary, $N\Psi'_j$, j=1,2, . . ., are the connected components of the set of transverse pseudo-equilibrium points on the stability boundary and $N\Xi'_l$, l=1,2, . . . , are the connected components of the set of semi-saddles on the stability boundary, then:

$$\partial\overline{A(x_s)} = \overline{\bigcup_i W^s(z_i) \bigcup_i W^s(N\Psi'_j) \bigcup_l W^s(N\Xi'_l) \cup (S \cap \partial\overline{A(x_s)})}. \tag{7.52}$$

Proof Assumption (A3) guarantees that

$$\partial\overline{A(x_s)} \subseteq \overline{\bigcup_i W^s(z_i) \bigcup_i W^s(N\Psi'_j) \bigcup_l W^s(N\Xi'_l) \cup (S \cap \partial\overline{A(x_s)})}.$$

Theorem 7-9, Theorem 7-10 and Theorem 7-11 prove the other inclusion and the theorem is proven.

Example 7-6 Consider again the DAE system (7.31). This system possesses an asymptotically stable equilibrium point $(x_s, y_s) = (1, 2, 1)$. The stability boundary of this equilibrium point is illustrated in Figure 7.14. The stability boundary is composed of three main pieces, the stable manifold of a pseudo-saddle, the stable manifold of a semi-saddle and a piece of set T on the singular surface.

7.8 Concluding remarks

A comprehensive theory of stability regions and of stability boundaries for constrained continuous dynamical systems has been developed in this chapter. Two approaches have been used in the development of a complete characterization for both the stability boundary and the stability regions of constrained nonlinear dynamical systems. The first approach explores an approximation of the stability region via the singular perturbation theory while the second approach is based on a regularization of the vector field on the singular surface.

It has been shown that under certain conditions, the stability boundary of DAE systems is composed of two parts: the first part is the stable manifolds of the equilibrium

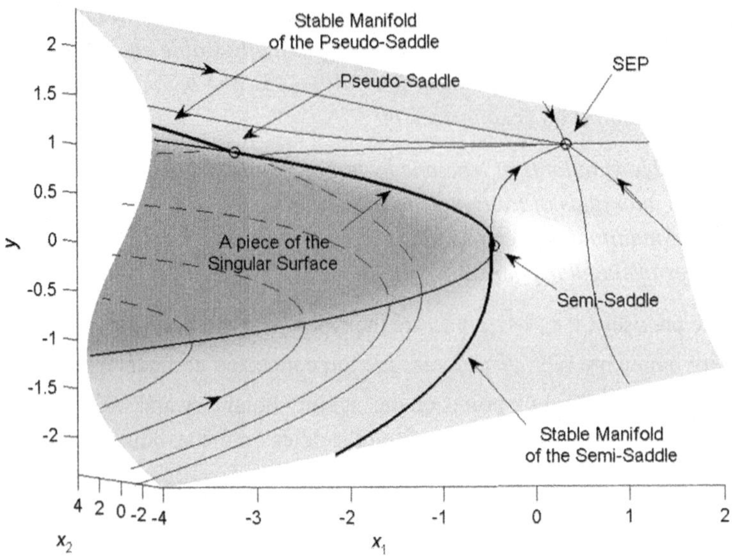

Figure 7.14 The stability boundary of the asymptotically stable equilibrium point of the DAE system (7.31). The stability boundary, the thick black curve, is composed of three main pieces: the stable manifold of the pseudo-saddle, a piece of the set T in the singular surface and the stable manifold of a semi-saddle. Points of the constraint manifold to the right of this curve belong to the stability region of the SEP.

points on the stability boundary while the second part contains points whose trajectories reach singular surfaces. The second part can be further delineated as a union of the stable manifolds of pseudo-equilibrium points and semi-singular points on the stability boundary and parts of the singular surface.

8 Relevant stability boundary of continuous dynamical systems

Knowledge of stability regions is essential for stability analysis of nonlinear systems under large disturbances. In many applications, however, it is desirable to characterize only the relevant portion of a stability region instead of the entire stability region. In this situation, characterization of the relevant stability boundary toward which the disturbed system trajectory is heading is more important than the characterization of the entire stability boundary. One important application of the relevant portion of the stability boundary is in the stability assessment of power grids.

Electric grids typically experience large disturbances. These disturbances include loss of transmission components (lines, transformers) due to short-circuits caused by lightning, high winds, or failures such as incorrect relay operations or insulation breakdowns. To protect power systems from damage due to disturbances, protective relays are placed strategically throughout the system to detect the disturbances and to trigger circuit breakers necessary to isolate the disturbances. As a result of the action of these protective relays, a power system subject to disturbances can be thought of as going through three stages of network configuration changes: a pre-disturbance system (Stage I), a disturbance-on system (Stage II), and a post-disturbance system (Stage III).

When a disturbance occurs, the system moves from a pre-disturbance system, which is in a stable equilibrium point (SEP), into a disturbance-on system before the disturbance is removed and the system enters the stage of post-disturbance. A fundamental stability question can be roughly stated as follows: given a pre-disturbance SEP and a disturbance-on system, will the post-disturbance system be stable (i.e. will the post-disturbance trajectory settle down to an SEP)? The answer to this question depends on whether or not the system state, when the disturbance is removed, lies inside the stability region of a (desired) post-disturbance SEP. If it does, then the post-disturbance trajectory will settle down to the post-disturbance SEP; otherwise, the post-disturbance trajectory may be unstable. Hence, knowledge of the relevant stability boundary toward which the disturbance-on system trajectory is heading, instead of knowledge of the entire stability boundary, plays an important role in the stability assessment of the post-disturbance trajectory.

In this chapter, we will present a rigorous introduction to the relevant stability boundary and its characterization for continuous nonlinear dynamical systems and constrained nonlinear dynamical systems. In particular, we will present the concept of a controlling UEP, the controlling UEP method, and develop a theoretical basis for the controlling UEP method. Both dynamic and geometric characterizations of the controlling UEP will be derived. These characterizations are useful to the development of solution methodologies for computing controlling UEPs. We then present the concept of the controlling unstable limit cycle (ULC) and the controlling ULC method. Finally, we illustrate the controlling UEP method numerically. The estimation of the relevant stability region will be presented in Chapter 14.

8.1 Relevant stability boundary

We first consider the following autonomous nonlinear dynamical system:

$$\dot{x} = f(x), \quad x \in R^n. \tag{8.1}$$

It is natural to assume the function (i.e. the vector field) $f : R^n \to R^n$ satisfies a sufficient condition for the existence and uniqueness of the solution.

Usually, the system is operating at an asymptotically stable equilibrium point of the pre-disturbance system:

$$\dot{x} = f_{pre}(x), \ x \in R^n, \ t \in [0, t_0^-] \tag{8.2}$$

Suppose that the system (8.2) undergoes a disturbance at time t_0. The resulting system, termed the *disturbance-on system*, is modeled by the following:

$$\dot{x} = f_{dis}(x), \ x \in R^n, \ t \in [t_0^+, t_{cl}^-]. \tag{8.3}$$

One special situation is when the disturbance function is additive so that the disturbance-on system is described by the following equation:

$$\dot{x} = f(x) + d(t), \ x \in R^n \tag{8.4}$$

where the disturbance function is modeled as $d(t)$.

Suppose the disturbance is removed at time t_{cl} and the post-disturbance system is described by the following:

$$\dot{x} = f(x), \ x \in R^n \quad \text{for} \quad t \geq t_{cl}. \tag{8.5}$$

One key question regarding the stability of the post-disturbance system is the following: will the post-disturbance system settle down to a stable equilibrium point? If the disturbance is quickly removed, then it is expected that the post-disturbance trajectory will settle down to an SEP. But, in this case the question is how can we quantify the "quickness"? In other words, what is the largest time period that the disturbance can stay in the system before the post-disturbance system becomes unstable. The largest time period is termed the *critical removal time* of the disturbance (function), which depends on the "relevant" stability boundary of the post-disturbance system toward which the disturbance-on trajectory is heading.

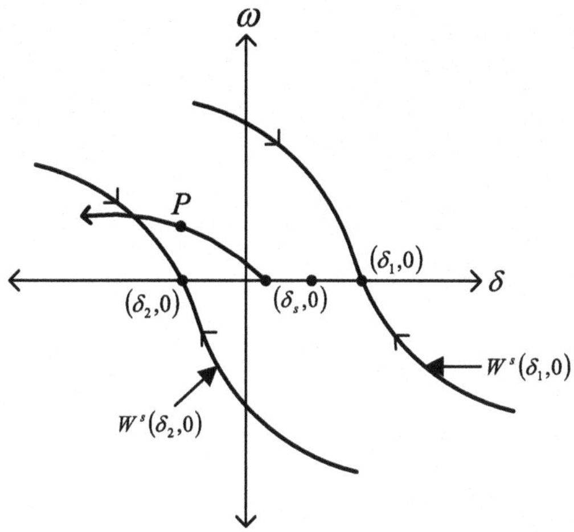

Figure 8.1 The exact stability region $A(\delta_s, 0)$ is completely characterized by its stability boundary $\partial A(\delta_s, 0)$ which is composed of the stable manifold of the UEP $(\delta_1, 0)$ and the stable manifold of the UEP $(\delta_2, 0)$. The disturbance-on trajectory crosses the stability boundary $\partial A(\delta_s, 0)$ through, $W^s(\delta_2, 0)$ then the UEP $(\delta_2, 0)$ is termed the CUEP relative to the disturbance-on trajectory and the stable manifold of the UEP $(\delta_2, 0)$ is termed the relevant stability boundary. The post-disturbance trajectory starting from the state P which lies inside the stability region $A(\delta_s, 0)$ will converge to $(\delta_s, 0)$.

We consider the following equations:

$$\begin{aligned} \dot{\delta} &= \omega \\ M\dot{\omega} &= -D\omega - P_0 \sin\delta + P_m \end{aligned} \tag{8.6}$$

which model a one-machine-infinite-bus power system for stability analysis.

There are three equilibrium points lying within the range of $\{(\delta, \omega) = -\pi < \delta < \pi, \omega = 0)\}$, and they are $(\delta_s, 0) = (\sin^{-1}(P_m/P_0), 0)$ which is a stable equilibrium point, and $(\delta_1, 0) = (\pi - \sin^{-1}(P_m/P_0), 0)$, $(\delta_2, 0) = (-\pi - \sin^{-1}(P_m/P_0), 0)$ which are unstable equilibrium points.

The system is two dimensional (2-D). Hence, the stability region of $(\delta_s, 0)$, shown in Figure 8.1, is two dimensional. The exact stability region $A(\delta_s, 0)$ is completely characterized by its stability boundary $\partial A(\delta_s, 0)$ which is composed of the stable manifold of the UEP $(\delta_1, 0)$ and the stable manifold of the UEP $(\delta_2, 0)$ (see Figure 8.1). If a disturbance-on trajectory crosses the stability boundary $\partial A(\delta_s, 0)$ through $W^s(\delta_2, 0)$, then the UEP $(\delta_2, 0)$ is termed the CUEP relative to the disturbance-on trajectory and the stable manifold of the UEP $(\delta_2, 0)$ is termed the relevant stability boundary. The post-disturbance trajectory starting from the state P, which lies inside the stability region $A(\delta_s, 0)$, will converge to $(\delta_s, 0)$ and hence it is stable. This simple example illustrates that knowledge of the relevant stability boundary relative to the disturbance-on trajectory instead of knowledge of the entire stability boundary is sufficient to determine the stability of the post-disturbance trajectory.

Let x_s^{pre} be an SEP of system (8.2) which is subject to a disturbance. Let $A(x_s)$ denote the stability region of the post-disturbance SEP x_s. The following assumption is needed for the concept of the relevant stability region to hold:

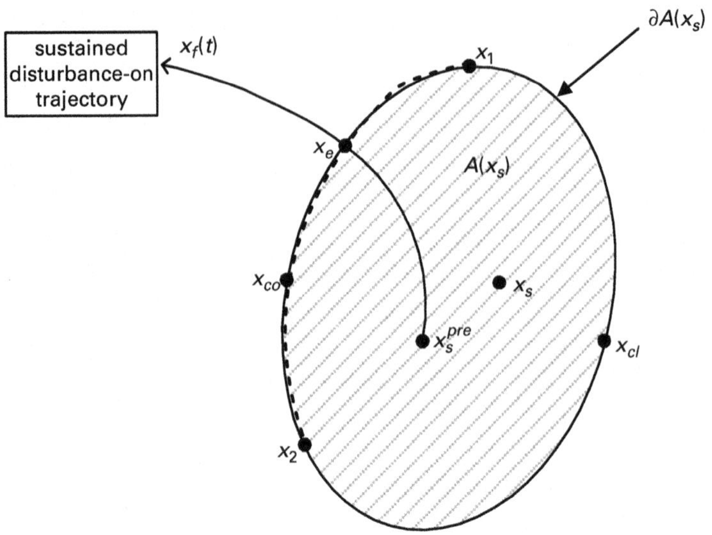

Figure 8.2 The sustained disturbance-on trajectory $x_f(t)$ moves toward the stability boundary $\partial A(x_s)$ and intersects it at the exit point x_e. The exit point lies on the stable manifold of the controlling UEP x_{co}. The relevant stability boundary is the stable manifold of x_{co}, the dashed line on the stability boundary.

$$\text{(D) } x_s^{pre} \in A(x_s).$$

It is generically true that the disturbance-on trajectory $x_f(t)$, i.e. the solution of (8.3) starting at x_s^{pre}, transversely intersects with the quasi stability boundary $\partial A_q(x_s)$ of the post-disturbance SEP x_s.

We next present a formal definition of relevant stability region for a class of nonlinear system (8.1) which satisfies the following assumption:

(A3) Every trajectory on the stability boundary converges to an equilibrium point as $t \to \infty$.

Before we further explain the relevant stability boundary, the following concept of an exit point is useful (see Figure 8.2).

DEFINITION **(Exit point)**
The point at which a (sustained) disturbance-on trajectory first intersects with the stability boundary of the post-disturbance SEP and exits the closure of the stability region is called the exit point *of the disturbance-on trajectory (relative to the post-disturbance trajectory).*

From a mathematical viewpoint, the disturbance-on trajectory may have several intersections with the stability boundary; the exit point is the first intersection of the disturbance-on trajectory $x_f(t)$ with the stability boundary. The intersection of the disturbance-on trajectory with the stability boundary by itself does not characterize the exit point. To be an exit point, the disturbance-on trajectory, at the point of intersection, has to exit the closure of the stability region.

If a post-disturbance system with an SEP x_s satisfies assumption (A3), then the stability boundary $\partial A(x_s)$ is contained in the union of the stable manifolds of the UEPs on the stability boundary, i.e.

$$\partial A(x_s) \subseteq \bigcup_{x_i \in \{E \cap \partial A(x_s)\}} W^s(x_i),$$

where E is the set of equilibrium points of the post-disturbance system. Let x_s^{pre} be an SEP of system (8.1) which is subject to a disturbance. Let $x_f(t)$ be the corresponding disturbance-on trajectory. Let $A(x_s)$ denote the stability region of the post-disturbance SEP x_s. We next present a formal definition of the controlling UEP.

DEFINITION **(Controlling UEP)**
The controlling UEP *of a disturbance-on trajectory $x_f(t)$ is the UEP whose stable manifold contains the exit point of $x_f(t)$ (i.e. the controlling UEP is the first UEP whose stable manifold is intersected by the disturbance-on trajectory $x_f(t)$ at the exit point).*

The issue of existence and uniqueness of the controlling UEP defined above will be investigated in the next section. The concept of the CUEP can be generalized into the controlling unstable limit cycle (ULC) as described in the following.

DEFINITION **(Controlling unstable limit cycle)**
The controlling unstable limit cycle *of a disturbance-on trajectory $x_f(t)$ is the unstable limit cycle whose stable manifold contains the exit point of $x_f(t)$ (i.e. the controlling ULC is the first ULC whose stable manifold intersects the disturbance-on trajectory $x_f(t)$ at the exit point).*

We next present a complete characterization of the stability boundary, which is useful in characterizing the relevant stability boundary via the controlling UEP.

THEOREM 8-1 **(Characterization of post-disturbance stability boundary)**
If a post-disturbance system with an SEP x_s has an energy function $V(\cdot):R^n \to R$, then the stability boundary $\partial A(x_s)$ is contained in the union of the stable manifolds of the UEPs on the boundary, i.e.

$$\partial A(x_s) \subseteq \bigcup_{x_i \in \{E \cap \partial A(x_s)\}} W^s(x_i),$$

where E is the set of equilibrium points of the post-disturbance system.

Proof The existence of an energy function ensures that every trajectory of the post-disturbance system on the stability boundary $\partial A(x_s)$ converges to a UEP on $\partial A(x_s)$. As a result, assumption (A3) is satisfied and Theorem 8-1 follows.

Since the stability boundary of the SEP of a post-disturbance system is contained in a set that is the union of the stable manifolds of the UEPs on the stability boundary, the exit point must lie on the stable manifold of a UEP on the stability boundary. This UEP is the controlling UEP relative to the disturbance-on trajectory. This controlling UEP is then used to characterize the relevant stability boundary for the disturbance-on trajectory. We are now in a position to present a formal definition of the relevant stability boundary.

DEFINITION **(Relevant stability boundary)**
The relevant stability boundary *with respect to a disturbance-on trajectory $x_f(t)$ is the stable manifold of the corresponding controlling UEP (or controlling limit cycle) whose stable manifold contains the exit point of $x_f(t)$.*

The controlling UEP (or the controlling ULC) with respect to a disturbance-on trajectory exists and is always unique (see next section) under assumptions (D) and (A3) (or B3), thus the concept of a relevant stability boundary is well defined. The relevant stability boundary is illustrated in Figure 8.2. The disturbance-on trajectory exits the stability region via the relevant stability boundary. Hence, knowledge of the relevant stability boundary, instead of knowledge of the entire stability boundary, is sufficient to assess the stability of the post-disturbance trajectory.

8.2 Existence and uniqueness

The issues of existence and uniqueness of the controlling UEP with respect to a disturbance-on trajectory and a post-disturbance system are addressed in this section.

THEOREM 8-2 **(Existence and uniqueness)**
Consider a pre-disturbance SEP, a disturbance-on system trajectory, and a post-disturbance system with an SEP x_s. If assumptions (D) and (A3) hold, then the controlling UEP of the disturbance-on trajectory exists and is unique.

Proof According to assumption (D), the exit point x_e, with respect to the disturbance-on trajectory $x_f(t)$, exists and is unique. Assumption (A3) ensures that x_e belongs to the stable manifold of a UEP x_{co} on the stability boundary. As a result, the controlling UEP exists. Since stable manifolds of different equilibrium points do not intersect, then x_{co} exists and is unique. This completes the proof.

The following corollary combines the results of Theorem 8-1 and Theorem 8-2 to show the existence and uniqueness of the controlling UEP of nonlinear systems (8.1) that admit energy functions.

COROLLARY 8-3 **(Existence and uniqueness)**
Consider a pre-disturbance SEP, a disturbance-on system trajectory, and a post-disturbance system with an SEP x_s, having an energy function $V(\cdot):R^n \to R$. If assumption (D) holds, then the controlling UEP of the disturbance-on trajectory always exists and is unique.

The uniqueness of the controlling UEP makes the task of finding the controlling UEP very difficult, due to the following computational challenges.

- Challenge I: The controlling UEP is a particular UEP embedded in a large-degree state-space.
- Challenge II: The controlling UEP is the first UEP whose stable manifold has a non-empty intersection with the disturbance-on trajectory at the exit point.
- Challenge III: The task of computing the exit point is very involved; it usually requires an iterative time-domain approach.

Challenge IV: The task of computing the controlling UEP requires solving a large set of constrained nonlinear algebraic equations.

Challenge V: A good initial guess for computing the controlling UEP is difficult to provide.

Challenge VI: The size of the convergence region of a controlling UEP with respect to a numerical method can be very small and irregular.

Challenge VII: Starting from the exit point as an initial guess, a numerical method such as the Newton method may not converge to the controlling UEP.

8.3 Dynamic and geometric characterizations

In this section, dynamic and geometric characterizations of the controlling UEP will be derived. These characterizations will lead to several advances in computing and identifying controlling UEPs. The notion of the quasi-stability region and its complete characterization will be employed to derive these characterizations.

We have the following assumptions.

(A1) All the equilibrium points on $\partial A_q(x_s)$ are hyperbolic.

(A2) The stable and unstable manifolds of equilibrium points on $\partial A_q(x_s)$ satisfy the transversality condition.

(A3) Every trajectory on $\partial A_q(x_s)$ approaches one of the equilibrium points as $t \to \infty$.

Under these assumptions, a complete characterization of the quasi-stability boundary $A_q(x_s)$ of an asymptotically stable equilibrium point x_s was derived in Chapter 6. More precisely, the following complete characterization of a quasi-stability boundary was derived

$$\partial A_q(x_s) = \bigcup_{x_i \in \partial A_q(x_s)} W^s(\sigma_i) \tag{8.7}$$

where x_i, $i = 1, 2, \ldots$ are the equilibrium points lying on the quasi-stability boundary $\partial A_q(x_s)$ of the stable equilibrium point x_s. Moreover, if σ_i, $i = 1, 2, \ldots$ are the type-one equilibrium points lying on the quasi-stability boundary $\partial A_q(x_s)$ of the stable equilibrium point x_s, then

$$\partial A_q(x_s) = \bigcup_{\sigma_i \in \partial A_q(x_s)} \overline{W^s(\sigma_i)}. \tag{8.8}$$

We next proceed to derive dynamic and geometric characterizations of the controlling UEP.

THEOREM 8-4 **(Dynamic characterization of the controlling UEP)**
Consider a pre-disturbance SEP, a disturbance-on trajectory and a post-disturbance system with an SEP x_s. If assumptions (D), (A1)–(A3) are satisfied, then the following dynamic results hold:

(a) the controlling UEP, say x_{co} of the disturbance-on trajectory always exists;

(b) x_{co} belongs to the quasi-stability boundary, i.e. $x_{co} \in \partial A_q(x_s)$;

(c) the unstable manifold of x_{co} converges to the SEP x_s, i.e. $W^u(x_{co}) \cap A(x_s) \neq \emptyset$;
(d) its unstable manifold intersects the complement of the closure of the stability region, i.e. $W^u(x_{co}) \cap (\overline{A}(x_s))^c \neq \emptyset$.

Proof The existence of the controlling UEP is a direct consequence of Corollary 8-3; this completes the proof of part (a). Assumption (D) ensures that the exit point x_e belongs to the quasi-stability boundary $\partial A_q(x_s)$. As a consequence of the quasi-stability boundary characterization (8.7) and the definition of controlling UEP, one has that $x_{co} \in \partial A_q$, proving part (b). The inclusion $\partial A_q \subset \partial A$ and Theorem 4-3 of Chapter 4, which states that a hyperbolic equilibrium point $x^* \in \partial A(x_s)$ if and only if $W^u(x^*) \cap A(x_s) \neq \emptyset$, prove part (c). Part (d) follows from part (b) and the definition of the quasi-stability region.

We next present a geometric characterization of the controlling UEP. It is shown that the controlling UEP is a type-one UEP.

THEOREM 8-5 **(Geometric characterization of the controlling UEP)**
Consider a pre-disturbance SEP, a disturbance-on trajectory, and a post-disturbance system with an SEP x_s. If assumptions (D), (A1)–(A3) are satisfied, then the controlling UEP x_{co} is generically a type-one UEP.

Proof Theorem 8-4 and Eq. (8.8) assert that the exit point generically lies on the stable manifold of type-one equilibrium points on the stability boundary. Hence, the controlling UEP is a type-one UEP lying on the stability boundary. This completes the proof.

Theorems 8-4 and 8-5 assert that the one-dimensional unstable manifold of the controlling UEP converges to the post-disturbance SEP and also intersects the regions which are "outside the stability region" (see Figure 8.3). We next illustrate the theoretical developments derived so far with a simple example. The following set of six-dimensional differential equations models the 3-machine, 9-bus power system shown in Figure 8.4:

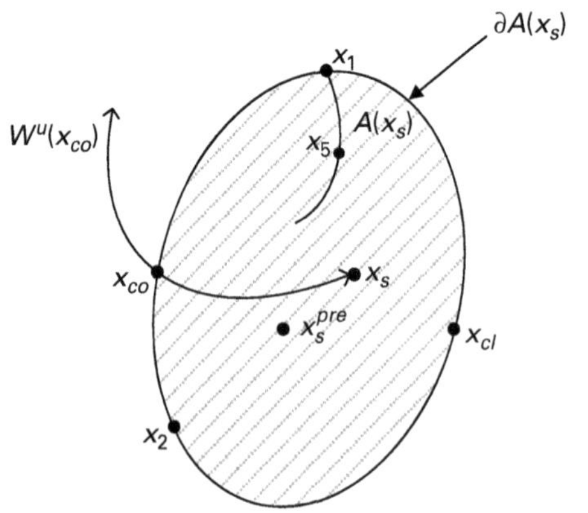

Figure 8.3 The controlling UEP is a type-one UEP lying on the stability boundary and its one-dimensional unstable manifold always converges to the post-disturbance SEP and also intersects with regions which are "outside the stability region."

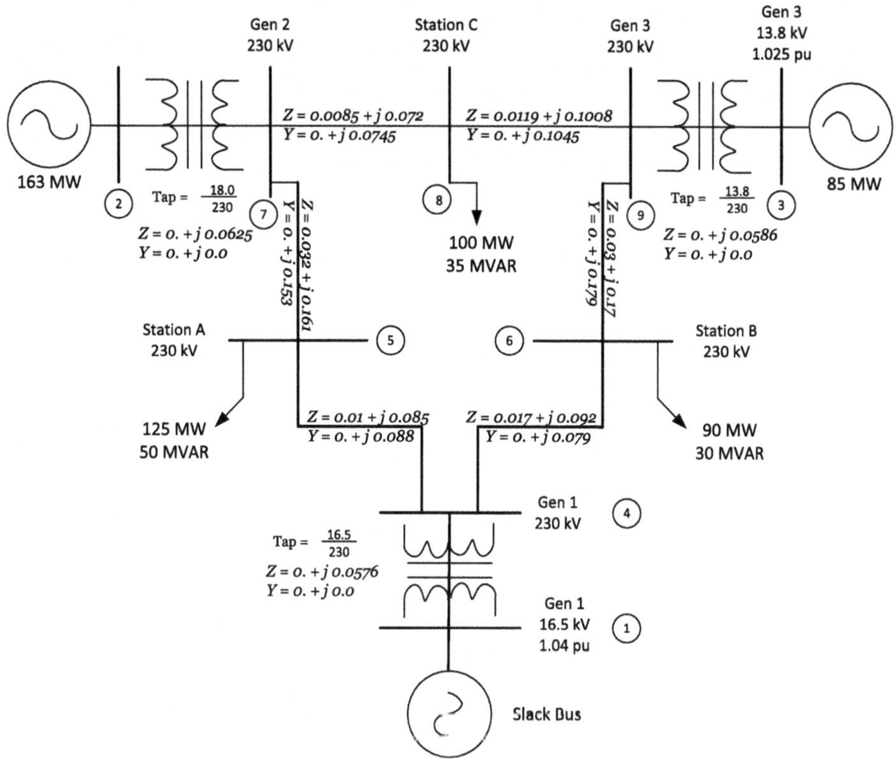

Figure 8.4 The on-line diagram of the (pre-fault) 3-machine, 9-bus power system. The value of Y is half the line charging.

$$
\begin{aligned}
\dot{\delta}_1 &= \omega_1 \\
\dot{\delta}_2 &= \omega_2 \\
\dot{\delta}_3 &= \omega_3 \\
m_1\dot{\omega}_1 &= -d_1\omega_1 + P_{m_1} - P_{e_1}(\delta_1, \delta_2, \delta_3) \\
m_2\dot{\omega}_2 &= -d_2\omega_2 + P_{m_2} - P_{e_2}(\delta_1, \delta_2, \delta_3) \\
m_3\dot{\omega}_3 &= -d_3\omega_3 + P_{m_3} - P_{e_3}(\delta_1, \delta_2, \delta_3)
\end{aligned} \tag{8.9}
$$

where

$$
P_{e_i}(\delta_1, \delta_2, \delta_3) = \sum_{j=1, j\neq 1}^{3} E_i E_j (B_{ij}\sin(\delta_i - \delta_j) + G_{ij}\cos(\delta_i - \delta_j)),
$$

$P_{m1} = 0.8980$, $P_{m2} = 1.3432$, $P_{m3} = 0.9491$, $E_1 = 1.1083$, $E_2 = 1.1071$, $E_3 = 1.0606$. The system admittance matrix, $G_{ij} + jB_{ij}$, of the pre-disturbance system is given by:

$$
G_{ij}^{pre} + jB_{ij}^{pre} = \begin{bmatrix} 0.845 - j2.988 & 0.287 + j1.513 & 0.210 + j1.226 \\ 0.287 + j1.513 & 0.420 - j2.724 & 0.213 + j1.088 \\ 0.210 + j1.226 & 0.213 + j1.088 & 0.277 - j2.368 \end{bmatrix}.
$$

We consider a uniform damping factor $d_i/m_i = 0.1$ with (d_1, d_2, d_3)= (0.0125, 0.0034, 0.0016). The coordinates of the pre-disturbance SEP are (−0.0482, 0.1252, 0.1124). To illustrate the computational procedure involved in the controlling UEP method, we numerically simulate the following dynamic objects for two different disturbances:

- pre-disturbance SEP and post-disturbance SEP,
- UEPs on the stability boundary of the post-disturbance system,
- the stability boundary of the post-disturbance system on the machine angle space,
- disturbance-on trajectory.

These dynamic objects are simulated with the consideration that the disturbance-on trajectory starting from the pre-disturbance SEP intersects the stability boundary of the post-disturbance system at the exit point. The exit point lies on the stable manifold of the controlling UEP. The relevant stability boundary toward which the disturbance-on trajectory heads is the stable manifold of the controlling UEP.

For disturbance #1, a short-circuit occurs near bus #8 and the line between bus #7 and bus #8 is tripped. The system admittance matrices, $G_{ij} + jB_{ij}$, of the disturbance-on and post-disturbance systems are given by:

$$G_{ij}^{dist} + jB_{ij}^{dist} = \begin{bmatrix} 0.685 - j3.685 & 0.077 + j0.523 & 0.0505 + j0.498 \\ 0.077 + j0.523 & 0.146 - 4.130 & 0.006 + j0.0548 \\ 0.0505 + j0.498 & 0.006 + j0.0548 & 0.120 - j3.127 \end{bmatrix}$$

$$G_{ij} + jB_{ij} = \begin{bmatrix} 0.838 - j0.015 & 0.203 + j1.464 & 0.236 + j1.202 \\ 0.203 + j1.464 & 0.326 - j1.908 & 0.066 + j0.324 \\ 0.263 + j1.202 & 0.066 + j0.324 & 0.505 - j1.814 \end{bmatrix}.$$

The coordinates of the SEP of the post-disturbance system are (−0.0655, 0.2430, −0.0024). The exact stability region of the post-disturbance system on the angle plane is highlighted in Figure 8.5. The disturbance-on trajectory intersects the stable manifold of the controlling UEP, a type-one UEP, the square in Figure 8.5. The coordinates of the CUEP are (−0.3495, 0.0745, 2.5864). There are two type-one UEPs and one type-two UEP lying on the stability boundary. The stable manifold of one type-one UEP, the controlling UEP, contains the exit point, which is the intersection between the stability boundary and the disturbance-on trajectory. This exit point is at some distance from the controlling UEP while it is very close to a type-two UEP lying on the stability boundary. It is, however, incorrect to use this type-two UEP as the controlling UEP.

Contrary to the claim made in the literature, this numerical example clearly shows that the controlling UEP is not the UEP closest to the disturbance-on trajectory. The controlling UEP is the UEP whose stable manifold contains the exit point.

For disturbance #2, a short-circuit occurs near bus #7 and the line between bus #7 and bus #8 is tripped. The system admittance matrices, $G_{ij} + jB_{ij}$, of the disturbance-on and post-disturbance systems are given by:

$$G_{ij}^{dist} + jB_{ij}^{dist} = \begin{bmatrix} 0.657 - j3.819 & 0 + j0 & 0.070 + j0.631 \\ 0 + j0 & 0 - j5.485 & 0 + j0 \\ 0.070 + j0.631 & 0 + j0 & 0.174 - j2.796 \end{bmatrix}$$

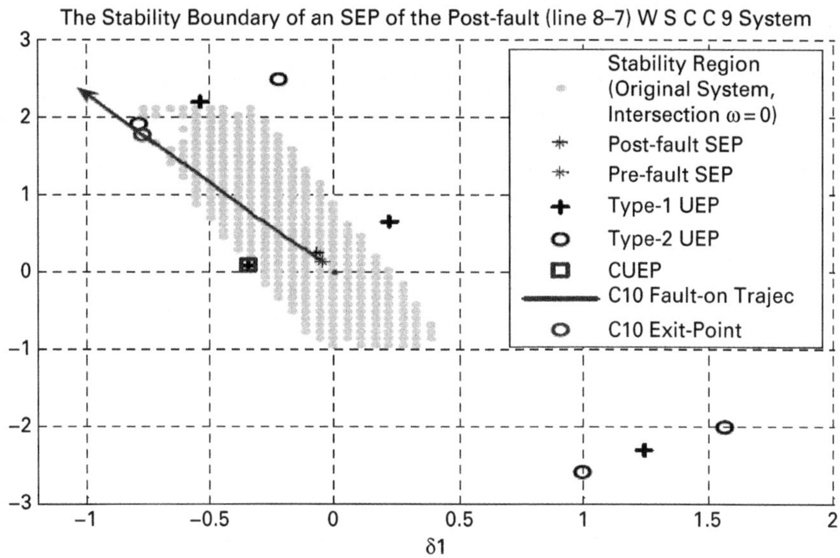

Figure 8.5 For disturbance #1, the stability region of the post-disturbance system on the angle plane is highlighted. The disturbance-on trajectory intersects the stable manifold of the controlling UEP at the exit point, as shown in the figure.

$$G_{ij} + jB_{ij} = \begin{bmatrix} 0.838 - j3.015 & 0.203 + j1.465 & 0.263 + j1.202 \\ 0.203 + j1.465 & 0.326 - j1.908 & 0.066 + j0.324 \\ 0.263 + j1.202 & 0.066 + j0.324 & 0.545 - j1.814 \end{bmatrix}.$$

The coordinates of the SEP of the post-disturbance system are (−0.0655, 0.2430, −0.0024). The stability region of the post-disturbance system on the angle plane is highlighted in Figure 8.6. The disturbance-on trajectory intersects the stable manifold of the controlling UEP at the exit point as shown in Figure 8.6. There are two type-one UEPs and one type-two UEP lying on the stability boundary. The stable manifold of one type-one UEP, the controlling UEP, contains the exit point, which is the intersection between the stability boundary and the disturbance-on trajectory. The coordinates of the CUEP are (−0.5424, 2.1802, −0.3755). This exit point is close to the controlling UEP, the small square in Figure 8.6, while it is at some distance from the other type-one UEP. It is interesting to note that the exit point is also close to a type-two UEP.

8.4 Relevant stability boundary of constrained dynamical systems

The concept of relevant stability boundary and the controlling UEP method, which were developed in the previous sections for dynamical systems modeled by ordinary differential equations, can be extended, with some adaptations, to constrained dynamical systems.

We consider a general class of nonlinear systems described by the following set of differential and algebraic equations (DAE):

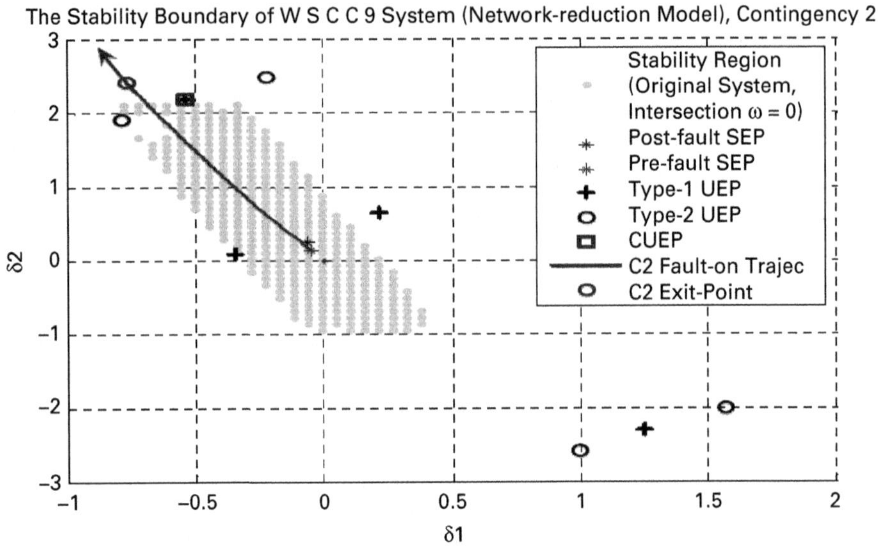

Figure 8.6 For disturbance #2, the stability region of the post-disturbance system on the angle plane is highlighted. The disturbance-on trajectory intersects the stable manifold of the controlling UEP, a type-one UEP, at the exit point, as shown in the figure.

$$\begin{aligned} \dot{x} &= f(x, y) \\ 0 &= g(x, y) \end{aligned} \tag{8.10}$$

where $x \in R^n$ and $y \in R^m$ are the corresponding dynamical and static variables of the system, respectively. Regarding this class of nonlinear systems, we describe a three-stage regime. In the pre-disturbance regime, the system model is represented as

$$\begin{aligned} \dot{x} &= f_{pre}(x, y) \\ 0 &= g_{pre}(x, y) \end{aligned} \quad t \in [0, t_0^-]. \tag{8.11}$$

Suppose that the system undergoes a disturbance at time t_0, which results in a structural change in the system. Suppose also that the disturbance duration is confined to the time interval $[t_0^+, t_{cl}^-]$. During this interval, the system is governed by the disturbance-on dynamics:

$$\begin{aligned} \dot{x} &= f_{dist}(x, y) \\ 0 &= g_{dist}(x, y) \end{aligned} \quad t \in [t_0^+, t_{cl}^-]. \tag{8.12}$$

After the disturbance is cleared at time t_{cl}, the system, termed the post-disturbance system, is henceforth governed by post-disturbance dynamics described by

$$\begin{aligned} \dot{x} &= f(x, y) \\ 0 &= g(x, y) \end{aligned} \quad t_{cl} \le t < \infty.$$

The post-disturbance system may or may not be the same as the pre-disturbance system. The DAE system (8.10) can be interpreted as an implicitly dynamical system defined on the constrained manifold Γ:

$$\Gamma = \{(x,y) : g(x,y) = 0\}.$$

All the system states and trajectories, including the equilibrium points, stable and unstable manifolds, and stability regions, must lie in the above constrained manifold. We note that the constrained manifold of the pre-disturbance, disturbance-on and post-disturbance systems are different. We will respectively denote them by Γ_{pre}, Γ_{dist} and Γ.

Usually the system is operating at an asymptotically stable equilibrium point (x_0, y_0) of the pre-disturbance system. At time t_0, the system undergoes a perturbation and the trajectory jumps instantaneously from $(x_0, y_0^-) \in \Gamma_{pre}$ to a point $(x_0, y_0^+) \in \Gamma_{dist}$. For the disturbance-on trajectory, along the disturbance constrained manifold Γ_{dist}, we will use the notation $z_{dist}(t) = (x_{dist}(t), y_{dist}(t))$. At the clearing time t_{cl}, the trajectory instantaneously jumps from $z(t_{cl}^-) = (x(t_{cl}), y(t_{cl}^-))$ on the disturbance manifold Γ_{dist} to a point $z(t_{cl}^+) = (x(t_{cl}), y(t_{cl}^+))$ on the post-disturbance manifold Γ.

These jump behaviors are a modeling problem that arises as a consequence of neglecting some parasitic or fast dynamics. We propose fixing this problem by using the singular perturbation approach. The singular perturbation approach treats the set of algebraic equations describing a DAE system as a limit of the fast dynamics: $\varepsilon\dot{y} = g(x,y)$. In other words, as ε approaches zero, the fast dynamics will approach its constrained manifold. In this way, the jumps will be associated with the limit, as ε approaches zero, of very fast dynamics.

Therefore, for the DAE system (8.10), we can define an associated singularly perturbed system

$$\begin{aligned} \dot{x} &= f(x,y) \\ \varepsilon\dot{y} &= g(x,y) \end{aligned} \tag{8.13}$$

where ε is a sufficiently small positive number. The state variables of system (8.13) have very different rates of dynamics and they can be separated into two distinct time scales: slow variable x and fast variable y. Note that trajectories of the singularly perturbed system (8.13) will not be confined to the constrained manifold Γ and are not exactly the same as those of the original DAE system (8.10). However, trajectories generated by the singularly perturbed system are still valid approximations of those of the DAE system. A theoretical justification ensuring that the difference of solution trajectories between the original DAE (8.10) and the singularly perturbed system (8.13) is uniformly bounded by the order of $O(\varepsilon)$ is provided by Tikhonov's theorem over the infinite time interval; see Chapter 16 for further details.

In Chapter 7, it was shown that a DAE system and its corresponding singularly perturbed system share several similar dynamical properties. Under some reasonable conditions, it was shown, for sufficiently small $\varepsilon > 0$, that $(\bar{x}, \bar{y})$ is a type-k equilibrium point of the DAE system (8.10) on a stable component Γ_s of Γ if and only if $(\bar{x}, \bar{y})$ is a type-k equilibrium point of the singularly perturbed system (8.13). Moreover, it was shown, for sufficiently small $\varepsilon > 0$ and under some conditions over the vector field, that (x_u, y_u) lies on the stability boundary $\partial A_0(x_s, y_s)$ of the DAE system (8.10) if and only if (x_u, y_u) belongs to the stable component Γ_s of Γ and lies on the stability boundary $\partial A_\varepsilon(x_s, y_s)$ of the singularly pesrturbed system (8.13).

These properties and relationships between equilibrium points on the stability boundary of the DAE system and equilibrium points on the stability boundary of the singularly perturbed system provide a theoretical basis for the controlling UEP method, developed for ODE systems, to be applied to DAE systems.

One key difficulty in defining the CUEP for the DAE system is that the final state of a disturbance-on trajectory will not lie on the constrained manifold Γ of its post-disturbance system. In other words, the disturbance-on trajectory will not intersect the stability boundary of its post-disturbance DAE system. Instead, the disturbance-on trajectory will intersect the stability boundary of the corresponding singularly perturbed post-disturbance system. Hence, the exit point of a disturbance-on trajectory must lie on the stable manifold of the controlling UEP of a singularly perturbed post-disturbance system. The singular perturbation approach hence allows us to extend the controlling UEP method developed for ODE systems to one applicable to DAE systems.

8.5 Controlling UEP for DAE systems

One key difficulty in applying the concept of the controlling UEP to DAE systems is that the final state of a disturbance-on trajectory will not lie on the constrained manifold of its post-disturbance DAE system, making the task of defining the controlling UEP for DAE systems difficult.

Two approaches have been developed to overcome this issue. One approach is to exploit the singularly perturbed system associated with the DAE system and define the controlling UEP of this DAE system as the controlling UEP of the associated singularly perturbed system. In this case, the disturbance-on trajectory will intersect the stability boundary of the corresponding singularly perturbed post-disturbance system and hence the controlling UEP of the singularly perturbed system is well defined.

The second approach is to define the controlling UEP of the DAE system via a projected disturbance-on trajectory on the constraint manifold of the post-disturbance DAE system. Even though these two approaches may appear quite different, we shall prove that these two approaches, under certain conditions, are equivalent.

DEFINITION **(Controlling UEP for singularly perturbed model)**
The controlling UEP of the singularly perturbed model (8.13) for a fixed small $\varepsilon > 0$ *with respect to a disturbance-on trajectory is the UEP on the stability boundary* $\partial A_\varepsilon(x_s, y_s)$ *of the singularly perturbed post-disturbance system (8.13), whose stable manifold contains the exit point of the disturbance-on trajectory.*

This definition is based on the fact that the exit point must lie on the stable manifold of some UEP on the stability boundary of the singularly perturbed post-disturbance model (8.13). By exploiting the complete characterization of the stability boundary of the singularly perturbed post-disturbance system (8.13), one can prove the existence and uniqueness of the controlling UEP of the singularly perturbed model (8.13) as shown below.

THEOREM 8-6 **(Existence and uniqueness of the controlling UEP)**
Consider a pre-disturbance SEP (x_0, y_0)*, a disturbance-on trajectory* $x_{f\varepsilon}(t)$ *and a post-disturbance system (8.11) with a fixed small* $\varepsilon > 0$ *and with an asymptotically SEP* $(x_{s\varepsilon}, y_{s\varepsilon})$*. If assumption (D) holds and the post-disturbance singularly perturbed system satisfies the following assumptions,*

(A1) all the equilibriums are hyperbolic, and
(A2) every trajectory on the stability boundary of the SEP $(x_{s\varepsilon}, y_{s\varepsilon})$ *converges to a UEP,*

then the controlling UEP with respect to the disturbance-on trajectory $x_{f\varepsilon}(t)$ *of the singularly perturbed system (8.13) exists and is unique.*

Proof According to assumption (D) and considering that the disturbance-on trajectory leaves the stability region $A_\varepsilon(x_s, y_s)$, the exit point $x_{e\varepsilon}$, with respect to the disturbance-on trajectory $x_{f\varepsilon}(t)$, exists and is unique. Assumptions (A1) and (A2) and Theorem 4.10 of Chapter 4 ensure that the stability boundary $\partial A_\varepsilon(x_s, y_s)$ of the singularly perturbed system is contained in the union of the stable manifolds of all equilibrium points on the stability boundary. As a consequence, the exit point must lie on one of these manifolds. Therefore, the controlling UEP exists. Since the intersection of stable manifolds of different equilibrium points must be an empty set, the controlling UEP is unique. This completes the proof.

Having shown the existence and uniqueness of the controlling UEP of the singularly perturbed system (8.13), we need to establish a relationship between the controlling UEP of the singularly perturbed system (8.13) and the controlling UEP of the DAE system (8.10). In other words, it is desirable to study the behavior of the controlling UEP for the singularly perturbed system (8.13) when $\varepsilon \to 0$. This requirement leads to the following definition of a uniform controlling UEP.

DEFINITION **(Uniform controlling UEP)**
Let $(x^\varepsilon_{co}, y^\varepsilon_{co})$ *be the controlling UEP of the singularly perturbed post-disturbance system (8.13) with respect to the disturbance-on trajectory* $(x^\varepsilon(t), y^\varepsilon(t))$*. Consider the map* $\varepsilon \to (x^\varepsilon_{co}, y^\varepsilon_{co})$*. If there exists an* $\varepsilon^* > 0$ *such that this map is constant for all* $\varepsilon \in (0, \varepsilon^*)$*, then* $(x^0_{co}, y^0_{co}) = (x^\varepsilon_{co}, y^\varepsilon_{co})$ *is a* uniform controlling UEP *with respect to the disturbance-on trajectory* $(x^\varepsilon(t), y^\varepsilon(t))$ *for all* $\varepsilon \in (0, \varepsilon^*)$.

The controlling UEP of the singularly perturbed system with respect to a disturbance-on trajectory may alter due to a change of ε. However, it will be shown that controlling UEPs of singularly perturbed systems are generically uniform. This property not only allows us to define the controlling UEP of DAE systems via the controlling UEP of the singularly perturbed system, but also provides a way to compute the controlling UEP of the DAE system with the aid of the associated singularly perturbed system.

DEFINITION **(Controlling UEP for DAE systems)**
The controlling UEP of a DAE system (8.10) with respect to a disturbance-on trajectory is the uniform controlling UEP of the associated singularly perturbed system (8.13).

The controlling UEP of the DAE system is well defined if the uniform controlling UEP of the associated singularly perturbed system lies on the stable component Γ_s of Γ that contains the asymptotically SEP (x_s, y_s). It might happen that the uniform controlling UEP of the singularly perturbed system lies in another component of Γ.

An alternative definition for the CUEP of DAE systems exploits the concept of projected disturbance-on trajectory. Since the disturbance-on trajectory does not belong to the constraint manifold of the post-disturbance DAE system, the disturbance-on trajectory does not reach the stability boundary of the post-disturbance DAE system. Thus, the exit point of the DAE trajectory cannot be defined as the point of intersection between the disturbance-on trajectory and the stability boundary of the post-disturbance system, which lies on a different constraint manifold.

We use the following approach to overcome this difficulty. For each point $(x(t), y(t))$ of the disturbance-on trajectory, we define a projected point on the post-disturbance constraint manifold denoted by $(x(t), y_p(t))$ such that $(x, y_p) \in \Gamma$ and $0 = g(x, y_p)$. Since the projected disturbance-on trajectory lies on the constrained manifold of the post-disturbance system, we can define the exit point of a DAE system with respect to a DAE disturbance-on trajectory as the point at which the projected disturbance-on trajectory intersects the stability boundary of the post-disturbance DAE system (see Figure 8.7). To this end, we consider the following assumption which is generically true and define the exit point for a DAE system.

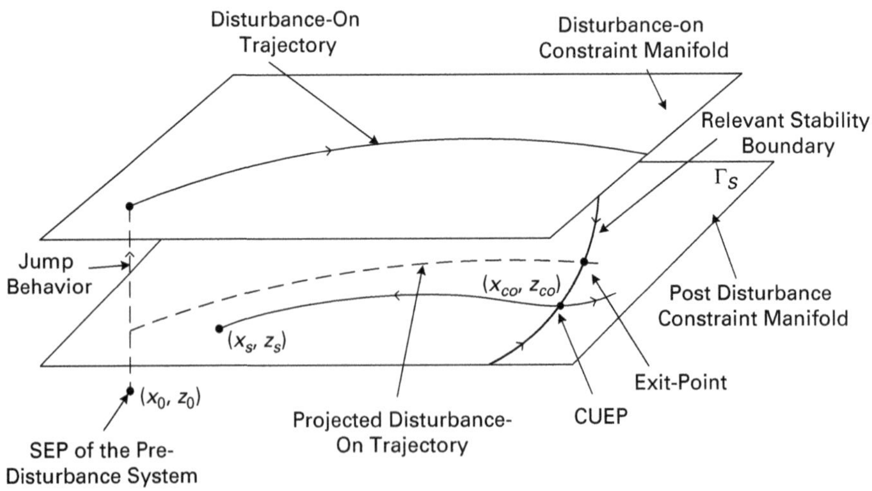

Figure 8.7 The controlling UEP of a disturbance-on DAE trajectory is the UEP of the post-disturbance DAE system whose stable manifold intersects the projected disturbance-on DAE trajectory at the exit point.

(D′) The projection of the pre-disturbance equilibrium point (x_0, y_0) on the stable component Γ_s of Γ that contains the asymptotically SEP (x_s, y_s) lies inside the stability region $A_0(x_s, y_s)$ of the DAE system.

Assumption (D′) ensures that the projected disturbance-on trajectory of the DAE system starts inside the stability region and we also assume that the projection of the sustained disturbance-on trajectory on Γ_s leaves the stability region in finite time.

DEFINITION **(Exit point)**
The point at which a (sustained) projected disturbance-on DAE trajectory intersects the stability boundary of a post-disturbance DAE system is termed the exit point of the disturbance-on DAE trajectory (relative to the post-disturbance DAE system).

With the definition of the exit point for a DAE system, we next present an alternative definition of the controlling UEP of DAE trajectories.

DEFINITION **(Controlling UEP of DAE systems – alternative definition)**
The controlling UEP of a disturbance-on DAE trajectory is the UEP of the post-disturbance DAE system whose stable manifold intersects the projected disturbance-on DAE trajectory at the exit point.

We now show that, under certain conditions, the two definitions of controlling UEP for DAE systems are equivalent. To this end, we assume that the jump behavior of the switched DAE systems can be approximately modeled by the fast dynamics of the associated singularly perturbed system.

THEOREM 8-7 **(Equivalence between the two definitions of CUEP)**
Suppose (x_{co}, y_{co}) is the controlling UEP of the DAE system (8.10), with respect to a projected disturbance-on trajectory $(x_{dist}(t), y_{dist_p}(t))$. If the singularly perturbed system (8.13) admits an energy function for every $\varepsilon > 0$, then (x_{co}, y_{co}) is also the uniform controlling UEP of the singularly perturbed system (8.13).

Proof We first show that the exit point of the singularly perturbed system lies on $W^s_\varepsilon(x_{co}, y_{co})$ for sufficiently small ε. Let t_{ex} be the time at which the projected disturbance-on trajectory $(x_{dist}(t), y_{dist_p}(t))$ intersects the stability boundary of the DAE system (8.10), that is, $(x^{DAE}_{ex}, y^{DAE}_{ex}) = (x_{dist}(t_{ex}), y_{dist_p}(t_{ex}))$. Using Tikonov's theorems and regular perturbation in the fast system, we can prove that the disturbance-on trajectory $\varphi_\varepsilon(t)$ of the singularly perturbed system intersects the stability boundary $\partial A_\varepsilon(x_s, y_s)$ of the post-disturbance singularly perturbed system in a neighborhood of the point $(x^*, y^*) = \varphi_0(t_{ex}, x_0, y_{sF})$ for sufficiently small ε. Since $(x^{DAE}_{ex}, y^{DAE}_{ex}) \in W^s_{DAE}(x_{co}, y_{co})$, $\varphi(t, (x^{DAE}_{ex}, y^{DAE}_{ex}))$ is a bounded solution of the DAE system for $0 \leq t < \infty$ that converges to (x_{co}, y_{co}) as $t \to \infty$. Let K be the closure of the segment of orbit of the DAE system that starts at (x_{ex}, y_{ex}) and tends toward (x_{co}, y_{co}) as $t \to \infty$, that is, K is the closure of the set $\{(x, y) \in \Gamma_s : (x, y) = \varphi(t, (x^{DAE}_{ex}, y^{DAE}_{ex})),\ 0 \leq t \leq \infty\}$. Set K is a compact set composed of equilibrium points of the fast system $dy/d\tau = g(x, y)$, where x is treated as a fixed parameter. Consider the extended system:

$$(\Pi_\varepsilon)\times 0\begin{cases}\dfrac{dx}{d\tau}=\varepsilon f(x,y)\\ \dfrac{dz}{d\tau}=g(x,y)\\ \dfrac{d\varepsilon}{d\tau}=0.\end{cases}$$

For each $(x, y) \in K$, let $E_{(x,y)}{}^s$, $E_{(x,y)}{}^u$ and $E_{(x,y)}{}^c$ denote the invariant subspaces of $R^{n+m} \times (-\varepsilon_0, \varepsilon_0)$ associated with the eigenvalues of the Jacobian matrix of the system $(\Pi_\varepsilon)\times 0$ calculated at $((x,z), \varepsilon=0)$. According to Theorem 9.1 of [93], there exists a center-stable manifold C^s for $(\Pi_\varepsilon)\times 0$ near K for sufficiently small ε. This center-stable manifold C^s satisfies the following properties: (i) $K \times \{\varepsilon = 0\} \subset C^s$; (ii) C^s is locally invariant[1] under the flow $(\Sigma_\varepsilon)\times 0$; (iii) C^s is tangent to $E_{(x,y)}{}^s \oplus E_{(x,y)}{}^c$ at $((x,y), \varepsilon=0)$ for all $(x, y) \in K$. Since (x^*, y^*) lies on $W^s_{fast}((x^{DAE}_{ex}, y^{DAE}_{ex}))$, given $\rho/2 > 0$, there exists $T > 0$ such that $\frac{\rho}{2} < \left\|\varphi_{fast}(T, x^*, y^*) - (x^{DAE}_{ex}, y^{DAE}_{ex})\right\| < \rho$.

Let P be the projection onto $E_{(x,z)}{}^s$. For a certain $\tau > T$, define $q = P(\varphi_{fast}(\tau, x^*, y^*) - ((x^{DAE}_{ex}, y^{DAE}_{ex})))$. According to [20], given $\rho > 0$, there exists a unique solution $\varphi_\varepsilon(t)$ of the singularly perturbed system, bounded for $t \geq 0$, such that the following hold

$$\begin{aligned} P(\varphi_\varepsilon(0) - (x^{DAE}_{ex}, y^{DAE}_{ex})) \quad &= q \\ \left\|\varphi_\varepsilon(t) - \varphi_o(t, (x^{DAE}_{ex}, y^{DAE}_{ex}))\right\| &\leq \rho \end{aligned}$$

for every $t \geq 0$ and ε sufficiently small. Since $\varphi_\varepsilon(t)$ is bounded for $t \geq 0$ and system (Σ_ε) has an energy function for every $\varepsilon > 0$, then $\varphi_\varepsilon(t) \to (x_{co}, z_{co})$ as $t \to \infty$. As a consequence, $\varphi_\varepsilon(0) \in C^s$ and $\varphi_\varepsilon(0) \in W^s_\varepsilon(x_{co}, y_{co})$. Since $(x^{DAE}_{ex}, y^{DAE}_{ex})$ is a hyperbolic equilibrium point of the fast system $(\Pi_F(x_{exS}))$, the stable manifold $W^s_{fast}(x^{DAE}_{ex}, y^{DAE}_{ex})$ is also tangent to $E_{(x,z)}{}^s$ at $(x^{DAE}_{ex}, y^{DAE}_{ex})$. As a consequence, given a number $\alpha > 0$, there exists $\rho > 0$ such that the following holds

$$\left\|(I-P)((x,y) - (x^{DAE}_{ex}, y^{DAE}_{ex})\right\| < \alpha\left\|(x,y) - (x^{DAE}_{ex}, y^{DAE}_{ex})\right\|$$

for all (x, y) lying on $W^s_{fast}(x^{DAE}_{ex}, y^{DAE}_{ex})$ or C^s satisfying $\left\|(x,y) - (x^{DAE}_{ex}, y^{DAE}_{ex})\right\| < \rho$. As a consequence

$$\begin{gathered} \left\|\varphi_{fast}(\tau, x^*, y^*) - \varphi_\varepsilon(0)\right\| = \\ \left\|P(\varphi_{fast}(\tau, x*, y*) - \varphi_\varepsilon(0)) - (I-P)(\varphi_{fast}(\tau, x^*, y^*) - \varphi_\varepsilon(0))\right\| = \\ \left\|(I-P)(\varphi_{fast}(\tau, x^*, y^*) - \varphi_\varepsilon(0))\right\| \leq 2\alpha\rho. \end{gathered}$$

Finally, we consider the regular perturbation in the system (Π_ε); given $\eta > 0$ there exists $\gamma > 0$ such that $\Phi_\varepsilon(-\tau, x, y) \in B_\eta(x^*, y^*)$ for every $(x, y) \in B_\gamma(\Phi_0(\tau, x^*, y^*))$

[1] A set V is invariant relative to U if orbit segments which leave V also leave U. A set V is locally invariant if it is invariant relative to some neighborhood U of V [93].

for sufficiently small ε. Choosing $\alpha < 1/2$ and ρ such that $2\alpha\rho < \gamma$, one guarantees that $\varphi_\varepsilon(0) \in B_\gamma(\Phi_0(\tau, x^*, y^*))$. As a consequence, there exists a point $p = \Phi_\varepsilon(-\tau, \varphi_\varepsilon(0)) \in W^s_\varepsilon(x_{co}, y_{co})$ that is η-close to (x^*, y^*). Since the disturbance-on trajectory has to intersect the stability boundary of the singularly perturbed system at a point on $W^s_\varepsilon(x_{co}, y_{co})$ for sufficiently small ε, (x_{co}, y_{co}) is a uniform CUEP of the singularly perturbed system. This completes the proof.

8.6 Numerical studies

The concept of relevant stability region and the theoretical results developed for DAE systems will be illustrated in the following two numerical examples.

Example 8-1 (A simple DAE system) The following dynamical system models a power system composed of a single generator and one load bus:

$$
\begin{aligned}
\dot{\omega} &= -\frac{1}{M_g} D_g \omega - \frac{1}{M_g} f(\delta, V) \\
\dot{\delta} &= -\frac{1}{D_l} f(\delta, V) + \omega \\
0 &= \frac{1}{V}(Q_l - B_{12} V \cos\delta - B_{22} V^2).
\end{aligned}
\tag{8.14}
$$

where $f(\delta, V) = B_{12} V \sin\delta - P_l$. For the following set of parameters, $M_g = 20$, $D_g = 9$, $D_l = 50$, $P_l = 4$, $Q_l = -0.5$, $B_{12} = 10$ and $B_{22} = -10$, system (8.14) possesses the stable equilibrium point (0, 0.4291, 0.9613) and the following two unstable equilibrium points (0, 1.2660, 0.4193) and (0, −5.0172, 0.4193). These equilibrium points lie on the constraint manifold $\Gamma = \{(\omega, \delta, V): g(\delta, V) = 0\}$. Figure 8.8 illustrates a stable component of this constraint manifold, which contains these three equilibrium points. Both unstable equilibrium points lie on the stability boundary of the stable equilibrium point. The stability boundary is composed of the union of the stable manifolds of these two UEPs, as indicated in Figure 8.8. The system is operating at the stable equilibrium point when it undergoes a short-circuit on the load bus.

The disturbance-on system is governed by the following equations:

$$
\begin{aligned}
\dot{\omega} &= -\frac{1}{M_g} D_g \omega - \frac{1}{M_g} f(\delta, V) \\
\dot{\delta} &= -\frac{1}{D_l} f(\delta, V) + \omega \\
0 &= V.
\end{aligned}
\tag{8.15}
$$

We consider the case that the post-disturbance system is equal to the pre-disturbance system. The constrained manifold of the disturbance-on system is the plane $\omega - \delta$. The disturbance-on trajectory instantaneously jumps on this manifold and evolves on it. Figure 8.9 shows the disturbance-on trajectory of this system.

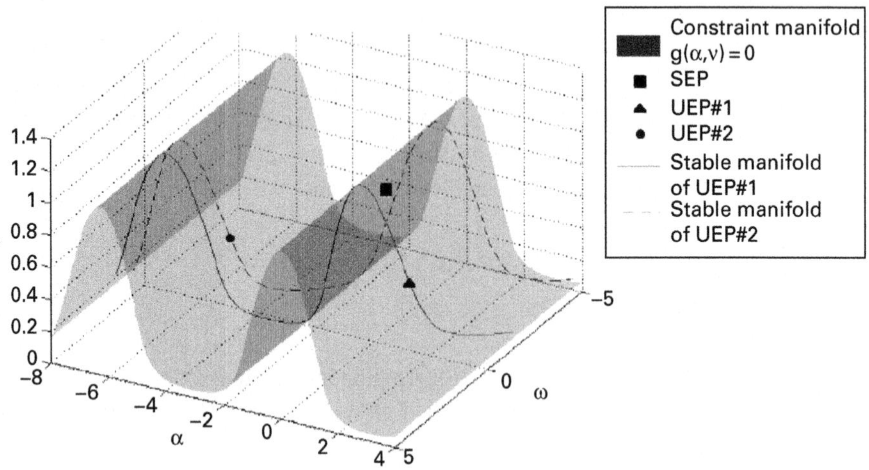

Figure 8.8 The constraint manifold of the DAE system (8.14). Both unstable equilibrium points (0, 1.2660, 0.4193) and (0, −5.0172, 0.4193), marked in black, lie on the stability boundary of the stable equilibrium point (0,0.4291,0.9613).

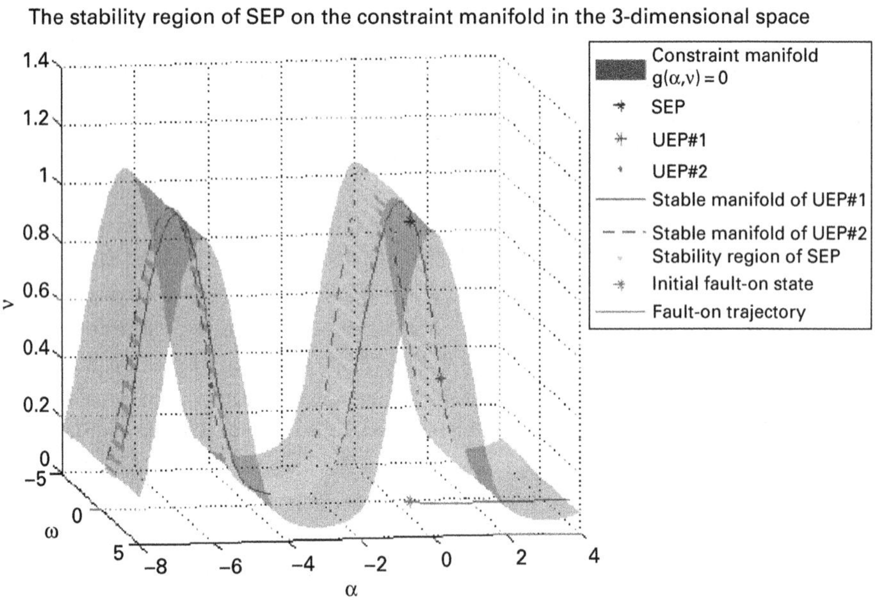

Figure 8.9 The disturbance-on trajectory of the DAE system (8.15). This trajectory evolves on the constraint manifold of the disturbance-on system, which differs from the constraint manifold of the post-disturbance system.

The projected disturbance-on trajectory of the DAE system (8.15) along the constraint manifold of the post-disturbance system (8.14) is depicted in Figure 8.10, as well as the corresponding exit point and the controlling UEP, which in this case is the UEP (0, 1.2660, 0.4193). The relevant stability region is also highlighted in this figure by a thick line. It is composed of the stable manifold of the controlling UEP.

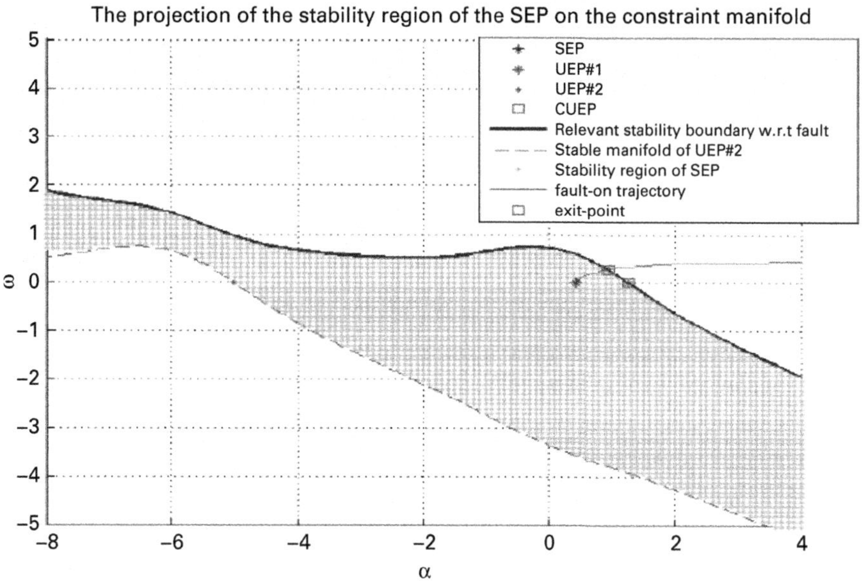

Figure 8.10 The projected disturbance-on trajectory of the DAE system (8.15) evolves on the constraint manifold of the post-disturbance system (8.14).

Example 8-2 (A higher dimension DAE system) We consider the following DAE system which represents the 9-bus power system model with three generators of Figure 8.4. This DAE system is composed of a set of six-dimensional differential equations in addition to a set of 18-dimensional nonlinear constrained equations:

$$
\begin{aligned}
\dot{\delta}_1 &= \omega_1 \\
\dot{\delta}_2 &= \omega_2 \\
\dot{\delta}_3 &= \omega_3 \\
m_1\dot{\omega}_1 &= -d_1\omega_1 + P_{m_1} - P_{G_1}(\delta_1, \theta_1, V_1) \\
m_2\dot{\omega}_2 &= -d_2\omega_2 + P_{m_2} - P_{G_2}(\delta_2, \theta_2, V_2) \\
m_3\dot{\omega}_3 &= -d_3\omega_3 + P_{m_3} - P_{G_3}(\delta_3, \theta_3, V_3) \\
0 &= \sum_{k=1} V_i V_k (G_{ik}\cos(\theta_i - \theta_k) + B_{ik}\sin(\theta_i - \theta_k)) - P_{Gi}(\delta_i, \theta_i, V_i) \\
0 &= \sum_{k=1}^{9} V_j V_k (G_{jk}\cos(\theta_j - \theta_k) + B_{jk}\sin(\theta_j - \theta_k)) \\
0 &= \sum_{k=1}^{9} V_i V_k (G_{ik}\sin(\theta_i - \theta_k) - B_{ik}\cos(\theta_i - \theta_k)) - Q_{Gi}(\delta_i, \theta_i, V_i) \\
0 &= \sum V_j V_k (G_{jk}\sin(\theta_j - \theta_k) - B_{jk}\cos(\theta_j - \theta_k))
\end{aligned}
\tag{8.16}
$$

for $i = 1, \ldots, 3$ and $j = 4, \ldots, 9$

Where

$$P_{G_i}(\delta,\theta,V)=\begin{cases}\dfrac{E_iV_i\sin(\delta_i-\theta_i)}{X'_{di}} & \text{if } i=1,\ldots,3\\ 0 & \text{if } i=4,\ldots,9\end{cases},$$

$$Q_{G_i}(\delta,\theta,V)=\begin{cases}-\dfrac{V_i^2}{X'_{di}}+\dfrac{E_iV_i\sin(\delta_i-\theta_i)}{X'_{di}} & \text{if } i=1,\ldots,3\\ 0 & \text{if } i=4,\ldots,9.\end{cases}$$

We consider a normal loading condition with a uniform damping factor $d_i/m_i = 0.1$ with (d_1, d_2, d_3)= (0.0125, 0.0034, 0.0016). The mechanical power input and the generator internal voltage of each generator are $P_{m1} = 0.8980$, $P_{m2} = 1.3432$, $P_{m3} = 0.9419$, $E_1 = 1.1083$, $E_2 = 1.1071$, $E_3 = 1.0606$, and $X_{d1}' = 0.0608$, $X_{d2}' = 0.1198$, $X_{d3}' = 0.1813$.

We consider a disturbance-on system which is a short-circuit near bus 8 on the line between bus 8 and bus 7. The post-disturbance system is obtained by tripping the transmission line between bus 7 and bus 8. The system admittance matrices (G and B) of the pre-disturbance system are summarized in Figure 8.11 and Figure 8.12. Matrices G and B of the post-disturbance system are identical to those of the pre-disturbance system, except matrix components G_{77}, G_{78}, G_{87}, G_{88}, B_{77}, B_{78}, B_{87} and B_{88}. Only these components are different because line 8–7 is tripped in the post-disturbance system to clear the disturbance. The new matrix components are $\hat{G}_{77} = 1.1876$, $\hat{G}_{78} = 0$, $\hat{G}_{87} = 0$, $\hat{G}_{88} = 1.9976$, $\hat{B}_{77} = -21.8221$, $\hat{B}_{78} = 0$, $\hat{B}_{87} = 0$ and $\hat{B}_{88} = -9.9746$.

$$\begin{bmatrix}0&0&0&0&0&0&0&0&0\\0&0&0&0&0&0&0&0&0\\0&0&0&0&0&0&0&0&0\\0&0&0&3,3074&-1,3652&-1,9422&0&0&0\\0&0&0&-1,3652&3,6411&0&-1,1876&0&0\\0&0&0&-1,9422&0&3,9895&0&0&-1,2820\\0&0&0&0&-1,1876&0&2,8047&-1,6171&0\\0&0&0&0&0&0&-1,6171&3,6147&-1,1551\\0&0&0&0&0&1,2820&0&-1,1551&2,4371\end{bmatrix}$$

Figure 8.11 The G matrix of the full admittance matrices of pre-disturbance system.

$$\begin{bmatrix}-17,3611&0&0&17,3611&0&0&0&0&0\\0&-16&0&0&0&0&16&0&0\\0&0&-17,0648&0&0&0&0&0&17,0648\\17,3611&0&0&-39,3089&11,6041&10,5107&0&0&0\\0&0&0&11,6041&-17,7735&0&5,9751&0&0\\0&0&0&10,5107&0&-16,0960&0&0&5,5882\\0&16&0&0&5,9751&0&-35,4456&13,6980&0\\0&0&0&0&0&0&13,6980&-23,5981&9,7843\\0&0&17,0648&0&0&5,5882&0&9,7843&-32,1539\end{bmatrix}$$

Figure 8.12 The B matrix of the full admittance matrices of pre-disturbance system.

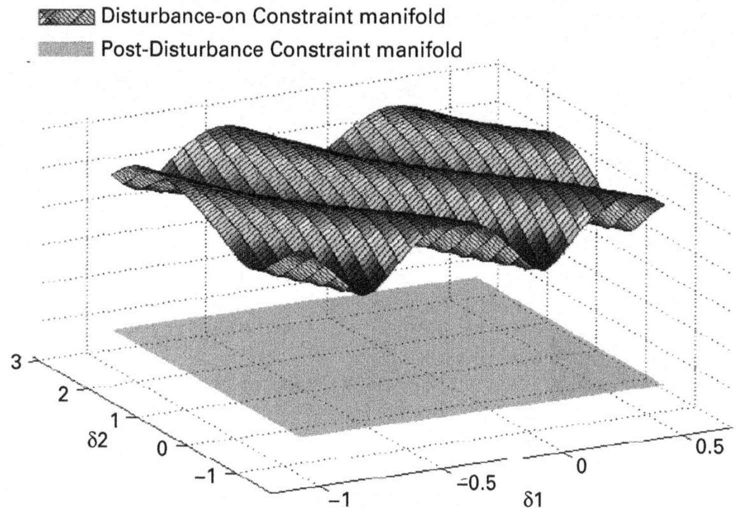

Figure 8.13 Constraint manifolds of disturbance-on and post-disturbance systems of DAE system (8.16).

The dynamical behaviors of the system during the disturbance-on period and the post-disturbance period are as follows.

[1] The constraint manifold of the disturbance-on system and that of the post-disturbance system are illustrated in Figure 8.13.

[2] At time 0.0 second, the system encounters a short circuit on the line 7–8 near bus 8. The system enters the disturbance-on period and the disturbance-on trajectory jumps from the pre-disturbance SEP onto the disturbance-on constraint manifold.

[3] During the disturbance-on period (i.e. before the disturbance is cleared), the disturbance-on trajectory moves on the disturbance-on constraint manifold (see Figure 8.14).

[4] The disturbance-on trajectory intersects the relevant stability boundary, which is the stable manifold of the CUEP. If the disturbance is cleared before the disturbance-on trajectory intersects the relevant stability boundary (i.e. the disturbance-on trajectory still lies inside the stability region of the post-disturbance system), then the disturbance-on trajectory will jump onto the post-disturbance constraint manifold and the corresponding post-disturbance trajectory lies inside the stability region of the post-disturbance and will converge to the SEP of the post-disturbance system (see Figure 8.15).

[5] If the disturbance is cleared after intersecting the relevant stability boundary, then the corresponding post-disturbance trajectory lies outside the stability region of the post-disturbance SEP and, hence it will diverge from the SEP (see Figure 8.15).

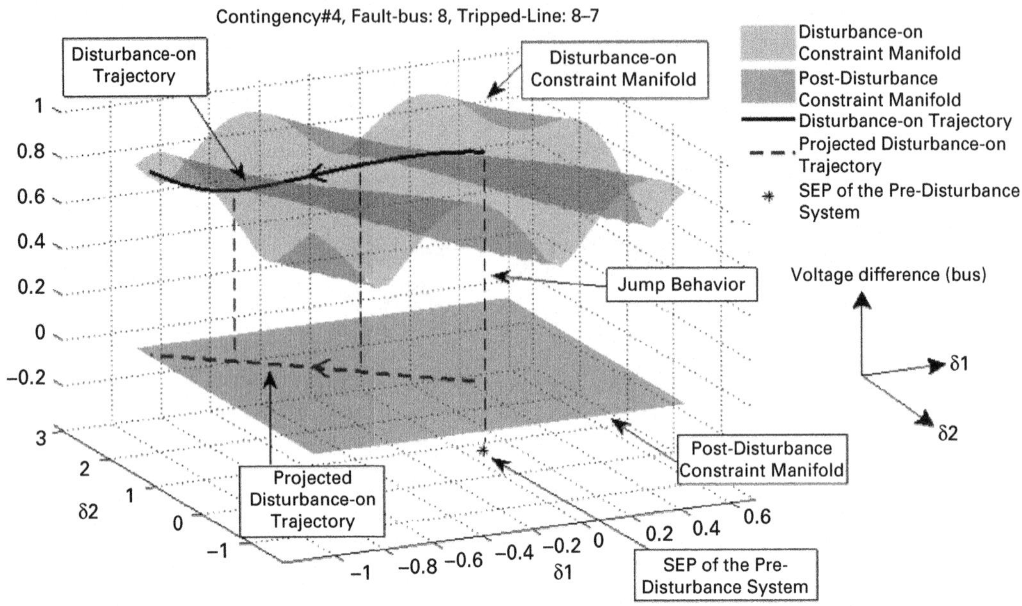

Figure 8.14 Several dynamic objects are shown: (1) projected disturbance-on trajectory, (2) SEP of the pre-disturbance system, (3) disturbance-on constraint manifold, (4) post-disturbance constraint manifold, (5) jump behavior, (6) disturbance-on trajectory of system (8.16).

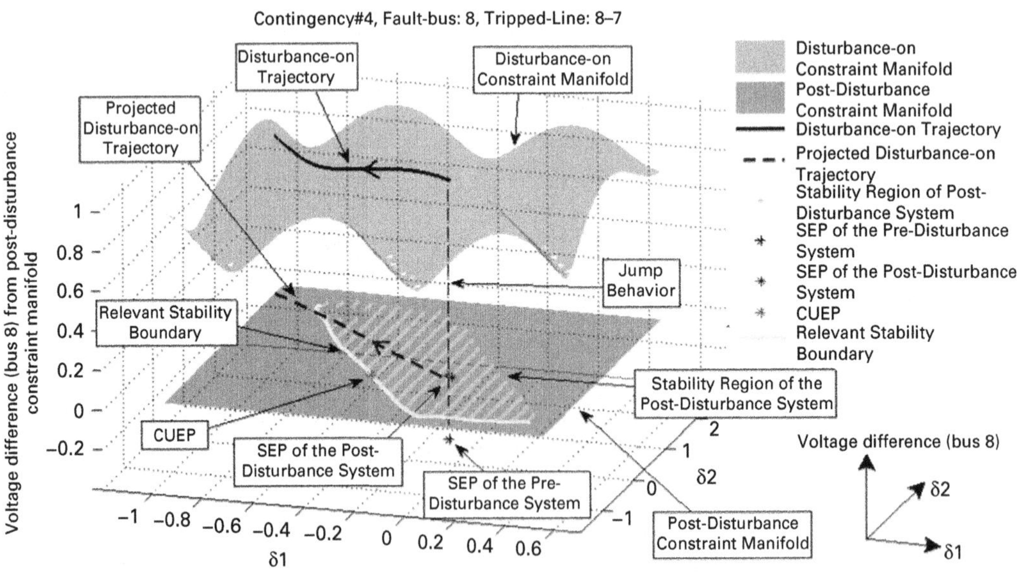

Figure 8.15 Several dynamic objects are shown: (1) controlling UEP relative to a disturbance-on trajectory, (2) stability region of the post-disturbance system, (3) disturbance-on constraint manifold, (4) post-disturbance constraint manifold, (5) jump behavior, (6) disturbance-on trajectory, (7) projected disturbance-on trajectory, (8) SEP of the pre-disturbance system, (9) SEP of the post-disturbance system, (10) relevant stability boundary of system (8.16).

8.7 Concluding remarks

This chapter has presented a rigorous introduction to the concept of the controlling UEP, the controlling limit cycle and the relevant stability boundary. The existence and uniqueness of the CUEP have been shown. A controlling UEP method has been developed along with its theoretical foundations. Some dynamic and geometric characterizations of controlling UEPs have also been derived. These characterizations are useful for the development of solution methodologies for computing controlling UEPs and estimating the relevant stability boundary.

The concept of a controlling UEP is well defined for nonlinear systems described by ODEs. One key difficulty in extending the concept of the controlling UEP of ODE systems to DAE systems is that the final state of a disturbance-on trajectory does not lie on the constrained manifold of its post-disturbance DAE system. Two approaches have been developed to overcome this difficulty. One approach is to exploit the singularly perturbed system associated with the DAE system and define the controlling UEP of this DAE system via the controlling UEP of the associated singularly perturbed system. With this approach, the disturbance-on trajectory always intersects the stability boundary of the corresponding singularly perturbed post-disturbance system and hence the controlling UEP of the singularly perturbed system is well defined. Another approach to define the controlling UEP is via the UEP of the post-disturbance DAE system whose stable manifold intersects the projected disturbance-on DAE trajectory at the exit point. We have shown that, under very mild conditions, the two definitions of controlling UEP for DAE systems are equivalent.

A comprehensive theory of controlling UEP method has been developed for nonlinear DAE systems. The concept of controlling UEP will be further explored in Chapter 14 to develop methods for optimally estimating the relevant stability boundary via energy functions.

9 Stability regions of discrete dynamical systems

A comprehensive theory of stability regions and of stability boundaries for continuous dynamical systems was presented in Chapters 4, 5 and 6. In this chapter, a comprehensive theory for both the stability boundary and the stability regions of nonlinear discrete dynamical systems will be developed.

The analytical results to be presented in this chapter can be viewed as the counterpart of the results derived in Chapter 4 for discrete systems. Although the theory of stability regions of discrete systems is parallel to the theory of continuous dynamical systems, discrete dynamical systems do possess some peculiarities. One fundamental difference is the connectivity of trajectories of continuous systems that do not exist in discrete systems. Another important difference is that backward trajectories are not defined or are not unique for many discrete dynamical systems. These peculiarities pose some challenges for the development of a theory of stability regions for discrete dynamical systems and justify an entire chapter devoted to this subject.

9.1 Introduction

Nonlinear discrete dynamical systems have been used to model a variety of practical nonlinear systems. Stability analysis and stability region characterizations are essential in many of these applications. For instance, the dynamics of power systems with LTCs (on Load Tap Changers) are modeled by difference equations [269], iterated-map neural networks are another example of discrete-time dynamical systems [174], discrete-time dynamical systems are used to study the dynamics of ecosystems [272] and economic models [283].

Discrete dynamical models are important in the stability analysis of sample-data systems. These models typically appear in the analysis of systems that are controlled by a computer [155]. Although the majority of physical systems have variables that continuously evolve in time, the employment of digital computers to simulate, control and interact with continuous systems has stimulated the development of stability theory for discrete systems [14,130,157,159]. Discrete dynamical models also appear in the analysis of systems that cannot be continuously measured. An application of discrete-time approximate models in the analysis of HIV dynamics and treatment schedules is presented in [87]. Although the physical variables of this system evolve continuously in time, measurements, blood examinations, are made periodically at discrete times. Also, it is impractical to employ continuous varying doses of medication during the treatment. Hence a discrete model is more suitable for the analysis of these systems.

The task of determining stability regions of nonlinear discrete dynamical systems is of fundamental importance. Knowledge of the stability boundary can lead to the development of more efficient and less conservative methods for estimating stability regions. Knowledge of the stability region is also important in controlled discrete systems [159]. Networked control systems, for example, can be modeled and analyzed by approximate discrete-time nonlinear models [72]. The control of engines with fuel injection is another application of nonlinear control of discrete-time dynamical systems [104].

In this chapter, a comprehensive theory for both the stability boundary and the stability regions of nonlinear discrete dynamical systems will be developed. In particular, theoretical developments of the following topics will be presented.

Topological properties and global behaviors:

(1) topological properties of stability regions of nonlinear discrete dynamical systems,
(2) global behavior of discrete-time trajectories,
(3) global behavior of discrete-time trajectories on the stability boundary.

Characterizations of the stability boundary:

(1) local characterizations of the stability boundary,
(2) complete characterizations of the stability boundary,
(3) sufficient conditions for an unbounded stability boundary.

A complete characterization of the stability boundary for a fairly large class of nonlinear discrete dynamical systems that admit energy functions will be developed. Energy functions for discrete system will be further explored in Chapter 12 in the development of algorithms to obtain optimal estimates of the stability region of discrete systems.

9.2 Discrete dynamical systems

Consider the following class of autonomous nonlinear discrete dynamical systems:

$$x_{k+1} = f(x_k) \tag{9.1}$$

where $k \in Z$, $x_k \in R^n$ and $f: R^n \rightarrow R^n$ is a vector-valued map. The solution of (9.1) starting from $x_0 \in R^n$ at $k = 0$, denoted by $\phi(x_0, \cdot): Z \rightarrow R_n$, is called an orbit (or trajectory) of (9.1). The solution of the discrete dynamical system (9.1) is an infinite sequence x_k that can be obtained by successive applications of the map f, i.e. $x_k = \phi(x_0, k) = f^k(x_0)$.

The existence and uniqueness of solutions of system (9.1) for times greater than zero ($k > 0$) is not an issue for discrete systems. If the map f is a well-defined function on R^n, then the solution starting at x_0 is defined in Z_+ and can easily be obtained by successive application of the map. On the other hand, solutions may not exist for negative times ($k < 0$), and when they exist the solution may not be unique. However, if function f is invertible, then solutions of (9.1) exist and are defined in Z.

A point x^* is a periodic point of period p if $f^p(x^*) = x^*$ and $f^k(x^*) \neq x^*$ for every k satisfying $0 < k < p$. If x^* is a periodic point of period p, the sequence $\gamma = \{x^*, f(x^*), \ldots, f^{p-1}(x^*)\}$ is a periodic orbit or closed orbit of system (9.1). If x^* has period one, i.e. $f(x^*) = x^*$, then x^* is called a fixed point of system (9.1). A state vector x is called a regular point if it is not a fixed point.

A fixed point x^* is stable if, for each $\varepsilon > 0$, there is $\delta = \delta(\varepsilon) > 0$ such that $\|x_0 - x^*\| < \delta$ implies $\|x_k - x^*\| < \varepsilon$, $\forall k \in Z_+$; it is asymptotically stable if it is stable and δ can be chosen such that $\|x_0 - x^*\| < \delta$ implies $\lim_{k\to\infty} x_k = x*$. A fixed point x^* is called unstable if it is not stable. A periodic point x^* of period p is (asymptotically) stable if x^* is an (asymptotically) stable fixed point of the map $f^p(\cdot)$.

When function f is continuous and differentiable, we say that a fixed point of (9.1) is hyperbolic if the Jacobian matrix at x^*, denoted by $Df(x^*)$, has no eigenvalues with modulus one. A hyperbolic fixed point is asymptotically stable if all the eigenvalues of its corresponding Jacobian have modulus less than one, while a hyperbolic fixed point is unstable if at least one eigenvalue of the Jacobian has modulus greater than one. A periodic or closed orbit γ of period p is hyperbolic if and only if every point of the orbit is a hyperbolic fixed point of f^p [228].

A set M is positively invariant with respect to the discrete system (9.1) if $f(M) \subset M$, which implies that every orbit x_k starting in M remains in M for all $k \geq 0$. *A* set M is invariant if $f(M) = M$ and negatively invariant if $f^{-1}(M) \subset M$. Solutions of discrete systems might be not defined in Z_-; however, if M is invariant, and x_0 is an initial condition in M, then there exists one solution x_k of system (9.1) satisfying the initial condition x_0, defined on Z, such that $x_k \in M$ for all $k \in Z$. In spite of this, and different from continuous systems, there may exist solutions starting outside M that enter M in finite time. Nevertheless, if f is injective, then invariance of M implies negative invariance. In this case, solutions starting outside of M cannot enter M in finite time.

Lemmas 9.1 and 9.2 establish a relationship between the invariance of a set with the invariance of its complement and its closure.

LEMMA 9-1 *If M is a positively invariant set with respect to the discrete system (9.1), then its complement $R^n - M$ is negatively invariant. Moreover, if f is a continuous function then:*

(a) the closure $\overline{M}$ is a positively invariant set;
(b) the complement of the closure $R^n - \overline{M}$ is a negatively invariant set.

LEMMA 9-2 *If M is an invariant set with respect to the discrete system (9.1) and f is a homeomorphism, then its closure $\overline{M}$ is also invariant.*

A point p is said to be in the ω-limit set (or α-limit set) of x_0 if for any given $\varepsilon > 0$ and $N > 0$ (or $N < 0$), there exists a $k > N$ (or $k < N$) such that $\|x_k - p\| < \varepsilon$. This is equivalent to the condition that there is a sequence $k_i \in Z$ with $k_i \to \infty$ (or $k_i \to -\infty$) as $i \to \infty$ such that $p = \lim_{i\to\infty} x_{k_i}$. The ω-limit set is closed and positively invariant. If function f is invertible, then the ω-limit set is closed and invariant. If f is continuous and the orbit $\{x_k\}$ is also bounded for $k > 0$, then the ω-limit set is non-empty, compact and invariant. Moreover,

$$x_k = f^k(x) \to \omega(x) \text{ as } k \to \infty.$$

In comparison with continuous dynamical systems, the ω-limit set of bounded orbits of discrete-time dynamical systems is usually not connected. This difference comes from the fact that orbits of discrete systems are not connected, while orbits of continuous systems are always connected. For instance, the α-limit and ω-limit of a periodic point of period p is the periodic orbit, which is formed of the union of p isolated points. Clearly, this is an example of a limit set that is not path connected.

The concept of invariantly connected sets relies on the dynamics to fix this problem, providing a concept of connectedness that is the counterpart of path connectedness in the continuous case. A closed and invariant set M is *invariantly connected* if it is not the union of two non-empty disjoint closed invariant sets [124,156,205,277]. If f is continuous and the orbit $\{x_k\}$ is bounded for $k > 0$, then the ω-limit set is also invariantly connected.

The hyperbolic fixed point x^* is called a *type-k fixed point* if the Jacobian $Df(x^*)$ has exactly k eigenvalues with modulus greater than one. The set of points whose ω-limit set is the fixed point x^* is called the stable set of x^* and is denoted $W^s(x^*)$. Similarly, the unstable set $W^u(x^*)$ is the set of points whose α-limit set is the fixed point x^*. Under the condition that the map f is a diffeomorphism, the stable and unstable sets have the structure of a manifold, similar to the case of stable and unstable manifolds of continuous dynamical systems, see Section 2.4 of Chapter 2. The function f is a diffeomorphism if it is differentiable and invertible and its inverse f^{-1} is differentiable. If f is a C^r-diffeomorphism, with $r \geq 1$, and x^* is a hyperbolic fixed point of (9.1), then the tangent space at x^*, $T_{x^*}(R^n)$, can be uniquely decomposed as a direct sum of two subspaces denoted by E^s and E^u, which are invariant with respect to the linear operator $Df(x^*)$:

$$E^s = \text{span}\{e_1, \ldots, e_s\},$$
$$E^u = \text{span}\{e_{s+1}, \ldots e_{s+u}\},$$

where $\{e_1, \ldots, e_s\}$ and $\{e_{s+1}, \ldots, e_{s+u}\}$ are the generalized eigenvectors of $Df(x^*)$ respectively associated with the eigenvalues of $Df(x^*)$ that have modulus less and greater than one. Vector subspaces E^s and E^u are respectively called stable and unstable subspaces or eigenspaces.

There are local manifolds $W^s_{loc}(x^*)$ and $W^u_{loc}(x^*)$ of class C^r, invariant with respect to (9.1) [123] that are tangent to E^s and E^u at x^*, respectively. $W^s_{loc}(x^*)$ and $W^u_{loc}(x^*)$ are respectively termed local stable and unstable manifolds. These local stable and unstable manifolds are unique. Every orbit x_k starting in $W^s_{loc}(x^*)$ approaches x^* as $k \to \infty$, while every orbit starting in $W^u_{loc}(x^*)$ approaches x^* as $k \to -\infty$.

The local unstable manifold can be extended via dynamics of system (9.1) to form the (global) unstable manifold:

$$W^u(x^*) = \bigcup_{k \geq 0} \phi(W^u_{loc}(x^*), k) = \bigcup_{k \geq 0} f^k(W^u_{loc}(x^*)).$$

The (global) stable manifold can also be obtained via backward iterations of system (9.1) to form the (global) stable manifold:

$$W^s(x^*) = \bigcup_{k\leq 0} \phi(W^s_{loc}(x^*), k) = \bigcup_{k\leq 0} f^k(W^s_{loc}(x^*))$$

The manifolds $W^s(x^*)$ and $W^u(x^*)$ are of class C^r. If x^* is a type-k fixed point, then the dimension of $W^u(x^*)$ is k and the dimension of $W^s(x^*)$ is $n-k$.

9.3 Stability regions and topological properties

The stability region of an asymptotically stable fixed point x_s of (9.1) is defined as

$$A(x_s) := \{x \in R^n : lim_{k\to\infty} f^k(x) = x_s\}.$$

Similar to the continuous case, the stability region of a discrete dynamical system is composed of a set of points whose trajectories approach the asymptotically stable fixed point as the discrete time k tends to infinity. In the following, we derive several topological characterizations of stability regions and their boundaries. We will start with a very general characterization and gradually move into classes of discrete vector fields to gain more refined topological properties of the stability region and the stability boundary.

PROPOSITION 9.3 **(Topological characterization 1)**
The stability region $A(x_s)$ of an asymptotically stable fixed point x_s of (9.1) is positively invariant and negatively invariant. The stability boundary $\partial A(x_s)$ is a closed set.

Proof In order to show that $A(x_s)$ is positively invariant, let $x \in A(x_s)$ be an arbitrary point in the stability region. We need to show that $x \in A(x_s) \Rightarrow f(x) \in A(x_s)$. Since $f^k(f(x)) = f^{k+1}(x)$, it is clear that $\lim_{k\to\infty} f^k(f(x)) = x_s$. Hence $f(x) \in A(x_s)$ and, as a consequence, $A(x_s)$ is a positively invariant set.

Let $y \in f^{-1}(x)$. Then $f^k(y) = f^{k-1}(x)$. Consequently, $f^k(y) \to x_s$ as $k \to \infty$. Therefore every point y that belongs to the inverse image of a point on the stability region belongs to the stability region. So $f^{-1}(A(x_s)) \subset A(x_s)$ and $A(x_s)$ is negatively invariant. Since the stability boundary $\partial A(x_s) = \overline{A(x_s)} \cap \overline{\{R^n - A(x_s)\}}$ and the intersection of two closed sets is closed, we conclude the stability boundary is a closed set. This completes the proof.

Proposition 9.3 provides general topological information regarding the stability region and stability boundary without imposing any condition on the map f. As we impose conditions on the map f, refined results regarding the topological properties of these sets can be obtained. The next proposition assumes that the map f is a continuous function.

PROPOSITION 9.4 **(Topological characterization 2)**
Let x_s be an asymptotically stable fixed point of (9.1) and suppose f is a continuous function. The stability region $A(x_s)$ is an open, positively and negatively invariant set. The stability boundary $\partial A(x_s)$ is a closed and positively invariant set formed by forward orbits. Moreover, the stability boundary $\partial A(x_s)$ is of dimension less than n and if $A(x_s)$ is not dense in R^n, then $\partial A(x_s)$ is of dimension $n-1$.

Proof In order to show that $A(x_s)$ is an open set, let p be an arbitrary point in $A(x_s)$. We will show that every point in a neighborhood of p belongs to $A(x_s)$. To that end, let $\varepsilon > 0$ be sufficiently small such that the set $\{x \in R^n : \|x - x_s\| < \varepsilon\}$ is contained in $A(x_s)$. This number ε always exists according to the definition of an asymptotically stable fixed point. Let $N > 0$ be large enough such that $\|f^N(p) - x_s\| < \varepsilon/2$.

Since f^N is a continuous function, we can choose δ small enough to ensure that for any point q in the neighborhood $\{x \in R^n : \|x - p\| < \delta\}$ of p, $\|f^N(q) - f^N(p)\| < \varepsilon/2$. Hence, $\|f^N(q) - x_s\| \leq \|f^N(q) - f^N(p)\| + \|f^N(p) - x_s\| < \varepsilon$. This shows that the point $f^N(q)$ is inside $A(x_s)$ and the set $A(x_s)$ is open. Invariance of the stability region trivially follows from Proposition 9.3. From Lemma 9.1, it follows that $\overline{A(x_s)}$ is positively invariant, i.e. $f(\overline{A(x_s)}) \subset \overline{A(x_s)}$. In order to show that $\partial A(x_s)$ is positively invariant, we must show, for every $x \in \partial A(x_s)$, that $f(x) \in \partial A(x_s)$. Suppose on the contrary that $f(x) \notin \partial A(x_s)$, then the positive invariance of $\overline{A(x_s)}$ implies $f(x) \in A(x_s)$. Since $x \in f^{-1}(A(x_s))$ and $A(x_s)$ is a negatively invariant set, we conclude that $x \in A(x_s)$. This leads to a contradiction because $x \in \partial A(x_s)$ and $A(x_s)$ is an open set. Therefore, $f(x) \in \partial A(x_s)$, and this shows that $\partial A(x_s)$ is a positively invariant set and the stability boundary $\partial A(x_s)$ is formed by forward trajectories of the points lying in $\partial A(x_s)$. We have already shown in Proposition 9.3 that the stability boundary is a closed set. The dimension of $\partial A(x_s)$ is a direct consequence of the results in [131]. This completes the proof.

The previous proposition asserts that the existence of at least two asymptotically stable fixed points is sufficient to guarantee that the dimension of the stability boundary of each stable fixed point is $n-1$. In the next proposition, the map f is assumed to be surjective. The stability region of surjective maps is shown to be invariant.

PROPOSITION 9.5 **(Topological characterization 3)**
Let x_s be an asymptotically stable fixed point of system (9.1) and suppose f is surjective, then the stability region $A(x_s)$ is an invariant set.

Proof From Proposition 9.3, the stability region is positively invariant, i.e. $f(A(x_s)) \subset A(x_s)$. In order to complete the proof, we have only to show that $A(x_s) \subset f(A(x_s))$. Let x be an arbitrary point in $A(x_s)$. Since f is surjective, there exists a y such that $x = f(y)$. Moreover, y belongs to $A(x_s)$. As a consequence, $x \in f(A(x_s))$ and the proof is complete.

Another topological property of the stability region of fixed points is that it may not be path connected. The following example illustrates a one-dimensional discrete system that contains a fixed point with a stability region that is not path connected.

Example 9-1 The following discrete system

$$x_{k+1} = x_k - 0.8\,x_k(x_k - 1)(x_k - 2)(x_k - 3) \tag{9.2}$$

possesses four fixed points, $x^1 = 0$, $x^2 = 1$, $x^3 = 2$ and $x^4 = 3$. These fixed points are indicated at the intersections of a graph of the discrete map and the straight line $x_{k+1} = x_k$. The fixed point $x^2 = 1$ is an asymptotically stable fixed point. A trajectory starting at $x_0 = 0.5$ and converging to the asymptotically stable fixed point x^2 is depicted in

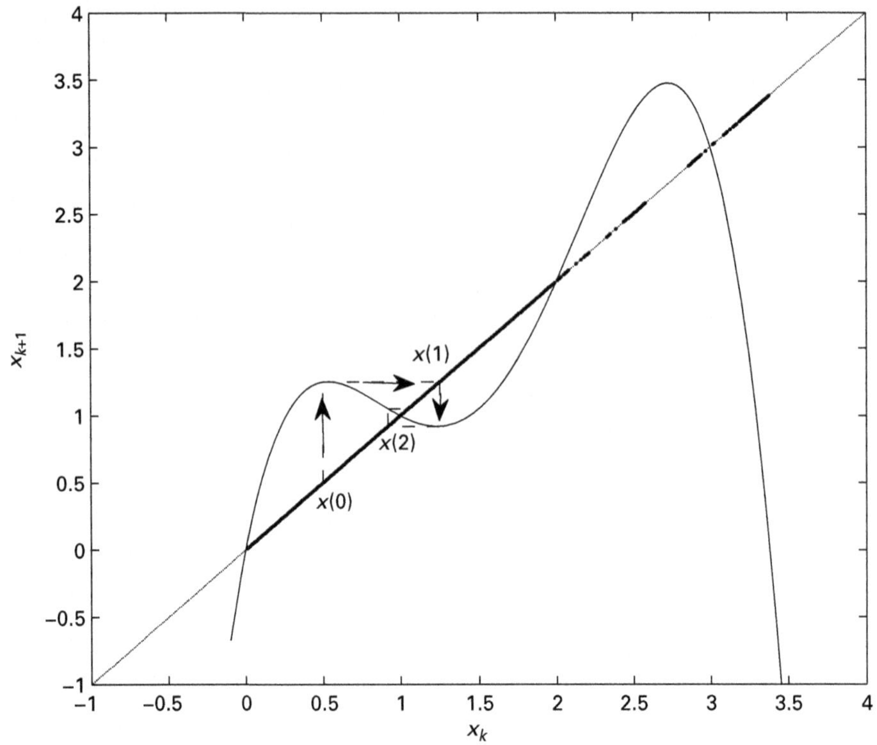

Figure 9.1 The stability region of system (9.2) is indicated by a thick diagonal line. The stability region is not a path connected set and it is not invariant.

Figure 9.1. Its stability region is highlighted by a thick black line on the diagonal line. The "shape" of this stability region is quite complex, even for this one-dimensional dynamical system. In particular, the stability region is not a connected set and it is not invariant.

However, if f is invertible and the inverse is a continuous function, then the stability region is path connected, as shown in Proposition 9.6.

PROPOSITION 9.6 **(Topological characterization 4)**
Let x_s be an asymptotically stable fixed point of the discrete system (9.1). If function f is invertible and its inverse is a continuous function, then the stability region is path connected.

Proof Since x_s is an asymptotically stable fixed point, there exists a path connected neighborhood U of x_s that is entirely contained in $A(x_s)$. The stability region can be written as $A(x_s) = \bigcup_{k>0} f^{-k}(U)$. Since f^{-k} is a continuous function and U is path connected, the set $f^{-k}(U)$ is also path connected. Moreover, set $f^{-k}(U)$ contains the fixed point x_s for every $k > 0$. Since arbitrary unions of open path connected sets sharing a common point are path connected, we conclude that the stability region is a path connected set. This completes the proof.

If function f is a homeomorphism, i.e. f is continuous, invertible and the inverse is also continuous, then all properties of the stability region and stability boundary that have been shown in Propositions 9.3–9.6 are valid. The next theorem asserts these properties and shows that the stability boundary is also an invariant set.

THEOREM 9-7 **(Topological characterization 5)**
Let x_s be an asymptotically stable fixed point of the discrete system (9.1). If function f is a homeomorphism, then the stability region $A(x_s)$ has the following topological and dynamic properties; it is

(a) open,
(b) positively and negatively invariant,
(c) invariant, and
(d) path connected;
and the stability boundary $\partial A(x_s)$ is:
(a) closed, and
(b) invariant.

Proof All the four properties of the stability region and property (a) of the stability boundary are a direct consequence of Propositions 9.3–9.6. The invariance of the stability boundary comes from the fact that the closure of an invariant set is also invariant (see Lemma 9.2). This completes the proof.

9.4 Characterization of stability regions

Our aim in this section is to present a comprehensive characterization of stability regions for the nonlinear discrete dynamical system (9.1). Our approach starts from a local characterization of the stability boundary and progresses towards a global characterization of the stability boundary.

We first derive a complete characterization for a fixed point lying on the stability boundary, which is a key step in the characterization of the stability region $A(x_s)$. We do this in two steps. First we impose only one generic assumption, namely, that fixed points are hyperbolic, and derive characterizations for a fixed point that lies on the stability boundary. These characterizations are expressed in terms of both the stable and unstable manifolds. Additional conditions are then imposed on the discrete dynamical system and the results are further sharpened.

Let x be a hyperbolic fixed point. Let U be a neighborhood of x in $W^s(x)$ whose boundary ∂U is transversal to the vector field f. We call ∂U a *fundamental domain* of $W^s(x)$. Any neighborhood $V \subset R^n$ of a fundamental domain ∂U that is disjoint from $W^u(x)$ is called a *fundamental neighborhood* for $W^s(x)$. Similarly we define a fundamental domain and a fundamental neighborhood for the unstable manifold $W^u(x)$.

If the vector field of the nonlinear discrete system (9.1) is a diffeomorphism, then we have already shown that the stability region is a path connected, open and invariant set and the stability boundary is a closed and invariant set. The following two theorems characterize all the fixed points lying on the stability boundary and prepare the terrain for developing a complete characterization of the stability boundary in terms of the invariant manifolds of the fixed points that lie on the stability boundary.

THEOREM 9-8 **(Fixed points on the stability boundary)**
Let $A(x_s)$ be the stability region of an asymptotically stable fixed point x_s of (9.1). Let $\hat{x} \neq x_s$ be a hyperbolic fixed point of system (9.1) and suppose that f is a diffeomorphism. Then,

(a) $\hat{x} \in \partial A(x_s)$ if and only if $\{W^u(\hat{x}) - \hat{x}\} \cap \bar{A}(x_s) \neq \varnothing$;
(b) if $\hat{x}$ is not a source, then $\hat{x} \in \partial A(x_s)$ if and only if $\{W^s(\hat{x}) - \hat{x}\} \cap \partial A(x_s) \neq \varnothing$.

Proof (a) We first prove the if condition. Suppose $\{W^u(\hat{x}) - \hat{x}\} \cap \bar{A}(x_s) \neq \varnothing$. Then, there exists $y \in \{W^u(\hat{x}) - \hat{x}\} \cap \bar{A}(x_s)$. Moreover $\lim_{n\to\infty} f^{-n}(y) = \hat{x}$. Since $\bar{A}(x_s)$ is an invariant set, it follows $f^n(y) \in \bar{A}(x_s)$ for all $n \in Z$. Consequently, $\hat{x} \in \bar{A}(x_s)$. Since $\hat{x} \notin A(x_s)$, one has $\hat{x} \in \partial A(x_s)$.

Next, we prove the only if condition. Suppose $\hat{x} \in \partial A(x_s)$. Then every neighborhood U of $\hat{x}$ has a non-empty intersection with the stability region $A(x_s)$. Let D be a fundamental domain of $W^u(\hat{x})$ and D_ε be an ε-neighborhood of D. According to the λ-lemma [201], there is a neighborhood U of $\hat{x}$ such that $\bigcup_{k>0} f^{-k}(D_\varepsilon)$ contains the set $\{U - W^s(\hat{x})\}$. Since $W^s(\hat{x}) \cap A(x_s) = \varnothing$ and $U \cap A(x_s) \neq \varnothing$, there exists a point $q \in \{U - W^s(\hat{x})\}$ such that $q \in A(x_s)$. Moreover, the positive invariance of $A(x_s)$ guarantees the existence of a point $p \in D_\varepsilon$ such that $p \in A(x_s)$. Since D_ε can be chosen arbitrarily small, we can find a sequence of points $\{p_j\}$ with $p_j \in A(x_s)$ for every $j = 1,2, \ldots,$ such that $d(p_j, D) \to 0$ as $j \to \infty$. Since the sequence $\{p_j\}$ is bounded, there exists a subsequence $\{p_{ji}\}$ that converges to p. By construction, $p \in \overline{A(x_s)} \cap \{W^u(\hat{x}) - \hat{x}\}$ and the proof of this part is completed.

(b) We prove the if condition. Now suppose $\{W^s(\hat{x}) - \hat{x}\} \cap \partial A(x_s) \neq \varnothing$. Then, there exists $y \in \partial A(x_s) \cap \{W^s(\hat{x}) - \hat{x}\}$. Moreover, $\lim_{n\to\infty} f^n(y) = \hat{x}$. Since $\partial A(x_s)$ is an invariant set, one has that $f^n(y) \in \partial A(x_s)$ for all $n \in Z$. Hence, $\hat{x} \in \partial A(x_s)$. The proof of the only if condition is very similar to that of part (a) and is hence omitted. This completes the proof.

The above characterization of a fixed point lying on the stability boundary can be extended to another critical element, i.e. closed orbit. The stable and unstable manifolds of a hyperbolic closed orbit γ are defined as follows:

$$W^s(\gamma) = \{x \in R^n : f^k(x) \to \gamma \text{ as } k \to \infty\}$$

$$W^u(\gamma) = \{x \in R^n : f^{-k}(x) \to \gamma \text{ as } k \to \infty\}.$$

A characterization of the closed orbit on the stability boundary is as follows.

THEOREM 9-9 **(Characterization of closed orbit on the stability boundary)**
Let $A(x_s)$ be the stability region of an asymptotically stable fixed point x_s of (9.1). Let γ be a hyperbolic closed orbit. Then

(a) $\gamma \subseteq \partial Ax_s$ if and only if $\{W^u(\gamma) - \gamma\} \cap \overline{A(x_s)} \neq \varnothing$;
(b) suppose $\{W^s(\gamma) - \gamma\} \neq \varnothing$, then $\gamma \subseteq \partial A(x_s)$ if and only if $\{W^s(\gamma) - \gamma\} \cap \partial A(x_s) \neq \varnothing$.

Proof The proof of Theorem 9-9 is very similar to the proof of Theorem 9-8 and will be omitted.

As a corollary to Theorem 9-8, if $\{W^u(\hat{x}) - \hat{x}\} \cap A(x_s) \neq \emptyset$, then $\hat{x}$ must be on the stability boundary. Since any orbit in $A(x_s)$ approaches x_s, a sufficient condition for $\hat{x}$ to be on the stability boundary is the existence of an orbit in $W^u(\hat{x})$ which approaches x_s. One favorable property of this sufficient condition is that it is numerically checkable. From a practical point of view, one would like to see when this sufficient condition is also necessary. We will show that this sufficient condition becomes necessary under two additional assumptions.

So far we have assumed only that the critical elements are hyperbolic. This is a generic property for dynamical systems. Roughly speaking, we say a property is generic for a class of systems if that property is true for *almost all* systems in the class. A formal definition is given in [203]. It has been shown that among C^r ($r \geq 1$) vector fields, the following properties are generic: (i) all fixed points and closed orbits are hyperbolic and (ii) the intersections of the stable and unstable manifolds of critical elements satisfy the transversality condition.

Theorem 9-8 can be sharpened under two conditions, one of which is generic for a nonlinear dynamical system (9.1). That is the transversality condition. The other condition requires that every orbit on the stability boundary approaches one of the fixed points. Hence, we consider the following assumptions.

(A1) All the fixed points on $\partial A(x_s)$ are hyperbolic.
(A2) The stable and unstable manifolds of fixed points on $\partial A(x_s)$ satisfy the transversality condition.
(A3) Every orbit on $\partial A(x_s)$ approaches one of the fixed points as $k \to \infty$.

Assumptions (A1) and (A2) are generic properties of dynamical discrete systems whose vector fields are diffeomorphisms, while assumption (A3) is not a generic property.

Now, we present the key theorem which characterizes a fixed point being on the stability boundary in terms of both its stable and unstable manifolds. From a practical point of view, this result is more useful in the numerical verification of fixed points on the stability boundary than Theorem 9-8.

THEOREM 9-10 **(Fixed points on the stability boundary)**
Let $A(x_s)$ be the stability region of an asymptotically stable fixed point x_s of system (9.1). Let $\hat{x}$ be a hyperbolic fixed point, suppose that f is a diffeomorphism and assumptions (A1)–(A3) are satisfied. Then, the following characterizations hold:

(a) $\hat{x} \in \partial A(x_s)$ if and only if $W^u(\hat{x}) \cap A(x_s) \neq \emptyset$;
(b) if $\hat{x}$ is not a source, then $\hat{x} \in \partial A(x_s)$ if and only if $W^s(\hat{x}) \subseteq \partial A(x_s)$.

Proof (a) We first prove the if condition. Suppose that $W^u(\hat{x}) \cap A(x_s) \neq \emptyset$. Since $\hat{x} \notin A(x_s)$, it follows that $\{W^u(\hat{x}) - \hat{x}\} \cap A(x_s) \neq \emptyset$. Moreover, $\{W^u(\hat{x}) - \hat{x}\} \cap A(x_s) \subset \{W^u(\hat{x}) - \hat{x}\} \cap \overline{A(x_s)}$, therefore, $\{W^u(\hat{x}) - \hat{x}\} \cap \overline{A(x_s)} \neq \emptyset$. By applying Theorem 9.8, it follows that $\hat{x} \in \partial A(x_s)$. We next prove the only if condition. Suppose $\hat{x} \in \partial A(x_s)$. Then, by Theorem 9.8, it follows that $\{W^u(\hat{x}) - \hat{x}\} \cap \overline{A(x_s)} \neq \emptyset$. We next show, under assumptions (A1)–(A3), that $\{W^u(\hat{x}) - \hat{x}\} \cap \overline{A(x_s)} \neq \emptyset$ implies $W^u(\hat{x}) \cap A(x_s) \neq \emptyset$. Let $p \in \{W^u(\hat{x}) - \hat{x}\} \cap \overline{A(x_s)}$. There are two possibilities: (1) $p \in A(x_s)$ and (2) $p \in \partial A(x_s)$. If $p \in A(x_s)$, then the result follows. Suppose that $p \in \partial A(x_s)$. From assumption (A3), there exists a fixed point $x^* \in \partial A(x_s)$ such that $f^k(p) \to x^*$ as $k \to \infty$, i.e. $p \in W^s(x^*)$. We note that both fixed points x^* and $\hat{x}$ are hyperbolic according to assumption (A1). Moreover, they are fixed points of type k with $k \geq 1$. The point p is a point of intersection between the manifolds $W^u(\hat{x})$ and $W^s(x^*)$ which, according to assumption (A2), is a point of transversal intersection.

Suppose that $\hat{x}$ is a hyperbolic fixed point of type one. Then, according to the transversality intersection, x^* has to be a fixed point of type zero. This leads to a contradiction since x^* must be a fixed point of type $k \geq 1$. Hence, it follows that $W^u(\hat{x}) \cap A(x_s) \neq \emptyset$ for all fixed points of type one on the stability boundary.

The proof will be completed by induction. Suppose that $W^u(x^*) \cap A(x_s) \neq \emptyset$ for all fixed points of type k or less on the stability boundary. Let $\hat{x}$ be a hyperbolic fixed point of type k+1 on the stability boundary. Then, according to the transversality intersection, $h = \dim\{W^u(x^*)\} \leq k$. Therefore, $W^u(x^*) \cap A(x_s) \neq \emptyset$. Let $y \in W^u(x^*) \cap A(x_s)$ and let $B_\varepsilon(y)$ be an open ball of radius ε centered at y, where ε is a sufficiently small real number. Since the stability region is an open set, $B_\varepsilon(y) \subset A(x_s)$ for ε sufficiently small. Let D_ε be a neighborhood of y in $W^u(x^*)$ (a disk of dimension h) induced by $B_\varepsilon(y)$, i.e. $D_\varepsilon = B_\varepsilon(y) \cap W^u(x^*)$. Let N be a neighborhood of p in $W^u(\hat{x})$. This neighborhood is immersed on a manifold of dimension k+1. This neighborhood contains a section of dimension h transversal to $W^s(x^*)$ at the point p. A direct application of the λ-lemma shows the existence of a point z in N and an integer $K > 0$ such that $f^K(z) \in B_\varepsilon(y)$. The invariance of $A(x_s)$ guarantees that $z \in A(x_s)$, therefore $W^u(\hat{x}) \cap A(x_s) \neq \emptyset$.

(b) We prove the if condition. Suppose that $W^s(\hat{x}) \subset \partial A(x_s)$. Since $\hat{x} \in W^s(\hat{x})$, it follows that $\hat{x} \in \partial A(x_s)$. We next prove the only if condition. Suppose that $\hat{x} \in \partial A(x_s)$. Then, from the proof of (a), one has that $W^u(\hat{x}) \cap A(x_s) \neq \emptyset$. Let $y \in \{W^u(\hat{x}) - \hat{x}\} \cap A(x_s)$ and $B_\varepsilon(y)$ be an open ball of radius ε, centered at y, where ε is an arbitrarily small number. Since the stability region is an open set, $B_\varepsilon(y) \subset A(x_s)$ for ε sufficiently small. Let D_ε be a neighborhood of y in $W^u(x^*)$ (a disk of dimension h) induced by $B_\varepsilon(y)$, i.e. $D_\varepsilon = B_\varepsilon(y) \cap W^u(x^*)$. Let p be an arbitrary point in $W^s(\hat{x})$ and let S be a section transversal to $W^s(\hat{x})$ at the point p. A direct application of the λ-lemma shows the existence of a point z in S and an integer K > 0 such that $f^K(z) \in B_\varepsilon(y)$. The invariance of $A(x_s)$ guarantees that $z \in A(x_s)$. Since ε and the section S can be chosen arbitrarily small, then there exist points of $A(x_s)$ arbitrarily close to p. This means $p \in \overline{A(x_s)}$. Since $W^s(\hat{x})$ cannot contain points in $A(x_s)$, $p \in \partial A(x_s)$. The arbitrariness of the choice of p in $W^s(\hat{x})$ guarantees that $W^s(\hat{x}) \subset \partial A(x_s)$. This completes the proof.

We are now in a position to develop a complete characterization of the stability boundary of a class of nonlinear discrete dynamical systems satisfying assumptions (A1)–(A3).

THEOREM 9-11 **(Stability boundary characterization)**
Let x_s be an asymptotically stable fixed point of the discrete system (9.1), suppose that f is a diffeomorphism and assumptions (A1)–(A3) are satisfied. Let $x_1, x_2, \ldots$ be the hyperbolic unstable fixed points on the stability boundary $\partial A(x_s)$. Then the stability boundary $\partial A(x_s)$ is completely characterized by the following:

$$\partial A(x_s) = \bigcup_i W^s(x_i)$$

Proof From Theorem 9-10,

$$\bigcup_i W^s(x_i) \subseteq \partial A(x_s). \tag{9.3}$$

From assumption (A3), we have

$$\partial A(x_s) \subset \bigcup_i W^s(x_i). \tag{9.4}$$

Combining Eqs. (9.3) and (9.4) leads to $\partial A(x_s) = \bigcup_i W^s(x_i)$ and the proof is complete.

Theorem 9-11 asserts that the stability boundary is composed of the union of the stable manifolds of all fixed points that lie on the stability boundary. Theorem 9-11 can be generalized to allow closed orbits to exist on the stability boundary.

9.5 Conceptual algorithms for exact stability regions

Theorem 9-11 leads to the following conceptual algorithm for determining the stability boundary of a stable fixed point of the nonlinear discrete dynamical systems (9.1) that satisfies assumptions (A1) to (A3).

Conceptual algorithm (To determine the stability boundary $\partial A(x_s)$)
Step 1: Find all the fixed points.
Step 2: Identify those fixed points whose unstable manifolds contain orbits approaching the stable fixed point x_s.
Step 3: The stability boundary of x_s is the union of the stable manifolds of the fixed points identified in Step 2.

Step 2 can be accomplished numerically. The following procedure is suggested.

(i) Find the Jacobian at the fixed point (say, $\hat{x}$).
(ii) Find many of the generalized unstable eigenvectors of the Jacobian that have unit length.
(iii) Find the intersection of each of these normalized, generalized, and unstable eigenvectors (say, y_i) with the boundary of an ε-ball of the fixed point (the intersection points are $\hat{x} + \varepsilon y_i$ and $\hat{x} - \varepsilon y_i$).
(iv) Iterate the vector field backward (reverse time) from each of these intersection points up to some specified time. If the orbit remains inside this ε-ball, then go to the next step. Otherwise, we replace the value ε by $\alpha\varepsilon$ and also the intersection points $\hat{x} \pm \varepsilon y_i$ by $\hat{x} \pm \alpha\varepsilon y_i$, where $0 < \alpha < 1$. Repeat this step.

(v) Numerically iterate the vector field starting from these intersection points.
(vi) Repeat steps (iii) through (v). If any of these orbits approaches x_s, then the fixed point is on the stability boundary.

For a planar system, the fixed point on the stability boundary is either a type-one fixed point or a type-two fixed point, which is a source. The stable manifold of a type-one fixed point in this case has dimension one, which can easily be determined numerically as follows.

(i) Find a normalized stable eigenvector y of the Jacobian at the fixed point $\hat{x}$.
(ii) Find the intersection of this stable eigenvector with the boundary of an ε-ball of the fixed point $\hat{x}$ (where the intersection points are $\hat{x}+\varepsilon y$ and $\hat{x}-\varepsilon y$).
(iii) Iterate the vector field from each of these intersection points after some specified time. If the orbit remains inside this ε-ball, then go to next step. Otherwise, we replace the value ε by $\alpha\varepsilon$ and also the intersection points $\hat{x}\pm\varepsilon y_i$ by $\hat{x}\pm\alpha\varepsilon y_i$, where $0<\alpha<1$. Repeat this step.
(iv) Numerically iterate the vector field backward (reverse time) starting from these intersection points.
(v) The resulting orbits are in the stable manifold of the fixed point.

This conceptual algorithm may have difficulty in finding exact stability regions of higher dimensional systems. This is because the numerical procedure similar to that described above can only provide a set of orbits on the stable manifold. Finding the stable manifold and unstable manifold of a fixed point is a nontrivial problem and advanced numerical methods for computing stable and unstable manifolds must be developed. We next use a simple example to illustrate this conceptual algorithm.

Example 9-2 Consider the nonlinear discrete system

$$\begin{aligned} x_{k+1} &= dx_k + x_k^3 + ey_k^2 \\ y_{k+1} &= cy_k \end{aligned} \tag{9.5}$$

with $c=d=e=0.5$. It is straightforward to show that $f(x_k, y_k)$ is a diffeomorphism. The fixed points of system (9.5) are (0, 0), which is an asymptotically stable fixed point, and $x_1=(\sqrt{1-d},0)$ and $x_2=(-\sqrt{1-d},0)$, which are type-one fixed points. All fixed points are hyperbolic and, as a consequence, assumption (A1) is satisfied. We make the following observations.

[1] The unstable manifolds of $x_1=(\sqrt{1-d},0)$ and $x_2=(\sqrt{1-d},0)$ converge to the asymptotically stable fixed point (0, 0). According to Theorem 9-10, these two type-one fixed points lie on the stability boundary of the stable fixed point (0, 0).
[2] Figure 9.2 displays the stability region $A(0,0)$ of the fixed point (0, 0). The unstable fixed points x_1 and x_2 lie on the stability boundary $\partial A(0,0)$. Assumptions (A2) and (A3) are also satisfied and therefore the stability boundary is composed of the union of the stable manifolds $W^s(x_1)$ and $W^s(x_2)$, in accordance with the results of Theorem 9-11.

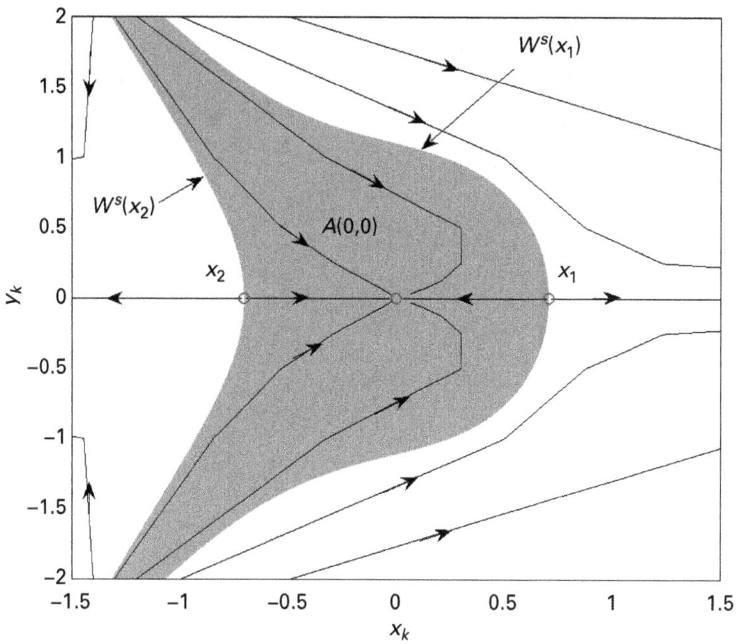

Figure 9.2 Stability region of the fixed point (0, 0) of system (9.5). The stability boundary is composed of the union of the stable manifolds $W^s(x_1)$and $W^s(x_2)$ of the type-one fixed points $x_1$and x_2.

Example 9.3 Consider the following discrete version of the nonlinear pendulum equation:

$$\begin{aligned} x_{k+1} &= x_k + hy_k \\ y_{k+1} &= (1 - dh)y_k - hc \sin x_k. \end{aligned} \tag{9.6}$$

For sufficiently small h, the vector field is a diffeomorphism. All the fixed points are hyperbolic and assumptions (A1), (A2) and (A3) are satisfied. Hence, the characterization of the stability boundary of Theorem 9-11 is valid. Figure 9.3 displays the stability region and stability boundary of the system for $h = 0.1$, $k = 1$ and $d = 0.5$. There are two hyperbolic fixed points whose unstable manifolds converge to the asymptotically stable fixed point (0,0). Hence, we have the following results.

[1] According to Theorem 9-10, one can conclude that these two hyperbolic fixed points lie on the stability boundary of the asymptotically stable fixed point (0, 0), as shown in Figure 9.3.

[2] It can be observed that the stability boundary of the asymptotically stable fixed point (0, 0) is composed of the union of the stable manifolds of these two fixed points lying on the stability boundary, confirming the complete characterization derived in Theorem 9-11.

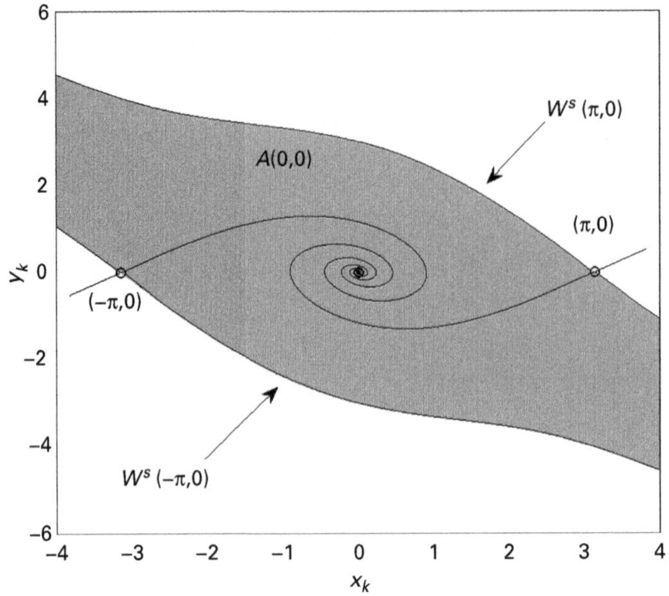

Figure 9.3 Stability region of system (9.6) for $k = 1$, $d = 0.5$ and $h = 0.1$. Two hyperbolic fixed points lie on the stability boundary of the asymptotically stable fixed point (0, 0), as asserted by Theorem 9-10. The stability boundary of the asymptotically stable fixed point (0, 0) is composed of the union of the stable manifolds of these two fixed points lying on the stability boundary, as asserted by Theorem 9-11.

While the stability boundary $\partial A(x_s)$ is of dimension $n-1$, the stable manifolds of non-type-one fixed points are thin sets on $\partial A(x_s)$. Exploring this property, the next theorem offers a complete characterization of the boundary of $\bar{A}$ in terms of the stable manifolds of type-one fixed points that lie on the stability boundary.

THEOREM 9-12 **(Another complete characterization)**
Let x_s be an asymptotically stable fixed point of the discrete system (9.1) and suppose that f is a diffeomorphism and assumptions (A1)–(A3) are satisfied. Let $x_1^1, x_2^1, \ldots$ be the type-one hyperbolic unstable fixed points on the stability boundary $\partial\bar{A}(x_s)$ of x_s. Then

$$\partial\overline{A}(x_s) = \bigcup_i \overline{W^s(x_i^1)}.$$

Proof Let $x_1^1, x_2^1, \ldots$be the type-one hyperbolic unstable fixed points on the stability boundary $\partial\bar{A}(x_s)$. The dimension of the stable manifold of a type-one fixed points is $n-1$, whereas the dimension of other fixed points is less than $n-1$, i.e. the interior of the stable manifolds of the fixed points that are not type one on $\partial A(x_s)$ is empty. The Baire category theorem [124] implies that $\partial A(x_s)$ cannot be written as a countable union of closed subsets having empty interiors, therefore we can affirm that $\partial\overline{A}(x_s) = \bigcup_i \overline{W^s(x_i^1)}$ where $x_1^1, x_2^1, \ldots$ are the type-one fixed points on $\partial\bar{A}(x_s)$. This completes the proof.

Theorem 9-12 asserts that a complete characterization of the stability boundary can be obtained by computing only the type-one fixed points that lie on the stability boundary. The following theorem gives an interesting result on the structure of the fixed points on the stability boundary. Moreover, it presents a necessary condition for the existence of certain types of fixed points on a *bounded* stability boundary.

THEOREM 9-13 **(Structure of fixed points on the stability boundary)**
Let x_s be an asymptotically stable fixed point of the discrete system (9.1) and suppose that f is a diffeomorphism and assumptions (A1)–(A3) are satisfied. If the stability region $A(x_s)$ is not dense in R^n then $\partial A(x_s)$ must contain at least one type-one fixed point. Moreover, if $A(x_s)$ is bounded, then $\partial A(x_s)$ must contain at least one type-n fixed point (i.e. a source).

Proof Since the stability region $A(x_s)$ is not dense in R^n, then Proposition 9.4 guarantees that the dimension of $\partial A(x_s)$ is $n-1$. Since $\partial A(x_s) = \underset{i}{\cup} W^s(x_i)$ where x_i, i=1,2, . . . are the fixed points on the stability boundary $\partial A(x_s)$, then at least one of the fixed points, say x_1, must be a type-one fixed point so that the dimension of $\underset{i}{\cup} W^s(x_i)$ is $n-1$. Repeating the same argument, if $\partial W^s(x_1)$ is non-empty, then the dimension of $\partial W^s(x_1)$ is $n-2$. The application of Theorem 9-11 yields $\partial W^s(x_1) = \underset{j}{\cup} W^s(x_j)$, where x_j are the fixed points in $\partial W^s(x_1)$. In order to guarantee that $\underset{j}{\cup} W^s(x_j)$ is of dimension $n-2$, at least one of the fixed points x_j on $\partial W^s(x_1)$ must be a type-two fixed point. If the stability region is bounded, the same argument can be repeated until we reach a type-n fixed point (a source). The proof is complete.

Theorem 9-13 offers a sufficient condition to check whether the stability region is unbounded. This condition is formally stated in the next corollary.

COROLLARY 9.14 **(Unbounded stability regions)**
If the stability boundary $\partial A(x_s)$ of the asymptotically stable fixed point x_s of the discrete system (9.1) has no source and assumptions (A1)–(A3) are satisfied, then $A(x_s)$ is unbounded.

We next illustrate Theorem 9-12, Theorem 9-13 and Corollary 9.14 on the simple test systems (9.5) and (9.6). We note that the stability boundaries of both system (9.5) and system (9.6) contain type-one fixed points on the stability boundaries as asserted by Theorem 9-13. We also note that both system (9.5) and system (9.6) contain no type-two fixed points (i.e. no source) on the stability boundaries of both systems. Hence, by Corollary 9.14, the stability region is unbounded. In addition, according to Theorem 9-12, the stability boundaries of these two simple systems (9.5) and (9.6) are composed of the union of the closure of the stable manifolds of two type-one fixed points.

9.6 Characterization via an energy function

In this section, we study energy functions which can be viewed as an extension of the Lyapunov functions. We focus on how to characterize the stability boundary of a class of nonlinear discrete systems that admit an energy function.

Before defining the concept of energy function, it is important to understand the concept of first difference, which plays the role of derivative for discrete systems.

Consider the scalar function $V: R^n \to R$. The first difference of V relative to (9.1), or to map f, at a point $x \in R^n$ is given by:

$$\Delta V(x) = V(f(x)) - V(x). \tag{9.7}$$

If x_k is a solution of (9.1) for $k \geq 0$, then the first difference of V along the solution x_k is given by:

$$\Delta V(x_k) = V(x_{k+1}) - V(x_k),\ k \geq 0. \tag{9.8}$$

DEFINITION **(Energy function)**
A continuous function $V: R^n \to R$ is called an energy function for the discrete system (9.1) if it satisfies the following conditions.

(E1) $\Delta V(x) \leq 0$ *for all* $x \in R^n$*;*
(E2) $\Delta V(x_k) = 0$ *implies* x_k *is a fixed point.*
(E3) If $V(x_k)$ *is bounded for* $k \in Z_+$*, then the orbit* x_k *is itself bounded for* $k > 0$.

Property (E1) indicates that the energy function is non-increasing along any orbit, but it alone does not imply that the energy function is strictly decreasing along nontrivial orbits. There may exist a discrete time k such that $\Delta V(x_k) = 0$. Properties (E1) and (E2) together imply that the energy function is strictly decreasing along any nontrivial system orbit. Property (E3) states that the energy function is a dynamic proper map along any system orbit but need not be a proper map for the entire state space. Recall that a proper map is a function $f: X \to Y$ such that for each compact set $D \in Y$, the set $f^{-1}(D)$ is compact in X. We note that if function V is proper or radially unbounded, then assumption (E3) is satisfied. From the above definition of energy function, it is obvious that an energy function may not be a Lyapunov function.

Energy functions are useful for global analysis of system orbits and for estimating stability regions and quasi-stability regions, among others. We next present a global analysis of system orbits of nonlinear discrete dynamical systems that have energy functions.

THEOREM 9-15 **(Energy functions and limit sets)**
If the nonlinear discrete dynamical system (9.1) admits an energy function and the map f is continuous with all fixed points being isolated, then every bounded trajectory x_k converges to a fixed point as $k \to \infty$.

Proof Let x_k be a bounded trajectory of (9.1) starting in x_0 at time $k=0$. Assumption (E1) implies that $V(x_k + 1) \leq V(x_k)$ for every $k \geq 0$. The non-increasing sequence $V(x_k)$ is bounded from below, since V is a continuous function. Hence, $V(x_k)$ converges to a certain value p as $k \to \infty$. On the other hand, the set $\omega(x_0)$ of a bounded orbit is non-empty and the orbit x_k approaches $\omega(x_0)$ as $k \to \infty$. Hence $V(x) = p$ for every $x \in \omega(x_0)$. The invariance of $\omega(x_0)$ implies that $\Delta V(x) = 0$ for every $x \in \omega(x_0)$. Suppose now that $x \in \omega(x_0)$ is not a fixed point. Then $\Delta V(x) = V(f(x)) - V(x) = p - p = 0$. This fact is in contradiction with assumption (E2). Therefore $\omega(x_0)$ is composed exclusively of fixed points. Since all fixed points are isolated and the limit set $\omega(x_0)$ is invariantly connected,

$\omega(x_0)$ is composed of a single fixed point. Therefore, every bounded trajectory converges to a fixed point as $k \to \infty$. This completes the proof.

Theorem 9-15 asserts that nonlinear discrete systems admitting an energy function do not present complex behavior, i.e. their limit sets are exclusively composed of fixed points, in particular, the system has no periodic orbits and consequently no limit cycles. The state space of this class of nonlinear systems does not admit nontrivial periodic solutions, quasi-periodic solutions and chaos. It will be shown in the next theorem that orbits on the stability boundary of nonlinear discrete dynamical systems admitting an energy function are bounded for $k > 0$, although the stability boundary itself can be unbounded.

THEOREM 9-16 **(Boundedness of orbits on the stability boundary)**
Let x_s be an asymptotically stable fixed point of the discrete system (9.1) that admits an energy function and suppose that the map f is continuous. Then every trajectory x_k on the stability boundary $\partial A(x_s)$ is bounded for $k > 0$.

Proof For an arbitrary $x \in \partial A(x_s)$, let $\{x_i\}$ be a sequence of points in $A(x_s)$ converging to x as $i \to \infty$. The trajectory x_{i_k} of system (9.1) starting in x_i converges to x_s as $k \to \infty$. Assumption (E1) implies $V(x_i) \geq V(x_s)$ for every x_i, $i=1,2,\ldots$. The continuity of V implies that $V(x) \geq V(x_s)$. Hence, $V(x_s)$ is a lower bound of V on the stability boundary. The positive invariance of $\partial A(x_s)$ and assumption (E3) imply that x_k, the solution of system (9.1) starting in x, is bounded for $k > 0$. This completes the proof.

Combining Theorem 9-15 and Theorem 9-16 leads to a sufficient condition for the satisfaction of assumption (A3) as shown in the following.

COROLLARY 9.17 **(Sufficient condition for (A3))**
Let x_s be an asymptotically stable fixed point of the discrete system (9.1) that admits an energy function, suppose that f is continuous and all fixed points are isolated. Then assumption (A3) is satisfied, i.e. every trajectory on $\partial A(x_s)$ converges to a fixed point.

Proof Theorem 9-16 asserts that every orbit on the stability boundary $\partial A(x_s)$ is bounded and Theorem 9-15 proves that every bounded orbit converges to a fixed point. This concludes the proof.

We are now in a position to present a complete characterization of the stability boundary for a class of discrete nonlinear dynamical systems that admits energy functions.

THEOREM 9-18 **(Stability boundary characterization 1)**
Let x_s be an asymptotically stable fixed point of the nonlinear discrete system (9.1) that admits an energy function. Suppose that f is a continuous map satisfying assumption (A1) and let $x_1, x_2, \ldots$ be the unstable fixed points on the stability boundary $\partial A(x_s)$. Then, the stability boundary is contained in the union of the stable sets of all the fixed points on the stability boundary

$$\partial A(x_s) \subseteq \bigcup_i W^s(x_i)$$

where $W^s(x_i)$ is the stable set of the unstable fixed point x_i.

Proof Assumption (A1) requires that every fixed point on the stability boundary is hyperbolic and therefore isolated. Corollary 9-17 guarantees that assumption (A3) is satisfied, i.e. every trajectory on the stability boundary converges to a fixed point on $\partial A(x_s)$. Hence, every trajectory on the stability boundary must lie in the stable set of a fixed point on the stability boundary.

The characterization of the stability boundary of systems that admit energy functions, asserted in Theorem 9-18, only requires continuity of the vector field. A sharper characterization of the stability boundary of systems having energy functions can be obtained by assuming that all fixed points are hyperbolic and f is a diffeomorphism. We can gain more structure on the stable set that in this case is a manifold at the expense of more conditions on the vector field.

THEOREM 9-19 **(Stability boundary characterization 2)**
Let x_s be an asymptotically stable fixed point of the nonlinear discrete system (9.1) that admits an energy function. Suppose that f is a diffeomorphism satisfying assumption (A1). Let $x_1, x_2, \ldots$ be the hyperbolic unstable fixed points on the stability boundary $\partial A(x_s)$ of x_s. Then

$$\partial A(x_s) \subseteq \bigcup_i W^s(x_i)$$

Moreover, if assumption (A2) is satisfied, then

$$\partial A(x_s) = \bigcup_i W^s(x_i)$$

where $W^s(x_i)$ is the stable manifold of the hyperbolic unstable fixed point x_i.

Proof The first part trivially follows from Theorem 9-18. Since assumptions (A1) and (A2) are satisfied and the existence of an energy function implies (A3), then the result follows from a direct application of Theorem 9-11. This proof is completed.

Example 9-4 Consider the following two-dimensional nonlinear discrete system:

$$\begin{aligned} x_{k+1} &= x_k^3 + \frac{3}{4}x_k \\ y_{k+1} &= y_k^3 + \frac{3}{4}y_k. \end{aligned} \tag{9.9}$$

The vector field is a diffeomorphism and the system possesses nine hyperbolic fixed points: (0,0), an asymptotically stable fixed point, and $\left(0, \pm\frac{1}{2}\right)$, $\left(\pm\frac{1}{2}, 0\right)$, $\left(\pm\frac{1}{2}, \pm\frac{1}{2}\right)$, which are unstable fixed points. Consider the following candidate energy function:

$$V(x,y) = \begin{cases} |x| + |y| & \text{if} \quad |x| \le \frac{1}{2} \quad \text{and} \quad |y| \le \frac{1}{2} \\ 1 - |x| + |y| & \text{if} \quad |x| > \frac{1}{2} \quad \text{and} \quad |y| \le \frac{1}{2} \\ 1 + |x| - |y| & \text{if} \quad |x| \le \frac{1}{2} \quad \text{and} \quad |y| > \frac{1}{2} \\ 2 - |x| - |y| & \text{if} \quad |x| > \frac{1}{2} \quad \text{and} \quad |y| > \frac{1}{2}. \end{cases}$$

It is straightforward to show that $\Delta V(x, y) \le 0$ for any (x,y) and $\Delta V(x, y) = 0$ if and only if (x,y) is a fixed point. Hence, assumptions (E1) and (E2) are satisfied. Assumption (E3) is also satisfied because $V(x,y)$ is a proper function. We conclude that V is an energy function for system (9.9) and all the conditions of Theorem 9-19 are satisfied. Thus the stability boundary $\partial A(0, 0)$ is composed of the union of the stable manifolds of every fixed point that lies on the stability boundary. Figure 9.4 illustrates the stability region and stability boundary of system (9.9). The stability boundary contains eight unstable fixed points since their unstable manifolds converge to the asymptotically stable fixed point, as asserted by Theorem 9-10. The stability boundary $\partial A(0, 0)$ equals the union of the stable manifolds of eight unstable fixed points, as asserted by Theorem 9-19.

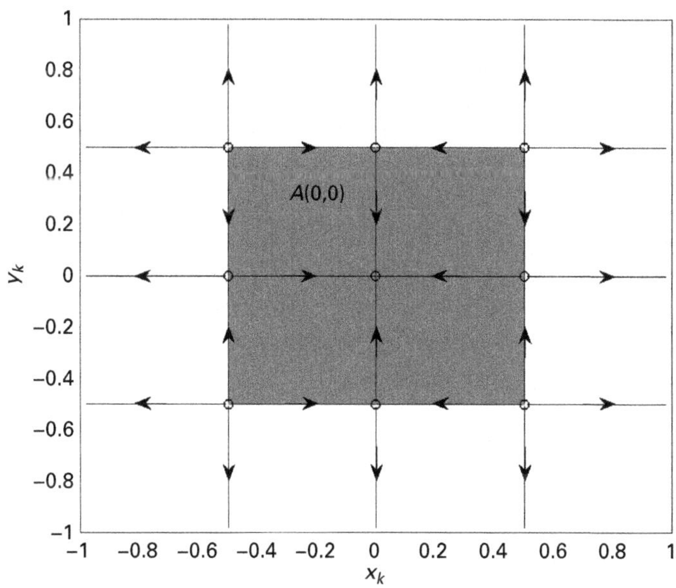

Figure 9.4 Stability region of (0, 0) which is an asymptotically stable fixed point of system (9.9). The stability boundary contains eight unstable fixed points since their unstable manifolds converge to (0, 0) (cf. Theorem 9-10). The stability boundary of (0, 0) is composed of the union of the stable manifolds of the eight unstable fixed points lying on the boundary.

9.7 Concluding remarks

A comprehensive theory of stability regions of general nonlinear autonomous discrete dynamical systems has been developed in this chapter. Several topological properties of the stability boundary and characterizations of limit sets lying on the stability boundary for general nonlinear discrete dynamical systems have been derived. Our approach starts from a local characterization of the stability boundary and progresses towards a global characterization of the stability boundary.

We have derived a complete characterization for a fixed point lying on the stability boundary, which is a key step in the characterization of the stability region $A(x_s)$. These characterizations are expressed in terms of both the stable and unstable manifolds. A complete characterization of stability boundaries for a fairly large class of nonlinear discrete dynamical systems has been obtained. For this class of nonlinear discrete systems, it was shown that the stability boundary of an asymptotically stable fixed point consists of the stable manifolds of all the fixed points on the stability boundary. Several necessary and sufficient conditions were derived to determine whether a given fixed point (or closed orbit) lies on the stability boundary.

A method to determine the exact stability region based on these results was proposed. The method, when feasible, will find the exact stability region, rather than a subset of the stability region as in the Lyapunov theory approach. For high-dimensional systems, an optimal scheme for estimating the stability region will be developed in Chapter 13.

Regarding the topological properties of stability regions of nonlinear discrete dynamical systems, the following properties have been derived.

- The stability region $A(x_s)$ is positively invariant and negatively invariant. The stability boundary $\partial A(x_s)$ is a closed set.
- If the vector field f is surjective, then the stability region $A(x_s)$ is an invariant set.
- If the vector field f is a continuous function, then the stability region $A(x_s)$ is an open, positively and negatively invariant set. The stability boundary $\partial A(x_s)$ is a closed and positively invariant set formed by forward orbits. Moreover, the stability boundary $\partial A(x_s)$ is of dimension less than n and if $A(x_s)$ is not dense in R^n, then $\partial A(x_s)$ is of dimension $n-1$.
- If the vector field f is a homeomorphism, then the stability region $A(x_s)$ has the following topological and dynamic properties. It is (i) open, (ii) positively and negatively invariant, (iii) invariant, and (iv) path connected; the stability boundary $\partial A(x_s)$ is (i) closed and (ii) invariant.

There has been significant work on the analysis of stability and asymptotic behavior of discrete-time dynamical systems in the literature. LaSalle proved an invariance principle for discrete-time systems in [156,157]. An extension of the invariance principle for discrete dynamical systems was independently derived in [219] and [5]. LaSalle theory seems to be the first practical tool to estimate the stability region of nonlinear discrete dynamical systems in the form of positive invariance sets. A survey on the theory of positively invariant sets in the analysis and control of discrete-time nonlinear dynamical systems can be found in [25]. The relationship between stability and the existence of smooth Lyapunov functions was studied in [140]. In spite of the enormous amount of work done in analysis of the asymptotic behavior of solutions of discrete-time nonlinear dynamical systems, only a few papers on the theory or estimation of stability regions of discrete dynamical systems exist [188].

The theory presented in this chapter is an extension of the theory of the stability region of continuous dynamical systems and provides a complete characterization of the stability boundary of these systems. This complete characterization allows the determination of the exact stability region and will provide the basis for the development of algorithms for estimating stability regions in Chapter 13.

Part II

Estimation

10 Estimating stability regions of continuous dynamical systems

In this chapter we discuss how to estimate the stability region of nonlinear continuous dynamical systems using an energy function, which can be a local energy function or a (global) energy function. Several methods for estimating stability regions described in this book are suitable for high-dimensional nonlinear systems. We consider a general nonlinear autonomous dynamical system described by the following equation:

$$\dot{x}(t) = f(x(t)). \tag{10.1}$$

It is natural to assume that the function (i.e. the vector field) f: $R^n \to R^n$ satisfies a sufficient condition for the existence and uniqueness of the solution. In general, the nonlinear continuous dynamical system (10.1) can be classified into one of the following two groups of nonlinear dynamical systems:

(i) the group of nonlinear systems which admit an energy function, and
(ii) the group of nonlinear systems which do not admit an energy function.

Despite not having a (global) energy function, those systems belonging to group (ii) do admit local energy functions around stable equilibrium points.

Depending on whether or not the nonlinear dynamical system under study has an energy function, there are two different procedures for estimating its stability regions. For systems having energy functions, the procedure starts with an energy function $V(.)$ associated with the system under study and proceeds in the following three steps.

Procedure A (For nonlinear dynamical systems having an energy function)

Step 1: Construct an energy function (or Lyapunov function) for system (10.1) whose stability regions are to be estimated.

Step 2: Determine the critical level value of the constructed energy function with respect to an SEP whose stability region is to be estimated.

Step 3: Approximate the stability boundary of the SEP via the constant energy surface with the critical level value.

It will be shown that the stability region estimated via Procedure A is the optimal in a certain sense. For systems belonging to group (ii) that do have local energy functions, the procedure starts with a local energy function $V(.)$ and proceeds in the following three steps.

Procedure B (For nonlinear dynamical systems only having a local energy function)

Step 1: Construct a local energy function around an SEP whose stability region is to be estimated.

Step 2: Determine the critical level value of the constructed local energy function with respect to the SEP.

Step 3: Approximate the stability boundary using the constant surface defined by the constructed local energy function with the critical level value determined in Step 2.

We note that both Procedure A and Procedure B have similar steps. The main difference lies in the issue of how to determine critical values in Step 2. Regarding Step 1 of Procedure B, there are several methods available for constructing local energy functions; some of them are based on linear matrix inequalities. These methods are usually not capable of dealing with high-dimensional nonlinear systems.

This chapter is focused on both theoretical and computational aspects behind Step 2 and Step 3 in Procedures A and B. This chapter also establishes a relationship between the structures of a quasi-stability boundary and constant energy surface at different level values.

10.1 Level set

Recall that a function $V{:}R^n \to R$ is an energy function for system (10.1) if the following three conditions are satisfied.

(i) The derivative of the energy function $V(x)$ along any system trajectory $x(t)$ is non-positive, i.e.

$$\dot{V}(x(t)) \leq 0.$$

(ii) If $x(t)$ is a nontrivial trajectory (i.e. $x(t)$ is not an equilibrium point), then along the nontrivial trajectory x(t) the set

$$\{t \in R : \dot{V}(x(t)) = 0\}$$

has measure zero in R.

(iii) That a trajectory $x(t)$ has a bounded value of $V(x(t))$ for $t \in R^+$ implies that the trajectory $x(t)$ is also bounded for $t \in R^+$. Stating this in brief: that $V(x(t))$ is bounded implies $x(t)$ itself is also bounded.

Property (i) indicates that the energy function is non-increasing along its trajectory, but does not imply that the energy function is strictly decreasing along its trajectory. Property (ii) states that there does not exist a time interval $[t_1, t_2]$ such that $\dot{V}(x(t)) = 0$ for $t \in [t_1, t_2]$. Properties (i) and (ii) together imply that the energy function is strictly decreasing along any system trajectory. Property (iii) states that the energy function is a proper map along any system trajectory, i.e. it is a dynamic proper map, but need not be a proper map for the entire state space.

Given a C^r-energy function $V(.):R^n \rightarrow R$, we consider the following set

$$S_v(k) = \{x \in R^n : V(x) < k\}. \tag{10.2}$$

Sometimes, we drop the subscript v of $S_v(k)$, simply writing $S(k)$, when it is clear from the context. We shall call the boundary of set (10.2), $\partial S(k)\!: = \{x \in R^n\!:\! V(x) = k\}$ the *level set* (or *constant energy surface*) and k the *level value*. If k is a regular value (i.e. $\nabla V(x) \neq 0$, for all $x \in V^{-1}(k)$), then by the inverse function theorem, $\partial S(k)$ is a C^r $(n-1)$-dimensional submanifold of R^n. Moreover, if $r > n - 1$, then by the Morse–Sard theorem, the set of regular values of V is residual; in other words "almost all" level values are regular. In particular, for almost all values of k, the level set $\partial S(k)$ is a C^r $(n-1)$-dimensional submanifold.

Generally speaking, this set $S(k)$ can be very complicated with several connected components even for the two-dimensional case. Let $S^i(k)$, $i = 1, 2, \ldots, m$ represent these disjoint connected component such that

$$S(k) = S^1(k) \cup S^2 \cup \ldots \cup S^m(k) \tag{10.3}$$

with $S^i(k) \cap S^j(k) = \emptyset$ when $i \neq j$. Since $V(.)$ is continuous, $S(k)$ is an open set. Because $S(k)$ is an open set, the level set $\partial S(k)$ is of $(n-1)$ dimensions. Furthermore, each component of $S(k)$ is a positively invariant set.

In spite of the possibility that a constant energy surface can contain several disjoint connected components, there is an interesting relationship between the constant energy surface and the stability boundary. This relationship reveals that there is at most one connected component of the set $S(r)$ which has a non-empty intersection with the stability region $A(x_s)$, as shown in Theorem 10-1 below.

THEOREM 10-1 **(Constant energy surface and stability region)**
Consider a general nonlinear autonomous dynamical system (10.1) which admits an energy function. Let x_s be an asymptotically stable equilibrium point of the system and let $A(x_s)$ be its stability region. Then the set $S(r)$ contains only one connected component which has a non-empty intersection with the stability region $A(x_s)$ if and only if $r > V(x_s)$.

Motivated by Theorem 10-1, we shall use the notation $S(r, x_s)$ to denote the connected set of $S(r)$ (whose level value is r) containing the stable equilibrium point x_s. We drop the notation x_s of $S(r, x_s)$ when it is clear from the context. In Figure 10.1, the relation between the constant energy surfaces at different level values and the stability region $A(x_s)$ is shown. It is observed in this figure that the connected set $S(r, x_s)$ with a level value r smaller than the critical value is very conservative in the approximation of the stability boundary $\partial A(x_s)$. As the set $S(r, x_s)$ is expanded with an increasing level value r, the approximation improves until this constant energy surface hits the stability boundary $\partial A(x_s)$ at some point. This point will be shown below to be a UEP. We call this point the *closest UEP* of the SEP x_s with respect to the energy function $V(.)$.

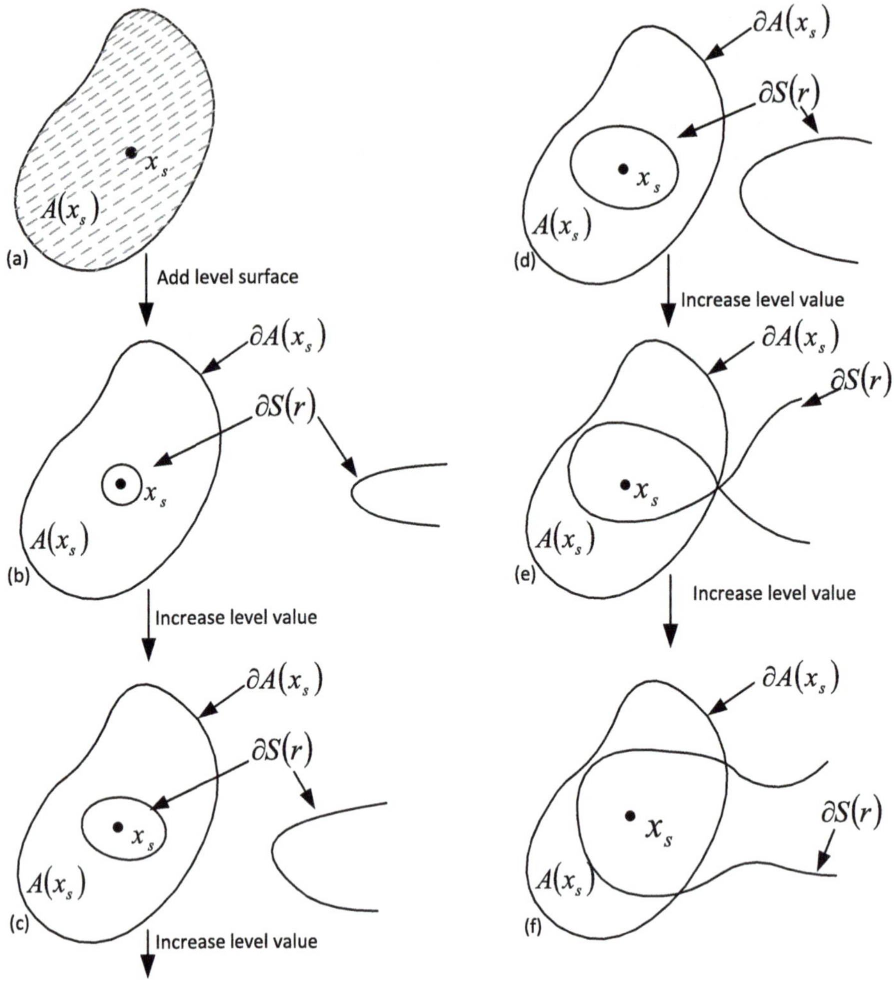

Figure 10.1 The relation between the constant energy surfaces $S(r)$ at different level values and the stability region $A(x_s)$. The connected set $S(r, x_s)$ with a level value r smaller than that at the closest UEP is very conservative in the approximation of the stability boundary $\partial A(x_s)$. As the set $S(r, x_s)$ is expanded with an increasing level value r, the approximation improves until this constant energy surface hits the stability boundary $\partial A(x_s)$ at the closest UEP. The connected set $S(r, x_s)$ with a level value r greater than that at the closest UEP is not suitable for approximating the stability boundary $\partial A(x_s)$.

DEFINITION **(Closest UEP)**
A UEP is termed the closest UEP of an SEP with respect to an energy function if the UEP is, among all the points of the stability boundary of the SEP, the point with the lowest energy function value.

The issue of the existence and uniqueness of the closest UEP will be investigated later on. In addition, the connected set $S(r)$ passing through the closest UEP will be shown to be the optimal estimation of the stability region $A(x_s)$ that can be obtained in the form of an energy function. Indeed, as we increase the level value r, the connected set $S(r, x_s)$ will contain points which lie outside the stability region $A(x_s)$. It is therefore inappropriate to approximate the stability region $A(x_s)$ by the connected set $S(r, x_s)$ with a level value higher than that of the lowest value of energy function on the stability boundary $\partial A(x_s)$. From Figure 10.1, it is obvious that among the several connected sets of the constant energy surface, there is only one connected component of the constant energy surface that has a non-empty intersection with the stability region $A(x_s)$.

10.2 Characterization of the closest UEP

A topological characterization and a dynamical characterization for the closest UEP (i.e. the point on the stability boundary with the minimum value of the energy function) will be derived. These characterizations will prove useful in identifying the closest UEP. Throughout this chapter and this book, the term asymptotically stable equilibrium point and stable equilibrium point will be used interchangeably.

THEOREM 10-2 **(Topological characterization)**
Consider a general nonlinear autonomous dynamical system (10.1) which admits an energy function. Let x_s be a stable equilibrium point and $A(x_s)$ be the stability region of system (10.1). If the stability region $A(x_s)$ is not dense in R^n, then the point with the minimum value of the energy function over the stability boundary $\partial A(x_s)$ exists and the point must be an unstable equilibrium point.

Proof By the conditions of energy functions, x_s is a stable equilibrium point if and only if x_s is a local minimum of any energy function $V(x)$ associated with the system (10.1). Hence, given any energy function $V(x)$, there exists a positive number n such that the set $S(n, x_s)$, which is defined to be the connected component of the level set $\{x: V(x_s) \le V(x) < n\}$ containing the stable equilibrium point, is bounded, and $S(n, x_s) \subset A(x_s)$. We denote $\partial S(n, x_s)$ the level surface of the level set $S(n, x_s)$.

Next, we investigate how the set $S(n, x_s)$ changes as the level value n is varied. Suppose that the level value n is a regular value of $V(\cdot)$, then as a consequence of the implicit function theorem the set $S(n, x_s)$ is a manifold. Furthermore, as we increase the level value n to a new value, say m with $m > n$, such that there are no critical points in the set $\{S(m, x_s) - S(n, x_s)\}$, then the set $S(m, x_s)$ is diffeomorphic to $S(n, x_s)$ [96]. In particular, $S(m, x_s)$ is bounded, and $S(m, x_s) \subset A(x_s)$ because the set $S(m, x_s)$ is a positively invariant set and contains only one ω-limit point.

Therefore, as we increase the level value n, the set $S(n, x_s)$ remains bounded and contained in $A(x_s)$ until it intersects the stability boundary at a certain point (we will show that the intersection cannot be multiple). The existence of this point is ensured by the fact that the energy function values over the stability boundary are bounded below by the energy function at the SEP x_s. Now we show that this point, say $\hat{x}$, must be a UEP lying on the stability boundary $\partial A(x_s)$. By contradiction, suppose that $V(\hat{x}) = \hat{m}$ and this point $\hat{x} \in \overline{A(x_s)}^c$; here $\overline{A(x_s)}^c$ denotes the complement of $\overline{A(x_s)}$. It is easy to see that the set $S(\hat{m}, x_s)$ has a non-empty intersection with the stability boundary $\partial A(x_s)$. Let $S := \partial S(\hat{m}, x_s) \cap \partial A(x_s)$. To show that S must contain at least an equilibrium point, we note that: (i) $\partial A(x_s)$ is a closed and invariant set of system (10.1); and (ii) $\partial S(\hat{m}, x_s)$ is a compact, positively invariant set of system (10.1); hence the set S is a compact, positively invariant set. Since any compact, positively invariant set must contain at least an ω-limit set and, according to Theorem 3-1 and Theorem 3-2, the ω-limit set of system (10.1) consists of equilibrium points, then the set S must contain at least one equilibrium point, which is a contradiction and we complete the proof.

Theorem 10-2 asserts that the point with the minimum value over the stability boundary must be an unstable equilibrium point. Theorem 10-3 gives a dynamical characterization of this point in terms of its unstable manifold. Note that this theorem holds without the transversality condition.

THEOREM 10-3 **(Dynamical characterization)**
Consider a general nonlinear autonomous dynamical system (10.1) satisfying assumption (A1) and which admits an energy function. Let x_s be a stable equilibrium point of system (10.1) and $A(x_s)$ be its stability region. If the stability region $A(x_s)$ is not dense in R^n, and $\hat{x}$ is the point with the minimum value of the energy function over the stability boundary $\partial A(x_s)$, then $W^u(\hat{x}) \cap A(x_s) \neq \emptyset$.

Proof We prove this theorem by contradiction. Suppose that the unstable manifold of $\hat{x}$ does not converge to the SEP x_s. Then, by Theorem 4-1, if $\hat{x}$ is a hyperbolic equilibrium point on the stability boundary $\partial A(x_s)$ of the stable equilibrium point x_s, then its unstable manifold has a non-empty intersection with the closure of the stability region, i.e. $W^u(\hat{x}) \cap \bar{A}(x_s) \neq \emptyset$. Suppose there exists a trajectory $x(t)$ in the unstable manifold $\{W^u(\hat{x}) - \hat{x}\}$ such that $x(t) \in \partial A(x_s)$ and $\lim_{t \to -\infty} x(t) = \hat{x}$. Since the energy function value is strictly decreasing along any nontrivial trajectory of system (10.1), there exist other points on the stability boundary with a lower energy function value than $V(\hat{x})$. This contradicts the fact that $\hat{x}$ possesses the lowest energy function value within the stability boundary $\partial A(x_s)$. This completes the proof.

10.3 Quasi-stability region and energy function

In this section, we will establish a relationship between the structures of a quasi-stability boundary and constant energy surface at different level values. Recall that at most one component of the constant energy surface can intersect the quasi-stability region. The constant energy surface with a small level value (of an energy function) is very conservative in approximating the quasi-stability boundary. As the level value increases, the constant energy surface expands until this constant energy surface intersects the stability boundary at the closest UEP (see Figure 10.2). Two scenarios can occur. The first scenario is that the closest UEP lies in the quasi-stability region and not on the quasi-stability boundary (see Figure 10.2), and the second scenario is that the closest UEP lies on the quasi-stability boundary.

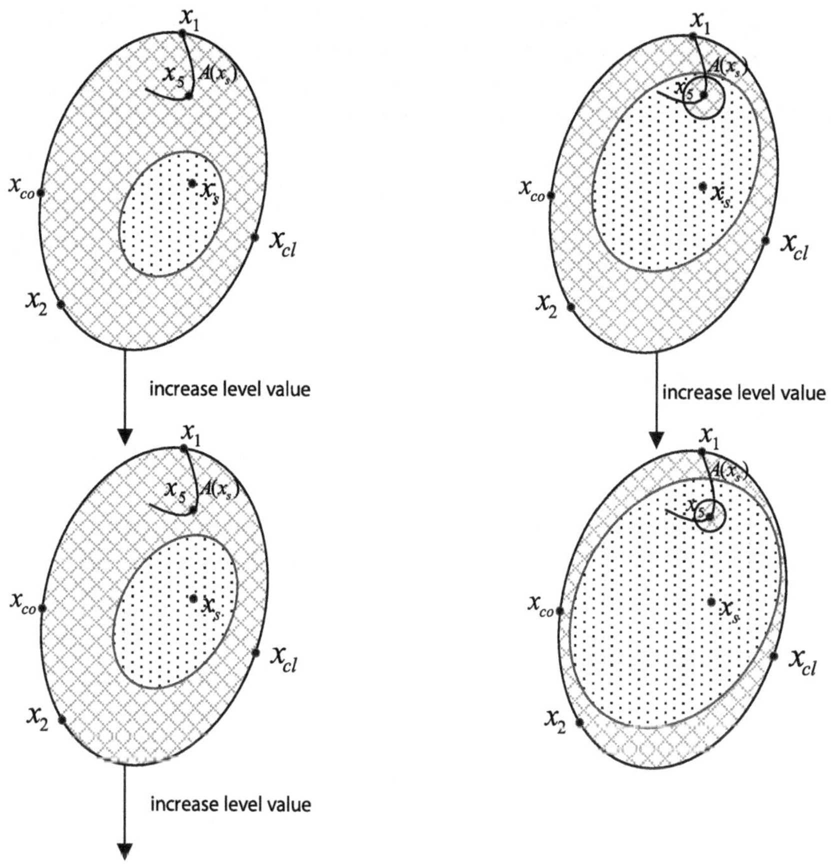

Figure 10.2 The structure of a constant energy surface changes with the increase of its level value. The constant energy surface contains several connected components; some of them are simply connected while others are not simply connected. In the beginning, for a low value of the energy function, only one connected constant energy surface intersects with the quasi stability region and the constant energy surface is conservative in approximating the quasi-stability boundary. As the level value increases, a hole appears in the connected constant energy surface which is then not simply connected. With the increase of the level value, the constant energy surface eventually intersects with a type-one UEP lying on the quasi-stability boundary. (A set is said to be connected if the line connecting two points in the set lies in the set. A set is said to be simply connected if the set is connected and the (topological) boundary of the set is also connected.)

If the closest unstable equilibrium point does not lie on the quasi-stability boundary, then it must lie in the quasi-stability region. As we increase the level value from that at the closest UEP, the constant energy surface continues to expand but still lies inside the quasi-stability region $A_q(x_s)$. It will be shown in the following theorem that the constant energy surface is still a connected set, but not a simply connected set.

THEOREM 10-4 **(Bounded property)**
Consider the nonlinear dynamical system described by (10.1) satisfying assumption (A1) and which admits an energy function $V(x)$. Let A and A_q be the stability region and the quasi-stability region of x_s, respectively. Let $c = \min_{x \in \partial A \cap E} V(x)$, where E is the set of

equilibrium points of (10.1). If the value c is attained at an equilibrium point $\hat{x} \in A_q$, *then for any* $\varepsilon > 0$ *sufficiently small, the following relation holds*

$$\overline{S(c+\varepsilon, x_s)} \subset A_q.$$

Proof Let k be the type of $\hat{x}$. Since $\hat{x}$ is a hyperbolic equilibrium point, by the Morse lemma, there exist local coordinates $(x_1, \ldots, x_k, y_1, \ldots, y_{n-k})$ defined in the neighborhood U centered at $\hat{x}$ such that $V(x, y) = c - |x|^2 + |y|^2$ in these coordinates. In these local coordinates, let $H(\varepsilon)$be a rectangular neighborhood of $\hat{x}$ defined as

$$H(\varepsilon) = \{(x, y) \in U : |y|^2 \leq \varepsilon \text{ and } |x|^2 \leq 2\varepsilon\}.$$

Set H is called a handle of the equilibrium point $\hat{x}$. Since $\overline{S(c+\varepsilon, x_s)}$ is homeomorphic to $\overline{S(c-\varepsilon, x_s)} \cup H(\varepsilon)$ and $\overline{S(c-\varepsilon, x_s)}$ and $H(\varepsilon)$ are compact sets, $\overline{S(c+\varepsilon, x_s)}$ is also compact because the image of a compact set under a homeomorphism is compact. Consequently, $\overline{S(c+\varepsilon, x_s)}$ is bounded.

Now, since $\hat{x} \in A_q = \text{int}\bar{A}$, it follows that $H(\varepsilon) \subset \text{int } \bar{A}$ for small ε. Moreover, $\overline{S(c-\varepsilon, x_s)} \subset \text{int}\bar{A}$. This implies that $H(\varepsilon) \cup \overline{S(c-\varepsilon, x_s)} \subset \text{int}\bar{A}$. For sufficiently small ε, the continuity of the homeomorphic transformation implies $\overline{S(c+\varepsilon, x_s)} \subset \text{int}\bar{A}$. This completes the proof.

Theorem 10-4 asserts that if one keeps on increasing the level value from $c + \varepsilon$ to a higher value, say c', such that the corresponding constant energy surface does not intersect ∂A_q, then $S(\bar{c}, x_s)$ is still inside A_q and bounded. In addition, it can be shown that if $a < b$ are regular values and there are no equilibrium points in $S(b, x_s) - S(a, x_s)$, then $S(a, x_s)$ and $S(b, x_s)$ are diffeomorphic.

Hence, the stability region approximated by a suitable constant energy level set always has a nice geometrical structure because it is homeomorphic to an open n-dimensional sphere. On the other hand, the quasi-stability region approximated by a constant energy level set can exhibit a very complicated shape, as shown in the following theorem.

THEOREM 10-5 **(Connected but not simply connected property)**
Consider the nonlinear dynamical system described by (10.1) satisfying assumption (A1) and that has an energy function V(x). Let A and A_q *be the stability region and the quasi-stability region of* x_s. *Let* $c = \min_{x \in \partial A \cap E} V(x)$, *where E is the set of equilibrium points of (10.1). If the value c is attained at an equilibrium point* $\hat{x} \in A_q$, *then* $S(c + \varepsilon, x_s)$ *is not simply connected.*

Proof We will first review the concept of deformation retraction. Let B be a subset of X. A deformation retraction of X onto B is a continuous map $F: X \times I \to X$, where I is the unit interval, such that $F(x, 0) = x$ for $x \in X$, $F(x, 1) \in B$ for $x \in X$, and $F(b, t) = b$ for $b \in B$. If such a function F exists, then B is called a deformation retraction of X. One can visualize a deformation retraction as a gradual collapsing of the space X onto the subspace B, such that each point of B remains fixed during the shrinking process. This concept is used in the following lemma.

LEMMA 10-6 *Let n_u be the type of $\hat{x}$ and let $W^u_\varepsilon(\hat{x})$ be an n_u-dimensional neighborhood of $\hat{x}$ in $W^u(\hat{x}) \cap H(\varepsilon)$, where $H(\varepsilon)$ is the rectangular neighborhood defined in the proof of Theorem 10.5. Then $S(c+\varepsilon, x_s)$ has $S(c-\varepsilon, x_s) \cup W^u_\varepsilon(\hat{x})$ as a strong deformation retraction.*

Lemma 10-6 states that by changing the level value from c to $c + \varepsilon$, the level set $S(c + \varepsilon, x_s)$ can be continuously transformed into $S(c-\varepsilon, x_s) \cup W^u_\varepsilon(\hat{x})$. Therefore, any topological invariant of the first set is preserved after the shrinking process.

Now we prove Theorem 10-5. Since $\hat{x} \in A_q = \text{int}\bar{A}$, it follows that $\{W^u(\hat{x}) - \hat{x}\} \subset A(x_s)$, which implies $W^u(\hat{x}) \subset \text{int}\bar{A}$. Hence, $W^u_\varepsilon(\hat{x}) \subset \text{int}\bar{A}$. From Lemma 10-6, it follows that since $S(c + \varepsilon, x_s)$ has $S(c-\varepsilon, x_s) \cup W^u_\varepsilon(\hat{x})$ as a strong deformation retraction, $S(c + \varepsilon, x_s)$ must "self-intersect". If not, then $W^u_\varepsilon(\hat{x})$ would connect $S(c - \varepsilon, x_s)$ which lies in int$\bar{A}$ to another component $C \subset (\bar{A})^c$ of the level set corresponding to the level value $c - \varepsilon$. So before reaching the level value $c - \varepsilon$, the component C must intersect $\partial\bar{A}$ and $S(c - \varepsilon, x_s) \cap C \neq \varnothing$. This contradicts the fact that for a given level value, exactly one component intersects $\bar{A}$, and this component must be $S(c + \varepsilon, x_s)$ respectively. Since $S(c - \varepsilon, x_s)$ is homeomorphic to an n-dimensional disk D^n, it can be retracted to a line segment. $W^u_\varepsilon(\hat{x})$ would connect $S(c - \varepsilon, x_s)$ which must self-intersect. So we conclude that $S(c-\varepsilon, x_s) \cup W^u_\varepsilon(\hat{x})$ can be retracted to a circle. Since the circle is not simply connected and $S(c + \varepsilon, x_s)$ has the same homotopy type as the circle, we conclude that $S(c + \varepsilon, x_s)$ is not simply connected. This completes the proof.

We have analyzed some behaviors of the constant energy surface and the relationship between constant energy surfaces and stability regions. We next discuss how to optimally approximate a stability region via a constant energy surface. Indeed, given an energy function, the key step of the energy function approach in estimating the stability region of a stable equilibrium point is the determination of the critical level value for the underlying energy function, i.e. determination of the minimum level of the energy function which provides the largest level set inside the stability region.

10.4 Optimal schemes for estimating stability regions

We discuss in this section how to optimally determine the critical level value of an energy function for estimating the stability boundary $\partial A(x_s)$. We use the notation $\bar{A}^c(x_s)$ to denote the complement of the set $\bar{A}(x_s)$. We use the notation $S(\hat{m}, x_s)$ to denote the only component of the several disjoint connected components of $S(\hat{m})$ that contains the stable equilibrium point x_s.

Theorem 10-7 asserts that the connected component $S(\hat{c}, x_s)$, where $\hat{c} = \min_{x_i \in \partial A(x_s) \cap E} V(x_i)$, is the best candidate to approximate the stability region $A(x_s)$ in the form of a level surface. Part (b) of Theorem 10-7 asserts that the scheme of choosing $\hat{c} = \min_{x_i \in \partial A(x_s) \cap E} V(x_i)$ to estimate the stability region $A(x_s)$ is optimal because the estimated stability region characterized by the corresponding constant energy surface is the largest one within the entire stability region.

THEOREM 10-7 **(Optimal estimation)**

Consider the nonlinear system (10.1) satisfying assumption (A1) and which has an energy function V(x). Let x_s be an asymptotically stable equilibrium point whose stability region $A(x_s)$ is not dense in R^n. Let E be the set of equilibrium points and $\hat{c} = \min_{x_i \in \partial A(x_s) \cap E} V(x_i)$, then

(a) $S(\hat{c}, x_s) \subset A(x_s)$, and
(b) the set $\{S(b, x_s) \cap \bar{A}^c(x_s)\}$ is non-empty for any number $b > \hat{c}$.

Theorem 6-11 asserts that the quasi-stability boundary equals the union of the closure of the stable manifolds of all type-one critical elements on the boundary. It follows that, without loss of generality, the closest UEP is a type-one equilibrium point. In addition, the unstable manifolds of closest UEP must converge to the stable equilibrium point. These properties are explored to propose a scheme to estimate the stability region $A(x_s)$. Based on Theorem 8-11, Theorems 10-2, 10-3 and 10-7, we propose the following scheme to estimate the stability region $A(x_s)$ of a nonlinear dynamical system (10.1) via an energy function $V(\cdot)$.

Scheme (Optimal estimation of the stability region $A(x_s)$ via an energy function $V(.)$)

A Determining the critical level value of the energy function.

Step A1: Find all the type-one equilibrium points.

Step A2: Order the equilibrium points whose corresponding values $V(.)$ are greater than $V(x_s)$ in terms of their energy function values.

Step A3: Of these, identify the one with the lowest value of energy function and whose unstable manifold converges to the stable equilibrium point x_s (let this one be $\hat{x}$).

Step A4: The value of the energy function at $\hat{x}$ gives the critical level value of this energy function (i.e. $V(x)$).

B Estimating the stability region $A(x_s)$.

Step B1: The connected component of $\hat{x}$ containing the stable equilibrium point x_s gives the estimated stability region.

The computation associated with Step A1 can be very involved. Efficient numerical methods in conjunction with utilizing special properties of the system under study are needed to implement this step. The above scheme is general and applicable to a nonlinear dynamical system (10.1) which admits an energy function. The computation efficiency of this scheme, however, can be significantly improved by exploring special structures of the underlying nonlinear systems, as will be shown in Chapter 18.

For the purpose of illustration, we consider the following simple example:

$$\begin{aligned} \dot{x}_1 &= -\sin x_1 - 0.5\sin(x_1 - x_2) + 0.01 \\ \dot{x}_2 &= -0.5\sin x_2 - 0.5\sin(x_2 - x_1) + 0.05. \end{aligned} \tag{10.4}$$

It is easy to show that the following function is an energy function for system (10.4):

Table 10.1 The coordinates of two type-one equilibrium points and their energy function values

Type-one EP	x_1	x_2	$V(.)$
1	0.04667	3.11489	−0.31329
2	−3.03743	0.33413	2.04547

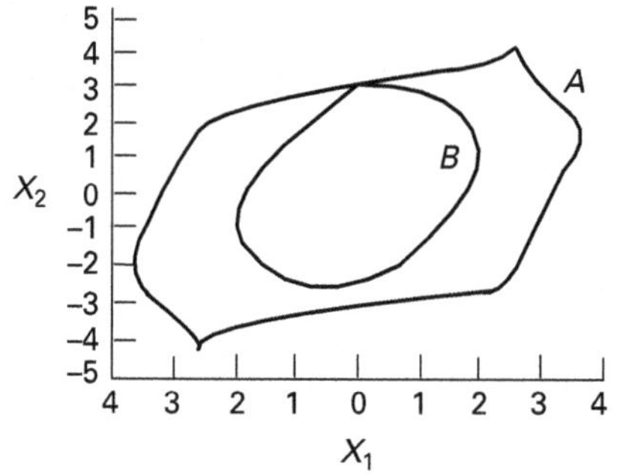

Figure 10.3 Curve A is the exact stability boundary $\partial A(x^s)$ of system (10.4). Curve B is the stability boundary estimated by the connected component (containing the SEP x_s) of the constant energy surface of the energy function (10.5) passing through −0.31329. The optimality of this estimation is also shown in the figure.

$$V(x_1, x_2) = -2\cos x_1 - \cos x_2 \quad \cos(x_1 - x_2) - 0.02x_1 - 0.1x_2. \tag{10.5}$$

The point $x^s := (x_1^s, x_2^s) = (0.02801, 0.06403)$ is the stable equilibrium point whose stability region is to be estimated. Applying the optimal scheme to system (10.4), we have the following.

Steps A1 and A2: There are two type-one equilibrium points within the region $\{(x_1, x_2) : x_1^s - \pi < x_1 < x_1^s + \pi,\ x_2^s - \pi < x_2 < x_2^s + \pi\}$, see Table 10.1.

Step A3: The type-one equilibrium point (0.04667, 3.11489) is the one with the lowest energy function value among all the UEPs on the stability boundary $\partial A(x^s)$ and its unstable manifold converges to the SEP (0.02801, 0.06403).

Step A4: The critical level value for the system (10.4) is −0.31329.

Curve A in Figure 10.3 is the exact stability boundary $\partial A(x^s)$ of system (10.4) while curve B is the stability boundary estimated by the connected component (containing the SEP x^s) of the constant energy surface of the energy function (10.5) passing through −0.31329. It can be seen from Figure 10.3 that the critical level value −0.31329 determined by the optimal estimation scheme is indeed the optimal value of the energy function (10.5) for estimating the stability region $A(x^s)$. The estimate by curve B, however, is still conservative compared to the exact stability boundary.

The optimal estimation scheme presented in this section requires the computation of the closest UEP. One issue arises: is it possible that two equilibrium points of system

(10.1) can have the same energy function value $V(\cdot)$ (i.e. $V(x_1) = V(x_2)$ for $x_1, x_2 \in E$)? This issue has been rigorously addressed in [60, Lemma 3-2], which states that, generically, all the equilibrium points of system (10.1) have distinct energy function values. Consequently, the closest UEP is generically unique. This method of estimating the stability region by picking the constant energy surface with the energy value of the closest UEP is called the closest UEP method.

10.5 Estimating the stability region via local energy functions

We will present a methodology for estimating the stability region for general nonlinear dynamical systems which do not have an energy function and analyze this methodology. The methodology is then applied to some numerical examples.

If there exists a function $V: R^n \to R$ such that it satisfies the following two conditions over some open set W containing a stable equilibrium point $x_s \in W$,

(i) the derivative of the energy function $V(x)$ along any system trajectory $x(t)$ is non-positive, i.e.

$$\dot{V}(x(t)) \leq 0$$

(ii) if $x(t)$ is a nontrivial trajectory (i.e. $x(t)$ is not an equilibrium point), then along the nontrivial trajectory $x(t)$ the set

$$\{t \in R : \dot{V}(x(t)) = 0\}$$

has measure zero in R,

then this function is then termed a *local energy function* over the open set W.

Let $V(x)$ be a local energy function around an asymptotically SEP, say x_s. Let the set $S(d, x_s)$ be the connected component of the set $\{x \in R^n: V(x) \leq d\}$ containing x_s. If the set $S(d, x_s)$ is contained entirely in the open set W, then $S(d, x_s)$ is contained in the stability region of x_s. If in particular $\hat{d}$ is the largest constant such that the above is true, then the set $S(\hat{d}, x_s)$ is the largest estimate of the stability region which one can obtain via this specific function $V(\cdot)$ and $\hat{d}$ is the optimal critical value with respect to the local energy function $V(\cdot)$.

We note that the level value $\hat{d}$ may not be the optimal critical value with respect to another local energy function. A number of algorithms for constructing local Lyapunov functions have been proposed in the literature. For example, the algorithms for constructing quadratic Lyapunov functions by solving a set of Lyapunov equations and for determining the corresponding critical level value appeared in [80]. Another class of algorithms for constructing local energy functions via normed Lyapunov functions can be found in [182]. The subject of how to find an optimal local energy function such that the largest estimate of the stability region can be obtained is attracting significant research. Recent results along this direction include the LMI approach [13,40,41,42].

A methodology for estimating stability regions of general nonlinear dynamical systems not having a global energy function is presented below.

Step 1: Construct a local energy function, say $V(\cdot)$ for the stable equilibrium point x_s. (Let $S(d, x_s)$ be the connected component containing x_s of the set $\{x \in R^n : V(x) \leq d\}$.)

Step 2: Determine the critical level value of $V(\cdot)$ for x_s such that conditions (i) and (ii) are satisfied for $x \in \{S(d, x_s)\} - \{x_s\}$. (Let $\hat{d}$ be the largest number for this part to be true.)

Step 3: Estimate the stability region $A(x_s)$ via the local energy function $V(\cdot)$. The connected component $S(d, x_s)$ of $\{x : V(x) \leq \hat{d}\}$ containing the stable equilibrium point x_s gives the estimated stability region for $A(x_s)$.

There are several approaches proposed for constructing local energy functions. One well-known method for constructing a local energy function (i.e. a local Lyapunov function) $V(x)$ for a stable equilibrium point, say x_s, of (10.1) is as follows.

Step 1: Solve the following Lyapunov matrix equation for matrix B

$$J^T B + BH = -C \tag{10.6}$$

where J is the Jacobian matrix at x_s and C is a symmetric, positive definite matrix and is usually chosen to be the identity matrix.

Step 2: Construct a local Lyapunov function which is a quadratic function, $V(x) = x^T B_x$.

This quadratic Lyapunov function can be used to estimate the stability region of x_s. Suppose that there is an open set W such that

$$\dot{V}(x) = 2f(x)^T Bx < 0 \text{ for } x \in \{W - x_s\}. \tag{10.7}$$

Step 3: Determine the critical value for the local Lyapunov function.

Find the critical value $\hat{d}$ such that it is the largest value for which the set $S(\hat{d}, x_s) := \{x \in R^n : x^T Bx \leq \hat{d}\}$ is contained in the set W.

Step 4: Estimate the stability region via the level set of the constructed Lyapunov function. The set $S(\hat{d}, x_s) := \{x \in R^n : x^T Bx \leq \hat{d}\}$ is contained in the stability region of x_s.

We illustrate the above method on several low-dimensional nonlinear systems.

Example 10-1 This is the well-known Vanderpol equation; here (0,0) is a stable equilibrium point of this equation whose stability region is to be estimated.

$$\begin{aligned}\dot{x}_1 &= -x_2\\ \dot{x}_2 &= x_1 - (1 - x_1^2)x_2.\end{aligned}$$

Applying the above methodology we have the following.

Steps 1 and 2: The constructed local Lyapunov function is

$$V(x_1, x_2) = (x_1, x_2)^t \begin{bmatrix} 0,593 & -0,182 \\ -0,182 & 0,437 \end{bmatrix} (x_1, x_2).$$

Step 3: The critical level value of $V(x_1, x_2)$ for (0,0) is 1.0, obtained by applying an optimization procedure for finding the largest value d.

Step 4: The connected component of $\{x\colon V(x_1, x_2) \leq 1{,}0\}$ containing (0,0) is the estimated stability region.

The estimated stability region and the exact stability region are all shown in Figure 10.4. It can be seen that the estimated stability region is contained in the entire stability region of (0,0).

Example 10-2 This is an example studied in [80,182]. The goal is to estimate the stability region of the origin of the system

$$\dot{x}_1 = -x_1 + 2x_1^2 x_2$$
$$\dot{x}_2 = -x_2.$$

Steps 1 and 2: The local Lyapunov function is

$$V(x_1, x_2) = (x_1, x_2)^t \begin{bmatrix} 0.330 & 0.249 \\ 0.249 & 0.376 \end{bmatrix} (x_1, x_2).$$

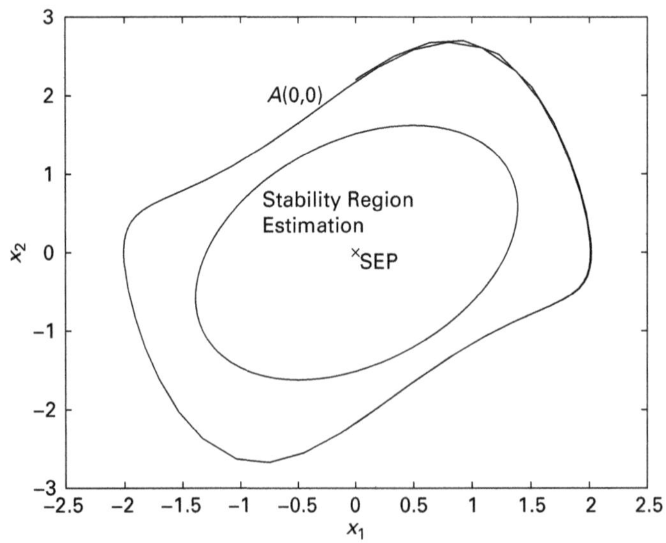

Figure 10.4 The stability boundary and the estimated stability region of the origin are highlighted with the estimated stability region contained in the entire stability region of (0,0).

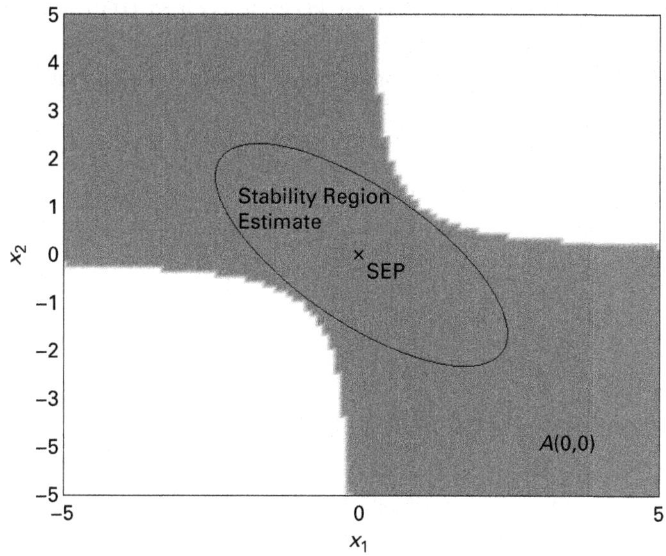

Figure 10.5 The estimated stability region is contained in the entire stability region of (0,0) and it is an optimal estimation with respect to the constructed local Lyapunov function.

Step 3: The critical level value of $V(x_1, x_2)$ for (0,0) is 1.0, obtained by applying an optimization procedure for finding the largest value d.

Step 4: The connected component of $\{x: V(x_1, x_2) \leq 1.0\}$ containing (0,0) is the estimated stability region.

The exact stability region and the estimated stability region are shown in Figure 10.5. Indeed, the estimated stability region is contained in the entire stability region of (0,0) and it is an optimal estimation with respect to the constructed local Lyapunov function.

Example 10-3 Consider the following three-dimensional system which has a stable equilibrium point at the origin and a limit cycle. The task of estimating the stability region of the origin proceeds as follows:

$$\begin{aligned}
\dot{x}_1 &= -x_2 \\
\dot{x}_2 &= -x_3 \\
\dot{x}_3 &= -0.915x_1 + (1 - 0.915x_1^2)x_2 - x_3.
\end{aligned}$$

Steps 1 and 2: The local Lyapunov function is

$$V(x_1, x_2, x_3) = (x_1, x_2, x_3)^t \begin{bmatrix} 12.5 & -8.1 & 3.0 \\ -8.1 & 20.8 & -8.5 \\ 3.0 & -8.5 & 13.4 \end{bmatrix} (x_1, x_2, x_3).$$

Step 3: The critical level value of $V(x_1, x_2, x_3)$ for (0,0,0) is 1.0, obtained by applying an optimization procedure for finding the largest value d.

Step 4: The connected component of $\{x : V(x_1, x_2, x_3) \leq 1.0\}$ containing (0,0,0) is the estimated stability region.

The stability region estimated using this method is colored dark gray in Figure 10.6 and Figure 10.7 respectively with different angles of view, while the exact stability boundary is highlighted in a light gray color. It can be seen that the estimated stability region is contained in the entire stability region of (0,0,0).

10.6 Optimal estimation of Lyapunov functions

In this section, we discuss a class of methods used to find Lyapunov functions for specific classes of nonlinear dynamical systems for estimating stability regions. This class of methods formulates the problem of finding a Lyapunov function in the form of an optimization problem, with the aim of maximizing the size of the estimated stability region. In this section, we admit, without loss of generality, that the origin is the asymptotically stable equilibrium point of interest. The estimated stability region is based on the following theorem.

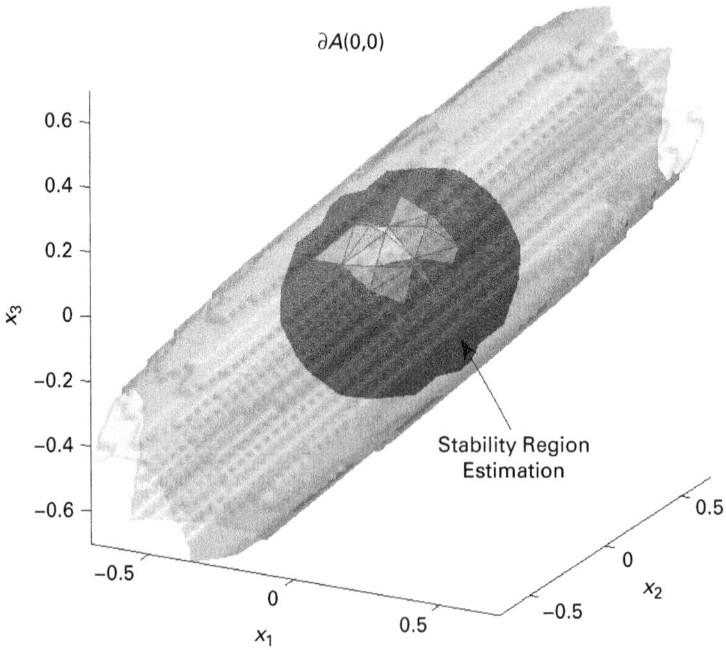

Figure 10.6 The estimated stability region is shown in dark gray while the exact stability boundary is a highlighted light gray transparent surface.

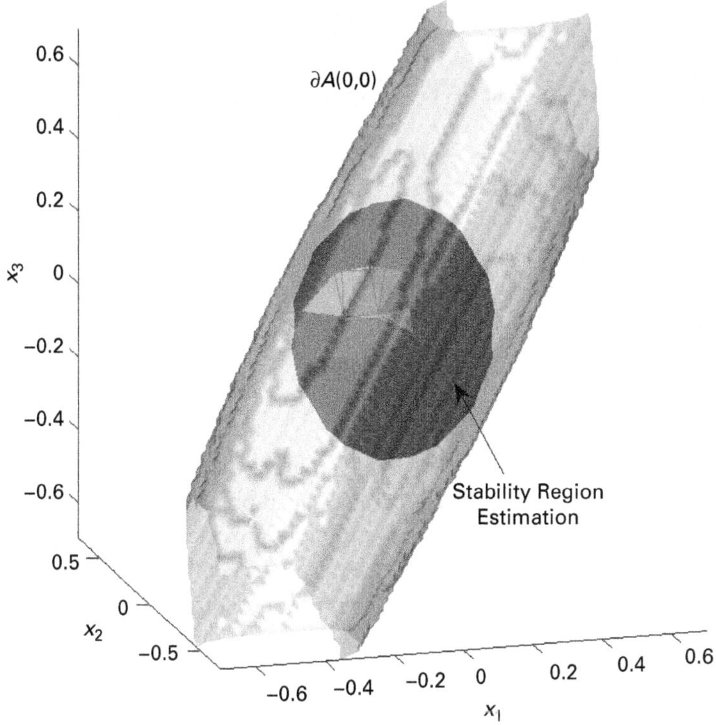

Figure 10.7 The estimated stability region is shown in dark gray while the exact stability region is a highlighted light gray transparent surface.

THEOREM 10-8 *Consider the nonlinear dynamical system $\dot{x} = f(x)$, with the origin as an asymptotically stable equilibrium point. Let $V: R^n \rightarrow R$ be a scalar C^1-function and let G be a domain containing the origin such that $\dot{V}(x) < 0$ for any $x \in G$. Let c be a real number such that the connected component $S(c, 0)$ of the level set $\{x \in R^n: V(x) < c\}$ containing the origin is bounded and contained in G, then $S(c, 0)$ is contained in the stability region of the origin.*

Theorem 10-8 justifies the task of estimating a stability region in the form of the level set of a Lyapunov function. However, Lyapunov functions are not unique and some Lyapunov functions may lead to very conservative estimates of stability regions. With the aim of estimating stability regions, a great effort has been made to develop computational methods to determine "optimal" Lyapunov functions for certain classes of nonlinear dynamical systems to maximize the estimated stability region.

Based on Theorem 10-8, the task of estimating stability regions can be divided into two steps. Step 1 seeks a function V defined in an open set G, containing the origin, such that the conditions of Theorem 10-8 are satisfied. Maximizing the set G is one scheme to obtain larger estimates of stability regions in Step 2. Consequently, Step 1 can be modeled as the following maximization problem:

$$\begin{aligned} &\max_V \text{ size of } G \\ &\text{such that } V(0) = 0 \\ V(x) > 0 \quad &\text{for all} \quad x \in G \backslash \{0\} \\ \dot{V}(x) = \frac{\partial V}{\partial x} f(x) < 0 \quad &\text{for any} \quad x \in G \backslash \{0\}. \end{aligned} \tag{10.8}$$

Once a function $V(\cdot)$ and a set G are determined, then the largest estimation of the stability region that can be obtained with $V(\cdot)$ is determined by solving a maximization problem in Step 2. This step is to determine the largest level set of $V(\cdot)$ which is entirely contained in the set G:

$$\begin{aligned} &\max \mathrm{L} \\ &\text{such that } \{x \in R^n : V(x) < L\} \subset G. \end{aligned} \tag{10.9}$$

There are several challenges to solving problems (10.8) and (10.9). Problem (10.9) is usually a nonconvex nonlinear programming problem. Theorem 10-8 relies on the existence of a Lyapunov function on a set G to provide estimates of stability regions, but does not offer any clue on how to find them. The combination of problems (10.8) and (10.9) leads to a "double" nonconvex nonlinear optimization problem.

We expect that maximizing the set G in Step 1 would lead to larger estimates of stability regions in Step 2; however, the choices of set G and function $V(\cdot)$ are interrelated. A general approach to maximize the stability region estimate is via solving the following general optimization problem:

$$\begin{aligned} &\max \text{ size of } \{x \in R^n : V(x) < L\} \\ &\text{such that } V(0) = 0 \\ V(x) > 0 \quad &\text{for all} \quad x \in G \backslash \{0\} \\ \dot{V}(x) = \frac{\partial V}{\partial x} f(x) < 0 \quad &\text{for all} \quad x \in G \backslash \{0\} \\ &\{x \in R^n : V(x) < L\} \subset G. \end{aligned} \tag{10.10}$$

The optimization problem (10.10) is very difficult to solve. Usually, it can be efficiently solved for certain classes of low-dimensional vector fields, certain forms of set G and certain classes of Lyapunov functions $V(\cdot)$.

A popular class of Lyapunov functions is the class of quadratic functions with the form $V(x)=x^T Px$, with matrix P positive definite. Without loss of generality, we set $L=1$ in (10.10). Since $V(\cdot)$ is a quadratic function, it is well known that the set $\{x \in R^n\colon V(x) < 1\}$ is an ellipsoid centered at the origin and that the volume of this ellipsoid, according to the principal axis theorem [239], is inversely proportional to the following scalar

$$\beta(P) = \left(\prod_1^n \lambda_i(P)\right)^{1/2}$$

where $\lambda_i(P)$ are the eigenvalues of matrix P. Hence, the problem (10.10) can be rewritten as:

$$\begin{gathered}\min_{P,G} \beta^2(P) \\ P > 0 \\ \dot{V}(x) = 2x^T P f(x) < 0 \quad \text{for any} \quad x \in G \backslash \{0\} \\ \{x \in R^n : x^T P x < 1\} \subset G.\end{gathered} \tag{10.11}$$

Davison and Kurak [80] and Michel *et al.* [182], for example, proposed methods to solve problem (10.11) by numerically checking on a grid of the state space the satisfaction of the nonconvex constraint: $\dot{V}(x) = 2x^T P f(x) < 0 \, for \, any \, x \in G \backslash \{0\}$ with $G = \{x \in R^n \colon x^T P x < 1\}$.

With the purpose of maximizing the size of set G and developing a systematic procedure for finding the scalar function $V(\cdot)$, set G is usually written in terms of a level set of a function g as $G = \{x \in R^n \colon g(x) < 0\}$. Hence, the problem of ensuring the derivative of function $V(\cdot)$ is negative definite inside a set G can be formulated as:

$$\dot{V}(x) - \alpha g(x) < 0 \Rightarrow \dot{V}(x) < 0 \quad \text{for any} \quad x \in G \tag{10.12}$$

where α is a positive real constant. Condition (10.12) is sufficient for $\dot{V}(x) < 0$ for any $x \in G$ and the problem of seeking a Lyapunov function can thus be formulated as follows:

$$\begin{gathered}\min_{P,G,\alpha} \beta^2(P) \\ P > 0, \alpha < 0 \\ 2x^T P f(x) - \alpha g(x) < 0 \\ \{x \in R^n : x^T P x < 1\} \subset G.\end{gathered} \tag{10.13}$$

One can choose $g(x) = c - V(x)$. In this case, set G will be a level set of function $V(\cdot)$. If $c \geq 1$, then the last constraint set of problem (10.13) is automatically satisfied and the problem of maximizing the size of the stability region estimate can be rewritten as:

$$\begin{gathered}\min_{P,\alpha} \beta^2(P) \\ P > 0, \alpha > 0 \\ 2x^T P f(x) - \alpha(1 - x^T P x) < 0.\end{gathered} \tag{10.14}$$

Several variations of formulation (10.14) have been presented in the literature with the aim of optimally searching for Lyapunov functions. Genesio and Vicino [98] studied this problem for the particular class of second-order systems with quadratic nonlinearity. Amato *et al.* [13] developed a method based on linear matrix inequalities (LMIs) to check whether a polytope is inside the stability region of quadratic systems. Chesi *et al.* [40,41,42] studied this problem for the particular class of polynomial systems. When the vector field f is of polynomial type, then the problem (10.14) can be written in terms of LMIs. It is not the objective of this book to introduce the theory of LMIs. For further reading on this subject, see for example [27].

The approach described of searching for Lyapunov functions by solving an optimization problem has the merit of making the process of finding the Lyapunov function automatic. These methods are known in the literature as having the property of selecting "the optimal Lyapunov function." The optimality, however, is achieved only for a particular class of Lyapunov functions, usually the quadratic ones. There is no guarantee that other classes of Lyapunov functions can yield larger sizes of estimated stability region. In addition, solving the optimization problems is very difficult due to the large number of variables and the nonlinear nature of the problem. These optimization problems can be efficiently solved only for particular classes of nonlinear systems and for low-dimensional dynamical systems. Hence, one key problem with this LMI approach is its scalability. Another problem with this approach is that the Lyapunov functions derived by these procedures are just numerical and are not given in general forms, i.e. changes in parameters require solving the optimization problem again.

10.7 Concluding remarks

In this chapter, we have presented schemes to optimally estimate stability regions of continuous nonlinear systems using energy functions. A theoretical basis for the optimal estimation schemes has also been developed. The relationship between the structures of the quasi-stability boundary and the constant energy surface at different level values has been established. Optimal schemes to estimate stability regions and quasi-stability regions have been presented and these schemes form the basis for the closest UEP method.

A topological characterization and a dynamical characterization for the closest UEP of an SEP with respect to an energy function have been derived. Generically speaking, the closest UEP is a type-one equilibrium point and the one-dimensional unstable manifold of the closest UEP converges to the SEP.

Depending on whether the nonlinear dynamical system under study has an energy function or not, the task of estimating the stability region requires two different estimation procedures. For systems having energy functions, the task starts with an energy function $V(\cdot)$ associated with the system under study and proceeds in the following three steps.

Procedure A (For nonlinear dynamical systems having an energy function)

Step 1: Construct an energy function (or energy function).

Step 2: Determine the closest UEP of an SEP with respect to the constructed energy function.

Step 3: Approximate the stability boundary of the SEP via the constant energy surface passing through the closest UEP.

For systems having local energy functions, the task of estimating the stability region starts with a local energy function $V(\cdot)$ associated with the system under study and proceeds in the following three steps.

Procedure B (For nonlinear dynamical systems only having a local energy function)

Step 1: Construct a local energy function around the SEP whose stability region is to be estimated.

Step 2: Determine the critical level value of the constructed local energy function with respect to the SEP.

Step 3: Approximate the stability boundary using the constant surface defined by the constructed local energy function with the critical level value determined at Step 2.

Several numerical methods have been developed to implement Procedure B. The subject of how to efficiently compute the closest UEP of an SEP with respect to an energy function will be presented in a later chapter.

11 Estimating stability regions of complex continuous dynamical systems

A comprehensive theory for estimating stability regions of a fairly large class of nonlinear dynamical systems admitting energy functions was developed in Chapter 10. In this chapter, we discuss how to estimate the stability region of an even larger class of dynamical systems, the class of general nonlinear dynamical systems introduced in Chapter 5. In Chapter 5, we concluded that general nonlinear dynamical systems can exhibit complex behaviors such as periodic or quasi-periodic orbits and chaos, making estimation of the stability region of these systems a challenge.

General nonlinear dynamical systems may exhibit complex behavior; therefore, they do not admit energy functions. As a result, the schemes developed in Chapter 10 cannot be applied to estimate the stability regions of this class of system. Generalized energy functions will be exploited in this chapter as a practical alternative for estimating stability regions of this class of general nonlinear dynamical system. The methods developed in this chapter provide estimates of the stability region in the form of a level set of the generalized energy function.

We first investigate the properties of generalized energy functions on the stability boundary of these general nonlinear dynamical systems. Afterwards, we explore these analytical results to develop optimal schemes for estimating the stability regions of general dynamical systems, including those that exhibit complex behavior such as closed orbits, quasi-periodic orbits and chaos.

11.1 Stability region estimation

We consider a general class of nonlinear autonomous dynamical system described by the following differential equation:

$$\dot{x}(t) = f(x(t)). \tag{11.1}$$

We are interested in estimating the stability region of an attracting set γ of system (11.1). For the purpose of estimating the stability region, we assume system (11.1) admits a generalized energy function V. Recall that a C^r-function $V:R^n \rightarrow R$, with $r > 1$ is said to be a generalized energy function if it satisfies the following conditions.

(G1) The number of connected components C_i of the set $C := \{x \in R^n : \dot{V}(x) \geq 0\}$, where the derivative of V is non-negative, is finite.

(G2) Every component C_i is bounded.

(G3) That a trajectory $x(t)$ has a bounded value of $V(x(t))$ for $t \in R^+$ implies that the trajectory $x(t)$ is also bounded for $t \in R^+$. The set where the derivative of the generalized energy function equals zero will be denoted M, i.e.

$$M = \{x \in R^n : \dot{V}(x) = 0\}. \tag{11.2}$$

It is clear that the set M is contained in C. Even though computation of set C is a challenging task, it is relatively easy to compute a set D which is an "upper" bound of C (i.e. $C \subset D$). All the analytical results to be presented in the following still hold if we substitute C by its estimate D. The estimates of attractors and stability regions will be less conservative, as the estimates D of C are less conservative.

We have shown in Chapter 3 that generalized energy functions provide useful information regarding the limit sets of general nonlinear systems, including those limit sets exhibiting complex behavior on the stability boundary. Moreover, we have shown in Chapter 5 that a generalized energy function is bounded below over a stability boundary, even though the stability boundary may be unbounded (see Lemma 5-7). This leads us to another useful property: every trajectory on the stability boundary is bounded and hence converges to one of the ω-limit sets. In addition, this property leads to a comprehensive characterization of the stability boundary as shown in Chapter 5.

The characterization of the stability region developed in Chapter 5 offers a way to determine the exact stability region. This characterization, however, requires the computation of complex invariant sets on the stability boundary and their associate stable manifolds. Although the computation of complex invariant sets and their stable manifolds can be very challenging, it offers a practical method to estimate the stability region of general nonlinear dynamical systems, as will be shown in this chapter.

In this chapter, we discuss various methods to determine critical level values of generalized energy functions for optimally estimating the stability regions of general dynamical nonlinear systems, including those that admit complex behavior such as closed orbits, quasi-periodic orbits and chaos. To this end, we next develop a topological characterization of points with a local minimum of generalized energy function values over the stability boundary of a general nonlinear dynamical system (11.1).

PROPOSITION 11-1 **(Topological characterization)**
Consider the nonlinear dynamical system (11.1) admitting a generalized energy function $V: R^n \to R$. Let M denote the set where the derivative of the generalized energy function equals zero and let γ be an attracting set whose stability region is not dense in R^n. Then, the following results hold.

(a) The point with the minimum value of the generalized energy function over the stability boundary $\partial A(\gamma)$ exists.

(b) The points which are local minima of the generalized energy function $V(x)$ on the stability boundary $\partial A(\gamma)$ are attained on the set

$$M \cap \partial A(\gamma).$$

Proof The existence of a global minimum value of the generalized energy function is a direct consequence of Lemma 5-7 of Chapter 5, which guarantees the existence of a lower bound of $V(x)$ on $\partial A(\gamma)$. Suppose $\hat{x} \in \partial A(\gamma)$ is a local minimum of V on $\partial A(\gamma)$, and suppose by contradiction that $\hat{x} \notin M$. Then $\dot{V}(\hat{x}) \neq 0$, and since all equilibrium points lie in the set $M, f(\hat{x}) \neq 0$. Consider the trajectory $\phi(t, \hat{x})$ passing through $\hat{x}$. Since $\partial A(\gamma)$ is an invariant set, $\phi(t, \hat{x}) \in \partial A(\gamma)$ for $t \in R$. If $\dot{V}(\hat{x}) < 0$, then there exists a time $t^* > 0$ such that $V(\phi(t, \hat{x})) < V(\hat{x})$ for every $0 < t < t^*$. This implies that arbitrarily close to $\hat{x}$ there exists $\hat{y} \in \partial A(\gamma)$ such that $V(\hat{y}) < V(\hat{x})$. But this is a contradiction. A similar argument applies if $\dot{V}(\hat{x}) > 0$ for some negative time $t^* < 0$. Therefore, every local minimum belongs to M. This concludes the proof.

The global minimum of a generalized energy function $V(x)$ on the stability boundary plays an important role in the estimation of the stability region. According to the above proposition, the search for the global minimum has to be made on M. Set M is contained in C and, therefore, M is composed of isolated components $M_i \subseteq C_i$. Usually, M_i coincides with the boundary of set C_i. As a result, the global minimum of the generalized energy function has to be searched for on every component M_i of those components C_i of C with a non-empty intersection with the stability boundary $\partial A(\gamma)$. Since every C_i is compact and V is a continuous function, the minimum L_i of V in each connected component M_i exists. Obviously, the following inequality holds

$$L_i \leq \min_{x \in M_i \cap \partial A(\gamma)} V(x). \tag{11.3}$$

That is, the minimum L_i of V on M_i is a local lower bound for the minimum of V on the stability boundary. It is important to highlight that each number L_i can be computed without knowing the stability boundary. Moreover, the L_i can be computed without knowing the nature of invariant sets and without computing invariant sets on the stability boundary.

Since there is a finite number of sets C_i on the stability boundary, there exists a number L satisfying $L := \min_i L_i$. Generically, there exists a unique connected component C_k such that $L = L_k$. This numerical value L can be used to provide an estimate of the stability region via a level set of the generalized energy function V. The next theorem presents sufficient conditions to guarantee that the connected component $S(L, \gamma)$ of the level set $\{x \in R^n : (x) < L\}$ containing γ is an estimate of the stability region ($S(L, \gamma) \subset A(\gamma)$).

THEOREM 11-2 **(Estimation of the stability region)**
Let γ be an attracting set of system (11.1) admitting a generalized energy function $V(x)$. Let C_i, $i = 1, 2, \ldots, m$ be the connected components of set C with a nonzero intersection with the stability boundary $\partial A(\gamma)$. Let $L_i = \min_{x \in M_i} V(x)$ and define $L = \min_i L_i$. Then, $S(L,\gamma)$, the connected component of the level set $\{x \in R^n : (x) < L\}$ containing γ, is an estimate of the stability region in the sense that $S(L,\gamma) \subset A(\gamma)$.

Proof Suppose by contradiction that $S(L,\gamma) \not\subset A(\gamma)$. Since $S(L,\gamma)$ is a connected set and γ is contained in $S(L,\gamma)$, there must exist at least one point $\hat{y}$, such that $\hat{y} \in \partial A(\gamma)$ and $\hat{y} \in S(L, \gamma)$. Since L is a lower bound for V on $\partial A(\gamma)$, one has $V(\hat{y}) \geq L$. But by the definition of $S(L, \gamma)$ it follows that $V(\hat{y}) < L$; hence, we reach a contradiction. This completes the proof.

Theorem 11-2 provides an analytical foundation for estimating stability regions of general nonlinear dynamical systems. It offers a scheme to derive these estimates via level sets of generalized energy functions. This method resembles the one developed in Chapter 10 for estimating stability regions of nonlinear dynamical systems admitting energy functions. Instead of searching for equilibrium points on the stability boundary, one searches for the components C_i of C intersecting the stability boundary. In the particular case when every set C_i is a hyperbolic equilibrium point on the stability boundary, then the scheme proposed in this section for estimating stability regions degenerates to the scheme proposed in Chapter 10. Therefore, the scheme proposed in this section can be viewed as a generalization of the scheme for estimating stability regions of nonlinear dynamical systems admitting energy functions.

For general nonlinear dynamical systems, the invariant sets lying on the stability boundary can be very complex. They can include quasi-periodic solutions and chaotic behaviors. In spite of that, the analytical results developed in Theorem 11-2 for estimating the stability region do not rely on the computation of these invariant sets, unlike the scheme of Chapter 10, which relies on the computation of unstable equilibrium points. Alternatively, the estimation can be obtained simply from knowledge of the generalized energy function. This feature is important because the computation of invariant sets, such as unstable closed orbits on the stability boundary, is still computationally challenging. The advantage of not requiring computation of invariant sets comes with one disadvantage, which is related to the optimality of the estimation. The estimate $S(L, \gamma)$ of the stability region $A(\gamma)$ may not be the optimal estimate of the stability region one can obtain via a generalized energy function. There may exist a higher level value $L_2 > L$ such that $S(L, \gamma) \subset S(L_2, \gamma) \subset A(\gamma)$. However, the value L may be the best computational estimate one can calculate without the need for computing invariant sets lying on the stability boundary and locating the point with the global minimum of a generalized energy function on the stability boundary.

It is obvious that the task of computing invariant sets lying on the stability boundary and the task of locating the point with the global minimum of a generalized energy function on the stability boundary are rather computationally challenging. To illustrate this difficulty, we show the relationship between different level surfaces and the exact stability region in Figure 11.1. As the level value of V increases, the corresponding estimated stability region also grows in size. The estimated stability region certainly remains inside the exact stability region until the estimated stability region intersects the connected components of C with a nonzero intersection with the stability boundary. Beyond this point, there is no guarantee that the level set will be entirely contained in the stability region.

It is also important to note that there is no guarantee that $S(L, \gamma)$ is a positively invariant set. Figure 11.2 shows a situation where the set $S(L, \gamma)$ is not positively invariant. We next derive conditions to guarantee the positive invariance of $S(L, \gamma)$.

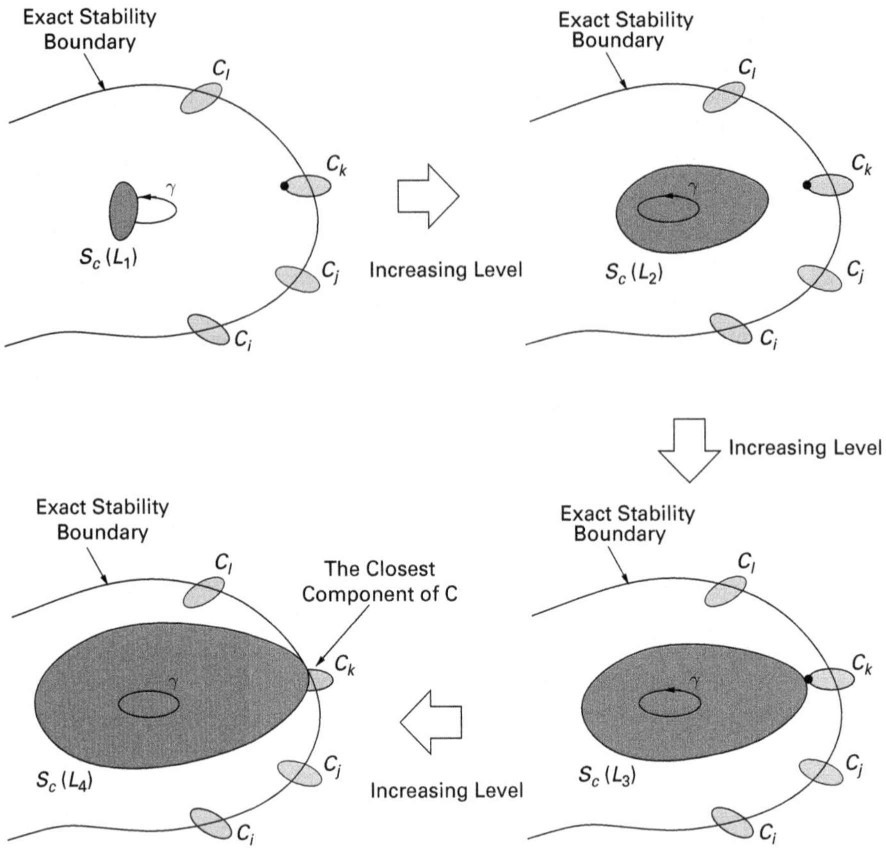

Figure 11.1 The relationship between different level surfaces of a generalized energy function and the exact stability region. As the level value is increased, the corresponding estimated stability region enlarges but remains inside the exact stability region until the estimated stability boundary intersects the connected components of C with a nonzero intersection with the stability boundary.

THEOREM 11-3 **(Positive invariance)**
Let γ be an attracting set of system (11.1) admitting a generalized energy function $V(x)$. Suppose $S(L, \gamma)$ is an estimate of the stability region $A(\gamma)$ and define $l = \sup_{x \in s(L,\gamma) \cap C} V(x)$. *If $l<L$, then $S(L, \gamma)$ is a positively invariant subset of $A(\gamma)$.*

Proof From Theorem 11-2, one has $S(L, \gamma) \subset A(\gamma)$. To finish the proof, we just need to show that $S(L, \gamma)$ is positively invariant. Suppose by contradiction that $x_0 \in S(L, \gamma))$ and that there exists a time t^* such that $\phi(t^*, x_0) \notin S(L, \gamma)$. Then, $V(x_0) < L$. The continuity of solutions guarantees the existence of a first time $\hat{t}$, satisfying $\hat{t} < t*$ such that $V(\phi(\hat{t}, x_o)) = L$. On the other hand, the condition $l = \sup_{x \in s(L,\gamma) \cap C} V(x) < L$ guarantees the existence of a time $\tilde{t}$ satisfying $0 \leq \tilde{t} \leq \hat{t}$ such that $V(\phi(\tilde{t}, x_o)) < L$ and $\dot{V}(\phi(t, x_o)) < 0$ for all $t \in (\tilde{t}, \hat{t})$. This is absurd, and consequently, the theorem is true.

A relationship between two level sets $S(L, \gamma)$ and $S(l, \gamma)$ of Theorem 11-3 is illustrated in Figure 11.3. The set $S(L, \gamma)$ is now a positively invariant subset of $A(\gamma)$.

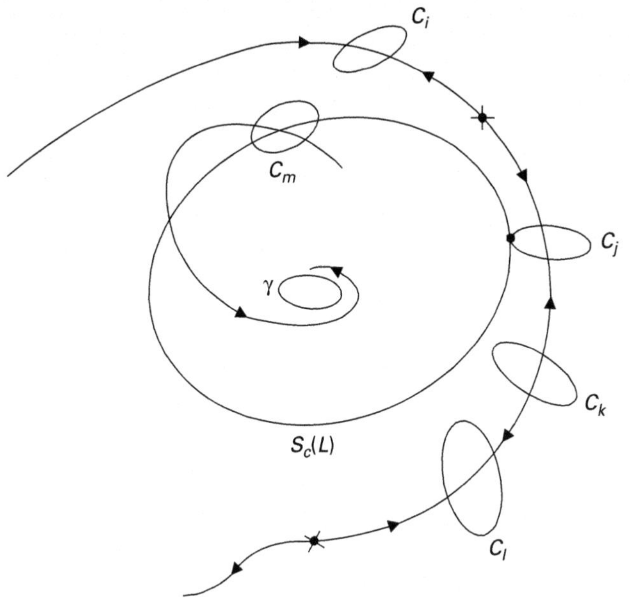

Figure 11.2 Set $S(L, \gamma)$ is an estimate of the stability region of γ, that is, $S(L, \gamma) \subset A(\gamma)$. However, there is no guarantee that it is a positively invariant set.

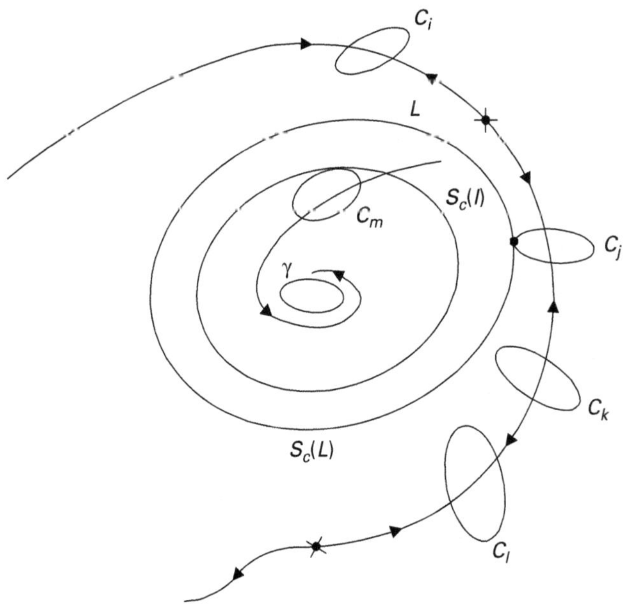

Figure 11.3 For $l < L$, set $S(L, \gamma)$ is a positively invariant subset of $A(\gamma)$.

Based on the above analytical results, we present the following conceptual scheme to estimate the stability region of a complex nonlinear system via a generalized energy function.

Scheme

Step 1: Find all the components C_i of set C intersecting the stability boundary of γ.

Step 2: Compute the minimum L_i of V on every component $M_i \subseteq C_i$ and calculate $L = \min_i L_i$.

Step 3: The connected component $S(L, \gamma)$ of the level set $\{x \in R^n: V(x) < L\}$ containing γ is an estimate of the stability region.

The above computational scheme for using the generalized energy function to estimate the stability region of general nonlinear dynamical systems extends previous results for estimating stability regions in the following senses.

(i) For complex nonlinear systems, an energy function does not exist and therefore cannot be used to estimate the stability regions of complex nonlinear systems. The generalized energy function and the previous proposed scheme allow for the estimation of the stability regions of general nonlinear dynamical systems.

(ii) The scheme does not require the calculation of critical elements such as closed orbits or chaotic orbits on the stability boundary, but it requires the location of sets C_i.

The proposed scheme may encounter the following numerical difficulties: (i) a systematic procedure is needed to check whether a component C_i has a non-empty intersection with the stability boundary, and (ii) the task of calculating L is a constrained minimization problem which can be difficult to solve. Additional work to develop robust numerical methods for estimating stability regions is still needed.

11.2 Numerical examples

We illustrate the proposed scheme for estimating stability regions of general dynamical systems in the following simple examples.

Example 11-1 We consider the same three-dimensional nonlinear dynamical system of Example 5-1 of Chapter 5 described by the following set of differential equations:

$$\begin{cases} \dot{x} = \dfrac{(\sigma - \alpha\rho^2)x - \sigma y}{1 + \alpha\rho^2} \\[2ex] \dot{y} = \dfrac{(1 - \alpha\rho^2)y + xz - rx}{1 + \alpha\rho^2} \\[2ex] \dot{z} = \dfrac{(b - \alpha\rho^2)z - xy}{1 + \alpha\rho^2}. \end{cases} \tag{11.4}$$

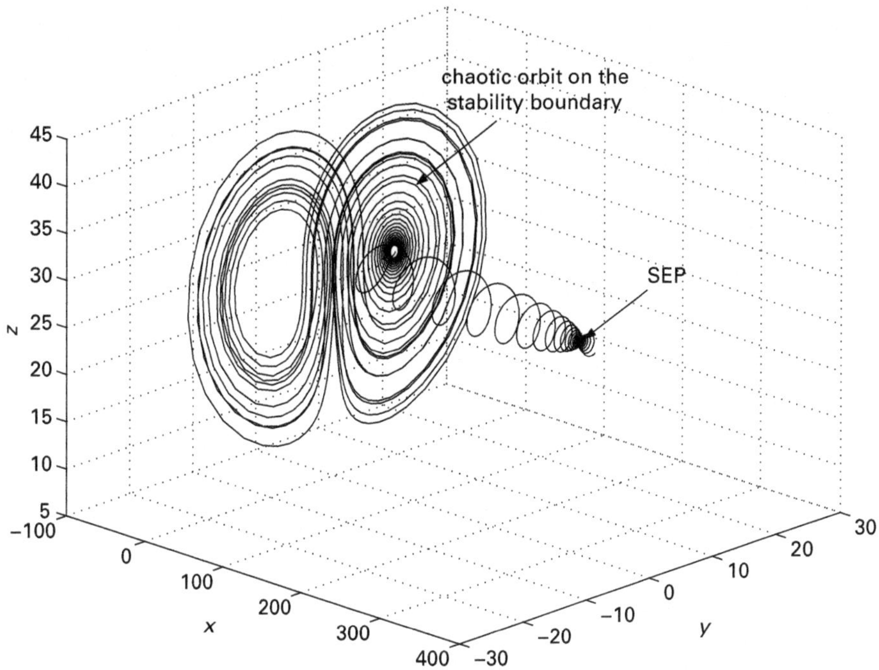

Figure 11.4 Complex behavior lying on the stability boundary of an asymptotically stable equilibrium point of system (11.4).

Where $\rho^2 = x^2 + y^2 + z^2$, $\sigma = 10$, $r = 28$ and $b = 8/3$. This system has an asymptotically stable equilibrium point at approximately (315, −0.65, 28). Numerical integration suggests the presence of very complex behavior on the stability boundary of this system, see Figure 5.6 in Chapter 5 and Figure 11.4. As a consequence, this system does not admit an energy function.

We employ a generalized energy function to estimate its stability region. Consider the following function:

$$V(x,y,z) = ax^4 - bx^2 + cy^4 + d(z-\beta)^4 \tag{11.5}$$

where $a = 1/4$, $b = 315^2/2$, $c = d = 8$ and $\beta = 28$. Here, function $V(x,y,z)$ is shown to be a generalized energy function for system (11.4).

Figure 11.5 shows the set C. Set C is composed of a single bounded component. This component intersects the stability boundary of the stable equilibrium point. Invariance of the stability boundary and Theorem 3-6 of Chapter 3 assert that every trajectory on the stability boundary is bounded and converges to a limit set that has a nonzero intersection with C.

For this example, the number $L = -4.43 \times 10^8$ is the minimum of V on the set M. Figure 11.5 shows the connected component of the level set $S(L, \gamma)$ containing the stable equilibrium point. This level set is an estimate of the stability region. This numerical result confirms the theoretical results of Theorem 11-2.

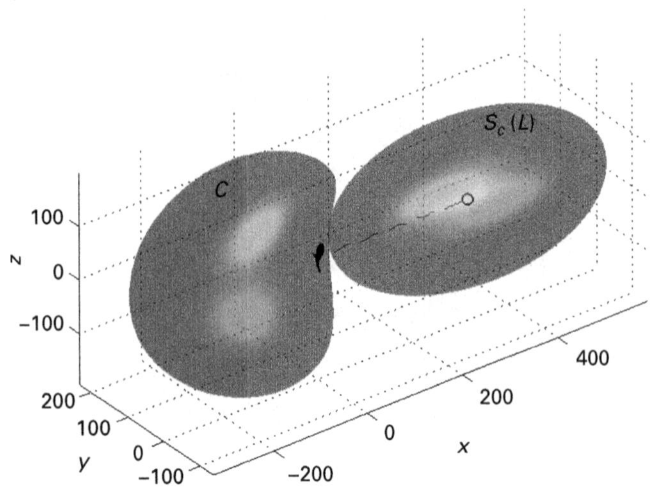

Figure 11.5 The complex invariant set on the stability boundary intersecting set C and the stability region estimate obtained via a generalized energy function.

Example 11-2 The following system of equations was obtained from the power system literature and was proposed to model the dynamical behavior of a two-generator system versus an infinite bus [29,47,233] considering transfer conductances:

$$\begin{cases} \dot{x}_1 = y_1 \\ \dot{x}_2 = y_2 \\ M_1\dot{y}_1 = P_1 - G_1 \sin x_1 - B_1 \cos x_1 - \\ \qquad - G_{12} \sin(x_1 - x_2) - \varepsilon \cos(x_1 - x_2) - D_1 y_1 \\ M_2\dot{y}_2 = P_2 - G_2 \sin x_2 - B_2 \cos x_2 - \\ \qquad - G_{12} \sin(x_2 - x_1) - \varepsilon \cos(x_2 - x_1) - D_2 y_2 \end{cases}$$

Parameter ε is used to represent the so-called transfer conductance. It has been shown that a general energy function does not exist for this system when $\varepsilon \neq 0$ [47]. Therefore, the traditional energy function methods for estimating and characterizing the stability region cannot be applied. For this reason, we have provided a generalized energy function for this system in Chapter 3 satisfying assumptions (G1)–(G3). Using this function, we show that an estimate of the stability region can be obtained.

Figure 11.6 shows the level sets of V for the following parameters: $P_1 = 1.78$, $P_2 = 3.83$, $G_1 = 3.16$, $G_2 = 7.85$, $B_1 = 0.28$, $B_2 = 0.255$, $G_{12} = 0.9$, $\varepsilon = 0.1$, $D_1 = D_2 = 0.1$, $M_1 = 0.053$, $M_2 = 0.079$, $\alpha = 13.017$ and $\beta = 0.005$. These sets were drawn in the plane $y_1 = y_2 = 0.8$ rad/s. The region where the derivative of V is positive

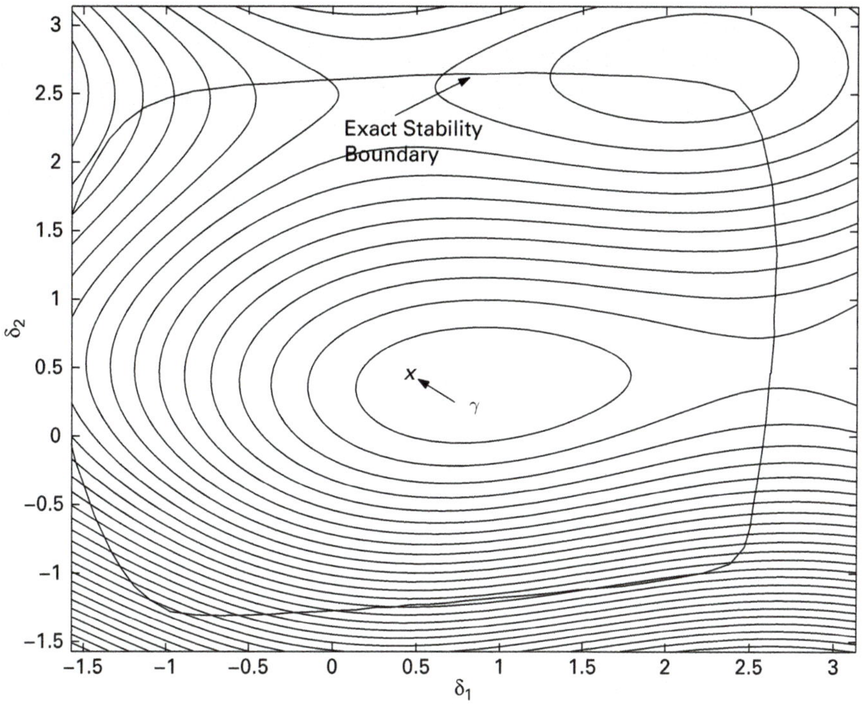

Figure 11.6 Different level surfaces associated with function V and their relationship with the exact stability boundary.

is composed of two small bounded sets C_1 and C_2. These sets are close to the unstable equilibrium points and their intersections with the subset $\{(x_1, x_2, y_1, y_2): x_1 \in R, x_2 \in R, y_1 = y_2 = 0.8\}$ are depicted in Figure 11.7. The number $L = 1.7619$ was calculated as the minimum of V on the set $M_2 \subset C_2$.

The intersection between the estimated stability region $\{x \in R^4: V(x) < L\}$ with the subset $\{(x_1, x_2, y_1, y_2) \in R^4\, y_1 = y_2 = 0.8\}$ is shown in Figure 11.8. Figure 11.9 shows that the choice of L is not optimal. However, if the sets C_i (in this case the set C_2) are small, then the amount of conservativeness introduced in the estimate is not significant.

11.3 Concluding remarks

In this chapter, we have presented a theoretical basis for using generalized energy functions to estimate the stability regions of complex nonlinear systems. The relationship between the structures of the stability boundary and the constant generalized energy

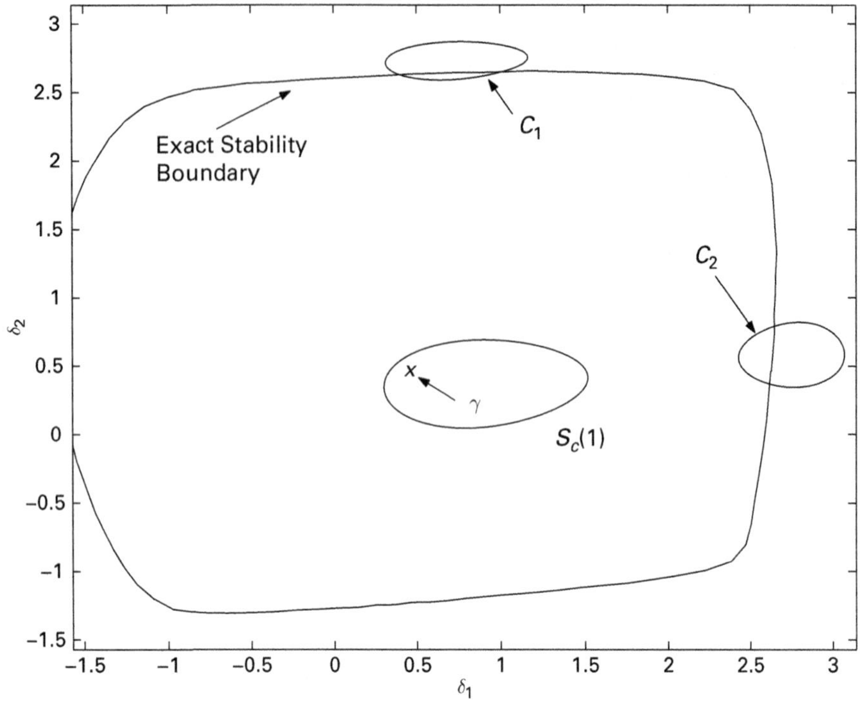

Figure 11.7 The connected level surface $S(r, \gamma)$, for $r = 1.0 < L = 1.76$, and its relationship with sets C_i and the stability boundary.

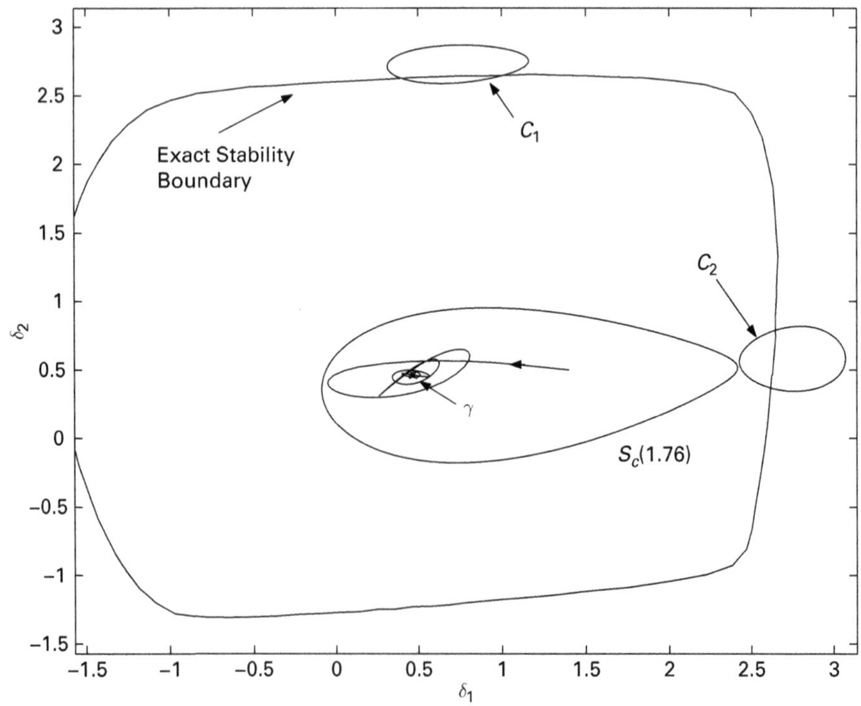

Figure 11.8 The estimated stability region via the connected level surface $S(r, \gamma)$, $r = L = 1.76$, and its relationship with sets C_i and the stability boundary.

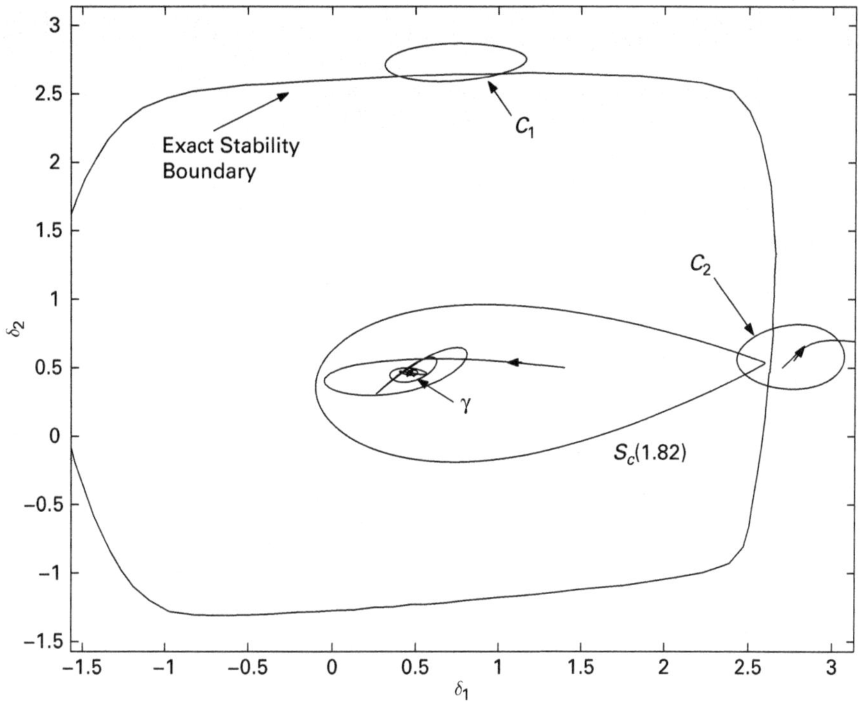

Figure 11.9 The connected level surface $S(r, \gamma)$, for $r = 1.82 > L = 1.76$, is the optimal estimate that can be obtained using the generalized energy function V.

surfaces at different level values has been established. In this context, we have presented optimal schemes for estimating the stability region. In spite of this, it is still numerically challenging to estimate the stability region of general nonlinear dynamical systems that exhibit complex behaviors. Further development of efficient numerical methods to estimate the stability regions of high-dimensional complex nonlinear systems is still needed.

12 Estimating stability regions of discrete dynamical systems

In this chapter, a theoretical basis for using energy functions to estimate stability regions of general nonlinear discrete-time dynamical systems is developed. The results developed in this chapter are the counterpart, for discrete systems, of the results developed in Chapter 10. On the basis of the theoretical developments, an optimal scheme for estimating the stability region of a large class of nonlinear discrete systems is developed. This computational scheme explores the stability boundary characterization and the tool of energy function (a scalar fuction) to optimally estimate stability regions of discrete dynamical systems in the form of level sets of a given energy function. More precisely, we develop an algorithm to obtain the largest level set of that energy function that is entirely contained in the stability region.

It is important to mention that most of the existing methods for estimating stability regions also search for a positively invariant level set contained inside the stability region. This invariant level set is usually characterized by a scalar function (usually a Lyapunov function). However, they do not explore the stability boundary characterization. As a consequence, their results might be conservative and not optimal in the sense that the given level set might not be the largest one entirely contained in the stability region.

12.1 Energy functions and the stability boundary

We will develop the foundations for estimating stability regions of the following class of autonomous nonlinear discrete dynamical systems:

$$x_{k+1} = f(x_k) \tag{12.1}$$

where $k \in Z$, $x_k \in R^n$ and $f: R^n \to R^n$ is a continuous map. Let E denote the set of fixed points of (12.1). We will assume that system (12.1) admits an energy function. Recall that an energy function for the discrete system (12.1) is a continuous scalar function $V: R^n \to R$ that satisfies the following conditions.

(E1) $\Delta V(x) \leq 0$ for all $x \in R^n$.
(E2) $\Delta V(x_k) = 0$ implies x_k is a fixed point.
(E3) If $V(x_k)$ is bounded for $k \in Z_+$, then the orbit x_k is itself bounded for $k > 0$.

Here $\Delta V(x_k) = V(x_{k+1}) - V(x_k)$ is the first difference of V along the solution x_k.

Given a C^r-energy function $V:R^n \to R$ and a real number k, we consider the following set:

$$S(k) = \{x \in R^n: V(x) < k\}. \tag{12.2}$$

We shall call the boundary of set (12.2), $\partial S(k) := \{x \in R^n: V(x) = k\}$ the *level set* (or *constant energy surface*) and k the corresponding *level value*. If k is a regular value of V (i.e. $\nabla V(x) \neq 0$, for all $x \in V^{-1}(k)$), then by the inverse function theorem, $\partial S(k)$ is a C^r $(n-1)$-dimensional submanifold of R^n. Moreover, if $r > n-1$, then by the Morse–Sard theorem, the set of regular values of V is residual; in other words "almost all" level values are regular. In particular, for almost all values of k, the level set $\partial S(k)$ is a C^r $(n-1)$-dimensional submanifold.

Generally speaking, set $S(k)$ can be very complicated with several connected components even for the two-dimensional case. Let $S^i(k)$, $i = 1, 2, \ldots, m$, be these connected components such that

$$S(k) = S^1(k) \cup S^2(k) \cup \ldots \cup S^m(k) \tag{12.3}$$

with $S^i(k) \cap S^j(k) = \emptyset$ when $i \neq j$. Each of these components is connected and disjoint from each other. Since $V(\cdot)$ is continuous, $S(k)$ is an open set. Because $S(k)$ is an open set, the level set $\partial S(k)$ is of $(n-1)$ dimensions. Actually, every connected component $S^i(k)$ is an open set whose boundary $\partial S^i(k)$ is an $(n-1)$-dimensional set.

Set $S(k)$ is positively invariant, however, unlike continuous systems, the connected components of the set $S(k)$ are not necessarily positively invariant. Moreover, unlike continuous systems, which admit a single connected component of the level set inside the stability region, discrete systems might admit more than one connected component of the level set entirely contained on the stability region.

As we increase the level value k, at least one connected component of the set $S(k)$ will appear inside the stability region. However, unlike continuous systems, multiple connected components might appear inside the stability region, even when the stability region is connected. By continually increasing the level value k, the connected components of the set S(k) inside the stability region enlarge until one of them touches the stability boundary. We will see, in the next theorem, that the point on the stability boundary that is first reached by the connected components of set S(k), as we increase k, must be an unstable fixed point (UFP).

THEOREM 12-1 **(Energy functions and fixed points on the stability boundary)** *Consider the nonlinear discrete dynamical system (12.1) that admits an energy function. Suppose that f is a continuous function and let x_s be an asymptotically stable fixed point of (12.1) whose stability region $A(x_s)$ is not dense in R^n. If $E \cap \partial A(x_s)$ is a bounded set, then the point with the minimal value of the energy function over the stability boundary $\partial A(x_s)$ exists and must be an unstable fixed point.*

Proof If the stability region of x_s is not dense in R^n, then the stability boundary is a nonempty set of dimension $n-1$. We have also proven, in the proof of Theorem 9-16 of Chapter 9, that $V(x_s)$ is a lower bound of the energy function V over the stability boundary. Hence, we claim that the global minimum of V over the stability boundary exists. In order to prove this, suppose, on the contrary, the non-existence of a global minimum of V over the stability boundary. Then $\partial A(x_s)$ has to be unbounded and $V(x_s) \leq \inf_{x \in \partial A(x_s)} V(x) = b \leq \infty$. Then there is a sequence of points $x_k \in \partial A(x_s)$ with $\|x_k\| \to \infty$ as $k \to \infty$ such that $V(x_k)$ is a decreasing function in k satisfying $V(x_k) \to b$ as $k \to \infty$. On the other hand, every trajectory $\phi(t, x_k)$ is bounded and converges to a fixed point x_k^{eq} on the stability boundary. Assumptions (E1) and (E2) guarantee that $V(x_k^{eq}) \leq V(x_k)$. This implies the existence of a sequence of fixed points such that $V(x_k^{eq}) \to b$ as $k \to \infty$. The condition that $E \cap \partial A(x_s)$ is bounded ensures that the sequence of fixed points x_k^{eq} is bounded; therefore there is a convergent subsequence that converges to a fixed point p on the stability boundary with $V(p) = b$. But this contradicts the fact that the global minimum over the stability boundary does not exist. Suppose the minimum of V over $\partial A(x_s)$ is attained at a regular point x. The existence of an energy function implies that $V(f(x)) < V(x)$. But $f(x) \in \partial A(x_s)$ because $\partial A(x_s)$ is a positively invariant set (see Proposition 7-9). This contradicts the fact that the minimum of V over $\partial A(x_s)$ is attained at x. As a consequence, the minimum of the energy function over the stability boundary must be attained at an unstable fixed point. This completes the proof.

12.2 Closest unstable fixed point and characterization

Theorem 12-1 shows that the minimum of an energy function on the stability boundary of a discrete dynamical system exists and must be attained at an unstable fixed point. The point of minimum energy on the stability boundary may not be unique. However, since the property that all equilibrium points of system (12.1) have distinct energy function values is generic [46], we can affirm that the point of minimum energy on the stability boundary is generically unique. In other words, the uniqueness of the minimum point is almost always guaranteed. We call the fixed point on the stability boundary with a minimum value of energy the *closest unstable fixed point.*

DEFINITION **(Closest unstable fixed point)**
A fixed point $\bar{x}$ is the closest unstable fixed point *of a stable fixed point x_s with respect to an energy function V, if $\bar{x} \in \partial A(x_s)$ and $V(\bar{x}) = \min_{x \in E \cap \partial A(x_s)} V(x)$.*

As in the case of the closest UEP for continuous nonlinear systems having an energy function, the closest UFP for discrete nonlinear systems exists and is generically unique. Theorem 12-2 gives a geometrical and dynamic characterization of the closest fixed point. Note that this theorem holds without the transversality condition.

THEOREM 12-2 **(Geometrical and dynamical characterization)**
Consider the nonlinear discrete dynamical system (12.1) satisfying assumption (A1) and that admits an energy function. Suppose that f is a diffeomorphism. Let x_s be an asymptotically stable fixed point of (12.1) and let $\hat{x}$ be the closest UFP on $\partial A(x_s)$ with respect to the energy function V. If $\hat{x}$ is hyperbolic, then $W^u(\hat{x}) \cap A(x_s) \neq \emptyset$.

Proof The closest UFP $\hat{x}$ is a hyperbolic equilibrium point lying on the stability boundary $\partial A(x_s)$. Then, by Theorem 9-8, $\{W^u(\hat{x}) - \hat{x}\} \cap \bar{A}(x_s) \neq \emptyset$. Suppose, by contradiction, that the unstable manifold $W^u(\hat{x})$ does not intercept $A(x_s)$. Then, there exists a trajectory $x(t)$ in the unstable manifold $W^u(\hat{x})$ such that $x(t) \in \partial A(x_s)$ for all t. As a consequence of Theorem 9-15, trajectory $x(t)$ converges to a fixed point on the stability boundary. In addition, the energy function value is strictly decreasing along any non-trivial trajectory of system (12.1). These two facts imply the existence of other points on the stability boundary with an energy function value lower than $V(\hat{x})$. This contradicts the fact that $\hat{x}$ possesses the lowest energy function value within the stability boundary $\partial A(x_s)$ and the proof is complete.

12.3 Optimal scheme for estimating the stability region

In this section, we explore the concept of closest UFP to develop a computational algorithm to obtain the optimal estimate of the stability region in the form of level sets of a given energy function.

THEOREM 12-3 **(Optimal estimation of the stability region)**
Consider the nonlinear discrete dynamical system (12.1) that admits an energy function $V(x)$. Suppose that f is a continuous function satisfying assumption (A1). Let x_s be an asymptotically stable fixed point of (12.1) and $\hat{x}$ be the closest UFP on $\partial A(x_s)$ with respect to $V(x)$. If $L = V(\hat{x})$, then:

(a) *every connected component $S^i(L)$ of the set $S(L)$ has an empty intersection with the stability boundary $\partial A(x_s)$;*
(b) *there exists a non-empty collection of connected components of the set $S(L)$ entirely contained in the stability region $A(x_s)$; this collection is positively invariant, in particular, there always exists a connected component in this collection containing the fixed point x_s;*
(c) *there exists a connected component of the set $S(B)$ that has a non-empty intersection with the stability boundary $\partial A(x_s)$ for any number $B > L$.*

Proof

(a) We shall prove this theorem by contradiction. Suppose the existence of a connected component $S^i(L)$ of $S(L)$ with a non-empty intersection with the stability boundary $\partial A(x_s)$. Let $q \in S^i(L) \cap \partial A(x_s)$. Therefore, $V(q) < L$. But this contradicts the fact that L is the minimum value of V on $\partial A(x_s)$. Consequently, every connected component $S(L)$ has an empty intersection with the stability boundary.
(b) Certainly, $V(x_s) < L$, therefore there exists one connected component of the set $S(L)$ that contains the asymptotically stable fixed point x_s. Since every connected component of $S(L)$ has an empty intersection with the stability boundary, either a connected component is entirely contained on the stability region or it is entirely contained on the interior of its complement. Let $S_A(L)$ be the collection of connected

components of $S(L)$ that are entirely contained in the stability region $A(x_s)$. Set $S_A(L)$ is non-empty because it contains the connected component of $S(L)$ that contains the asymptotically stable fixed point x_s. If $x \in S_A(L)$, then $x \in A(x_s)$ and $f(x) \in A(x_s)$ because $A(x_s)$ is positively invariant. Moreover, property (E1) of an energy function ensures that $f(x) \in S(L)$. Then $f(x) \in S(L) \cap A(x_s) = S_A(L)$. Consequently, $f(S_A(L)) \subset S_A(L)$.

(c) For any $B > L$, the closest UFP $\hat{x} \in S(B)$. In particular, $\hat{x}$ belongs to one of the connected components of $S(B)$. Therefore $S(B) \cap \partial A(x_s) \neq \varnothing$. This completes the proof.

In practice, Theorem 12-3 ensures that by calculating the energy function value of all fixed points on the stability boundary, we can obtain the optimal estimate of the stability region, in the form of a level set of the energy function $V(x)$, by picking the level set with a level value that equals the value of the energy of the fixed point on the stability boundary which has the lowest value of energy over the stability boundary.

The next example illustrates a discrete dynamical system that admits an energy function and whose optimal estimation of the stability region is composed of two connected components of the level set $S(L)$.

Example 12-1 Consider the nonlinear discrete dynamical system:

$$x_{k+1} = f(x_k) \tag{12.4}$$

with

$$f(x) = \begin{cases} bx + c & \text{if} \quad x < -1 \\ ax & \text{if} \quad -1 \leq x \leq 1 \\ dx + e & \text{if} \quad x > 1 \end{cases} \tag{12.5}$$

where $a = 1/2, b = 2, c = 1, d = -2, e = 3$. This system possesses two fixed points. The origin is an asymptotically stable fixed point while $\text{UPF} = -c/(b-1) = -1.5$ is an unstable fixed point on the stability boundary of the origin. The stability region and stability boundary are shown in Figure 12.1. The stability boundary is composed of the stable set of the unstable fixed point, UFP, which is composed of the UFP union with the point indicated by an 'X' in Figure 12.1.

This system admits an energy function:

$$V(x) = \begin{cases} x + f & \text{if} \quad x < \text{UFP} \\ -x & \text{if} \quad \text{UFP} \leq x \leq 0 \\ 2x & \text{if} \quad 0 < x \leq 1 \\ gx + h & \text{if} \quad 1 < x \leq -\dfrac{e}{d} \\ -gx + j & \text{if} \quad -\dfrac{e}{d} < x \leq -\dfrac{c}{d(b-1)} - \dfrac{e}{d} \\ -x + n & \text{if} \quad x > -\dfrac{c}{d(b-1)} - \dfrac{e}{d} \end{cases} \tag{12.6}$$

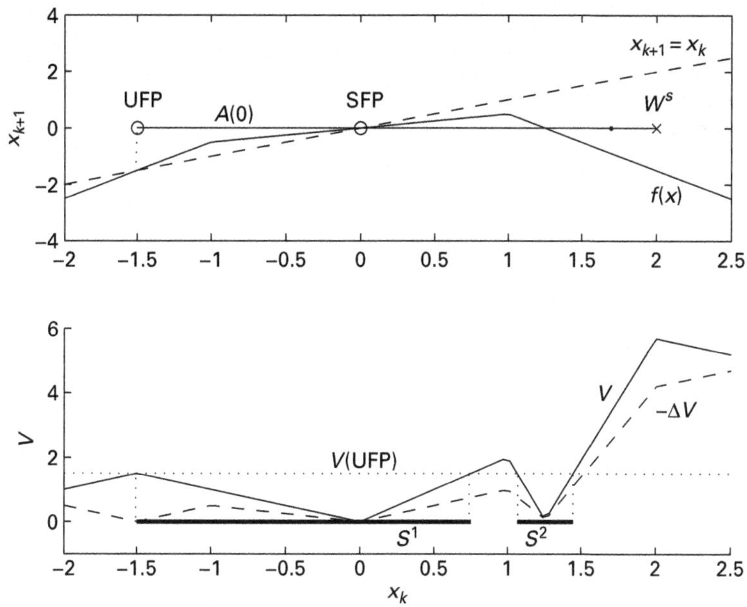

Figure 12.1 The stability region and stability region estimation of system (12.4).

where $f = 2c/(b-1)$, $g = 1.9/(1 + e/d)$, $h = 2 - g$, $j = -ge/d$, $n = (-g + 1)(-c/d/(b-1) - e/d) + j$. The graphic of function $V(x)$ and its first difference ΔV is plotted in Figure 12.1. Clearly, $\Delta V = 0$ only at the fixed points. Since the UFP is the only unstable fixed point on the stability boundary, UFP is also the closest UFP. Using the value of the energy function calculated at the closest UEP we obtain the level curve set depicted in Figure 12.1. This set is composed of two connected components S^1 and S^2 entirely contained on the stability region of the origin. The connected component S^2 is not positively invariant. But both S^1 and the union $S^1 \cup S^2$ are positively invariant.

Based on Theorem 12-3, we propose the following scheme to estimate the stability region $A(x_s)$ of the nonlinear discrete dynamical system (12.1) that admits an energy function.

Scheme (Optimal estimation of the stability region $A(x_s)$ via an energy function $V(\cdot)$)

A Determining the type-one critical level value of an energy function.

Step 1: Find all fixed points.

Step 2: Identify those whose unstable manifold (or unstable set) intersects the stability boundary $\partial A(x_s)$, say $x_1, x_2, \ldots$.

Step 3: Compute the value of the energy at these fixed points, $V(x_1)$, $V(x_2)$, ... and select the one with the smallest value of energy, say $L = \min_j\{V(x_j)\}$.

B Estimating the stability region $A(x_s)$.

Step 4: The connected component of the set $S(L)$ containing the fixed point x_s is an estimate of the stability region.

We comment on the estimation scheme. The analytical basis for Steps 1 and 2 is Theorem 12-2 while the analytical basis for Steps 3 and 4 is Theorem 12-3.

12.4 Numerical studies

Example 12-2 This example illustrates the proposed optimal scheme. Consider the two-dimensional nonlinear discrete dynamical system:

$$\begin{aligned} x_{k+1} &= x_k^3 + \frac{3}{4}x_k \\ y_{k+1} &= \alpha\frac{y_k + \beta y_k^2 + y_k^3}{y_k^2 + 1}. \end{aligned} \tag{12.7}$$

It is not difficult to check that the vector field is a diffeomorphism if $\beta^2 < \alpha(3 - \alpha)$. System (12.7) possesses six hyperbolic fixed points. The origin is a hyperbolic asymptotically stable fixed point while the other five fixed points lie on the stability boundary of (0, 0). Consider the following function:

$$V(x,y) = \begin{cases} |x| + |y| & \text{if } |x| \le \frac{1}{2} \text{ and } y \le \frac{1-\alpha}{\beta} \\ 1 - |x| + |y| & \text{if } |x| > \frac{1}{2} \text{ and } y \le \frac{1-\alpha}{\beta} \\ |x| + 2\left(\frac{1-\alpha}{\beta}\right) - y & \text{if } |x| \le \frac{1}{2} \text{ and } y > \frac{1-\alpha}{\beta} \\ 1 + 2\left(\frac{1-\alpha}{\beta}\right) - |x| - y & \text{if } |x| > \frac{1}{2} \text{ and } y > \frac{1-\alpha}{\beta}. \end{cases} \tag{12.8}$$

It is straightforward to verify that function $V(x, y)$ satisfies conditions (E1), (E2) and (E3). As a consequence, $V(x, y)$ is an energy function for system (12.7) and all conditions of Theorem 9-18 are satisfied. Thus the stability boundary $\partial A(0, 0)$ is composed of the union of the stable manifolds of every unstable fixed point that lies on the stability boundary. Figure 12.2 illustrates the fixed points and the stability boundary $\partial A(0, 0)$ for $\alpha = 1/3$ and $\beta = 2/3$. Applying Steps 1 and 2 of the optimal estimation scheme presented in the previous section, one can conclude that the fixed points (0.5, 0), (0.5, 1), (0, 1), (−0.5, 1) and (−0.5, 0) lie on the stability boundary $\partial A(0, 0)$.

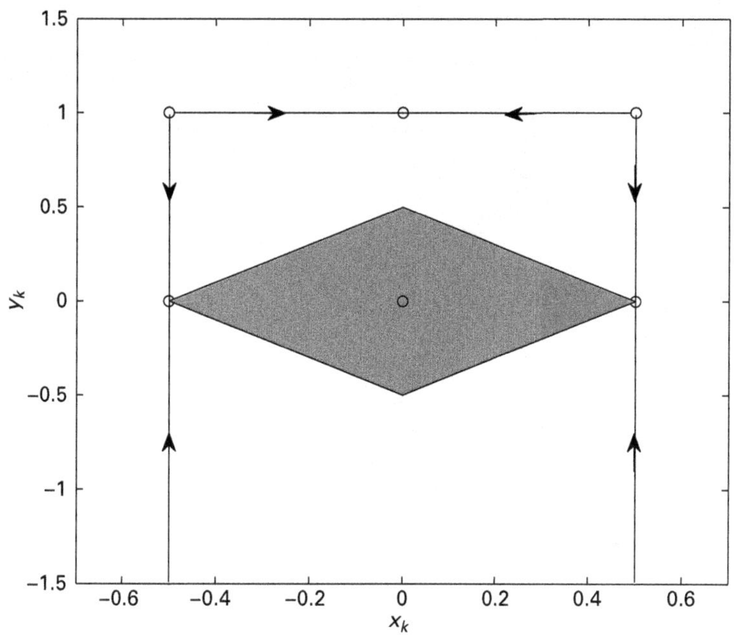

Figure 12.2 Stability region and stability region estimate of system (12.7) for $\alpha = 1/3$ and $\beta = 2/3$.

Computing the value of the energy function at these fixed points, as suggested in Step 3, one obtains $L = 0.5$. This minimum value of energy is attained at the fixed points $(0.5, 0)$ and $(-0.5, 0)$. Set $S(L)$, the optimal estimate of the stability region in the form of a level set of V, is the gray area of Figure 12.2. It can be seen from this figure that the critical level 0.5 determined by the proposed scheme is indeed the optimal one for estimating the stability region $A(0, 0)$.

Example 12-3 Consider the following discrete-time nonlinear system that resembles the model of a recurrent neural network:

$$y_i(k+1) = \sigma\left(\sum_{j=1}^{3} \mu_i \omega_{ij} y_j(k) + \mu_i s_i\right) \quad i = 1, 2, 3. \tag{12.9}$$

Function σ is the activation function of the neural network. Matrix $W = [w_{ij}]_{3 \times 3}$ is the synaptic weight matrix. The vector $s = [s_i]_{3 \times 1}$ is the input of the network and $\mu = [\mu_i]_{3 \times 1}$ is the activation gain of the network. The following energy function has been suggested in [174]:

$$E(y) = -\frac{1}{2}\sum_{i=1}^{3}\sum_{j=1}^{3} \omega_{ij} y_i y_j - \sum_{i=1}^{3} s_i y_i + \sum_{i=1}^{3} \frac{1}{\mu_i} \int_0^{y_i} \sigma^{-1}(\tau) d\tau. \tag{12.10}$$

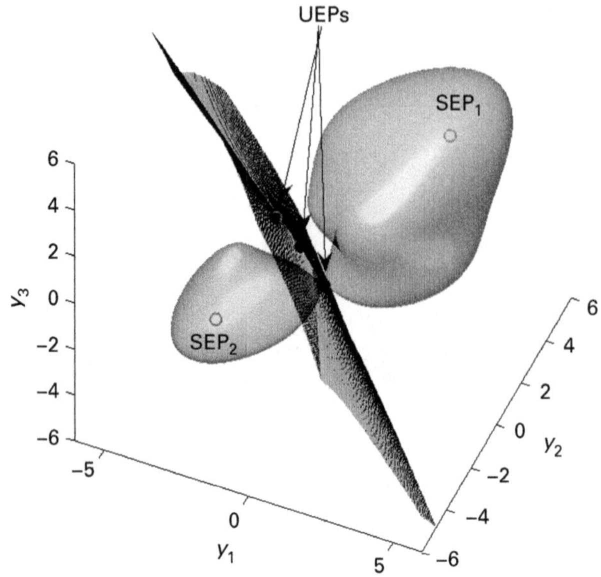

Figure 12.3 Stability region estimation of system (12.9). The dark surface containing the UEPs is the stability boundary while the surfaces involving the SEPs represent the optimal stability region estimations obtained with energy function (12.10).

In this example, $\sigma(z) = z^{1/3}$. For the following set of parameters, $\mu_1 = \mu_2 = \mu_3 = 10$, $\omega_{11} = 0.3$, $\omega_{22} = 0.2$, $\omega_{33} = 0.4$, $\omega_{12} = \omega_{21} = 0.4$, $\omega_{13} = \omega_{31} = 0.2$, $\omega_{23} = \omega_{322} = 0.1$, $s_1 = 0.5$, $s_2 = 0.2$ and $s_3 = 0.1$, the network possesses two asymptotically stable fixed points, they are SEP$_1$ =[3.15, 2.85, 2.77] and SEP$_2$ =[−2.63, −2.53, −2.58]. There are three fixed points on the stability boundary of these fixed points. The circles of Figure 12.3 display the location of these fixed points. The exact stability boundary is the black surface in this figure. By applying the optimal scheme for obtaining optimal estimates of the stability region of these asymptotically stable fixed points, we conclude that the unstable fixed point [0.17, −0.43, −1.88] is the one with the lowest energy function value on the stability boundary. The optimal stability region estimate $S(L)$, with L=−0.2152, of both fixed points is depicted in Figure 12.3.

12.5 Conclusions

A theoretical basis for using energy functions to estimate stability regions of general nonlinear discrete dynamical systems has been presented in this chapter. In addition, an optimal scheme for estimating the stability region was developed. A topological characterization and a dynamical characterization for the closest fixed point of a stable fixed point with respect to an energy function have been derived. Generically speaking, the closest fixed point is a type-one fixed point and the one-dimensional unstable manifold of the closest fixed point converges to the stable fixed point. This computational scheme can optimally estimate stability regions in the form of level sets of an energy function. Several examples have been given to illustrate the optimal estimation scheme.

13 A constructive methodology to estimate stability regions of nonlinear dynamical systems

13.1 Introduction

We consider the following (autonomous) nonlinear dynamical system

$$\dot{x} = f(x), \ x \in R^n. \tag{13.1}$$

It is natural to assume the function (i.e. the vector field) $f: R^n \rightarrow R^n$ satisfies a sufficient condition for the existence and uniqueness of the solution. In this chapter we present a constructive methodology for estimating the stability regions of nonlinear dynamical systems with the following characteristics:

- the ability to reduce the conservativeness in estimating the stability regions;
- computational efficiency;
- adaptability;
- sound theoretical basis.

By "adaptability" we mean the ability either to stand by itself or to accept an estimate from some existing method as an input.

In general, nonlinear dynamical systems can be classified into two groups:

(1) the systems that admit a function $V(\cdot)$ such that $\dot{V}(\cdot) < 0$ along any nontrivial trajectory (we shall refer to this function as an energy-like function throughout this chapter), and
(2) the systems that do not have an energy-like function.

The constructive methodology contains slightly different procedures for each group of nonlinear dynamical systems. For systems having a (global) energy-like function, the methodology starts with an energy-like function $V(\cdot)$ associated with the system under study and proceeds in the following three steps.

Procedure A (For nonlinear dynamical systems having a global energy-like function)

Step 1: Optimally determine the critical level value of the energy-like function with respect to a stable equilibrium point.

Step 2: Construct a new energy-like function from the present energy-like function.

Step 3: Estimate the stability region via the new energy-like function constructed in Step 2 and via the critical level value determined at Step 1 (and go to Step 2).

Steps 2 and 3 are then applied iteratively to improve the estimated stability region (in reducing its conservativeness). It should be stressed that the procedure only requires simple algebraic operations at Step 2 without invoking complex optimization techniques which many existing algorithms rely on. The critical level value determined at Step 1 is also the critical level value for every newly constructed energy-like function at Step 2, and this critical level value will also be shown to be optimal for every newly constructed energy-like function at Step 2. (In contrast, existing algorithms need to determine the critical level value for every newly constructed energy-like function.) The stability regions estimated by repeating Steps 2 and 3 will be shown to form a sequence of sets with strictly monotonic increasing size and yet each set is contained in the entire stability region. Hence, at every iteration, this methodology yields a larger estimated stability region than the previous iteration, thus reducing the conservativeness in estimating the stability regions.

For systems having local energy-like functions, the methodology starts with a local energy-like function $V(\cdot)$ associated with the system under study and proceeds in the following three steps.

Procedure B (For nonlinear dynamical systems only having a local energy-like function)

Step 1: Determine the critical level value of the given local energy-like function.

Step 2: Construct a new local energy-like function from the present local energy-like function.

Step 3: Estimate the stability region using the newly constructed local energy-like function and the critical level value determined at Step 1.

Steps 2 and 3 are then applied iteratively to improve the estimated stability region (in reducing its conservativeness). Basically, the above two procedures are similar and the merits of Procedure A are carried over to Procedure B. The only difference between the above two procedures lies in the way that the critical value is determined. Chapter 10 presented numerical implementation of Step 1 in Procedure A and Step 1 in Procedure B.

13.2 First-order expansion scheme

In this section, we present two schemes for constructing a sequence of functions from a given energy-like function $V(\cdot)$ for the nonlinear system (13.1). We show that the sequence of functions can be applied to estimate the stability region of (13.1). In addition, we show that the level sets defined by these energy-like functions form a sequence of sets with strictly monotonic increasing size and yet each set is contained in the entire stability region. These two schemes are then incorporated into a constructive algorithm which, at every iteration, offers a larger (hence less conservative) estimate of the stability region $A(x_s)$. This constructive algorithm is discussed in the next section.

Given an energy-like function $V(\cdot)$ for the nonlinear system (13.1), we construct a sequence of functions via the following scheme:

$$\begin{aligned} V_1(x) &= V(x + d_1 f(x)) \\ V_2(x) &= V_1(x + d_2 f(x)) \\ &\cdots \\ V_n(x) &= V_{n-1}(x + d_n f(x)) \end{aligned} \tag{13.2}$$

where $f(x)$ is the vector field of (13.1) and d_i, $i = 1, 2, \ldots, n$ are positive numbers.

Theorem 13-1 below shows that the sequence of functions constructed by (13.2) are all energy-like functions.

THEOREM 13-1 **(Energy-like function)**
Let $V(\cdot): R^n \to R$ be an energy-like function for the nonlinear system (13.1) and let K be a compact set, containing no other equilibrium points, in the state space of (13.1). Then, there exists a $\hat{d} > 0$ such that, for $d < \hat{d}$, the function $V_1(x) = V(x + df(x))$ is also an energy-like function on the compact set K for the nonlinear system.

Proof The term $\dot{V}_1(x)$ can be written as

$$\dot{V}_1(x) = \left\langle \frac{\partial V}{\partial x}\bigg|_{x=x+df(x)}, f(x) + d\left[\frac{\partial f}{\partial x}\right] f(x) \right\rangle.$$

We use two steps to prove that $\dot{V}_1(x) < 0$, for $x \in K$. In the first step we shall show that $\left\langle \frac{\partial V}{\partial x}\big|_x, f(x) \right\rangle < 0$, for all $x \in K$, implies that $\left\langle \frac{\partial V}{\partial x}\big|_{x=x+df(x)}, f(x + df(x)) \right\rangle < 0$, for all $x \in K$. In the second step we shall show that $\left\langle \frac{\partial V}{\partial x}\big|_{x=x+df(x)}, f(x + df(x)) \right\rangle < 0$, for all $x \in K$, implies that $\left\langle \frac{\partial V}{\partial x}\big|_{x=x+df(x)}, f(x) + d\left[\frac{\partial f}{\partial x}\right] f(x) \right\rangle < 0$, for all $x \in K$.

First, since $\dot{V}(x) = \left\langle \frac{\partial V}{\partial x}\big|_x, f(x) \right\rangle < 0$ and $f(x)$ is uniformly bounded for $x \in K$, it follows by the continuity argument that, for $x \in K$, there exists a $\hat{d}_1 > 0$ such that if $d_1 < \hat{d}_1$, then $\left\langle \frac{\partial V}{\partial x}\big|_{x=x+df(x)}, f(x + df(x)) \right\rangle < 0$. Moreover, because $\dot{V}_1(x)$ is a continuous function, there exists a neighborhood $U(x)$ of x such that $\dot{V}_1(\hat{x}) < 0$ for all $\hat{x} \in U(x)$. Hence, we have shown that for any point $x \in K$, there exist a positive number $\hat{d}_1$ and a neighborhood $U(x)$ such that, for $d_1 < \hat{d}_1$ and $y \in U(x)$, the following holds:

$$\left\langle \frac{\partial V}{\partial x}\big|_{x=y+df(y)}, f(y + df(y)) \right\rangle < 0 \text{ for } y \in U(x) \text{ and } d_1 < \hat{d}_1. \tag{13.3}$$

Since the set K is compact, there are finitely many open sets that cover K. Therefore, there are finitely many points $p_1, \ldots, p_n$ in K, and there exist a neighborhood U_{p_i} and a positive number $\hat{d}_{1i}$, $i = 1, 2, \ldots, n$ such that Eq. (13.3) is satisfied for $y \in U_{p_i}$, $d_{1i} < \hat{d}_{1i}$, and the following relationship holds:

$$K \subset U_{p_1} \cup \ldots \cup U_{p_n}.$$

By taking $\hat{d}_1 = \min\{d_{11}, d_{12}, \ldots, d_{1n}\}$ we notice that

$$\left\langle \frac{\partial V}{\partial x}\bigg|_{x=x+df(x)}, f(x+df(x)) \right\rangle < 0 \text{ for all } x \in K \text{ and } d < \hat{d}_1. \tag{13.4}$$

Hence, the first step is completed.

Second, suppose the function $f(\cdot)$ has continuous partial derivatives of order r, $r > 1$. Then, for $x \in R^n$, it follows that

$$\begin{aligned} f(x+df(x)) &= f(x) + d\left\langle \frac{\partial f}{\partial x}, f(x) \right\rangle + \text{higher order terms} \\ &= f(x) + d\left\langle \frac{\partial f}{\partial x}, f(x) \right\rangle + d^2(\alpha_1(x) + \alpha_2(x)d \\ &\quad + \ldots + \alpha_{r-2}(x)d^{r-3} + \alpha_{r-1}(z)d^{r-2}) \end{aligned} \tag{13.5}$$

where $z = x + tdf(x)$, for some $t \in [0, 1]$. Since every continuous function is bounded on any compact set, it follows from the above equation and the compactness argument as well as the continuity argument that, given any positive number α, there exists a number $\tilde{d}_2 > 0$ such that, for $d < \tilde{d}_2$, the following is true:

$$\left| f(x+df(x)) - f(x) - d\left\langle \frac{\partial f}{\partial x}, f(x) \right\rangle \right| < d^2\tilde{\alpha} \text{ for all } x \in K \tag{13.6}$$

Similarly, there exist $\alpha_1 > 0$ and $\tilde{d}_3 > 0$ such that, for $d < \tilde{d}_3$, it follows that

$$\left| \frac{\partial V}{\partial x}\bigg|_{x=x+df(x)} \right| < \alpha_1, \text{ for all } x \in K. \tag{13.7}$$

On the other hand, there exists a number $\beta_1 > 0$ such that, for $d < \hat{d}_1$,

$$\left\langle \frac{\partial V}{\partial x}\bigg|_{x=x+df(x)}, f(x+df(x)) \right\rangle < -\beta_1 \text{ for all } x \in K. \tag{13.8}$$

Thus, from equations (13.7) and (13.8) we have that, for $d < \min\{\tilde{d}_2, \tilde{d}_3\}$,

$$\left| \left\langle \frac{\partial V}{\partial x}\bigg|_{x=x+df(x)}, f(x+df(x)) - f(x) - d\left[\frac{\partial f}{\partial x}\right] f(x) \right\rangle \right| > \alpha_1 d^2 \hat{\alpha} \text{ for all } x \in K. \tag{13.9}$$

Therefore, for $\hat{d}_2 = \min\{\tilde{d}_2, \tilde{d}_3, (\beta_1/\alpha_1\hat{\alpha})^{1/2}\}$, it follows

$$\left\langle \frac{\partial V}{\partial x}\bigg|_{x=x+df(x)}, f(x) + d\left[\frac{\partial f}{\partial x}\right] f(x) \right\rangle < 0 \text{ for all } x \in K \text{ and } d < \hat{d}_2. \tag{13.10}$$

By taking $\hat{d} = \min\{\hat{d}_1, \hat{d}_2\}$ and combining the first and second steps we complete the proof.

Next, we show that the sizes of the sets which are the intersections between the level sets of the functions $V_i(\cdot)$ and any compact set are monotonically increasing.

THEOREM 13-2 **(Expansion)**

Let $V(\cdot):R^n \rightarrow R$ be an energy-like function for the nonlinear system (13.1) and let K be a compact set, containing no equilibrium points, in the state space of (13.1). Suppose that the set $S_V(c)$: = $\{x$: $V(x) \leq c$ and $x \in K\}$ is non-empty for some constant c. Then, there exists a $\tilde{d} > 0$ such that for the set characterized by $S_{V_1}(c): \; = \{x : V_1(x) \leq c \text{ and } x \in K\}$, where $V_1(x) = V(x + df(x))$ and $d < \tilde{d}$, the following is true:

$$S_V(c) \subset S_{V_1}(c).$$

Proof Suppose the function $V(\cdot)$ has continuous partial derivatives of order r, $r \geq 1$, at each point x of R^n. Without loss of generality, we assume that $r > 1$. Then, for $x \in R^n$ we have the following:

$$\begin{aligned} V_1(x) &= V(x + df(x)) \\ &= Va(x) + d\langle \frac{\partial V}{\partial x}, f(x)\rangle + d^2(\alpha_1(x) + \alpha_2(x)d + \ldots \\ &+ \alpha_{r-2}(x)d^{r-3} + \alpha_{r-1}(z)d^{r-2}) \end{aligned} \tag{13.11}$$

where $z = x + tdf(x)$, for some $t \in [0, 1]$. Since every continuous function is bounded on any compact set, it follows that there exist positive numbers d_i, $i = 1, 2, \ldots, r - 1$, such that the functions $\alpha_i(x)$ in (13.11) are bounded, i.e. $\alpha_i(x) < d_i$ for $x \in K$. Thus, the sum of higher order terms in (13.11) is bounded above. Therefore, given any $\alpha > 0$, there exists a number $\tilde{d} > 0$ such that for $d < \tilde{d}$, the sum of higher order terms is bounded by $d^2\alpha$. Thus, we have

$$V_1(x) < V(x) + d\langle \frac{\partial V}{\partial x}, f(x)\rangle + d^2\alpha \text{ for } x \in K,\ d < \tilde{d} \tag{13.12}$$

or

$$V_1(x) = V(x) + d\left(\langle \frac{\partial V}{\partial x}, f(x)\rangle + d\alpha\right). \tag{13.13}$$

Since $V(\cdot)$ is an energy-like function, we notice that $\langle \partial V/\partial x, f(x)\rangle < 0$; and also that for $x \in K$, there exists a number $\beta > 0$ such that $\langle \partial V/\partial x, f(x)\rangle < -\beta$, for $x \in K$. Consequently, for $0 < d < \min\{\tilde{d}, \beta/\alpha\}$, it follows that

$$V_1(x) < V(x) + d(d\alpha - \beta) < V(x). \tag{13.14}$$

This suffices to show that $S_V(c) \subset S_{V_1}(c)$. This completes the proof.

13.3 Second-order expansion schemes

In this section, we present another scheme for constructing a sequence of functions from a given energy-like function $V(\cdot)$ for the nonlinear system (13.1). We show that these functions are also energy-like functions for the nonlinear system (13.1), and that the level sets defined by these functions form a monotonically increasing sequence.

Given an energy-like function $V(\cdot)$ for the nonlinear system (13.1), we construct a sequence of energy-like functions through the following second-order expansion scheme:

$$\begin{aligned} V_1(x) &= V\left(x + \frac{d}{2}(f(x + df(x)) + f(x)\right) \\ V_2(x) &= V_1\left(x + \frac{d}{2}(f(x + df(x)) + f(x))\right) \\ &\cdots \\ V_n(x) &= V_{n-1}\left(x + \frac{d}{2}(f(x + df(x)) + f(x))\right). \end{aligned} \tag{13.15}$$

The analysis of the proposed scheme (13.15) parallels the analysis of the scheme (13.2). In other words, we first show that every function $V_i(\cdot)$, $i = 1, 2, \ldots,$ in (13.15) is an energy-like function on a compact set for system (13.1), and then show that the sequence formed by the level sets defined by the sequence of functions (13.15) is monotonically increasing.

THEOREM 13-3 **(Energy-like function)**
Let $V(\cdot)$: $R^n \to R$ be an energy-like function for the nonlinear system (13.1) and let K be a compact set, containing no equilibrium points, in the state space of (13.1). Then, there exists a $\hat{d} > 0$ such that, for $d < \hat{d}$, the function $V_1(x) = V(x + (d/2)(f(x + df(x)) + f(x)))$ is also an energy-like function on the compact set K for the nonlinear system (13.1).

Proof The term $\dot{V}_1(x)$ is written as

$$\begin{aligned} \dot{V}_1(x) &= \dot{V}\left(x + \frac{d}{2}(f(x + df(x)) + f(x))\right) \\ &= \left\langle \frac{\partial V}{\partial x}\bigg|_{x+\frac{d}{2}(f(x+df(x))+f(x))}, \left(I + \frac{d}{2}\left(\frac{\partial f}{\partial x}\bigg|_{x+df(x)}\left(I + d\frac{\partial f}{\partial x}\bigg|_x\right) + \frac{\partial f}{\partial x}\bigg|_x\right)\right)\dot{x}\right\rangle. \end{aligned}$$

To show $\dot{V}_1(x) < 0$, for $x \in K$, we first show that there exists a $d_1 > 0$ such that $\langle \partial V/\partial x|_{x+\frac{d}{2}(f(x+df(x))+f(x))}, \dot{x}\rangle < 0$ for $x \in K$ and $d < d_1$; using this result we then show that $\dot{V}_1(x) < 0$, for $x \in K$.

Since $\dot{V}(x) = \langle \partial V/\partial x|_x, \dot{x}\rangle < 0$, for $x \in K$ there exists a positive number $\beta > 0$ such that

$$\dot{V}(x) < -\beta \text{ for } x \in K. \tag{13.16}$$

Also, because $\partial V/\partial x$ is continuous and $f(x)$ is continuous, it follows that, given any $\beta_1 > 0$, there exists a positive number d_1 such that for $d < d_1$ the following holds:

$$\left|\frac{\partial V}{\partial x}\bigg|_{x+\frac{d}{2}(f(x+df(x))+f(x))} - \frac{\partial V}{\partial x}\bigg|_x\right| < \beta_1 \text{ for } x \in K. \tag{13.17}$$

Furthermore, it follows that

$$\begin{aligned}
&\left\langle \frac{\partial V}{\partial x}\bigg|_{x+\frac{d}{2}(f(x+df(x))+f(x))}, \dot{x} \right\rangle \\
&= \left\langle \frac{\partial V}{\partial x}\bigg|_{x}, \dot{x} \right\rangle + \left\langle \frac{\partial V}{\partial x}\bigg|_{x+\frac{d}{2}(f(x+df(x))+f(x))} - \frac{\partial V}{\partial x}\bigg|_{x}, \dot{x} \right\rangle \\
&< \left\langle \frac{\partial V}{\partial x}\bigg|_{x}, \dot{x} \right\rangle + \beta_1 \quad [\text{from (13.17)}] \\
&< -\beta + \beta_1 \quad [\text{from (13.16)}] \\
&< 0 \quad (\text{by choosing}\, \beta_1 < \beta).
\end{aligned} \tag{13.18}$$

Thus, we have shown that there exists a positive number d_1 such that for $d < d_1$,

$$\left\langle \frac{\partial V}{\partial x}\bigg|_{x+\frac{d}{2}(f(x+fd(x))+f(x))}, \dot{x} \right\rangle < 0, \text{ for } x \in K. \tag{13.19}$$

Next, we observe that the matrices $[\partial f/\partial x]$ and $[\partial f/\partial x]^2$ are bounded for $x \in K$, then applying the continuity argument and compactness reasoning, it follows that there exists a positive number $d_2 > 0$, such that, for $d < d_2$,

$$\left\langle \frac{\partial V}{\partial x}\bigg|_{x+\frac{d}{2}(f(x+df(x))+f(x))}, \left(I + \frac{d}{2}\left(\frac{\partial f}{\partial x}\bigg|_{x+df(x)}\left(I + d\frac{\partial f}{\partial x}\bigg|_{x}\right) + \frac{\partial f}{\partial x}\bigg|_{x}\right)\right)\dot{x} \right\rangle < 0$$

which implies that

$$\dot{V}_1(x) = \dot{V}\left(x + \frac{d}{2}(f(x+fd(x)) + f(x))\right) < 0 \,\text{for}\, d < d_2 \text{ and } x \in K.$$

Taking $\hat{d} = \min\{d_1, d_2\}$ we conclude the proof of this theorem.

THEOREM 13-4 **(Expansion)**
Let $V(\cdot){:}R^n \rightarrow R$ be an energy-like function for the nonlinear system (13.1) and let K be a compact set, containing no other equilibrium points, in the state space of (13.1). Suppose that the set $S_V(c){:} = \{x{:}V(x) \geq c$ and $x \in K\}$ is non-empty for some constant c. Then, there exists a $\hat{d} > 0$ such that for the set characterized by $S_{V_1}(x) : = \{x : V_1(x) \leq c$ and $x \in K\}$, where $V_1(x) = V(x + d(f(x + df(x)) + f(x)))$ and $d < \hat{d}$, the following is true:

$$S_V(c) \subset S_{V_1}(c).$$

Proof The procedure of this proof is similar to that of Theorem 13-2. First, since

$$
\begin{aligned}
V_1(x) &= V\left(x+\frac{d}{2}(f(x+df(x))+f(x))\right) \\
&= V(x)+\left\langle\frac{\partial V}{\partial x},\frac{d}{2}(f(x+df(x))+f(x)\right\rangle + \text{higher order terms} \\
&= V(x)+\left\langle\frac{\partial V}{\partial x},\frac{d}{2}(2f(x)+d\left[\frac{\partial V}{\partial x}\right]f(x)+ \quad + \text{higher order terms}\right\rangle \\
&= V(x)+\left\langle\frac{\partial V}{\partial x},df(x)\right\rangle + \text{higher order terms.}
\end{aligned}
\tag{13.20}
$$

Next, since $\langle \partial V/\partial x, df(x)\rangle < 0$ for $x \in K$ and the higher order terms in (13.20) are bounded above, one can employ the continuity property and use the compactness argument, as in the proof of Theorem 13-2, to show that there exists a positive number $\tilde{d}$ such that for $d < \tilde{d}$,

$$V_1(x) < V(x) \text{ for } x \in K$$

or

$$S_V(c) \subset S_{V_1}(c).$$

This completes the proof.

It is noted that there is a close relationship between the expansion schemes described above and the integration schemes used for solving ordinary differential equations. Indeed, the first-order expansion scheme A is related to the backward Euler integration scheme while the second-order expansion scheme B is related to the trapezoidal integration scheme. Using this analogy, a variety of other expansion schemes can be developed in a similar way.

13.4 Constructive methodology with energy-like functions

A constructive methodology for estimating the stability region of system (13.1) which has an energy-like function is described in this section. The methodology is iterative in character and computationally efficient. At each iteration, this method offers a larger estimated stability region than the previous iteration and yet it still lies inside the exact stability region, thus reducing the conservativeness in estimating the stability regions. The theoretical basis of the methodology is also presented.

The constructive methodology for estimating the stability region $A(x_s)$ starts with a given energy-like function $V(\cdot)$ and proceeds as follows.

Step A: Compute the critical energy.

Determine the critical level value of $V(\cdot)$ via the closest UEP method. Let $\hat{x}$ be the closest UEP relative to the SEP x_s.

Step B: Estimate the stability region.

Estimate the stability region $A(x_s)$ via the function $V(\cdot)$. The connected component of $\{x : V(x) < V(\hat{x})\}$ containing the stable equilibrium point x_s gives the estimated stability region for $A(x_s)$.

Step C: Modify the energy-like function.

Modify the energy-like function $V(\cdot)$. Replace $V(x)$ by either $V(x + df(x))$ or $V(x + (d/2)(f(x + df(x)) + f(x)))$, $d > 0$, and go to Step B.

This constructive method yields a sequence of estimated stability regions for $A(x_s)$. In addition, for a finite number of iterations, there exists a positive number $\hat{d}$ such that, for $d < \hat{d}$, this sequence of estimated stability regions is a strictly increasing sequence contained in the entire stability region $A(x_s)$.

Remarks

[1] One of the appealing features of this methodology is that it only requires simple algebraic operations in Step C without invoking optimization techniques, which many existing methods rely upon. Another salient feature is that the critical level value determined at Step A is employed as the critical level value for each newly constructed function at Step C without the need to recompute the critical values, while many existing algorithms need to determine the critical level value for each newly constructed function.

[2] In terms of the terminology from decision theory, Step A of this methodology determines the *threshold value* for the *performance index* $V(\cdot)$ while Step C improves the performance index. Thus, each iteration of this methodology improves the performance index in such a way that the threshold value remains constant, and yet this threshold value is guaranteed to be optimal.

[3] This constructive methodology is very general; it can stand by itself or it can cooperate with existing optimization-based methods by using the energy-like functions produced by these methods as the starting energy-like function.

[4] The choice of the "step size" d is dependent upon the nonlinear dynamical system under study. It is suggested that the step size d for $V(x + df(x))$ be chosen to be the step size associated with the backward Euler method for the system, and the step size d for $V(x + (d/2)(f(x + df(x)) + f(x)))$ be chosen to be the step size associated with the trapezoidal method for the system.

13.5 Analysis of the constructive methodology

Chapter 10 presented a method to determine the optimal critical level value for estimating the stability region $A(x_s)$ for the nonlinear dynamical system (13.1) using an energy-like function $V(\cdot)$. Specifically, using the following value as the critical level value for the energy-like function $V(\cdot)$ is optimal in the sense that the estimated stability region characterized by the energy-like function is the largest one within the entire stability region $A(x_s)$:

$$k = \min_{x_i \in \partial A(x_s) \cap E} V(x_i). \tag{13.21}$$

In this section it will be shown that the value k determined in (13.21) is also the optimal critical level value of the energy-like functions generated by the expansion schemes.

THEOREM 13-5 **(Optimal estimation for the first-order scheme)**
Consider the nonlinear dynamical system (13.1) which admits an energy-like function $V(\cdot)$. Let x_s be an SEP and E be the set of UEPs of (13.1) and $A(x_s)$ be the associated stability region. Let $V_1(x) = V(x + df(x))$ and $S_{V_1}(k)$ denote the connected component containing x_s of the set $\{x \in R^n: V_1(x) \leq k\}$. If $k = \min_{x_i \in \partial A(x_s) \cap E} V(x_i)$, then there exists a $\hat{d}$ such that for $d \leq \hat{d}$:

(a) the component $S_{V_1}(k - \varepsilon)$, with ε being an arbitrarily positive small number, lies inside the stability region $A(x_s)$; and
(b) the set $\{S_{V_1}(b) \cap A^c(x_s)\}$ is non-empty for any number $b > k$.

Proof From Theorem 13-2 and the fundamental theorem of the closest UEP method, there exists a $d_1 > 0$ such that for $d < d_1$, the set $S_{V_1}(k - \varepsilon)$ is a positive invariant set of (13.1). Also, from the continuity argument, there exists a $d_2 > 0$ such that for $d < d_2$, the set $S_{V_1}(k - \varepsilon)$ is also bounded. Thus, let $\hat{d} = \min\{d_1, d_2\}$; we notice that, for $d < \hat{d}$, the set $S_{V_1}(k - \varepsilon)$ is a bounded and positive invariant set. Now, we prove part (a) by contradiction. Suppose $S_{V_1}(k - \varepsilon) \not\subset A(x_s)$, then $S_{V_1}(k - \varepsilon) \cap \partial A(x_s) \neq \emptyset$.

Let $B = S_{V_1}(k - \varepsilon) \cap \partial A(x_s)$. Since the stability boundary $\partial A(x_s)$ is an invariant set, B is also a bounded, closed, positively invariant set. Since every compact positively invariant set contains the ω-limit set of every trajectory in the set, it follows that B contains some equilibrium points on the stability boundary $\partial A(x_s)$, which implies that $S_{V_1}(k - \varepsilon)$ contains some equilibrium points on the stability boundary (see Theorem 3.1 and 3.2). This is a contradiction because: (1) if x is an equilibrium point, then $V_1(x) = V(x + df(x)) = V(x)$ and (2) $k = \min_{x_i \in \partial A(x_s) \cap E} V(x_i)$. Thus, part (a) follows. Next, we prove part (b). It is clear that the set $S_{V_1}(b)$ has a non-empty intersection with $\overline{A}(x_s)$ and contains the equilibrium point which is on the stability boundary with the lowest value of $V(\cdot)$. Since a point p is on the stability boundary $\partial A(x_s)$ if and only if, for every $\varepsilon > 0$, the ball $B(p, \varepsilon)$ contains points of the stability region $A(x_s)$ and of $A^c(x_s)$, it follows from a continuity argument that the set $S_{V_1}(b)$ also contains points which do not belong to the stability region $A(x_s)$ and part (b) holds. This completes the proof.

Theorem 13-5 can be extended to the second-order expansion scheme, as described in the following theorem.

THEOREM 13-6 **(Optimal estimation)**
Consider the nonlinear dynamical system (13.1) having an energy-like function $V(\cdot)$. Let x_s be an SEP and E be the set of UEPs of (13.1) and $A(x_s)$ be the associated stability region. Let $V_1(x) = (x + (d/2)(f(x + df(x)) + f(x)))$ and $S_{V_i}(k)$ denote the connected

component containing x_s of the set $\{x \in R^n: V_1(x) \le k\}$. If $k = \min_{x_i \in \partial A(x_s) \cap E} V(x_i)$, then there exists a $\hat{d}$ such that for $d < \hat{d}$:

(a) the component $S_{V_1}(k - \varepsilon)$ lies inside the stability region $A(x_s)$, for ε an arbitrarily positive small number; and
(b) the set $\{S_{V_1}(b) \cap A^c(x_s)\}$ is non-empty for any number $b > k$.

Proof This proof is similar to that of Theorem 13-5; hence it is omitted.

Let $V_i(\cdot)$ be the functions generated by the first-order expansion scheme described in (13.15) and $S_{\hat{V}_i}(k)$ denote the connected component containing x_s of the set $\{x \in R^n: V_i(x) \le k\}$ with $k = \min_{x_i \in \partial A(x_s) \cap E} V(x_i)$, then according to Theorem 13-2 and Theorem 13-5, we obtain the following relationship:

$$\begin{aligned} S_V(k) \subset S_{V_1}(k) \subset S_{V_2}(k) \subset \ldots \subset S_{V_n}(k) \\ \subset S_{V_{n+1}}(k) \subset \ldots \subset A(x_s). \end{aligned} \tag{13.22}$$

Hence, the level sets defined by these energy-like functions form a sequence of sets with strictly monotonic increasing size and yet each set is contained in the entire stability region. The constructive methodology with the first-order expansion scheme iteratively offers a larger (hence less conservative) estimate of the stability region $A(x_s)$.

Similarly, let $\hat{V}_i(\cdot)$ be the functions generated by the second-order expansion scheme and $S_{\hat{V}_i}(k)$ denote the connected component containing x_s of the set $\{x \in R^n : \hat{V}_i(x) \le k\}$ with $k = \min_{x_i \in \partial A(x_s) \cap E} V(x_i)$, then according to Theorem 13-4 and Theorem 13-6, we obtain the following relationship:

$$\begin{aligned} S_{\hat{V}_1}(k) \subset S_{\hat{V}_2}(k) \\ \subset \ldots \subset S_{\hat{V}_n}(k) \subset S_{\hat{V}_{n+1}}(k) \subset \ldots \subset A(x_s). \end{aligned} \tag{13.23}$$

Hence, the level sets defined by these energy-like functions form a sequence of sets with strictly monotonic increasing size and yet each set is contained in the entire stability region.

For illustrative purposes, we reconsider the following simple example.

Example 13-1

$$\begin{aligned} \dot{x}_1 &= -\sin x_1 - 0.5 \sin(x_1 - x_2) + 0.01 \\ \dot{x}_2 &= -0.5 \sin x_2 - 0.5 \sin(x_2 - x_1) + 0.5. \end{aligned}$$

One SEP of this simple system is $x_s = (0.02801, 0.06403)$ whose stability region is to be estimated. The following function is an energy-like function associated with the system:

$$V(x_1, x_2) = -2\cos x_1 - \cos x_2 - \cos(x_1 - x_2) - 0.02x_1 - 0.1x_2. \tag{13.24}$$

Applying Step A of the constructive methodology, the optimal level value of the energy-like function, for estimating the stability region of $x_s = (0.02801, 0.06403)$, is −0.31329. The stability boundary estimated by the connected component (containing the SEP x_s) of

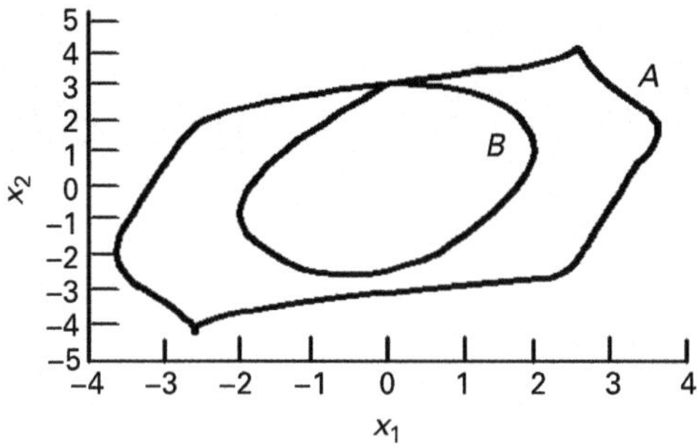

Figure 13.1 Curve A is the exact stability boundary based on the complete characterization of the stability boundary while curve B is the optimal estimation of the stability region.

the level surface passing through −0.3132 is shown in Figure 13.1 and the estimated stability region lies inside the exact stability region and is a conservative estimate.

Next, we apply the constructive solution methods by iteratively applying Steps B and C of the constructive methodology to obtain the sequence of estimated stability regions shown in Figure 13.2. As we have examined in Theorem 13-5, the estimated stability regions of this system indeed form a strictly increasing sequence and are contained in the entire stability region of the system.

The constructive method is general and can be applied to any finite-dimension nonlinear dynamical systems which admit energy functions. This methodology was applied to a high-order system (40 dimensions), which is a power system dynamic model for transient stability analysis, with very promising results [55].

13.6 Constructive methodology with local energy-like functions

A constructive methodology for estimating stability regions of nonlinear dynamical systems which do not have an energy-like function will be presented in this section. Theoretical analysis of the constructive method will also be performed.

We consider the nonlinear dynamical system (13.1) with a function $V{:}R^n \to R$ such that the derivative of $V(\cdot)$ along the trajectory of (13.1) is negative definite over some open set W except at an asymptotically stable equilibrium point $x_s \in W$. This function is termed a *local energy-like function* of this equilibrium point. As described in Chapter 10, if the set $S_V(d)$, defined as the connected component containing x_s of the set $\{x \in R^n{:}\ V(x) \le d\}$, is contained entirely in the open set W, then it follows that

$$S_V(d) \subset A(x_s).$$

If in particular $\hat{d}$ is the largest constant such that the above is true, then the set $S_V(\hat{d})$ is the largest estimate of the stability region which one can obtain via this specific local energy-like function $V(\cdot)$ and $\hat{d}$ is the corresponding optimal threshold value.

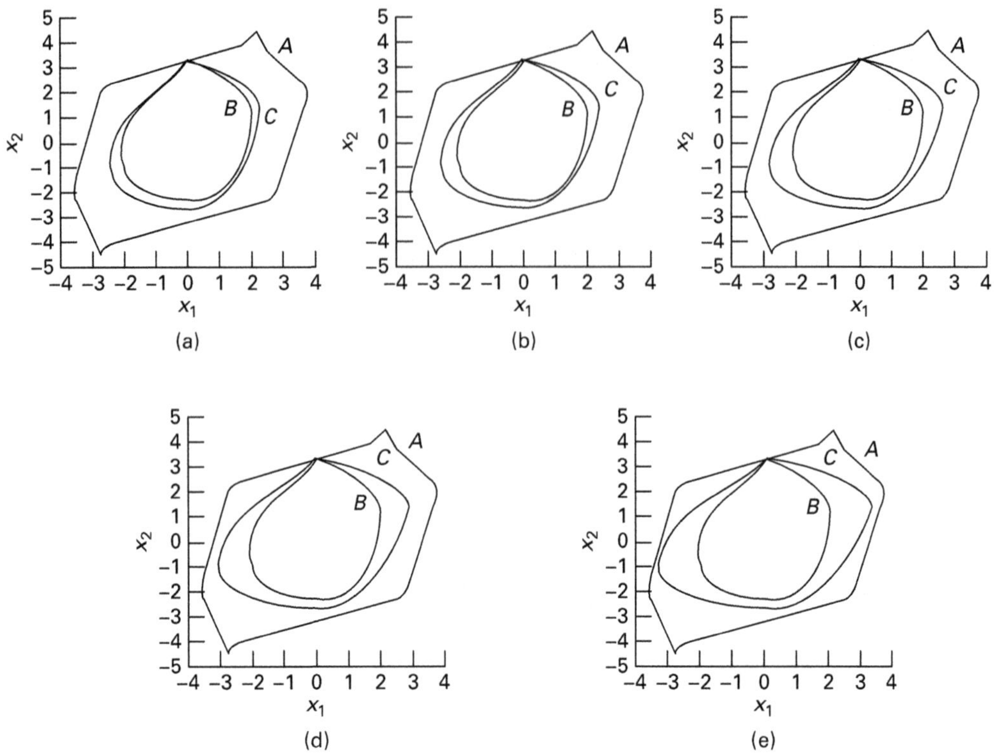

Figure 13.2 Curve A is the exact stability boundary while curve B is the stability boundary estimated by the connected component (containing the SEP of interest) of the level surface passing through the critical level value. Curve C is the stability boundary estimated by the proposed methodology after 2 iterations (a), after 4 iterations (b), after 6 iterations (c), after 8 iterations (d), after 10 iterations (e). All of the stability regions estimated by the proposed methodology lie inside the stability region.

One well-known method for constructing a local energy-like function $V(x)$ for the stable equilibrium point x_s of (13.1) is as follows.

Step 1: Solve the energy-like matrix equation for B

$$J^T B + BJ = -C \tag{13.25}$$

where J is the Jacobian matrix of (13.1) at x_s and C can be any symmetric, positive definite matrix and is usually chosen to be the identity matrix.

Step 2: Construct a local energy-like function, $V(x) = x^T Bx$.

This quadratic energy-like function can be used to estimate the stability region of x_s. Suppose that there is an open set W such that

$$\dot{V}(x) = 2f(x)^T Bx < 0 \text{ for } x \in \{W - x_s\} \tag{13.26}$$

and suppose that $\hat{d}$ is the largest value for which the set $S_V(\hat{d}) := \{x \in R^n : x^T Bx \le \hat{d}\}$ is contained in the set W. Then, $S_V(\hat{d})$ is contained in the stability region of x_s. Several schemes to determine the critical level value $\hat{d}$ of $V(x)$ can be found in [18,19,21].

It is clear that the set $S_V(\hat{d})$ still presents a conservative estimate of the stability region $A(x_s)$. In order to improve the estimate of the stability region $A(x_s)$ via the quadratic energy-like function $V(x) = x^T Bx$ and its associated critical value $\hat{d}$, several researchers have suggested modifying matrix B into B_i and its associated critical level value $\hat{d}$ into $\hat{d}_i$ in such a way that the sets

$$S_{V_i}(\hat{d}_i) := \{x \in R^n : V_i(x) = x^T B_i x \le \hat{d}_i \text{ and } \dot{V}_i(x) = 2f(x)^T B_i x < 0\} \tag{13.27}$$

satisfy the condition

$$S_V(\hat{d}_i) \subset S_{V_1}(\hat{d}_1) \subset S_{V_2}(\hat{d}_2) \subset S_{V_3}(\hat{d}_3) \subset \ldots. \tag{13.28}$$

The above task was formulated as a constrained optimization problem and solved by constrained optimization techniques. At each step of the optimization procedure, the form of quadratic energy-like function is preserved but the matrix B is modified over the parameter space of B and the critical level value is also updated. This requires considerable computation effort and achieves insignificant improvement.

We next present a constructive methodology that meets the condition (13.28) for any given local energy-like functions without use of optimization methods. In addition, the methodology can work with the existing optimization methods in the manner that they provide the initial local energy-like function and the associated critical level value for the constructive method.

A constructive methodology for general nonlinear dynamical systems

Step A: Construct a local energy-like function, say $V(\cdot)$, for the stable equilibrium point x_s. (Let $S_V(d)$ be the connected component containing x_s of the set $\{x \in R^n: V(x) \le d\}$.)

Step B: Determine the critical level value of $V(\cdot)$ for x_s such that $\dot{V}(x) < 0$ for $x \in \{S_V(d) - x_s\}$. (Let $\hat{d}$ be the largest number for this step to be true.)

Step C: Estimate the stability region $A(x_s)$ via the local energy-like function $V(\cdot)$.

The connected component of $\{x : V(x) \le \hat{d}\}$ containing the stable equilibrium point x_s gives the estimated stability region for $A(x_s)$.

Step D: Modify the local energy-like function $V(\cdot)$.

Replace $V(x)$ by either $V(x + df(x))$ or $V(x + (d/2)(f(x + df(x)) + f(x)))$, $d > 0$, and go to Step C.

Remarks

[1] A number of algorithms for constructing local energy-like functions can be applied to Step A; for example, the algorithms for constructing quadratic energy-like functions by solving a set of energy-like equations or the algorithms for constructing normed energy-like functions proposed in [18,19], which also contain efficient algorithms to determine the critical level value in Step B for quadratic energy-like functions as well as for normed energy-like functions.

[2] The solution methodology requires only simple algebraic operations in achieving the condition (13.28) without invoking complicated optimization techniques which the existing algorithms rely upon. Furthermore, the critical level value determined at Step B is utilized as the critical level value for every newly constructed local energy-like function at Step C, while many existing algorithms need to determine the critical level value for every newly constructed function.

We next show that this constructive methodology yields a sequence of estimated stability regions for $A(x_s)$. In particular, it is shown that for a finite number of iterations, there exists a positive number $\hat{d}$ such that, for $d < \hat{d}$, this sequence of estimated stability regions is a strictly increasing sequence (thus satisfying the condition (13.28)) contained in the entire stability region $A(x_s)$.

THEOREM 13-7 **(Optimal estimation)**
Let x_s be a stable equilibrium point of (13.1) and $A(x_s)$ be the associated stability region. Let $V(\cdot)$ be a local energy-like function of x_s. Let $V_1(x) = V(x + df(x))$ (or $V_1(x) = V(x + (d/2)(f(x + df(x)) + f(x))))$. $S_V(k)$ denotes the connected component containing x_s of the set $\{x \in R^n\colon V(x) \le k\}$, and $S_{V_1}(k)$ denotes the connected component containing x_s of the set $\{x \in R^n\colon V_1(x) \le k\}$. If $\hat{k}$ is the largest number such that $\dot{V}(x) < 0$ for $x \in \{S_V(\hat{k}) - x_s\}$, then there exists a $\hat{d}$ such that for $d < \hat{d}$ the following results hold:

(a) the component $S_{V_1}(\hat{k})$ lies inside the stability region $A(x_s)$; and
(b) $S_V(\hat{k}) \subset S_{V_1}(\hat{k})$.

Proof The proof of this theorem is similar to that of Theorem 13-7 and is therefore omitted.

Let $V(\cdot)$ be a local energy-like function around the stable equilibrium point x_s of (13.1) and $A(x_s)$ be the associated stability region. Let $V_i(\cdot)$ be the sequence of functions generated by the first-order scheme in (13.2) and $S_{\hat{V}_i}(k)$ denote the connected component containing x_s of the set $\{x \in R^n\colon V_i(x) \le k\}$. If $\hat{k}$ is the largest number such that $\dot{V}(x) < 0$ for $x \in \{S_V(\hat{k}) - x_s\}$, then, according to Theorem 13-3 and Theorem 13-7, the following relationship holds:

$$\begin{aligned} S_V(\hat{k}) &\subset S_{V_1}(k) \subset S_{V_2}(\hat{k}) \\ &\subset \ldots \subset S_{V_n}(\hat{k}) \subset S_{V_{n+1}}(\hat{k}) \subset \ldots \subset A(x_s). \end{aligned} \tag{13.29}$$

Similarly, let $\hat{V}_i(\cdot)$ be the functions generated by the second-order scheme in (13.15) and $S_{\hat{V}_i}(k)$ denote the connected component containing x_s of the set $\{x \in R^n : \hat{V}_i(x) \le k\}$. Then, according to Theorem 13-7, the following relationship holds:

$$\begin{aligned} S_{\hat{V}_1}(\hat{k}) &\subset S_{\hat{V}_2}(\hat{k}) \subset \ldots \subset S_{\hat{V}_n}(\hat{k}) \\ &\subset S_{\hat{V}_{n+1}}(\hat{k}) \subset \ldots \subset A(x_s). \end{aligned} \tag{13.30}$$

The constructive method described will be illustrated by three simple examples. The algorithm presented in [18,19] was used to implement Steps A and B of the constructive method. In each example we give several figures to compare the stability regions estimated by previous methods with the stability regions estimated by the constructive method. The phase portrait of each example will be used to verify the accuracy of the constructive method.

Example 13-2 This is the well-known Vanderpol equation; here (0,0) is a stable equilibrium point of this equation whose stability region is to be estimated.

$$\begin{aligned} \dot{x}_1 &= -x_2 \\ \dot{x}_2 &= x_1 - (1 - x_1^2)x_2. \end{aligned} \tag{13.31}$$

Applying the constructive method leads to the following.

Step A: A local energy-like function constructed by the algorithm [18,19] is

$$V(x_1, x_2) = (x_1, x_2)^T \begin{bmatrix} 0.593 & -0.182 \\ -0.182 & 0.437 \end{bmatrix} (x_1, x_2). \tag{13.32}$$

Step B: The critical level value of $V(x)$ for (0,0) is 1.0.

Step C: The connected component of $\{x: V(x) \leq 1.0\}$ containing (0,0) is the estimated stability region.

Step D: Replace $V(x_1, x_2)$ by $V(x_1 + dx_2, x_2 - dx_1 + d(1 - x_1^2)x_2)$, and go to Step C.

By repeating the procedure of Steps C and D one can obtain a sequence of estimated stability regions as shown in Figure 13.3. To show that the estimated stability regions are indeed subsets of the exact stability region, the phase portrait of this system is shown in Figure 13.4. It can be seen that the sequence of estimated stability regions is a strictly increasing sequence contained in the entire stability region of (0,0).

Example 13-3 We consider the task of estimating the stability region of (0,0), the origin of the following second-order system [18,19]:

$$\begin{aligned} \dot{x}_1 &= -x_1 + 2x_1^2 x_2 \\ \dot{x}_2 &= -x_2. \end{aligned}$$

By applying the constructive methodology, we have the following steps:

Step A: A local energy-like function is

$$V(x_1, x_2) = (x_1, x_2)^T \begin{bmatrix} 0.330 & 0.249 \\ 0.249 & 0.376 \end{bmatrix} (x_1, x_2).$$

Step B: The critical level value of $V(x)$ for (0,0) is 1.0.

Step C: The connected component of $\{x: V(x) \leq 1.0\}$ containing (0,0) is the estimated stability region.

Step D: Replace $V(x_1, x_2)$ by $V(x_1 + dx - 2dx_1^2, x_2 + dx_2)$, and go to Step C.

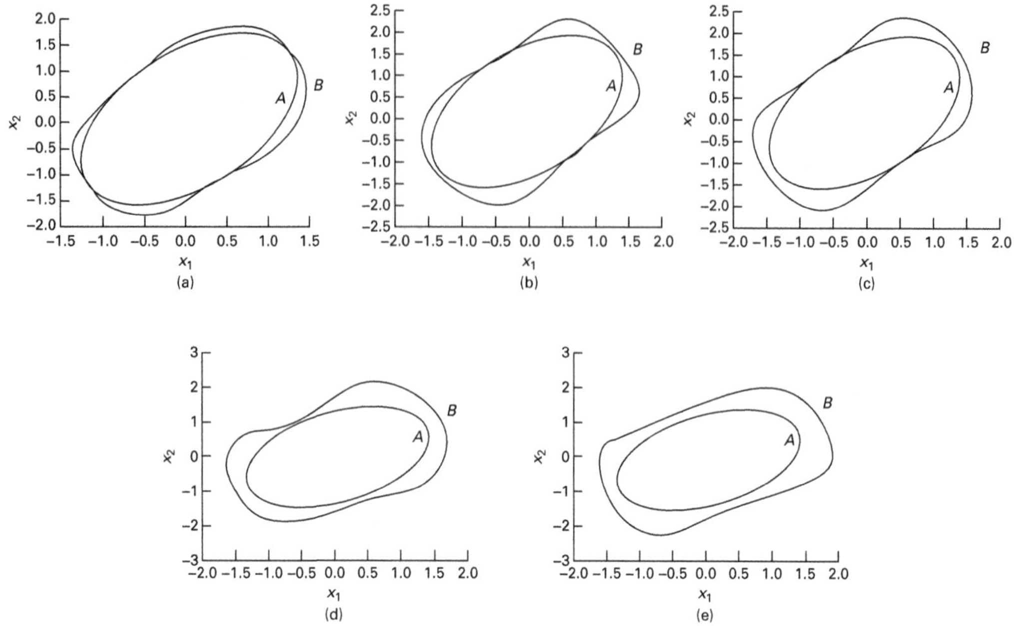

Figure 13.3 Curve A is the exact stability boundary $\partial A(x_s)$ of system (13.31). Curve B is the stability boundary estimated by the connected component (containing the SEP x_s) of the level surface of (13.32) passing through 1.0. Curve C is the stability boundary estimated by the constructive methodology: (a) after 2 iterations, (b) after 4 iterations, (c) after 6 iterations, (d) after 8 iterations, (e) after 10 iterations. All these stability regions estimated by the constructive methodology lie inside the stability region.

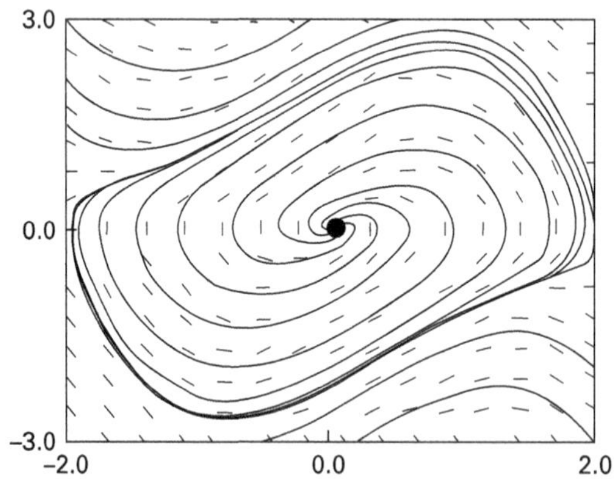

Figure 13.4 The phase portrait of Example 13-2. Note that all the estimated stability regions delimited by curve C in Figure 13.3 are inside the exact stability region.

By repeating the procedure of Steps C and D one can obtain a sequence of estimated stability regions as shown in Figure 13.5. Curve A in this figure is the estimated stability boundary, obtained by using the original local energy-like function at Step A, while the other curves are the estimated stability boundary obtained by using the sequence of local

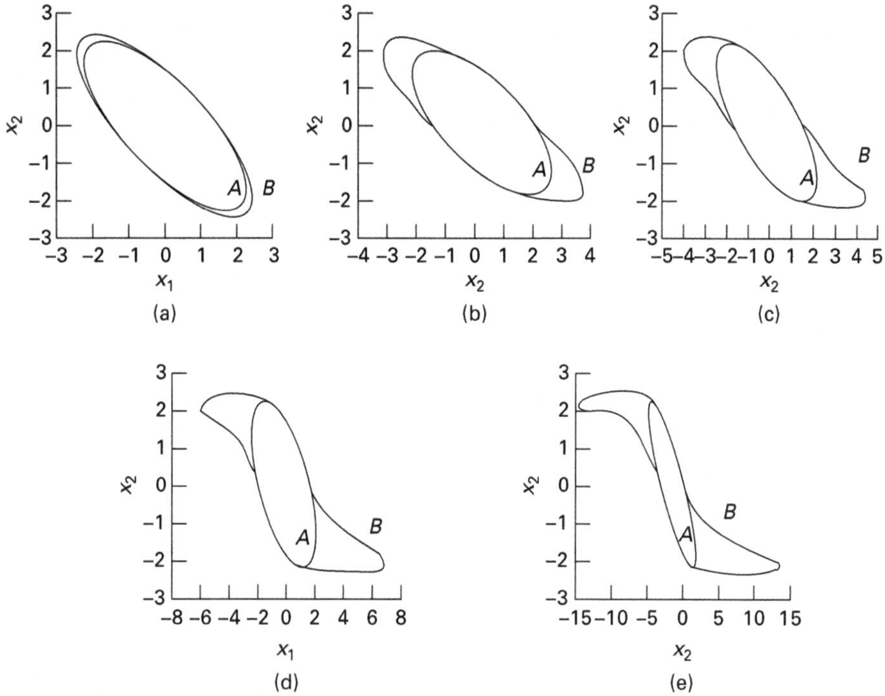

Figure 13.5 Curve A is the estimated stability boundary obtained by the local energy-like function obtained at Step A for Example 13-3. Curve B is the stability boundary estimated by the constructive methodology: (a) after 2 iterations, (b) after 4 iterations, (c) after 6 iterations, (d) after 8 iterations, (e) after 10 iterations.

energy-like functions at Step D. It can be seen that the sequence of estimated stability regions is a strictly increasing sequence contained in the entire stability region of (0,0).

Example 13-4 Consider the following three-dimensional system which has a stable equilibrium point at the origin (0,0,0) and a limit cycle. It is desired to estimate the stability region of the origin:

$$\begin{aligned} \dot{x}_1 &= -x_2 \\ \dot{x}_2 &= -x_3 \\ \dot{x}_3 &= -0.915x_1 + (1 - 0.915x_1^2)x_2 - x_3 \end{aligned} \tag{13.33}$$

By applying the constructive methodology, we have the following steps in which the algorithm presented in [18, 19] is used to implement Steps A and B

Step A: A local energy-like function is

$$V(x_1, x_2, x_3) = (x_1, x_2, x_3)^T \begin{bmatrix} 12.5 & -8.1 & 3.0 \\ -8.1 & 20.8 & -8.5 \\ 3.0 & -8.5 & 13.4 \end{bmatrix} (x_1, x_2, x_3).$$

Step B: The critical level value of $V(x)$ for (0,0,0) is 1.0.

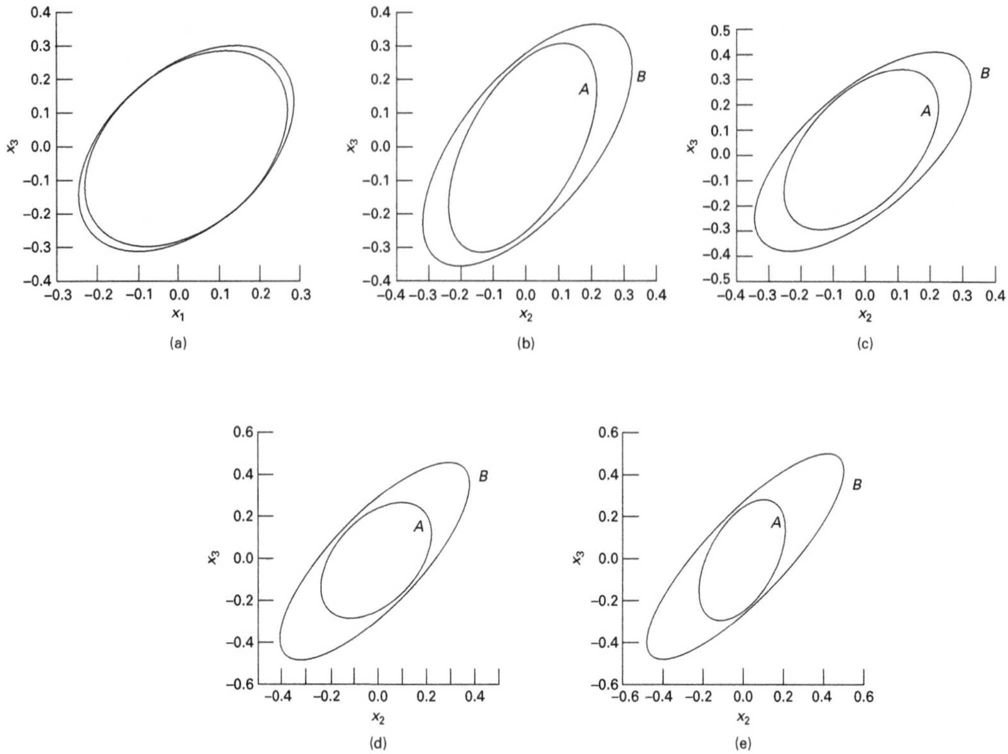

Figure 13.6 Curve A is the estimated stability boundary obtained by the local energy-like function obtained at Step A for Example 13-4. Curve B is the stability boundary estimated by the constructive methodology: (a) after 2 iterations, (b) after 4 iterations, (c) after 6 iterations, (d) after 8 iterations, (e) after 10 iterations.

Step C: The connected component of $\{x\colon V(x) \leq 1.0\}$ containing (0,0,0) is the estimated stability region.

Step D: Replace $V(x_1, x_2, x_3)$ by the following function

$$V(x_1 + dx_2, x_2 + dx_3, x_3 + 0.915dx_1 - d(1 - 0.915x_1^2)x_2 - dx_3),$$

and go to Step C.

By repeating the procedure of Steps C and D, one can obtain a sequence of estimated stability regions as shown in Figure 13.6. Curve A in this figure is the estimated stability boundary, obtained by the original local energy-like function at Step A, while the other curves are the estimated stability boundary obtained by using the sequence of local energy-like functions constructed at Step D. It can be seen from the phase portrait of the system that the sequence of estimated stability regions is a strictly increasing sequence contained in the entire stability region of (0,0,0).

13.7 Conclusion

General nonlinear dynamical systems can be classified into those admitting (global) energy-like functions and those admitting local energy-like functions. A constructive methodology for estimating stability regions of general nonlinear dynamical systems has been presented. For systems having a (global) energy-like function, a constructive methodology starts with an energy-like function $V(\cdot)$ associated with the system under study and proceeds in three steps. Step 1 is to optimally determine the critical level value of the energy-like function with respect to a stable equilibrium point. Steps 2 and 3 are designed to iteratively improve the estimated stability region (by reducing its conservativeness). The critical level value determined at Step 1 is also the critical level value for every newly constructed energy-like function at Step 2, and this critical level value has been shown to be optimal for every newly constructed energy-like function at Step 2. The stability regions estimated by repeating Steps 2 and 3 have been shown to form a sequence of sets with strictly monotonic increasing size, and yet each set is contained in the entire stability region; it thus reducing the conservativeness in estimating the stability regions.

For systems having local energy-like functions, the methodology starts with a local energy-like function $V(\cdot)$ associated with the system under study and proceeds with the same three steps as for the systems with energy-like functions. The only difference between the above two procedures lies in the way that the critical value is determined at Step 1.

One distinguishing feature of the constructive methodology is that it enables the existing methods for estimating stability regions reduce their conservativeness. The constructive methodology is rather general: it can either stand by itself to estimate stability regions or it can cooperate with the existing methods to estimate stability regions.

14 Estimation of relevant stability regions

14.1 Introduction

A rigorous introduction to the relevant stability region and its characterization for continuous nonlinear dynamical systems and constrained nonlinear dynamical systems was presented in Chapter 8. In this chapter, we will present effective energy-function-based methods to estimate the relevant stability regions.

We start with a one-machine-infinite-bus system described by the following equations:

$$\begin{aligned} \dot{\delta} &= \omega \\ M\dot{\omega} &= -D\omega - P_0 \sin\delta + P_m. \end{aligned} \tag{14.1}$$

There are three equilibrium points lying within the range of $\{(\delta, \omega) = -\pi < \delta < \pi, \omega = 0\}$, and they are $(\delta_s, 0) = (\sin^{-1}(P_m/P_0), 0)$ which is a stable equilibrium point, and $(\delta_1, 0) = (-\pi - \sin^{-1}(P_m/P_0), 0)$, $(\delta_2, 0) = (-\pi - \sin^{-1}(P_m/P_0), 0)$ which are unstable equilibrium points. The exact stability region $A(\delta_s, 0)$ is completely characterized by its stability boundary $\partial A(\delta_s, 0)$ which is equal to the union of the stable manifold of the UEP $(\delta_1, 0)$ and the stable manifold of the UEP $(\delta_2, 0)$ (see Figure 14.1). The intersection between the stability region $A(\delta_s, 0)$ and the angle space $\{(\delta, \omega): \delta \in R, \omega = 0\}$ is $A_\delta := \{(\delta, \omega): \delta \in [\delta_2, \delta_1], \omega = 0\}$. The boundary of this one-dimensional region A_δ is composed of two points δ_1 and δ_2, where $(\delta_1, 0)$ and $(\delta_2, 0)$ are the UEPs on the stability boundary $\partial A(\delta_s, 0)$. To estimate the stability region, we consider the following function which is an energy function for this simple system

$$V(\delta, \omega) = \frac{1}{2}M\omega^2 - P_m\delta - P_0 \cos\delta. \tag{14.2}$$

The energy function can be divided into kinetic energy $K(\omega)$ and potential energy $U(\omega)$ functions

$$V(\delta, \omega) = K(\omega) + U(\delta) \tag{14.3}$$

where $K(\omega) = \frac{1}{2}M\omega^2$ and $U(\delta) = -P_m\delta - P_0 \cos\delta$.

The closest UEP method uses the constant energy surface $\{(\delta, \omega): V(\delta, \omega) = U(\delta_1)\}$ passing through the closest UEP $(\delta_1, 0)$ to approximate the stability boundary $\partial A(\delta_s, 0)$

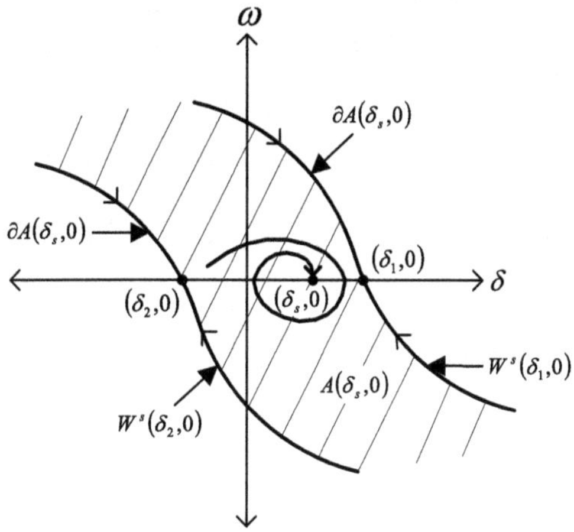

Figure 14.1 The position of the stable equilibrium point $(\delta_s, 0)$ along with its stability region $A(\delta_s, 0)$ (the shaded area). The stability boundary $\partial A(\delta_s, 0)$ is composed of the stable manifold of the UEP $(\delta_1, 0)$ and the stable manifold of the UEP $(\delta_2, 0)$.

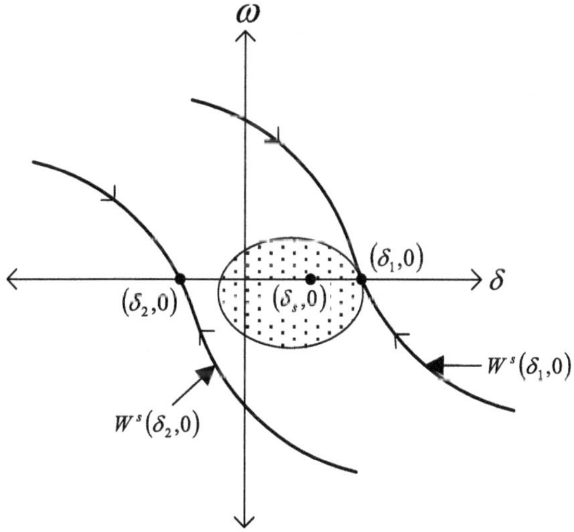

Figure 14.2 The closest UEP method uses the constant energy surface passing through the closest UEP $(\delta_1, 0)$ to approximate the (entire) stability boundary $\partial A(\delta_s, 0)$. The shaded area is the stability region estimated by the closest UEP method.

(see Figure 14.2). If a given state, say $(\delta_{cl}, \omega_{cl})$, has energy function value $V(\delta_{cl}, \omega_{cl})$ less than $U(\delta_1)$, then the state $(\delta_{cl}, \omega_{cl})$ is classified by the closest UEP method to be lying inside the stability region of $(\delta_s, 0)$. Thus one can assert, without numerical integration, that the resulting trajectory will converge to $(\delta_s, 0)$.

This closest UEP method gives very conservative stability assessments, especially for those disturbance-on trajectories crossing the stability boundary $\partial A(\delta_s, 0)$ through $W^s(\delta_2, 0)$ (see Figure 14.3(a)). For example, the post-disturbance trajectory starting from

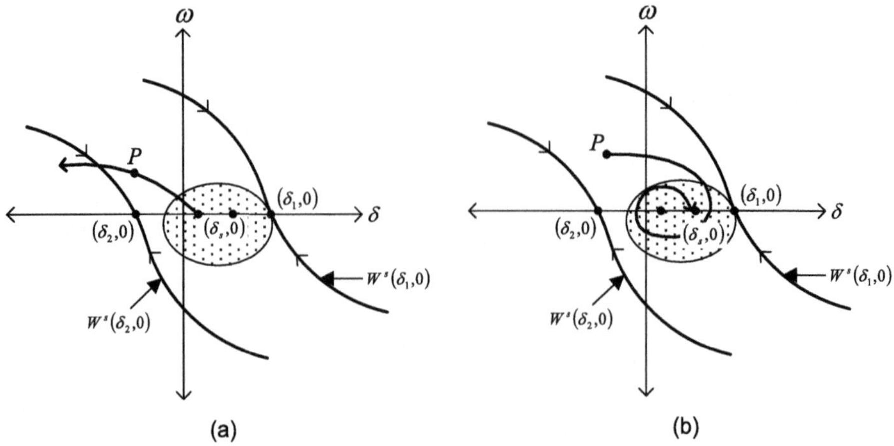

Figure 14.3 The post-disturbance trajectory starting from the state *P*, which lies inside the stability region $A(\delta_s, 0)$, is classified as unstable by the closest UEP method while in fact the resulting trajectory will converge to $(\delta_s, 0)$ and hence it is stable.

the state *P* which lies inside the stability region $A(\delta_s, 0)$, is classified as unstable by the closest UEP method. This is incorrect because the resulting trajectory will converge to $(\delta_s, 0)$ and hence it is stable (see Figure 14.3(b)).

The closest UEP method does provide a good approximation for the entire stability boundary. However, it does not provide an accurate approximation for the relevant stability boundary. This is due to the fact that the disturbance-on trajectory is not taken into account in the closest UEP method.

14.2 Controlling UEP method

The controlling UEP method aims to reduce the conservativeness of the closest UEP method in estimating the relevant stability region by taking the dependence of the disturbance-on trajectory into account. We illustrate this method using the simple system as an example. To assess the stability of the post-disturbance trajectory whose corresponding disturbance-on trajectory $(\delta(t), \omega(t))$ moves towards δ_1, the controlling UEP method uses the constant energy surface passing through the UEP $(\delta_1, 0)$ which is $\{(\delta, \omega): V(\delta, \omega) = U(\delta_1)\}$ as the local approximation for the relevant stability boundary of the post-disturbance system. In the same manner, for those disturbance-on trajectories $(\delta(t), \omega(t))$ whose $\delta(t)$ component moves towards δ_2, the controlling UEP method uses the constant energy surface passing through the UEP $(\delta_2, 0)$ which is $\{(\delta, \omega): V(\delta, \omega) = U(\delta_2)\}$ as the local approximation for the relevant stability boundary (see Figure 14.4). Therefore, for each disturbance-on trajectory, there exists a unique corresponding unstable equilibrium point whose stable manifold constitutes the relevant stability boundary. This unique unstable equilibrium point is the controlling UEP. The constant energy surface passing through the controlling UEP can be used to accurately approximate the relevant part of the stability boundary toward which the disturbance-on

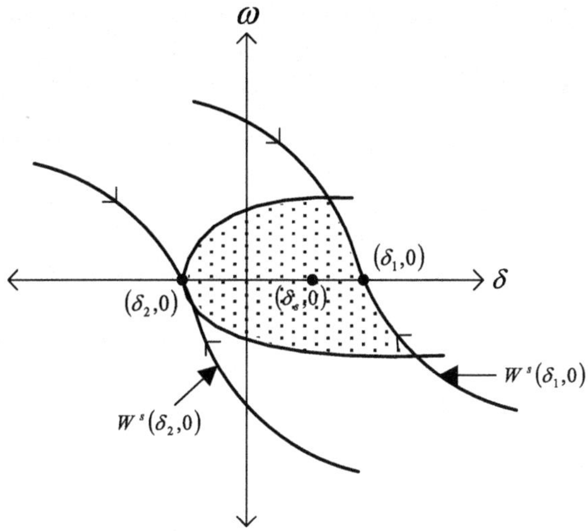

Figure 14.4 The shaded area is the stability region estimated by the controlling UEP method. The controlling UEP method does not provide an approximation for the entire stability boundary. However, it provides an accurate approximation for the relevant stability boundary.

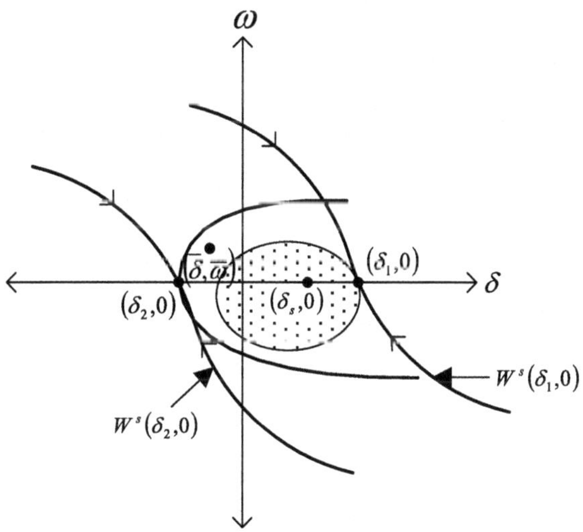

Figure 14.5 The post-disturbance trajectory starting from the state $(\bar{\delta}, \bar{\omega})$, which lies inside the stability region $A(\delta_s, 0)$, is correctly classified as stable by the controlling UEP method while it is classified as unstable by the closest UEP method.

trajectory is heading. If the energy function value of a given state is less than that of the controlling UEP, then the state is classified as lying inside the stability region of $(\delta_s, 0)$ by the controlling UEP method (see Figure 14.5).

Suppose a disturbance is cleared when the disturbance-on trajectory reaches the state $(\bar{\delta}, \bar{\omega})$ in Figure 14.5. The post-disturbance trajectory starting from the state $(\bar{\delta}, \bar{\omega})$, which lies inside the stability region $A(\delta_s, 0)$, will converge to the SEP $(\delta_s, 0)$.

Hence, the post-disturbance system is stable. This post-disturbance trajectory is correctly classified as stable by the controlling UEP method while it is classified as unstable by the closest UEP method. Again, this shows the conservativeness of the closest UEP method, which does not take into account the dependence of the disturbance-on trajectory.

The spirit of the controlling UEP method is that it does not provide an approximation for the entire stability boundary; instead, it provides an accurate approximation for the relevant stability boundary. The controlling UEP method, although more complex than the closest UEP method, gives much more accurate and less conservative stability assessments when compared to the closest UEP method in estimating relevant stability regions.

The controlling UEP method for estimating relevant stability regions of large-scale nonlinear systems proceeds as follows.

1. Determination of the critical energy
   Step 1.1: Find the controlling UEP, x_{co}, for a given disturbance-on trajectory $x_f(t)$.
   Step 1.2: The critical energy, V_{cr}, is the value of the energy function $V(\cdot)$ at the controlling UEP, i.e.

$$V_{cr} = V(x_{co}).$$

2. Approximation of the relevant stability boundary
   Step 2.1: Use the connected constant energy surface of $V(\cdot)$ passing through the controlling UEP x_{co} and containing the SEP x_s to approximate the relevant part of the stability boundary for the disturbance-on trajectory $x_f(t)$.

3. Check whether the disturbance-on trajectory at the disturbance clearing time (t_{cl}) is located inside the stability boundary characterized in Step 2.1. This is done as follows.
   Step 3.1: Calculate the value of the energy function $V(\cdot)$ at the time of disturbance clearance (t_{cl}) using the disturbance-on trajectory

$$V_f = V(x_f(t_{cl})).$$

   Step 3.2: If $V_f < V_{cr}$, then the point $x_f(t_{cl})$ is located inside the stability boundary and the post-disturbance system is stable. Otherwise, it may be unstable.

To ensure the conservative nature of the controlling UEP method, it is important that the disturbance-on trajectory passes through the constant energy surface containing the controlling UEP before it exits the stability region at the exit point. If the disturbance is cleared before the disturbance-on trajectory reaches the constant energy surface containing the controlling UEP, then the post-disturbance trajectory will converge to the SEP and become stable (see Figure 14.6). To check when the disturbance-on trajectory passes through the constant energy surface containing the controlling UEP, one can monitor the value $V(x_f(t))$ along the disturbance-on trajectory $x_f(t)$ and identify when $V(x_f(t)) = V(x_{co})$ (see Figure 14.6).

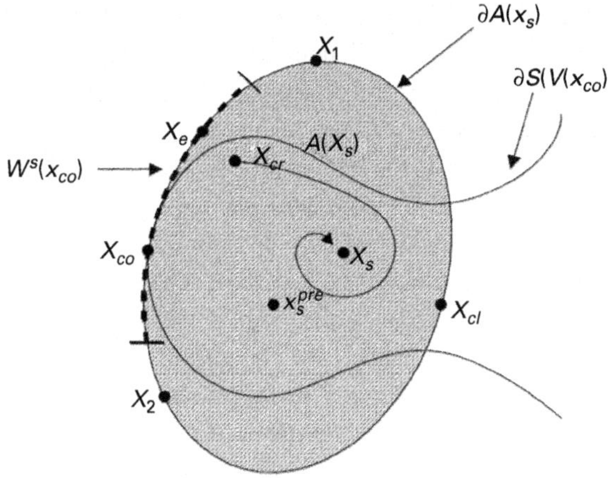

Figure 14.6 The controlling UEP method approximates the relevant stability boundary, the stable manifold of the controlling UEP $W^s(x_{co})$, by the constant energy surface, $\partial S(V(x_{co}))$, passing through the controlling UEP.

The above analysis suggests that the energy value at the controlling UEP can be used as the critical energy with respect to an energy function for the disturbance-on trajectory and that the constant energy surface containing the controlling UEP can be used to approximate the relevant part of the stability boundary. We will present a theoretical analysis of the controlling UEP method in the next section.

14.3 Analysis of the controlling UEP method

Theorem 14-1 below gives a rigorous theoretical justification for the controlling UEP method. We define the following two components: $S(r)$:= the connected component of the energy level set $\{x \in R^n: V(X) < r\}$ containing x_s, and $\partial S(r)$:= the (topological) boundary of $S(r)$, which is the constant energy surface with energy value r.

THEOREM 14-1 **(Controlling UEP method for approximating the relevant stability boundary)**
Consider a general post-disturbance system which has an energy function $V(\cdot)$: $R^n \to R$. Let x_{co} be an equilibrium point on the stability boundary $\partial A(x_s)$ of an SEP x_s of this system.
Then, the following results hold.

(a) The connected constant energy surface $\partial S(V(x_{co}))$ intersects with the stable manifold $W^s(x_{co})$ only at the point x_{co}; moreover, the set $S(V(x_{co}))$ has an empty intersection with the stable manifold $W^s(x_{co})$.
(b) Any connected path starting from a point $p \in \{S(V(x_{co})) \cap A(x_s)\}$ and intersecting with $W^s(x_{co})$ must also intersect with $\partial S(V(x_{co}))$.

Theorem 14-1 asserts that for any disturbance-on trajectory $x_f(t)$ starting from a point $x_s^{pre} \in A(x_s)$and $V(x_s^{pre}) < V(x_{co})$, if the exit point of this disturbance-on trajectory $x_f(t)$ lies on the stable manifold of x_{co}, then this disturbance-on trajectory $x_f(t)$ must pass through the connected constant energy surface $\partial S(V(x_{co}))$ before it passes through the stable manifold of x_{co} (thus exiting the stability boundary $\partial A(x_s)$). Therefore, the connected constant energy surface $S(V(x_{co}))$ is adequate for approximating the relevant part of the stability boundary.

Using the energy value at a different UEP instead of the controlling UEP as the critical energy can give an erroneous stability assessment. We show in Theorem 14-2 below that among all the UEPs on the stability boundary, the controlling UEP is the one that gives the most accurate critical energy. Indeed, there are usually multiple UEPs lying on the stability boundary. As such it is inappropriate to choose the energy values at UEPs other than the controlling UEP, as the critical energy.

THEOREM 14-2 **(Controlling UEP method for accurate critical energy)**
Consider a general post-disturbance system which has an energy function $V(\cdot)$: $R^n \rightarrow R$. Let x_{co} be an equilibrium point on the stability boundary $\partial A(x_s)$ of an SEP x_s of this system.
Then, the following results hold.

(a) *If x^u is a UEP and $V(x^u) > V(x_{co})$, then $S(V(x^u)) \cap W^s(x_{co}) \neq \emptyset$.*
(b) *If x^u is a UEP and $V(x^u) < V(x_{co})$, then $S(V(x^u)) \cap W^s(x_{co}) = \emptyset$.*
(c) *If $\hat{x}$ is a state vector on the stability boundary but not the closest UEP, then the constant energy surface passing through $\hat{x}$ has a non-empty intersection with the complement of the closure of the stability region; i.e. $\partial S(V(\hat{x})) \cap (\bar{A}(x_s))^c \neq \emptyset$.*

Results (a) and (b) of Theorem 14-2 assert that the following two situations may occur.

Case (1): The energy level set $S(V(x_1))$ contains only part of the stable manifold $W^s(x_{co})$.
Case (2): The energy level set $S(V(x_1))$ contains the entire stable manifold $W^s(x_{co})$ (see Figure 14.7).

In Case (1), the disturbance-on trajectory $x_f(t)$ may pass through the connected constant energy surface $\partial S(V(x_1))$ after it passes through the stable manifold $W^s(x_{co})$ (see Figure 14.8). In this situation, using the energy value at x_1 as the critical energy gives an inaccurate stability assessment. In Case (2), the disturbance-on trajectory $x_f(t)$ passes through the connected constant energy surface $\partial S(V(x_1))$ after it passes through the stable manifold $W^s(x_{co})$ (see Figure 14.9). In this situation, using the energy value at x_1 as the critical energy can give inaccurate stability assessments. In particular, it can classify the post-disturbance trajectory as stable when in fact it is unstable. Results (b) and (c) of Theorem 14-2 assert that the set $S(V(x_1))$ has an empty intersection with the stable manifold $W^s(x_{co})$. Under this situation, the connected constant energy surface $\partial S(V(x_1))$ does not approximate the relevant stability boundary.

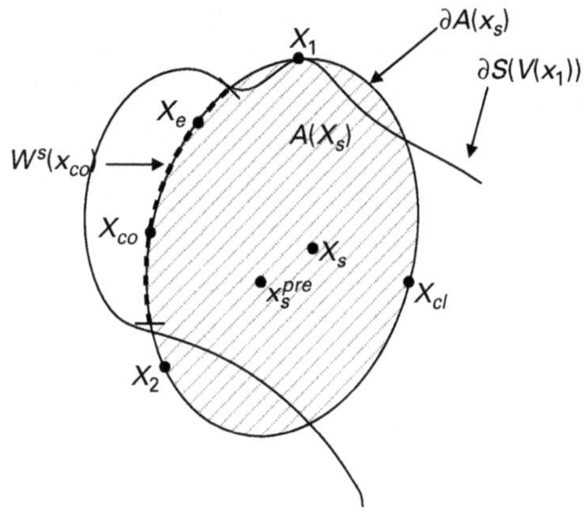

Figure 14.7 The constant energy surface passes through the UEP x_1 and the corresponding energy level set contains the entire stable manifold $W^s(x_{co})$.

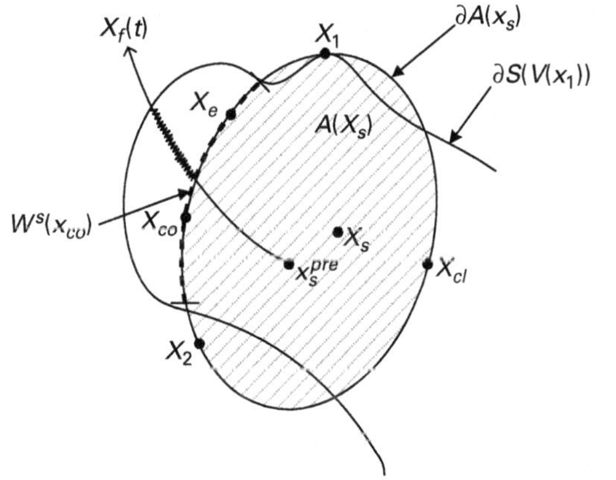

Figure 14.8 The energy level set $S(V(x_1))$ contains the whole stable manifold $W^s(x_{co})$. In this situation, the disturbance-on trajectory $x_f(t)$ always passes through the connected constant energy surface $\partial S(V(x_1))$ after it passes through the stability boundary. Using the energy value at $x_1$as the critical energy gives inaccurate stability assessments. It classifies an unstable contingency as stable when the disturbance is cleared in the highlighted segment of the disturbance-on trajectory.

In order to illustrate the controlling UEP method, we use the following simple numerical example, which closely represents a three-machine power system, with machine number 3 as the reference machine:

$$\begin{aligned}
\dot{\delta}_1 &= \omega_1 \\
\dot{\omega}_1 &= -\sin\delta_1 - 0.5\sin(\delta_1 - \delta_2) - 0.4\omega_1 \\
\dot{\delta}_2 &= \omega_2 \\
\dot{\omega}_2 &= -0.5\sin\delta_2 - 0.5\sin(\delta_2 - \delta_1) - 0.5\omega_2 + 0.05.
\end{aligned}$$

It is easy to show that the following function is an energy function for this system:

$$V(\delta_1, \delta_2, \omega_1, \omega_2) = \omega_1^2 + \omega_2^2 - 2\cos\delta_1 - \cos\delta_2 - \cos(\delta_1 - \delta_2) - 0.1\delta_2.$$

Table 14.1 Coordinates of type-two equilibrium points lying on the stability boundary of x_s

Type-two EP	δ_1	ω_1	δ_2	ω_1
1	3.60829	0	1.58620	0
2	2.61926	0	4.25636	0
3	−2.67489	0	1.58620	0
4	−3.66392	0	−2.02684	0
5	−2.67489	0	−4.69699	0
6	2.61926	0	−2.02684	0

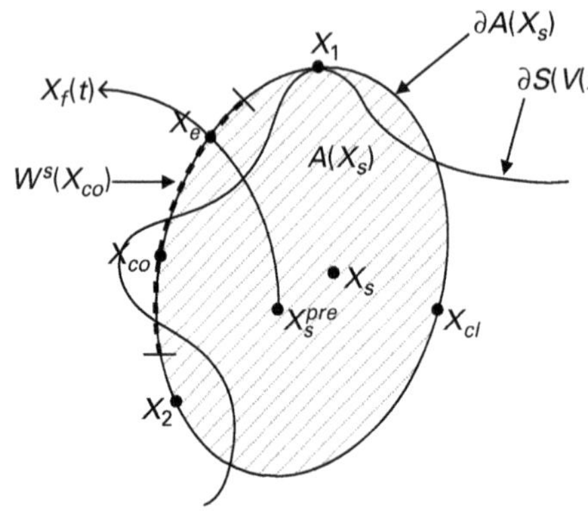

Figure 14.9 The energy level set $S(V(x_1))$ contains only part of the stable manifold $W^s(x_{co})$. The disturbance-on trajectory $x_f(t)$ passes through the connected constant energy surface $\partial S(V(x_1))$ before it passes through the stable manifold $W^s(x_{co})$. In this situation, incorrect use of the energy function value of $x_1$as the critical energy still gives a correct stability assessment.

The point $x_s = (\delta_1^{s}, \omega_1^{s}, \delta_2^{s}, \omega_2^{s}) = (0.02001, 0, 0.06003, 0)$ is an SEP of the above post-disturbance system. There are six type-two equilibrium points (see Table 14.1) and six type-one equilibrium points (see Table 14.2) lying on the stability boundary of x_s. The unstable manifold of each of these equilibrium points converges to the SEP, x_s.

The stability boundary, $\partial A(x_s)$, is contained in the set which is the union of the stable manifolds of these six type-one equilibriums and of these six type-two equilibriums. Figure 14.10 depicts the intersection between the stability boundary $\partial A(x_s)$ and the angle space $\{\delta_1, \delta_2 : \delta_1 \in R, \delta_2 \in R\}$. For illustrational purposes, we assume the disturbance-on trajectory, $x_f(t)$, shown in Figure 14.10, is due to the disturbance. The disturbance-on trajectory, $x_f(t)$, starting from the pre-disturbance SEP x_s first intersects with the constant energy surface passing through the type-one equilibrium point x_{co} = (0.03333, 0, 3.10823, 0) and then intersects with the stability boundary, $\partial A(x_s)$, by passing through the stable manifold of the type-one equilibrium point x_{co} at the exit point, x_e. Among the stable manifolds of the six type-one UEPs on the stability boundary, the disturbance-on trajectory $x_f(t)$ only passes through the stable manifold of the type-one equilibrium point, x_{co}. Hence, x_{co} is the controlling UEP relative to the disturbance-on trajectory $x_f(t)$.

Table 14.2 Coordinates of six type-one equilibrium points lying on the stability boundary of x_s

Type-one EP	δ_1	ω_1	δ_2	ω_1
1	3.24512	0	0.31170	0
2	3.04037	0	3.24387	0
3	0.03333	0	3.10823	0
4	−3.03807	0	0.3117	0
5	−3.24282	0	−3.03931	0
6	0.03333	0	−3.17496	0

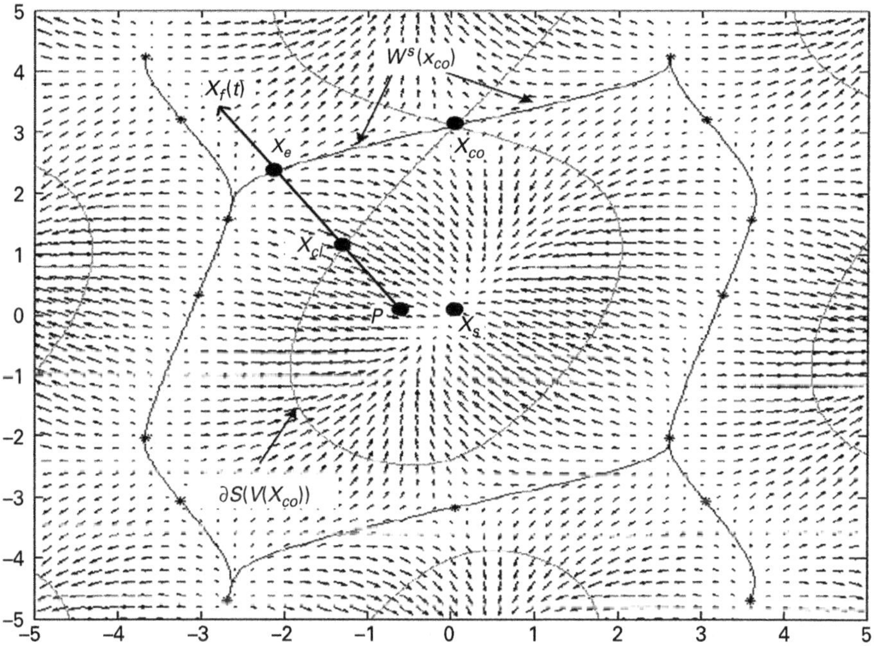

Figure 14.10 The intersection between the stability boundary $\partial A(x_s)$ and the angle subspace. The stability boundary $\partial A(x_s)$ is contained in the set which is the union of the stable manifolds of these six type-one equilibrium points and six type-two equilibrium points. The constant energy surface passing through one UEP is drawn. The vector field at many points is plotted in the figure, which can serve to confirm the location of the exact stability boundary.

Given a disturbance-on trajectory $x_f(t)$ with a preset disturbance clearing time, let x_{cl} be the state vector when the disturbance is cleared, which is also the initial point of the post-disturbance trajectory. Let $W^s(x_{co})$ be the stable manifold of the corresponding controlling UEP x_{co}. The task of determining whether or not the segment of the disturbance-on trajectory, $x_f(t)$, between x_s^{pre} and x_{cl} lies inside the stability region $A(x_s)$ based on the stable manifold $W^s(x_{co})$ is numerically very difficult, if not impossible. The main reason for this difficulty is due to the lack of an explicit expression for stable manifolds. However, the task becomes relatively easy if an energy function instead of the stable manifold $W^s(x_{co})$ is given.

14.4 Numerical studies

In order to illustrate the controlling UEP and the controlling UEP method presented in the previous section, we consider the following equation which closely represents the three-machine power system in absolute angle coordinates shown in Figure 14.11:

$$\begin{aligned}
\dot{\delta}_1 &= \omega_1 \\
\dot{\delta}_2 &= \omega_2 \\
\dot{\delta}_3 &= \omega_3 \\
m_1\dot{\omega}_1 &= -d_1\omega_1 + P_{m_1} - P_{e_1}(\delta_1, \delta_2, \delta_3) \\
m_2\dot{\omega}_2 &= -d_2\omega_2 + P_{m_2} - P_{e_2}(\delta_1, \delta_2, \delta_3) \\
m_3\dot{\omega}_3 &= -d_3\omega_3 + P_{m_3} - P_{e_3}(\delta_1, \delta_2, \delta_3)
\end{aligned}$$

where

$$P_{e_i}(\delta_1, \delta_2, \delta_3) = \sum_{j=1, j\neq 1}^{3} E_iE_j(B_{ij}\sin(\delta_i - \delta_j) + G_{ij}\cos(\delta_i - \delta_j))$$

$P_{m1} = 0.8980$, $P_{m2} = 1.3432$, $P_{m3} = 0.9419$, $E_1 = 1.1083$, $E_2 = 1.1071$, $E_3 = 1.0606$. The system admittance matrix, $G_{ij} + jB_{ij}$, of the pre-disturbance system is given by:

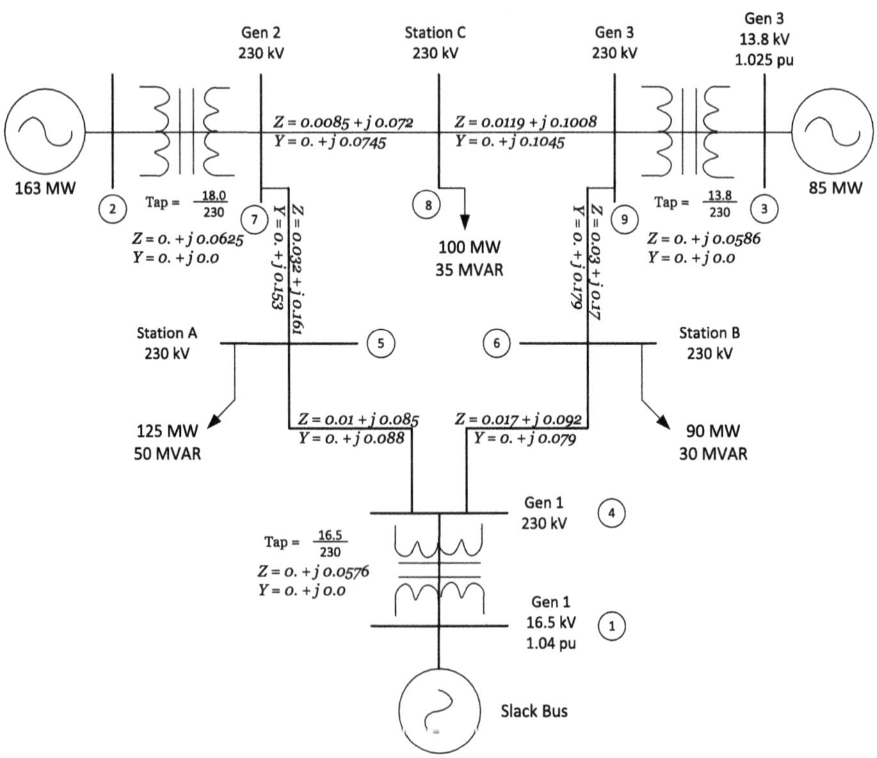

Figure 14.11 A 3-machine, 9-bus power system; the value of Y is half the line charging.

$$G_{ij}^{pre} + jB_{ij}^{pre} = \begin{bmatrix} 0.845 - j2.988 & 0.287 + j1.513 & 0.210 + j1.226 \\ 0.287 + j1.513 & 0.420 - j2.724 & 0.213 + j1.088 \\ 0.210 + j1.226 & 0.213 + j1.088 & 0.277 - j2.368 \end{bmatrix}.$$

We consider a normal loading condition with a uniform damping factor $d_i / m_i = 0.1$ with $[d_1, d_2, d_3] = [0.0125, 0.0034, 0.0016]$. The coordinate of the pre-disturbance SEP is $[-0.0482, 0.1252, 0.1124]$. The contingency list, the associate disturbance, and the post-disturbance SEP are summarized in Table 14.3.

The system admittance matrices $G_{ij} + jB_{ij}$ for the disturbance-on system due to a short-circuit at bus 7 and bus 8 are respectively given by

$$G_{ij}^{dist(7)} + jB_{ij}^{dist(7)} = \begin{bmatrix} 0.657 - j3.816 & 0 + j0 & 0.070 + j0.631 \\ 0 + j0 & 0 - j5.485 & 0 + j0 \\ 0.070 + j0.631 & 0 + j0 & 0.174 - j2.796 \end{bmatrix}$$

$$G_{ij}^{dist(8)} + jB_{ij}^{dist(8)} = \begin{bmatrix} 0.685 - j3.685 & 0.077 + j0.523 & 0.0505 + j0.498 \\ 0.077 + j0.523 & 0.146 - j4.130 & 0.006 + j0.0548 \\ 0.0505 + j0.498 & 0.006 + j0.0548 & 0.120 - j3.127 \end{bmatrix}$$

and the post-fault system admittance matrices $G_{ij} + jB_{ij}$ for the post-fault system for contingencies 1, 2, 3 and 4 are respectively given by

$$G_{ij}^{(1)} + jB_{ij}^{(1)} = \begin{bmatrix} 1.099 - j2.362 & 0.121 + j0.687 & 0.174 + j1.049 \\ 0.121 + j0.687 & 0.361 - j2.075 & 0.186 + j1.185 \\ 0.174 + j1.049 & 0.186 + j1.185 & 0.266 - j2.361 \end{bmatrix}$$

$$G_{ij}^{(2)} + jB_{ij}^{(2)} = \begin{bmatrix} 0.838 - j3.015 & 0.203 + j1.465 & 0.263 + j1.202 \\ 0.203 + j1.465 & 0.326 - j1.908 & 0.066 + j0.324 \\ 0.263 + j1.202 & 0.066 + j0.324 & 0.545 - j1.814 \end{bmatrix}$$

$$G_{ij}^{(3)} + jB_{ij}^{(3)} = \begin{bmatrix} 0.819 - j3.015 & 0.316 + j1.384 & 0.133 + j1.273 \\ 0.316 + j1.384 & 0.650 - j2.113 & 0.059 + j0.325 \\ 0.133 + j1.273 & 0.059 + j0.325 & 0.271 - j1.657 \end{bmatrix}$$

$$G_{ij}^{(4)} + jB_{ij}^{(4)} = \begin{bmatrix} 0.838 - j3.015 & 0.203 + j1.464 & 0.263 + j1.202 \\ 0.203 + j1.464 & 0.326 - j1.908 & 0.066 + j0.324 \\ 0.263 + j1.202 & 0.066 + j0.324 & 0.545 - j1.814 \end{bmatrix}.$$

The controlling UEP relative to each contingency is computed using the time-domain-based controlling UEP method and is summarized in Table 14.4.

Table 14.3 The contingency list, the associated disturbance and the post-disturbance SEP

Contingency number	Fault-bus	Disturbance clearing type	Description		Post-disturbance SEP $[\delta_1, \delta_2, \delta_3]$
			From bus	To bus	
1	7	Line trip	7	5	[−0.1204, 0.3394, 0.2239]
2	7	Line trip	8	7	[−0.0655, 0.2430, −0.0024]
3	8	Line trip	9	8	[−0.0462, 0.0728, 0.2082]
4	8	Line trip	8	7	[−0.0655, 0.2430, −0.0024]

Table 14.4 The controlling UEP relative to each contingency

Contingency number	Controlling UEP $[\delta_1, \delta_2, \delta_3]$
1	[−0.7589, 1.9528, 1.8079]
2	[−0.5424, 2.1802, −0.3755]
3	[−0.2910, −0.1011, 2.5008]
4	[−0.3495, 0.0745, 2.5864]

14.5 Estimation of constrained dynamical systems

We consider a constrained nonlinear dynamical system which is mathematically described by the following differential and algebraic equations (DAEs):

$$\begin{aligned} \dot{x} &= f(x, y) \\ 0 &= g(x, y) \end{aligned} \tag{14.4}$$

where $x \in R^n$ and $y \in R^m$ are the corresponding dynamic and static variables of the systems, respectively. The DAE system (14.4) can be interpreted as an implicitly dynamical system defined on the algebraic manifold Γ

$$\Gamma = \{(x, y) : g(x, y) = 0\}.$$

Let Γ_s be a stable component of Γ and $\phi(t, x, y)$ be a trajectory of the DAE system (14.4) starting from (x, y). The stability region of a given stable equilibrium point (x_s, y_s) of a DAE system (14.4) is defined as

$$A(x_s, y_s) = \{(x, y) \in \Gamma_s : \lim_{t\to\infty} \phi(t, x, y) = (x_s, y_s)\}. \tag{14.5}$$

Here we restrict the stability region to lying on the stable component Γ_s and exclude other stable components from which trajectories pass through a singular surface and converge to the stable equilibrium point (x_s, y_s).

The singular perturbation approach treats the set of algebraic equations describing a DAE system as a limit of the fast dynamics: $\varepsilon\dot{y} = g(x, y)$. In other words, as ε approaches zero, the fast dynamics will approach the algebraic manifold. Therefore, for the DAE system (14.4), we can define an associated singularly perturbed system

$$\begin{aligned} \dot{x} &= f(x, y) \\ \varepsilon\dot{y} &= g(x, y) \end{aligned} \tag{14.6}$$

where ε is a sufficiently small positive number. If f and g are both smooth functions and bounded for all $(x, y) \in R^{n+m}$, then the vector field is globally well defined. The state variables of system (14.4) have very different rates of dynamics and they can be separated into two distinct time scales: slow variable x and fast variable y. A DAE system and its corresponding singularly perturbed system share several similar dynamical properties. The controlling UEP relative to a DAE system can also be defined via the controlling UEP of the corresponding singularly perturbed system (14.6). We next present a controlling UEP method for DAE systems.

Given a disturbance-on DAE trajectory $(x_f(t), y_f(t))$, a post-disturbance DAE system with a post-disturbance SEP (x_s, y_s) and an energy function $W(x, y)$ for the post-disturbance DAE system, we have the following method.

A controlling UEP method for DAE systems

Step 1: Determination of the critical energy.

Step 1.1: Find the controlling UEP (x_{co}, y_{co}) relative to the disturbance-on DAE trajectory $(x_f(t), y_f(t))$.

Step 1.2: The critical energy value W_{cr} is the value of energy function $W(x, y)$ at the controlling UEP, i.e. $W_{cr} = W(x_{co}, y_{co})$.

Step 2: Approximation of the relevant stability boundary.

Step 2.1: Use the connected constant energy surface of $W(x, y)$ passing through the controlling UEP (x_{co}, y_{co}) to approximate the relevant part of stability boundary for the disturbance-on trajectory $(x_f(t), y_f(t))$.

We next present a fundamental theorem for the controlling UEP method for DAE system (14.4).

THEOREM 14-3 **(Fundamental theorem)**

Consider the DAE system (14.4) and its associated singularly perturbed system (14.6). Suppose the associated singularly perturbed system (14.6) has an energy function $V(\cdot)$: $R^n \rightarrow R$. Let $\hat{p}$ be an equilibrium point on the stability boundary $\partial A(p_s)$ of this system. Let $r > V(p_s)$ and

- $S(r)$:= the connected component of the set $\{x \in R^n: V(x) < r\}$ containing p_s, and
- $\partial S(r)$:= the connected component of the set $\{x \in R^n: V(x) = r\}$ containing p_s.

Then,

(a) the connected constant energy surface $\partial S(V(\hat{p}))$ intersects with the stable manifold $W^s(\hat{p})$ only at the equilibrium point $\hat{p}$; moreover, the set $S(V(\hat{p}))$ has an empty intersection with the stable manifold $W^s(\hat{p})$. In other words, $\partial S(V(\hat{p})) \cap W^s(\hat{p}) = \hat{p}$ and $S(V(\hat{p})) \cap W^s(\hat{p}) = \emptyset$,

(b) $S(V(p_u)) \cap W^s(\hat{p}) \neq \emptyset$ if p_u is an EP and $V(p_u) > V(\hat{p})$,

(c) $S(V(p_u)) \cap W^s(\hat{p}) = \emptyset$ if p_u is an EP and $V(\hat{p}) > V(p_u)$,

(d) if $\hat{p}$ is not the closest UEP, then $\partial S(V(\hat{p})) \cap (\overline{A}(p_s))^c \neq \emptyset$,

(e) any connected path starting from $p \in \{S(V(\hat{p})) \cap A(p_s)\}$ *and passing through* $W^s(\hat{p})$ *must intersect* $\partial S(V(\hat{p}))$ *first before the path intersects* $W^s(\hat{p})$.

The above theorem is similar to the fundamental theorem for the controlling UEP method for ODE systems. This theorem asserts that one can use the constant energy surface passing the controlling UEP of the singularly perturbed model as the relative stability boundary for the disturbance-on trajectory of the DAE system. Once the initial condition of the post-disturbance trajectory of the DAE system lies inside the constant energy surface, the ensuing post-disturbance trajectory will converge to the stable equilibrium point of the post-disturbance DAE system.

14.6 Numerical studies

In order to illustrate the CUEP and the CUEP method for DAE system in a simple context, we consider the same 9-bus power system of Figure 14.11, which is modeled, in absolute angle coordinates, by the following differential-algebraic equations (DAE) composed of 6 differential equations and 18 nonlinear algebraic equations:

$$\begin{aligned}
\dot{\delta}_1 &= \omega_1 \\
\dot{\delta}_2 &= \omega_2 \\
\dot{\delta}_3 &= \omega_3 \\
m_1\dot{\omega}_1 &= -d_1\omega_1 + P_{m_1} - P_{G_1}(\delta_1, \theta_1, V_1) \\
m_2\dot{\omega}_2 &= -d_2\omega_2 + P_{m_2} - P_{G_2}(\delta_2, \theta_2, V_2) \\
m_3\dot{\omega}_3 &= -d_3\omega_3 + P_{m_3} - P_{G_3}(\delta_3, \theta_3, V_3)
\end{aligned}$$

$$\begin{aligned}
0 &= \sum_{k=1}^{9} V_2V_k(G_{2k}\cos(\theta_2 - \theta_k) + B_{2k}\sin(\theta_2 - \theta_k)) - P_{G_2}(\delta_2, \theta_2, V_2) \\
0 &= \sum_{k=1}^{9} V_3V_k(G_{3k}\cos(\theta_3 - \theta_k) + B_{3k}\sin(\theta_3 - \theta_k)) - P_{G_3}(\delta_3, \theta_3, V_3) \\
0 &= \sum_{k=1}^{9} V_4V_k(G_{4k}\cos(\theta_4 - \theta_k) + B_{4k}\sin(\theta_4 - \theta_k)) \\
0 &= \sum_{k=1}^{9} V_5V_k(G_{5k}\cos(\theta_5 - \theta_k) + B_{5k}\sin(\theta_5 - \theta_k)) \\
0 &= \sum_{k=1}^{9} V_6V_k(G_{6k}\cos(\theta_6 - \theta_k) + B_{6k}\sin(\theta_6 - \theta_k)) \\
0 &= \sum_{k=1}^{9} V_7V_k(G_{7k}\cos(\theta_7 - \theta_k) + B_{7k}\sin(\theta_7 - \theta_k)) \\
0 &= \sum_{k=1}^{9} V_8V_k(G_{8k}\cos(\theta_8 - \theta_k) + B_{8k}\sin(\theta_8 - \theta_k))
\end{aligned}$$

$$0 = \sum_{k=1}^{9} V_2 V_k (G_{2k} \sin(\theta_2 - \theta_k) + B_{2k} \cos(\theta_2 - \theta_k)) - Q_{G_2}(\delta_2, \theta_2, V_2)$$
$$0 = \sum_{k=1}^{9} V_3 V_k (G_{3k} \sin(\theta_3 - \theta_k) + B_{3k} \cos(\theta_3 - \theta_k)) - Q_{G_3}(\delta_3, \theta_3, V_3)$$
$$0 = \sum_{k=1}^{9} V_4 V_k (G_{4k} \sin(\theta_4 - \theta_k) + B_{4k} \cos(\theta_4 - \theta_k))$$
$$0 = \sum_{k=1}^{9} V_5 V_k (G_{5k} \sin(\theta_5 - \theta_k) + B_{5k} \cos(\theta_5 - \theta_k))$$
$$0 = \sum_{k=1}^{9} V_6 V_k (G_{6k} \sin(\theta_6 - \theta_k) + B_{6k} \cos(\theta_6 - \theta_k))$$
$$0 = \sum_{k=1}^{9} V_7 V_k (G_{7k} \sin(\theta_7 - \theta_k) + B_{7k} \cos(\theta_7 - \theta_k))$$
$$0 = \sum_{k=1}^{9} V_8 V_k (G_{8k} \sin(\theta_8 - \theta_k) + B_{8k} \cos(\theta_8 - \theta_k))$$

where

$$P_{G_i}(\delta, \theta, V) = \begin{cases} \dfrac{E_i V_i \sin(\delta_i - \theta_i)}{X'_{di}} & \text{if } i = 1, \ldots, 3 \\ 0 & \text{if } i = 4, \ldots, 9 \end{cases},$$

$$Q_{G_i}(\delta, \theta, V) = \begin{cases} -\dfrac{V_i^2}{X'_{di}} + \dfrac{E_i V_i \sin(\delta_i - \theta_i)}{X'_{di}} & \text{if } i = 1, \ldots, 3 \\ 0 & \text{if } i = 4, \ldots, 9 \end{cases}$$

We consider a uniform damping factor $d_i/m_i = 0.1$ with $[d_1, d_2, d_3] = [0.0125, 0.0034, 0.0016]$. We consider two contingencies summarized in Table 14.5.

The post-disturbance SEP and the controlling UEP relative to each contingency are summarized in the following two tables.

To illustrate the computational issues involved in the CUEP method for DAE systems, we numerically simulate the following dynamical information for two contingencies:

- pre-disturbance SEP, SEP and controlling UEP of post-disturbance system,
- stability boundary of post-disturbance system intersecting with the machine angle space,
- disturbance-on trajectory and relevant stability boundary.

Table 14.5 The contingency list and the associated disturbance

Contingency number	Faulted bus	Disturbance clearing type	Description	
			From bus	To bus
1	8	Line trip	8	7
2	7	Line trip	8	7

Table 14.6 The post-disturbance SEP, the controlling UEP relative to contingency 1

	$[\delta_1, \delta_2, \delta_3]$	$[V_1, V_2, \ldots, V_9]$	$[\theta_1, \theta_2, \ldots, \theta_9]$
Post-disturbance SEP	[−0.1204, 0.3394, 0.2239]	[1.0781, 1.0845, 1.0669, 1.0511, 1.0066, 1.0402, 1.0802, 1.0657, 1.0740]	[−0.1584, 0.2091, 0.0756, −0.1963, −0.2816, −0.1621, 0.1396, 0.0589, 0.0284]
CUEP	[−0.7589, 1.9528, 1.8079]	[0.8652, 0.9073, 0.6730, 0.6349, 0.6080, 0.3222, 0.8113, 0.6945, 0.5598]	[−0.7557, 1.8228, 1.5978, −0.7504, −0.8357, −0.3298, 1.7305, 1.6101, 1.4697]

Table 14.7 The post-disturbance SEP, the controlling UEP relative to contingency 2

	$[\delta_1, \delta_2, \delta_3]$	$[V_1, V_2, \ldots, V_9]$	$[\theta_1, \theta_2, \ldots, \theta_9]$
SEP	[−0.0655, 0.2430, −0.0024]	[1.0930, 1.0940, 1.0538, 1.0801, 1.0547, 1.0615, 1.0947, 1.0235, 1.0568]	[−0.1048, 0.1129, −0.1534, −0.1430, −0.1361, −0.2089, 0.0443, −0.2823, −0.2022]
CUEP	[−0.3495, 0.0745, 2.5864]	[0.9318, 1.0088, 0.4284, 0.7647, 0.8074, 0.4451, 0.9633, 0.2576, 0.2660]	[−0.3387, −0.0384, 2.2596, −0.3239, −0.2914, −0.2163, −0.1060, 1.7530, 1.8332]

For contingency 1, a disturbance occurs near bus 8 and the line between bus 7 and bus 8 is tripped. The exact stability region of the post-disturbance system on the angle plane is highlighted in Figure 14.12. The disturbance-on trajectory intersects with the stable manifold of the controlling UEP, a type-one UEP, highlighted in the figure. There are two type-one UEPs and one type-two UEP lying on the stability boundary. The stable manifold of one type-one UEP, the CUEP, contains the exit point, which is the intersection between the exact stability boundary and the disturbance-on trajectory. This exit point is at some distance from the CUEP while it is very close to a type-two UEP lying on the stability boundary.

It is interesting to note that contrary to the claim made in the literature, the CUEP is not the UEP which is closest to the disturbance-on trajectory, for instance, the type-two UEP is closest to the disturbance-on trajectory, but this type-two UEP is not the controlling UEP.

For contingency 2, a disturbance occurs near bus 7 and the line between bus 7 and bus 8 is tripped. The exact stability region of the post-disturbance system on the angle plane is highlighted in Figure 14.13. There are two type-one UEPs and one type-two UEP lying on the stability boundary. The stable manifold of one type-one UEP, the CUEP, contains the exit point (of the original model), which is the intersection between the exact stability boundary and the disturbance-on trajectory. This exit point is close to the CUEP while it is at some distance from the other type-one UEP. On the other hand, the exit point is also close to a type-two equilibrium point.

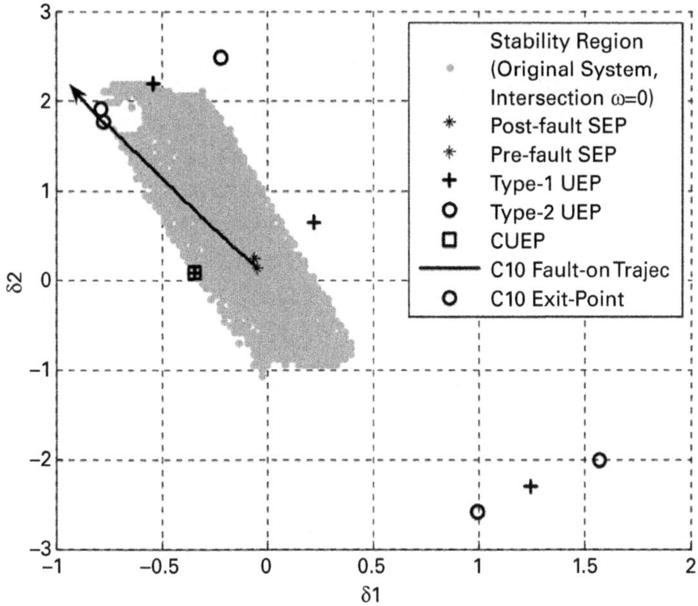

Figure 14.12 For contingency 1, the exact stability region of the post-disturbance system on the angle plane is identified. The disturbance-on trajectory intersects the stable manifold of the controlling UEP, a type-one UEP, at the exit point. Note that the exit point (of the original model) is at some distance from the CUEP while it is close to a type-two UEP, which is not the CUEP.

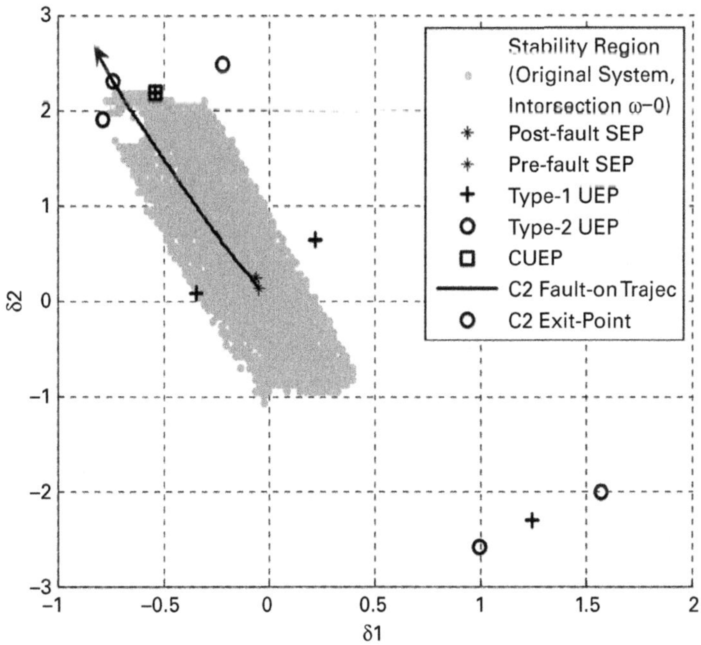

Figure 14.13 For contingency 2, a disturbance occurs near bus 7 and the line between bus 7 and bus 8 is tripped. The exact stability region of the post-disturbance system on the angle plane is highlighted. The disturbance-on trajectory intersects with the stable manifold of the controlling UEP, a type-one UEP, at the exit point. There are two type-one and one type-two UEPs.

14.7 Conclusion

We have presented effective energy-function-based controlling UEP methods to estimate relevant stability regions for nonlinear dynamical systems described by ODEs and by DAEs. An analytical basis for the energy-function-based controlling UEP methods has also been developed. It will be shown in the next chapter that the energy function provides the "best" metric to estimate stability regions of large-scale nonlinear systems.

To ensure the conservative nature of the controlling UEP method, it has been shown that the disturbance-on trajectory passes through the constant energy surface containing the controlling UEP before it exits the stability region at the exit point. If the disturbance is cleared before the disturbance-on trajectory reaches the constant energy surface containing the controlling UEP, then the post-disturbance trajectory will converge to the SEP and is stable. It is rather easy and straightforward to check when the disturbance-on trajectory passes through the constant energy surface containing the controlling UEP by simply monitoring the energy value along the disturbance-on trajectory and identifying when its value equals the energy value of the controlling UEP.

15 Critical evaluation of numerical methods for approximating stability boundaries

15.1 Introduction

There have been several methods proposed to approximate the stability boundaries of nonlinear dynamical systems. One key goal of this chapter is to show that the energy function approach described in this book is the best approach for approximating the stability boundaries of nonlinear systems. To this end, we provide a critical evaluation of different methods for approximating, without loss of generality, the relevant stability boundary of a class of nonlinear systems which admit energy functions. These methods include the energy-function-based method, the hyperplane-based method and methods employing the solution set (via a quadratic approximation) of a set of partial differential equations. The *relevant stability boundary* (of a post-disturbance SEP) relative to a disturbance-on trajectory was studied in Chapter 8 and it is the stable manifold of the controlling UEP intersected by the disturbance-on trajectory.

It is noted that the LMI approach, see for example [39,41,247], currently is suitable only for small-sized nonlinear dynamical systems, say 15-dimensional systems. It is hence not included in this evaluation study. Two test nonlinear dynamical models representing two simplified practical power systems, the Western System Control Council (WSCC) 9-bus system and the IEEE 145-bus system, will be employed in this evaluation study with focus on the following two aspects:

(i) the accuracy of each method in approximating relevant stability boundaries,
(ii) the accuracy of each method in computing the *critical clearing times* (CCTs) which are a measure of the distance from the current state to the exact stability boundary.

15.2 Methods for approximating the relevant stability boundary

We consider a nonlinear autonomous dynamical system described by the following equation:

$$\dot{x}(t) = f(x(t)). \tag{15.1}$$

We assume that this system admits an energy function, $V{:}R^n \rightarrow R$. Recall that the CUEP relative to a disturbance-on trajectory (or just a path) is the UEP lying on the stability boundary of a post-disturbance SEP and whose stable manifold intersects the trajectory. The stable manifold of the CUEP is the relevant stability boundary for this trajectory. We evaluate the following methods based on the accuracy with which they approximate the relevant stability boundary:

- CUEP-based hyperplane method,
- CUEP-based quadratic method,
- CUEP-based energy function method.

15.2.1 Hyperplane-based CUEP method

The hyperplane method seeks to use a hyperplane passing through the CUEP to approximate the relevant stability boundary. This approximation via a hyperplane containing the CUEP x_{co} can be described by the following equation

$$g(x) = \alpha^t(x - x_{co}) = 0 \tag{15.2}$$

where α is the eigenvector associated with the only unstable eigenvalue μ:

$$\alpha J(x_{co}) = \mu\alpha^t \tag{15.3}$$

where $J(x) = \partial f/\partial x$ is the Jacobian of f at the CUEP. The hyperplane method requires knowledge of the CUEP and will therefore be referred to as the *CUEP-based hyperplane method.*

Note that a hyperplane is of co-dimension one while the dimension of a type-one stable manifold is also of co-dimension one. The idea behind this hyperplane method is to approximate the stable manifold of a type-one UEP via the hyperplane passing through the type-one UEP. Intuitively, this approximation can be accurate in the neighborhood of the type-one UEP and may be inaccurate in regions away from the type-one UEP. The hyperplane-based CUEP method is summarized as follows.

Step 1: Compute the controlling unstable equilibrium point x_{co} relative to the disturbance-on trajectory.

Step 2: Determine the hyperplane passing through the controlling UEP x_{co} using Eq. (15.2).

Step 3: Approximate the relevant stability boundary by using the hyperplane determined at Step 2.

15.2.2 CUEP-based quadratic method

It can be shown that the function describing the stable manifold of a type-one unstable equilibrium point x_e^1 satisfies a set of nonlinear algebraic equations [108,255]

$$W^s(x_e^1) = h(x). \tag{15.4}$$

According to the CUEP theorem described in Chapter 9, the relevant stability boundary can therefore be implicitly expressed by the above set of equations $h(x) = 0$. The implicit expression of $h(x)$ can be approximately expressed by a power series expansion at an equilibrium point. Let x_e^1 be a type-one unstable equilibrium point with one unstable eigenvalue, say μ. The series expansion at x_e^1 is as follows

$$f(x) = f(x_e^1) + J(x_e^1)(x - x_e^1) + \frac{1}{2}\begin{pmatrix} (x - x_e^1)H_1(x_e^1)(x - x_e^1) \\ \cdots \\ (x - x_e^1)H_n(x_e^1)(x - x_e^1) \end{pmatrix} + \cdots \tag{15.5}$$

where $J(x) = \partial f / \partial x$ is the Jacobian of f and $H_i(x) = \partial^2 f_i / \partial x^2$ is the Hessian matrix of f_i. Then a quadratic approximation of the stable manifold $h(x)$ in a neighborhood of a controlling UEP, say x_{co} is the following:

$$h_Q(x) = \eta^T(x - x_{co}) + (x - x_{co})^T Q(x - x_{co})/2 \tag{15.6}$$

and

$$h(x) = h_Q(x) + O(\|x - x_e^1\|^3).$$

The coefficients η and Q of the quadratic expression can be determined as follows.

[1] The coefficient vector $\eta \in R^n$ of the linear term is the left eigenvector associated with the eigenvalue μ of the Jacobian J at the equilibrium point x_e^1

$$J^T\eta = \mu\eta, \quad \eta^T\eta = 1\,. \tag{15.7}$$

[2] The coefficient matrix Q of the quadratic term is the solution of the Lyapunov equation

$$CQ + QC^T = H \tag{15.8}$$

where $C = (\mu I/2 - J^T)$, I is the $n \times n$ identical matrix, and $H = \sum_{i=1}^{n} \eta_i H_i$. This method can be summarized as follows.

Step 1: Compute the controlling unstable equilibrium point x_{co} relative to the disturbance-on trajectory.

Step 2: Determine a quadratic approximation of the stable manifold $h_Q(x)$ in a neighborhood of the controlling UEP, x_{co}.

Step 3: Approximate the relevant stability boundary by using the connected set $h_Q(x)$ containing the CUEP x_{co}.

The relevant stability boundary is thus obtained by a quadratic approximation of the stable manifold at a controlling UEP (or a type-one UEP). We refer to this method as the *quadratic-surface-based CUEP method, or the quadratic-based CUEP method.*

15.2.3 CUEP-based energy function method

The CUEP-based energy function method for approximating the relevant stability boundary, as described in Chapter 9, can be summarized by the following steps.

Step 1: Compute the controlling unstable equilibrium point x_{co} relative to the disturbance-on trajectory.

Step 2: Determine an energy function for the post-disturbance system and compute its value at x_{co}, $V_c = V(x_{co})$ where $V: R^n \to R$ is an energy function.

Step 3: Approximate the relevant stability boundary by using the connected set $S = \{x \in R^n : V(x) = V_c\}$ containing the CUEP x_{co}.

The performance of each of the above methods in approximating the relevant stability boundary will be evaluated on two test systems: the WSCC 9-bus system and the IEEE 145-bus system. We consider the so-called classical nonlinear stability models in these two test systems. Consider a power system with n generators and loads interconnected by a transmission network. The classical power system dynamical model is used and the loads are modeled as constant impedances. The dynamics of the ith generator can hence be represented by the following so-called swing equations:

$$\begin{aligned} \dot{\delta}_i &= \omega_i \\ M_i\dot{\omega}_i &= P_{mi} - P_{ei}(\delta) - D_i\omega_i, \; i = 1, \ldots, n \end{aligned} \tag{15.9}$$

where D_i and M_i are the damping ratio and inertia constant of machine i. E_i is the constant voltage behind the direct axis transient reactance, P_{mi} is the mechanical power, $Y = (Y_{ij})_{n\times n} = (G_{ij}+jB_{ij})_{n\times n}$ is the reduced admittance matrix and the electric power output is expressed as

$$P_{ei} = \sum_{i=1}^{n} E_iE_jB_{ij}\sin(\delta_i - \delta_j) + \sum_{i=1}^{n} E_iE_jG_{ij}\cos(\delta_i - \delta_j).$$

If the center of inertia (COI) for all machines is used as the reference and uniform damping is assumed, then (15.9) can be transformed into the following system:

$$\begin{aligned} \dot{\tilde{\delta}}_i &= \tilde{\omega}_i \\ \dot{\tilde{\omega}}_i &= (P_{mi} - P_{ei}(\tilde{\delta}))/M_i - P_{COI}/M_T - d_0\omega_i \; i = 1, \ldots, n \end{aligned} \tag{15.10}$$

where

$$\begin{aligned} \tilde{\delta}_i &= \delta_i - \delta_0 \qquad \delta_0 = \left(\sum_k M_k\delta_k\right)/M_T \\ \tilde{\omega}_i &= \omega_i - \omega_0 \qquad \omega_0 = \left(\sum_k^n M_k\omega_k\right)/M_T \\ M_T &= \sum_k^n M_k \qquad d_0 = D_i/M_i, \; i = 1, \ldots, n \\ P_{COI} &= \sum^n P_{mk} - \sum^n\sum^n (E_kE_jG_{kj}\cos\tilde{\delta}_{kj} - E_kE_jB_{kj}\sin\tilde{\delta}_{kj}). \end{aligned}$$

The exact stability boundary of an SEP can be numerically derived via the time-domain approach. This time-domain approach is inefficient and computationally difficult for larger systems. It is used only as a benchmark for the evaluation.

15.3 Evaluation on small systems

The one-line diagram of the WSCC 9-bus system and its dynamic data can be found in [224, p.171]. A uniform damping of 0.1 is applied to every machine while loads are maintained at the original values. Two contingencies are applied to evaluate the performance of each method in approximating the relevant stability boundary. The first contingency has bus 9 as the faulted bus, which defines the disturbance-on system and line 9–6 is tripped at the clearing time, which defines the post-disturbance system. The second contingency has bus 8 as the faulted bus, which defines the disturbance-on system and line 8–7 is tripped, which defines the post-disturbance system.

Our evaluation is focused on the accuracy of each method in approximating the relevant stability boundary of the post-disturbance system relative to the disturbance-on system. Figure 15.1 shows the relevant stability boundary of the first contingency

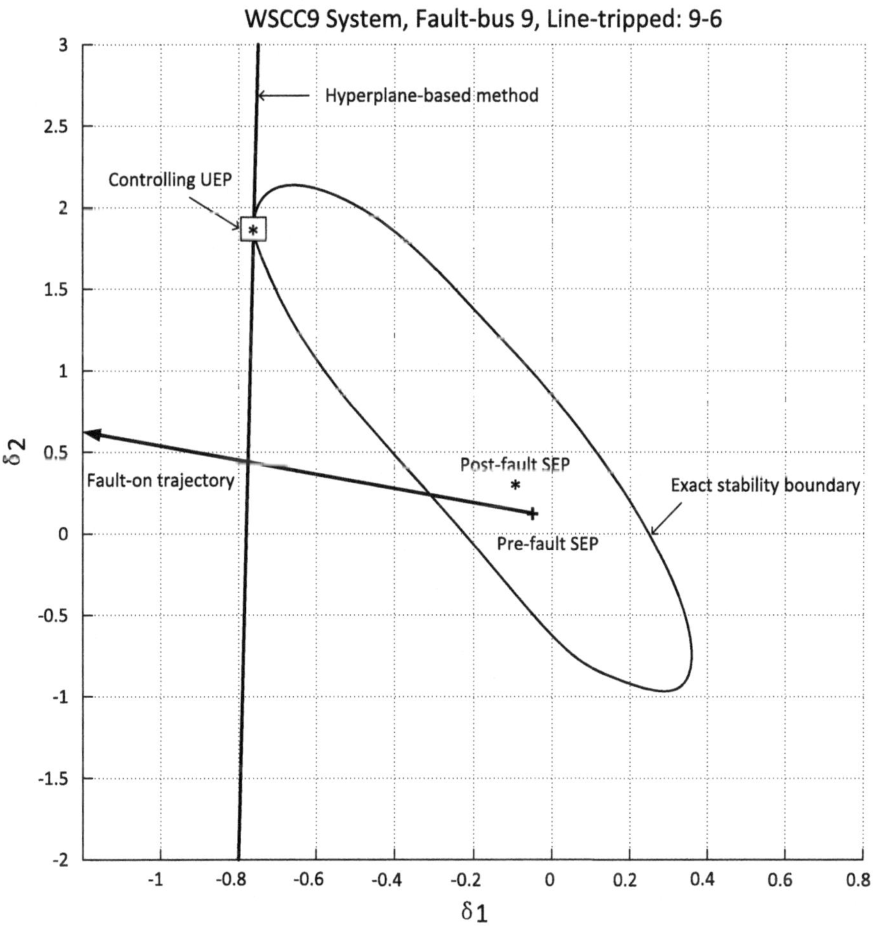

Figure 15.1 The exact stability boundary of the post-disturbance SEP and the approximation by the hyperplane-based CUEP method. The fault-bus is 9 and line 9–6 is tripped to clear the fault. The hyperplane-based method gives an over-approximation of the relevant stability boundary.

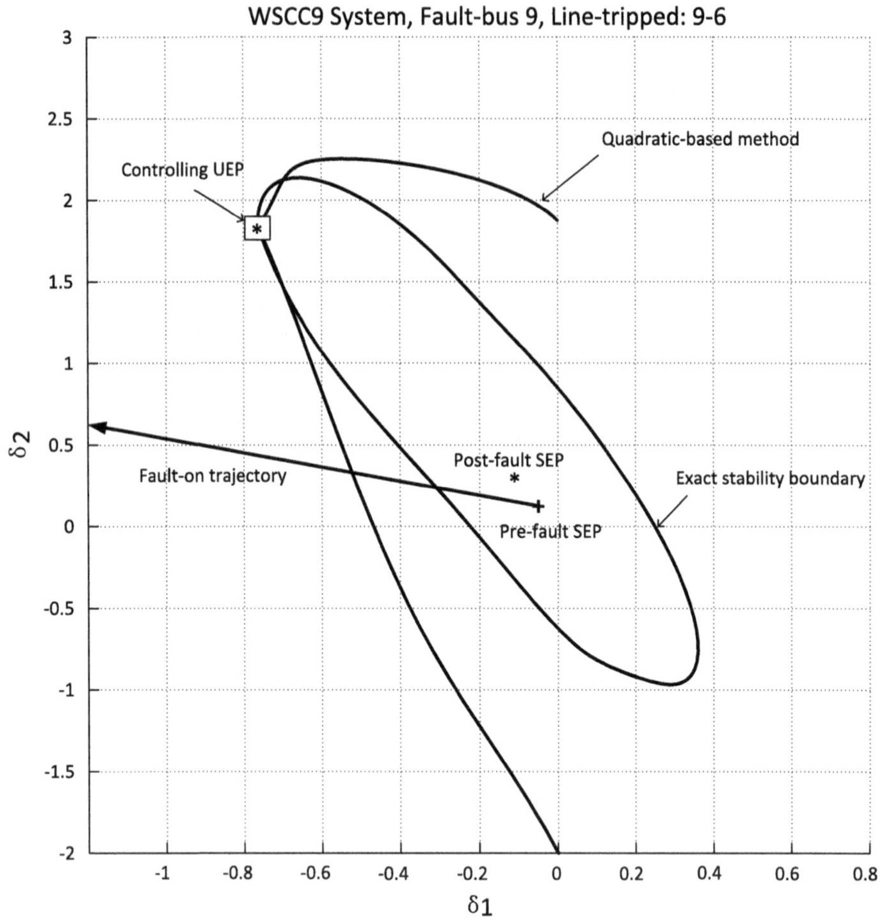

Figure 15.2 The exact stability boundary of the post-disturbance SEP and the approximation by the quadratic-based CUEP method. The fault-bus is 9 and line 9–6 is tripped to clear the fault. The quadratic-based method gives an over-approximation of the relevant stability boundary.

approximated by the hyperplane-based CUEP method while Figure 15.2 shows the relevant stability boundary approximated by the quadratic-based CUEP method. The exact stability boundary, displayed by the solid black line, is obtained via the time-domain simulation and is included as a benchmark. Since the hyperplane method employs only the hyperplane at the controlling UEP, the approximation is the straight line passing through the controlling UEP at $(-0.7576, 1.858)$. The line with an arrow is the projection of the actual fault-on trajectory onto the $\omega = 0$ plane and is presented in the figure as a fault-on trajectory example.

It can be seen that the disturbance-on trajectory intersects the exact stability boundary first before it reaches the boundary approximated by the hyperplane method. This large approximation error of the hyperplane-based CUEP method is caused by two factors: the first is the nonlinear characteristic of the exact stability boundary near the controlling UEP, and the second is the large distance between the controlling UEP and the exit point, i.e. the point at which the disturbance-on trajectory intersects the exact stability boundary.

The quadratic-based CUEP method alleviates the nonlinear characteristic issue by taking into account the second derivative at the controlling UEP. The approximation is illustrated by the parabola passing through the controlling UEP, and it appears to capture the characteristics of the stability boundary around the controlling UEP more accurately than the hyperplane-based CUEP method. However, the quadratic approximation still yields large approximation errors in areas far away from the controlling UEP. Moreover, both the hyperplane-based CUEP method and the quadratic-based CUEP method give over-shoots in approximating the relevant stability boundary.

The performance of the energy-function-based CUEP method is shown in Figure 15.3. The light gray line in Figure 15.3 represents the relevant stability boundary

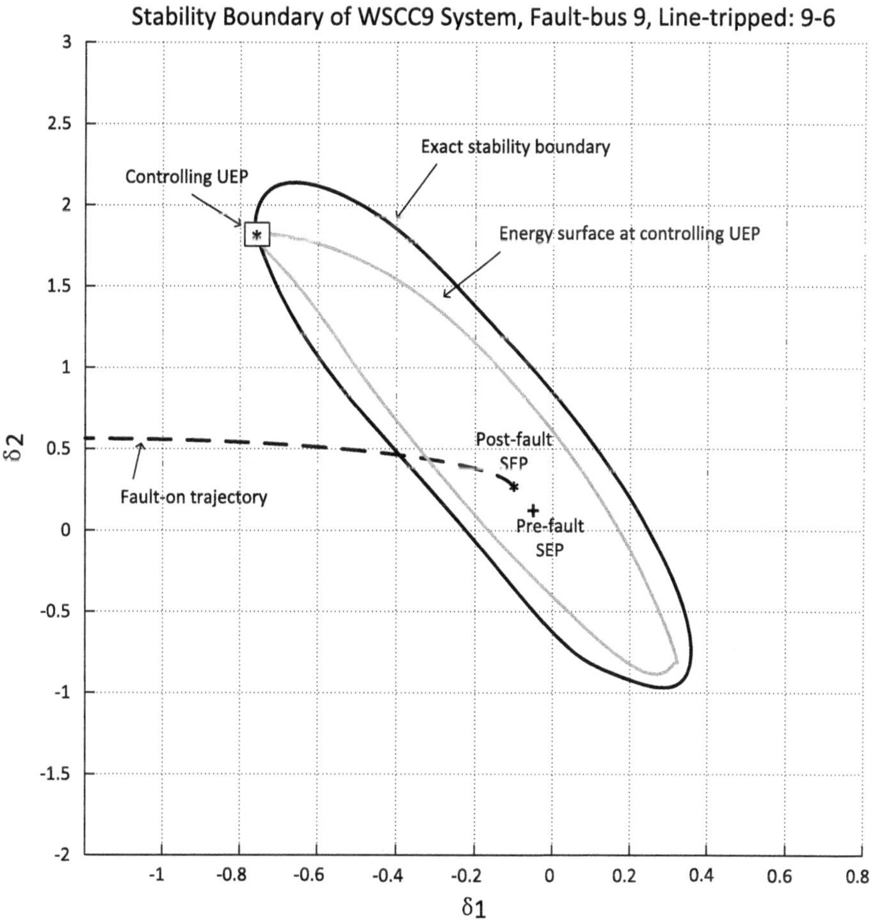

Figure 15.3 The exact stability boundary of the post-disturbance system and the approximation by the energy-function-based CUEP method. Because the controlling UEP is also the closest UEP in this case, this method gives a slightly conservative approximation of the entire stability boundary.

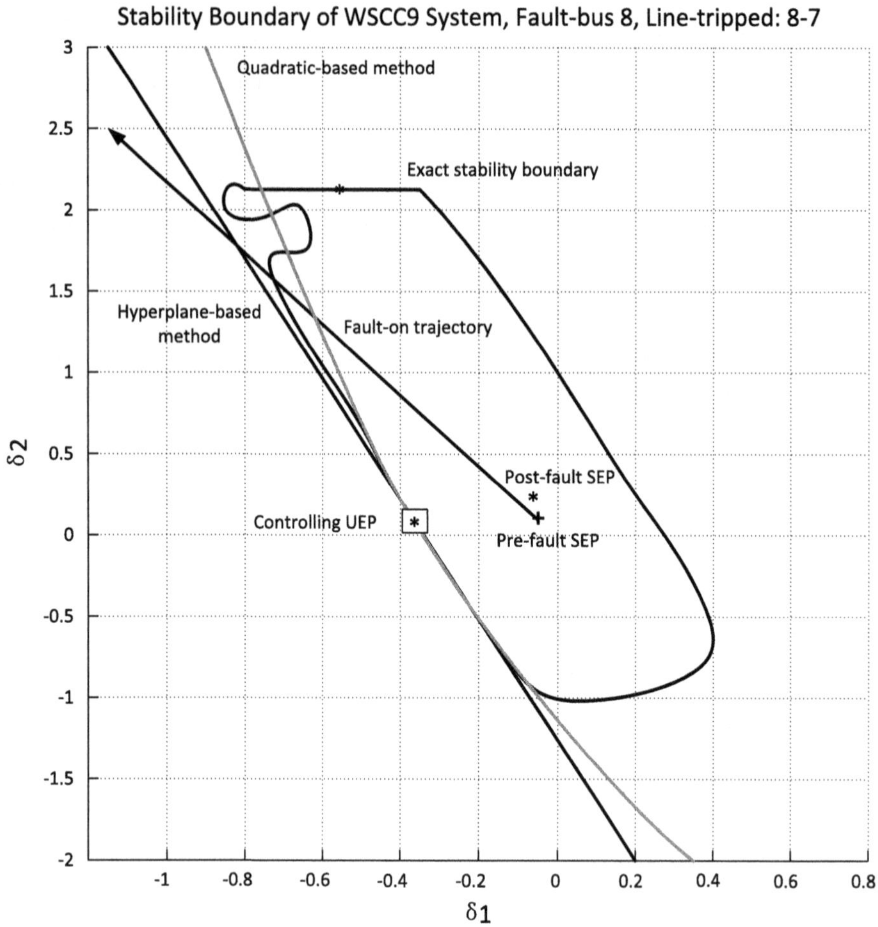

Figure 15.4 The exact stability boundary of the post-disturbance SEP, the approximations by the hyperplane-based CUEP and the quadratic-based CUEP methods. The fault-bus is 8 and line 8–7 is tripped to clear the fault.

approximated by the energy surface passing through the controlling UEP (i.e. energy-function-based CUEP method). In this situation, the controlling UEP is also the closest UEP, making the approximation strictly conservative in all the directions from the SEP.

The results of both methods for the second contingency (where bus 8 is the faulted bus and line 8–7 is tripped) are shown in Figure 15.4. It can be seen that both non-energy-function-based methods give much improved approximations of the relevant stability boundary compared with the first contingency. Even though the point where the disturbance-on trajectory intersects the exact stability boundary is farther away from the SEP than for the first contingency, the linear characteristic of the stability boundary in a neighborhood of the controlling UEP helps make the approximations by both methods quite accurate, even though the hyperplane method still gives an overestimate while the quadratic method gives a slight underestimate in this case. In Figure 15.5, it is

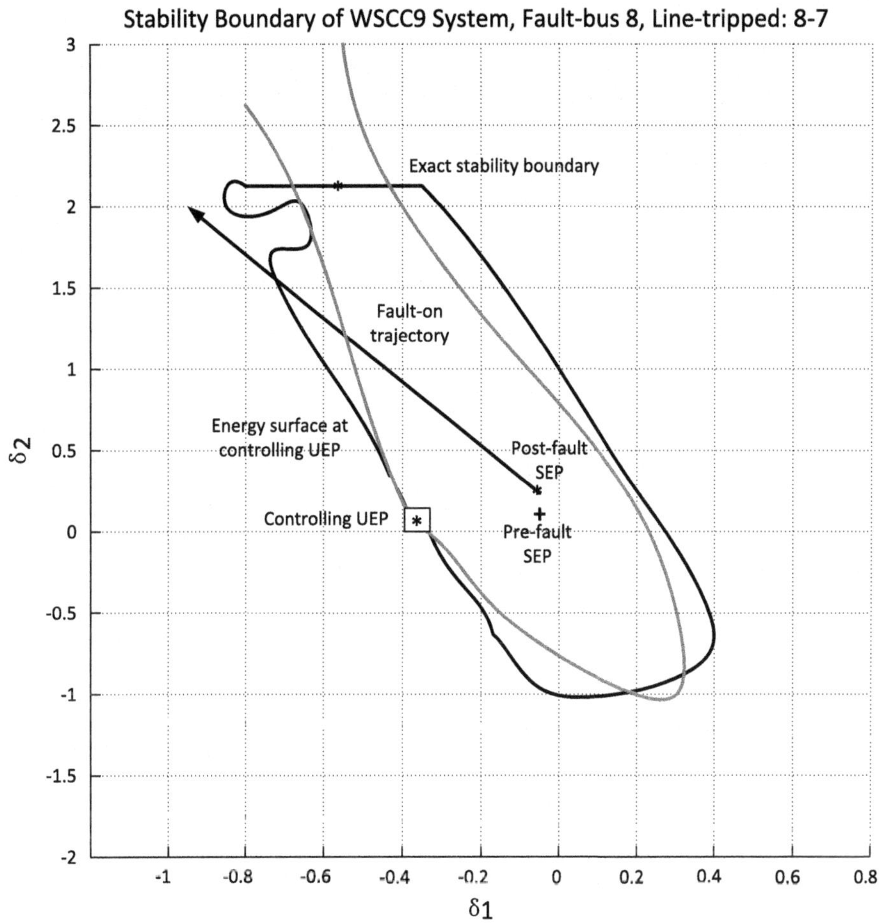

Figure 15.5 The exact stability boundary of the post-disturbance system and the approximation by the energy-function-based CUEP method. This method gives a slightly conservative approximation of the relevant stability boundary.

clear that the controlling UEP at $(-0.3765, 0.012)$ is not the same as the closest UEP at $(-0.561, 2.309)$. The approximation for the relevant stability boundary by the energy-function-based CUEP method is still slightly conservative.

From these numerical simulations, we have the following observations.

- Neither of the non-energy-function-based methods, the hyperplane-based method and the quadratic-based method, is able to give consistently accurate approximations.
- The hyperplane-based method always gives an overestimation of the relevant stability boundary while the quadratic-based method can give either an overestimation or conservative approximations.
- The energy-function-based method gives consistently accurate and yet conservative approximations of the relevant stability boundary, as asserted by the controlling UEP Theorems 14.1 and 14.2.

15.4 Evaluation on large systems

Our evaluation of the methods is next performed on the IEEE 145-bus test system, which contains fifty generators with the classical model. The system dimension is 99 (using the COI reference) and the relevant values and parameters are available in [1]. A uniform damping of 1.0 is applied to every generator. One contingency (59 as the fault-bus, and 59–72 as the fault-line) is presented.

Since the dimension of the relevant stability boundary of this test system is 99, we use the visualization scheme described as follows to display the stability boundary approximated by each method. The ability to visualize stability regions can provide insightful information on the dynamic behaviors and nature of stability of the study system. The visualization of stability regions makes observation of the variation of stability regions due to several factors, such as changes in loading conditions, damping factors and network topology, more effective. This task of developing effective visualization schemes is challenging. A scheme for visualizing two-dimensional portraits of a high-dimensional stability boundary while maintaining critical information of interest on the stability boundary, is described below. To determine a two-dimensional plane on which a stability region is to be projected, we choose the following three points as the reference points used in the visualization, as described Figure 15.6:

- the post-disturbance SEP,
- the controlling UEP,
- the exit point (i.e. the intersection between the disturbance-on trajectory and the stability boundary).

For the large test system, the relevant stability boundaries approximated by the hyperplane-based CUEP and quadratic-based CUEP methods are displayed in Figure 15.7. The relevant stability boundary approximated by the energy-function-based

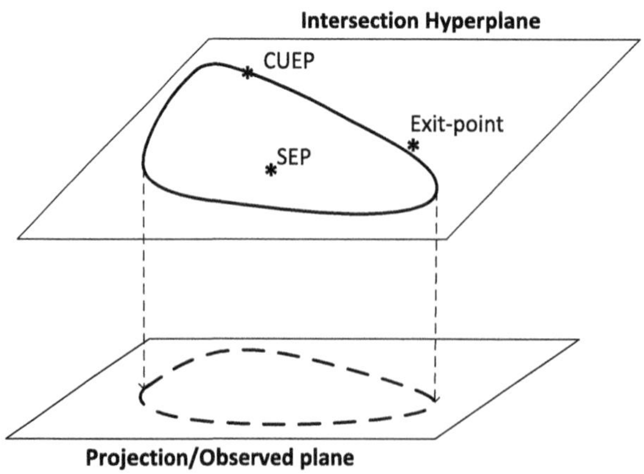

Figure 15.6 Use of the coordinates of three points to obtain an informative two-dimensional display of the stability regions of a high-dimensional stability boundary.

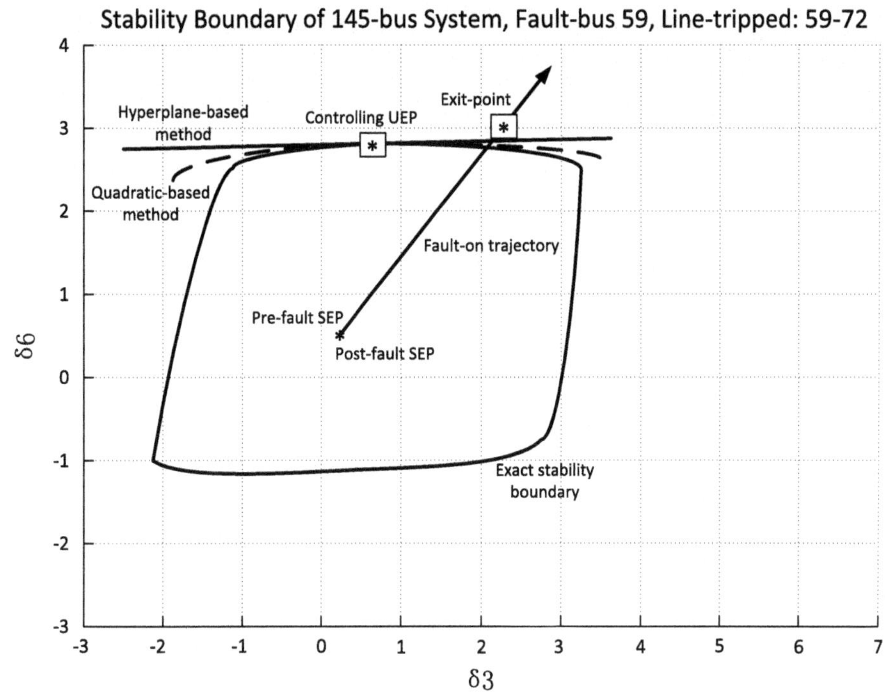

Figure 15.7 A two-dimensional portrait of the exact stability boundary of the IEEE 145-bus system and approximations by the hyperplane-based CUEP and the quadratic-based CUEP methods. The fault-bus is 59 and line 59–72 is tripped. Both methods give overestimations of the relevant stability boundary.

CUEP method is shown in Figure 15.8. The chosen angle coordinates are machine 3 at bus 79 and machine 6 at bus 89.

For this contingency (i.e. disturbance), the approximations of the relevant stability boundaries near the controlling UEP by the hyperplane-based CUEP method and by the quadratic-based CUEP method are nearly linear and quite accurate. Both methods, however, give overestimations of the relevant stability boundary in the direction of the disturbance-on trajectory. In Figure 15.8, the stability boundary approximated by the energy-function-based CUEP method is shown and, as asserted by the CUEP theorem, is contained inside the exact stability region. This numerical study again confirms that the energy-function-based CUEP method indeed always yields accurate and yet conservative approximations for the relevant stability boundary.

The numerical studies of the relevant stability boundary approximations by these three methods lead to the following observations.

- The energy function-based CUEP method always gives accurate and yet conservative approximations of the relevant stability boundary.
- The quadratic-based CUEP method gives a relatively accurate approximation of the relevant boundary but gives overestimation approximation.

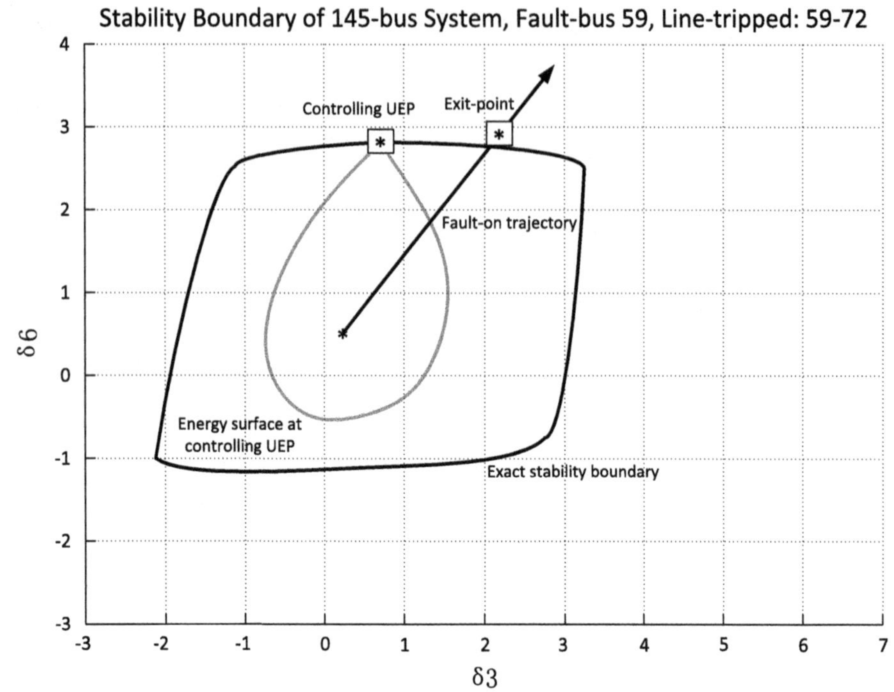

Figure 15.8 A two-dimensional portrait of the exact stability boundary of the IEEE 145-bus system and approximations by the energy-function-based CUEP method. The fault-bus is 59 and line 59–72 is tripped. This method gives a strictly conservative approximation of the relevant stability boundary, as asserted by the CUEP theorem.

- The hyperplane-based CUEP method always gives over-approximations of the relevant stability boundary. This may imply the convex characteristic of the stability region in the vicinity of the controlling UEP.

15.5 Evaluation of CCT estimations

The accuracy of each method in computing critical clearing times (CCTs) is evaluated. The CCT is the duration of time between the pre-disturbance SEP and the exit point of the disturbance-on trajectory. The time-domain approach is used as a benchmark. The three methods are evaluated for ten different contingencies (i.e. ten different faults) at five different loading conditions, hence a total of fifty nonlinear dynamical systems of both the WSCC 9-bus and IEEE 145-bus test systems. Moreover, we also study whether each method can give a consistent conservative estimate of the CCTs. Conservative assessment is a crucial property for the fast dynamic contingency screening of large-scale power grids in which only stable contingencies should be screened out.

Table 15.1 CCT estimates of the WSCC 9-bus system by four different methods, the unit of CCT is the second

	Time-domain method	Energy-function method	Hyperplane-based method	Quadratic-based method
Fault #1, Fault bus 5, line-tripped 7–5	0.5809	0.5802	> 0.5907	0.5732
Fault #2, Fault bus 7, line-tripped 7–5	0.2611	0.2604	> 0.2967	0.2560
Fault #3, Fault bus 7, line-tripped 8–7	0.2736	0.2429	> 0.2951	> 0.2835
Fault #4, Fault bus 8, line-tripped 8–7	0.4178	0.3415	> 0.4735	0.4083
Fault #5, Fault bus 4, line-tripped 4–6	0.4117	0.4112	> 0.4198	> 0.4155
Fault #6, Fault bus 6, line-tripped 4–6	0.7436	0.7433	> 0.7453	0.7426
Fault #7, Fault bus 6, line-tripped 6–9	0.6797	0.6752	> 0.6830	0.6785
Fault #8, Fault bus 9, line-tripped 6–9	0.2430	0.2276	> 0.3237	> 0.2749
Fault #9, Fault bus 8, line-tripped 9–8,	0.4210	0.3003	> 0.4650	0.3892
Fault #10, Fault bus 9, line-tripped 9–8	0.2273	0.2155	> 0.2357	0.2268
Degree of conservativeness	–	100%	0%	70%

15.5.1 WSCC 9-bus system

For the WSCC 9-bus system, the CCTs computed by each method for ten different contingencies are summarized in Table 15.1. The average relative errors and conservativeness percentages are also included. The CCTs computed by the time-domain approach are used as the benchmark. Regarding the percentage of conservative assessment, the energy-function-based CUEP method is 100%, the quadratic-based method is 70%, and the hyperplane-based method is 0%.

We next evaluate the relative errors of each method as loading conditions increase. The quadratic-based method performs much worse under heavy loading conditions, exhibiting the highest relative error of 40.6959% at 200% of the base loading condition. The hyperplane-based method has a high average relative error of 30.3171% at 200% of the base case loading condition, and it has 0% conservative assessment for all loading conditions. The energy-function-based CUEP method is the least affected in the average error at different loading conditions and gives 100% conservative estimates at all loading conditions. Its average relative error at 200% of the loading condition is the smallest among the three methods, which is 9.2614%.

Figure 15.9 summarizes the percentage of conservative estimation at different loading conditions.

15.5.2 IEEE 145-bus system

The simulation results for the IEEE 145-bus system are presented in the same format as for the 9-bus system. Table 15.2 contains the CCT estimation by each method for ten contingencies at the base loading condition. Figure 15.10 summarizes the average

Table 15.2 CCT estimates by the three methods on the IEEE 145-bus system

	Time-domain method	Energy-function method	Hyperplane-based method	Quadratic-based method
Fault #1	0.1610	0.1235	> 0.1860	> 0.2744
Fault #2	0.2009	0.0985	> 0.2498	> 0.4780
Fault #3	0.2591	0.1420	> 0.2881	> 0.2707
Fault #4	0.2699	0.2307	> 0.5035	0.1863
Fault #5	0.3169	0.3129	> 0.3195	> 0.3189
Fault #6	0.2998	0.2840	> 0.3043	> 0.3032
Fault #7	0.2698	0.2307	> 0.5034	0.1860
Fault #8	0.2850	0.2619	> 0.2935	> 0.2871
Fault #9	1.0439	1.0023	> 1.2981	> 1.0589
Fault #10	0.2095	0.1990	> 0.2153	> 0.2116
Average relative error	–	17.2108%	25.6618%	27.9817%
Degree of conservativeness	–	100%	0%	20%

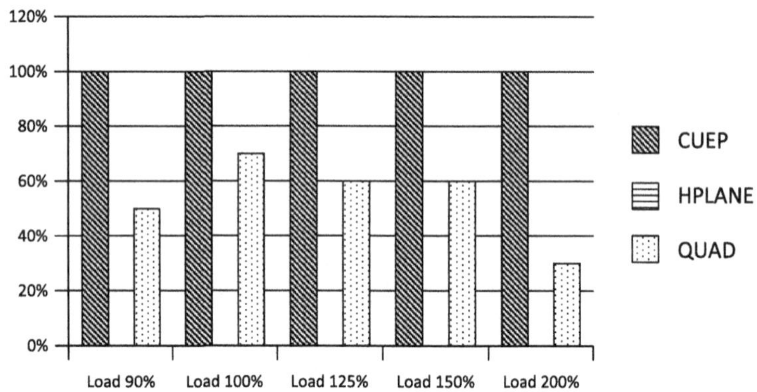

Figure 15.9 Percentages of conservative assessment for ten different contingencies of the WSCC 9-bus system. The energy-function-based method consistently gives conservative estimations.

relative errors while Figure 15.11 summarizes the percentage of conservative approximation at four different loading conditions.

The fault sequence is described as follows: Fault #1, Faulted bus 7, Line-tripped 7–6; Fault #2, Faulted bus 6, Line-tripped 7–6; Fault #3, Faulted bus 59, Line-tripped 59–72; Fault #4, Faulted bus 72, Line-tripped 59–72; Fault #5, Faulted bus 115, Line-tripped 115–116; Fault #6, Faulted bus 116, Line-tripped 115–116; Fault #7, Faulted bus 72, Line-tripped 100–72; Fault #8, Faulted bus 100, Line-tripped 100–72; Fault #9, Faulted bus 75, Line-tripped 91–75; Fault #10, Faulted bus 91, Line-tripped 91–75.

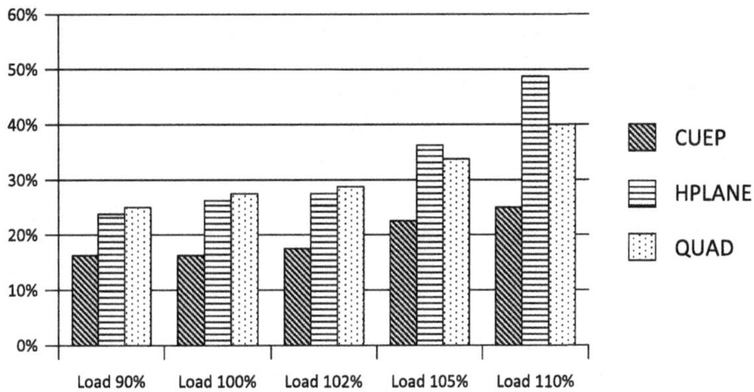

Figure 15.10 Average relative error percentages for ten different contingencies at five different loadings of the IEEE 145-bus system.

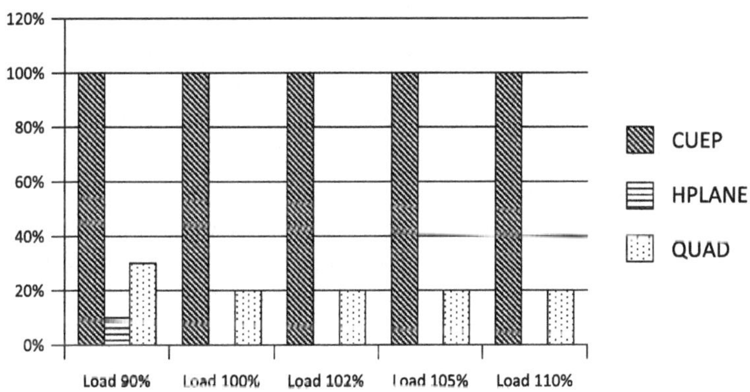

Figure 15.11 Percentages of conservative assessment for ten different contingencies of the IEEE 145-bus system. The energy-function-based method consistently gives (slightly) conservative estimations.

Table 15.2 shows that the energy-function-based CUEP method gives the smallest error of 17.2108% when compared to 25.6618% and 27.9817% for the other two methods. The energy-function-based CUEP method still maintains 100% conservative assessment. Similar to the results for the WSCC 9-bus system, the hyperplane-based method always overestimates the CCTs and hence shows 0% in its conservative assessment.

Compared to the WSCC9 system, the accuracy of the hyperplane-based method and the quadratic-based method is greatly reduced in this larger system as the average errors increase from 8.6550% to 25.6618% for the hyperplane-based method and from 3.1306% to 27.9817% for the quadratic-based method. It is expected that the performance of these two methods will deteriorate as the system dimension increases.

From the evaluation of CCT computations by the three methods, we have the following observations.

- The energy-function-based CUEP method is the most accurate method in estimating CCTs on the larger IEEE 145-bus system at all loading conditions.
- The energy-function-based CUEP method is the only method that guarantees conservative estimations of CCTs at all loading conditions.
- The accuracy in estimating CCTs by the energy-function-based CUEP method is the least affected by loading increases when compared to the other methods.
- The hyperplane-based method always gives overestimated CCTs while the quadratic-based method is inconsistent in estimating CCTs (it gives both underestimation and overestimation).
- The hyperplane-based CUEP method gives overestimated CCTs in every case regardless of the size of the system and loading conditions. This may imply the convex characteristic of the stability region in the vicinity of the controlling UEP.

15.6 Conclusions

A critical evaluation of three methods, the energy-function-based CUEP method, the hyperplane-based CUEP method and the quadratic-based CUEP method, is given in this chapter. A technique to visualize stability regions of high-dimensional systems is also presented, showing that it is possible to obtain a simple yet informative two-dimensional portrait of the stability region of large systems. Using the proposed visualization technique, we provide a graphical evaluation of the performance of each method in approximating the relevant stability boundary of two test systems. The evaluation of CCT estimations provides more insight regarding the conservative assessment capability of each method. The LMI approach is currently only suitable for small-sized nonlinear dynamical systems; hence it is not included in this evaluation.

A comparison of the simulation results with the time-domain approach reveals several interesting observations regarding the performance of the methods.

- The hyperplane-based method always gives overestimated CCTs while the quadratic-based method is inconsistent in estimating CCTs.
- The energy-function-based CUEP method is the most accurate method in estimating CCTs on the IEEE 145-bus system at most loading conditions.
- The energy-function-based CUEP method always gives conservative estimations of CCTs in both systems and at all loading conditions.
- The accuracy in estimating CCTs by the energy-function-based CUEP method is the least affected by the loading conditions.

From these numerical results and the theoretical foundation of the controlling UEP method, it can be concluded that the energy-function-based controlling UEP method excels in the following regard: accuracy in approximating the relevant stability boundary, ability to provide conservative approximation, lower error sensitivity at all loading conditions and least computational burden.

Part III

Advanced topics

16 Stability regions of two-time-scale continuous dynamical systems

In this chapter, we explore the special system structure of a class of continuous dynamical systems, the so-called singularly perturbed systems, to obtain improved and specialized results about the stability region and stability boundary characterization of these systems as well as about the theory of stability region estimation. In particular, the relationship between the stability regions and stability boundaries of the singularly perturbed system and those of two simpler systems, the fast and slow subsystems, will be developed.

16.1 Introduction

Many nonlinear physical systems possess two-time-scale features in their state variables, fast and slow dynamic variables coexist [6,67,204]. Electromechanical systems are good examples of systems with two-time-scale behaviors in which electrical variables are usually much faster than mechanical variables. Also, there is a close relationship between time-scale and simplified models.

One may face several numerical and analytical problems when applying traditional analytical techniques as well as numerical techniques to two-time-scale nonlinear systems. Two-time-scale systems usually have the same model structure as regular nonlinear systems modeled by a set of ordinary differential equations but, from a numerical point of view, the time step required to numerically integrate two-time-scale nonlinear dynamical systems has to be sufficiently small to accurately compute fast dynamics. This can become cumbersome since a great deal of computational effort would be required to accomplish the integration in the time range of slow variables. Indeed, the need for complex numerical integration algorithms, suitable for stiff nonlinear equations, arises. From both computational and analytical viewpoints, traditional techniques usually lead to very conservative estimations of the stability region of two-time-scale systems [81,125,223].

These difficulties have motivated the development of nonlinear tools that take into account the two-time-scale features of these systems. Several advantages in terms of computational speed and accurate results can be derived from time-scale features by decomposing the analysis of the original system into the analysis of two simpler systems: the *slow system* and the *fast system*. The advantages of decomposing the two-time-scale (TTS) system into slow and fast subsystems include: (i) fast estimation of solutions, (ii) well-conditioned numerical algorithms, (iii) insightful information about the two-time-scale system dynamics from the analysis of reduced simpler subsystems.

In this chapter, we study the characterization of the stability boundary of a class of nonlinear autonomous two-time-scale dynamical systems. The names *two-time-scale systems* and *singularly perturbed systems* are used interchangeably in this chapter. A brief introduction to the theory of singularly perturbed systems is presented, and results on the stability region of two-time-scale dynamical systems are developed. Conditions under which the stability region and stability boundary characterizations of singularly perturbed systems can be decomposed into the stability region and stability boundary characterizations of the slow and fast subsystems are derived. The approximation of the stability boundary of two-time-scale systems by the stability boundaries of the fast and slow subsystems is presented.

The theory of estimating stability regions of two-time-scale dynamical systems is also presented in this chapter. Several conditions, in terms of location and stability properties of equilibriums of the slow and fast subsystems, for ensuring that a type-one equilibrium point lies on the stability boundary of the two-time-scale system are derived. These conditions are further explored to develop schemes for obtaining optimal estimates of the stability region of two-time-scale dynamical systems and the relevant stability region of two-time-scale systems.

16.2 Singularly perturbed systems

In this chapter, we study the following particular class of two-time-scale system:

$$(\Sigma_\varepsilon)\begin{cases} \dot{x} = f(x, z) \\ \varepsilon\dot{z} = g(x, z) \end{cases} \tag{16.1}$$

where $x \in R^n$ and $z \in R^m$. Functions f and g are of class C^1 and ε is a small non-negative scalar. If f and g have the same magnitude, then ε small implies that variable z exhibits much faster variations compared to variable x. As a consequence, z represents *fast variables* while x represents *slow variables*.

A point (x^*, z^*) is an *equilibrium point* of (Σ_ε) if $f(x^*, z^*) = 0$ and $g(x^*, z^*) = 0$. Let E denote the set of equilibrium points of (Σ_ε), that is, $E = \{(x, z) \in R^n \times R^m: f(x, z) = 0, g(x, z) = 0\}$. An equilibrium point is *hyperbolic* if not all the eigenvalues of the Jacobian matrix $J_\varepsilon = \begin{bmatrix} D_x f & D_z f \\ \frac{1}{\varepsilon} D_x g & \frac{1}{\varepsilon} D_z g \end{bmatrix}$ calculated at the equilibrium point lie on the imaginary axis. A hyperbolic equilibrium point is a *type-k equilibrium point* if there exist k eigenvalues of J on the right half of the complex plane and $(m+n-k)$ on the left half-plane.

For this class of dynamical systems, we are primarily interested in studying the dynamical behavior and stability region as the parameter $\varepsilon > 0$ approaches zero. Changes in the parameter ε can be viewed as a perturbation and, for this reason, system (Σ_ε) is usually termed a *singularly perturbed model.* As $\varepsilon > 0$ approaches zero, we study the relationship between the dynamics of the singularly perturbed system (16.1) and the dynamics of two simplified systems, the slow and fast systems, which can be derived from (Σ_ε) by formally setting $\varepsilon = 0$. It will be shown, under some conditions, that analysis of dynamics, stability, and stability region of the singularly perturbed system can be decomposed into the analysis of dynamics, stability, and stability regions of the slow and fast systems.

16.2.1 The slow system

The slow system is obtained by setting $\varepsilon = 0$ in (Σ_ε):

$$(\Sigma_0) \begin{cases} \dot{x} = f(x, z) \\ 0 = g(x, z). \end{cases} \tag{16.2}$$

The algebraic equation $0=g(x, z)$ in (Σ_0) constrains the dynamics of the slow system to the set $\Gamma = \{(x, z) \in R^n \times R^m: g(x, z) = 0\}$. This manifold is known in the literature as the constraint manifold, algebraic manifold or equilibrium manifold.

The slow system aims to capture the long-term dynamics of the singularly perturbed system. In other words, assuming that the fast dynamics of variable z settle down quickly by approaching the constraint manifold Γ, the dynamics of the slow system approximate the dynamics of the singularly perturbed system, for small ε, in the slow time scale.

Figure 16.1 illustrates the set Γ and the relationship between the dynamics of the singularly perturbed system and the dynamics of the slow system on Γ. In the beginning, since the initial condition is far away from Γ, the solution of the singularly perturbed system is very different from the solution of the slow system. However, within a short interval of time, the trajectory of the singularly perturbed system approaches the set Γ and the solutions of both systems become close to each other.

16.2.2 The fast system

In order to explore the time-scale property of (Σ_ε) and capture the dynamics of the singularly perturbed system in the short term, we define the fast time scale $\tau = t/\varepsilon$. In this new time scale, system (Σ_ε) takes the form:

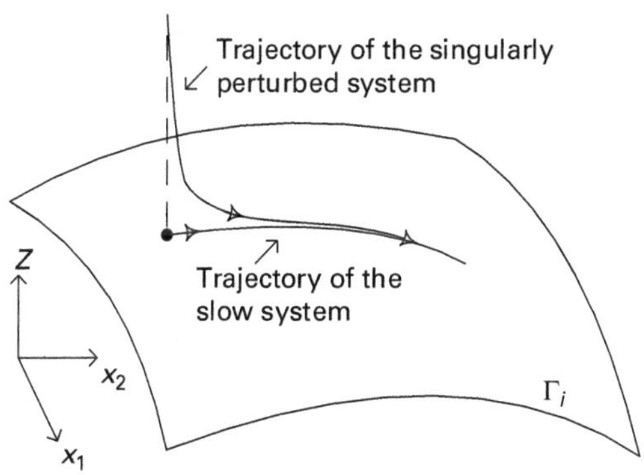

Figure 16.1 Relationship between the trajectory of the singularly perturbed system and the trajectory of the slow system.

$$(\Pi_\varepsilon)\quad \begin{cases} \dfrac{dx}{d\tau} = \varepsilon f(x, z) \\ \dfrac{dz}{d\tau} = g(x, z). \end{cases} \tag{16.3}$$

The fast system (Π_F) or (Π_0) is obtained by setting ε=0 in (Π_ε):

$$(\Pi_0)\begin{cases} \dfrac{dx}{d\tau} = 0 \\ \dfrac{dz}{d\tau} = g(x, z) \end{cases} \tag{16.4}$$

In the fast system, variable x is frozen. For this reason, one can think of the fast system (Π_0) as a family of dynamical systems parametrized by the variable x. These systems are also termed boundary-layer systems:

$$\left(\Pi_F(x)\right)\ \left\{\frac{dz}{d\tau} = g(x, z)\right. \tag{16.5}$$

where x is frozen and treated as a parameter. Figure 16.2 illustrates the dynamics of the slow system along the constraint manifold Γ and the dynamics of the family of fast systems. For each fixed x, a fast dynamical system $(\Pi_F(x))$ is defined. The intersections of the hyperplane $\{(x, z): x = x^*\}$ with Γ are equilibrium points of the fast system $(\Pi_F(x^*))$.

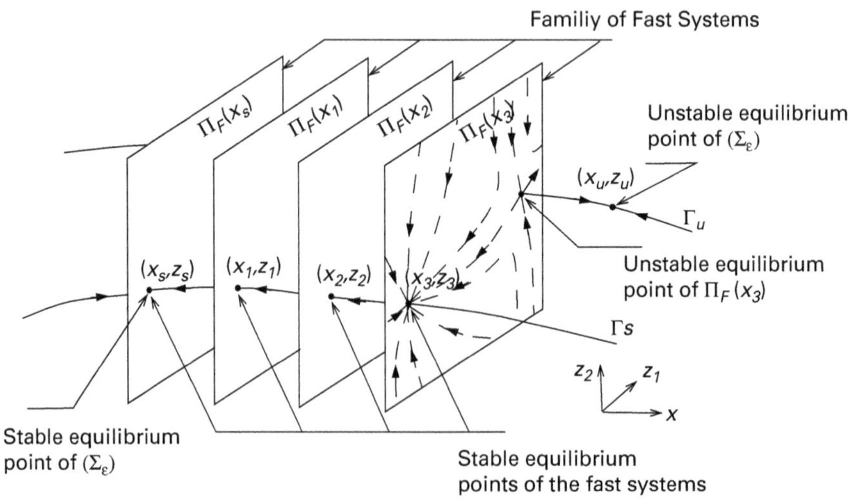

Figure 16.2 Illustration of the family of fast systems. For each frozen x, a fast system $\Pi_F(x)$ is defined. The dynamic of the slow system is constrained to the set Γ. The constrained manifold Γ is a set of equilibriums of the fast system (Π_0). Set Γ may contain several connected components. Every point on the stable component Γ_s is an asymptotically stable equilibrium point of one member of the family of fast systems. Every point on the type-one unstable component Γ_u is a type-one hyperbolic unstable equilibrium point of one member of the family of fast systems.

16.2.3 The constraint manifold

The constraint manifold Γ plays an essential role in the relationship between the stability region of the two-time-scale system (Σ_ε) and those of the simplified subsystems (Σ_0) and $(\Pi_F(x))$. The constraint manifold Γ is a set of equilibriums of the fast system (Π_0) because $g(x, z) = 0$ for every $(x, z) \in \Gamma$. The set $E = \{(x, z) \in R^n \times R^m: f(x, z) = 0, g(x, z) = 0\}$ of equilibriums of the singularly perturbed system (Σ_ε) is a subset of Γ.

Typically, the constraint set Γ is a smooth manifold composed of several disjoint connected components. Assuming that rank$[D_x g\ D_z g] = m$ for every $(x, z) \in \Gamma$, the local form for submersions assures that Γ is an n-dimensional smooth manifold in R^{n+m} [201]. Let $NH \subset \Gamma$ be the set of non-hyperbolic points on Γ, that is, the points of Γ where $D_z g$ has at least one eigenvalue on the imaginary axis. Set NH is a manifold of dimension $n-1$ that separates each of the components of Γ into smaller connected components Γ_i such that $\Gamma - NH = \bigcup_i \Gamma_i$ [264].

In each connected component Γ_i, the number of eigenvalues of $D_z g$ on the right half of the complex plane is constant [121]. Therefore, it makes sense to define a type of stability for each component Γ_i with respect to the family of fast systems $(\Pi_F(x))$.

DEFINITION **(Stability type of components of Γ)**
The connected set Γ_i is said to be a type-k component of Γ if the matrix $D_z g$, evaluated at every point of Γ_i, has k eigenvalues lying in the right complex half-plane and $m-k$ in the left half-plane. If all the eigenvalues of $D_z g$ calculated at points of Γ_i have a negative real part, then we call Γ_i a stable component of Γ. Otherwise, it is called an unstable component.

The stability of the equilibrium points of the fast system determines the stability of these components. Every point in Γ is an equilibrium point of the fast system. More precisely, if (x^*, z^*) lies on a type-k component Γ_i of Γ, then z^* is a type-k hyperbolic equilibrium point of $(\Pi_F(x^*))$. Figure 16.2 illustrates a stable component Γ_s of Γ and a type-one unstable component Γ_u of Γ. Points on Γ_s are asymptotically stable equilibrium points of the family of fast systems while points on Γ_u are type-one equilibrium points of the family of fast systems.

The vector field of the slow system is well defined in every connected component Γ_i of Γ. In the non-hyperbolic points of Γ, the dynamics of the slow system may not be well defined. These points are called impasse points. For more details on impasse points, see [69]. Solutions of the slow system may reach impasse points on Γ in finite time. The interval $[0, t_\omega)$ denotes the maximal interval of existence to the right of the solution $\varphi_0(t, x_0, z_0)$ of the slow system (Σ_0).

16.3 Tikhonov results

We will show, in this section, under certain conditions on the vector field, that the dynamics of the singularly perturbed system can be approximated by the dynamics of one fast system and by the dynamics of the slow system along Γ.

The first question which arises when studying singularly perturbed systems is why the perturbation is called singular. The answer to this question is at the core to understanding the difficulties that appear in the study of this class of dynamical systems. The singularly perturbed model can be rewritten in the form of an autonomous dynamical system $\dot{x} = h(x, \varepsilon)$, depending on the parameter ε, by dividing the second equation of (Σ_ε) by ε. In this case, function h takes the form:

$$h(x, y, \varepsilon) = \begin{bmatrix} f(x, y, \varepsilon) \\ \frac{1}{\varepsilon} g(x, y, \varepsilon) \end{bmatrix}. \tag{16.6}$$

Consider now the following basic result from the qualitative theory of ordinary differential equations.

THEOREM 16-1 **(Continuity of solutions)**
Let $D \subset R^n$ be an open set, C a compact set in R, and suppose $h: D \to R^n$ is a C^1 function. Then, for each initial condition $x_0 \in D$ and each fixed parameter ε, there exists a unique maximal solution $x(t, x_0, \varepsilon)$ of $\dot{x} = h(x, \varepsilon)$ defined on the maximal interval $(\omega_-(x_0, \varepsilon), \omega_+(x_0, \varepsilon))$. Indeed, the solution $x(t, x_0, \varepsilon)$ is continuous with respect to the variables.

Theorem 16.1 states that solutions of differential equations are continuous in the variable t. Indeed, these solutions vary continuously with respect to changes in the initial condition and with respect to changes in the parameter ε. To put it simply, small changes in the parameters cause small changes in the solutions. Therefore, the problem of regular perturbations is relatively simple since no drastic changes of trajectories are expected as a result of parameter variation. For equations (Σ_ε), the vector field is not C^1 when $\varepsilon = 0$. Therefore the conditions of Theorem 16.1 are violated. In reality, the vector field is not well defined at $\varepsilon = 0$ and, for this reason, we call this perturbation of ε, in the neighborhood of the origin, a singular perturbation.

One possibility for overcoming this difficulty is to make a change in the time scale. In the fast time scale, τ, system (Σ_ε) takes the form (Π_ε). The conditions of Theorem 16.1 are satisfied for (Π_ε), even for $\varepsilon = 0$. The limitation of this approach is that the limit system obtained by making $\varepsilon = 0$ captures only the behavior of the fast system, while the slow variable is kept frozen. That means we are unable to analyze the slow component of the dynamics and to determine a simplified model for analysis of the slow system.

The theory of singularly perturbed systems overcomes these problems by decomposing the analysis of the two-time-scale system (Σ_ε) into the analysis of two simpler systems: the slow and fast systems. Tikhonov derived conditions for ensuring the proximity between solutions of singularly perturbed systems and solutions of the slow and fast systems [173, p.46].

16.3.1 Reduced system and boundary-layer system

We expect that the fast variable z will, under certain conditions, quickly converge to a stable component Γ_s of the constraint manifold Γ, approaching a solution of the slow system on Γ_s. In order to study these conditions, let Γ_s be a stable component of Γ and suppose the existence of a function $h(x)$, satisfying $0 = g(x, h(x))$, such that $\Gamma_s = \{(x,z): z = h(x)\}$. Then, the following change of coordinates is convenient:

$$y = z - h(x). \tag{16.7}$$

The new variable y measures the distance of the fast variable z from the stable component Γ_s of the constraint manifold Γ. The singularly perturbed system (Σ_ε) in this new variable assumes the form:

$$(\Sigma'_\varepsilon)\begin{cases} \dot{x} = f(x, y + h(x)) \\ \varepsilon\dot{y} = g(x, y + h(x)) - \varepsilon\dfrac{\partial h(x)}{\partial x} f(x, y + h(x)). \end{cases} \tag{16.8}$$

System (Σ'_ε), in the fast time scale, is given by:

$$(\Pi'_\varepsilon)\begin{cases} \dfrac{dx}{d\tau} = \varepsilon f(x, y + h(x)) \\ \dfrac{dy}{d\tau} = g(x, y + h(x)) = -\varepsilon\dfrac{\partial h(x)}{\partial x} f(x, y + h(x)). \end{cases} \tag{16.9}$$

The boundary-layer system (BLS) or fast system is obtained by setting $\varepsilon = 0$ in the previous equation. That is:

$$(\Pi_{BLS})\ \frac{dy}{d\tau} = g(x, y + h(x)) \tag{16.10}$$

where x is frozen and treated as a parameter. The origin is an equilibrium of (Π_{BLS}) independently of x. The reduced system can be obtained by setting $\varepsilon = 0$ in (Σ'_ε) and substituting the equilibrium y=0 of the fast variable in (Σ'_0). The reduced system can be expressed as:

$$(\Sigma_{red})\ \dot{x} = f(x, h(x)). \tag{16.11}$$

System (Σ_{red}) is termed a *reduced model*, or *slow system*, or quasi-steady-state model. It is called reduced because the dimension has been reduced from $n + m$, the dimension of the original system, to the dimension n, of the variable x.

Assuming that the origin of system (Π_{BLS}) is exponentially stable, uniformly with respect to $x \in D$, and assuming some smoothness conditions on the vector field, then the next theorem guarantees, at least for sufficiently small ε, that solutions of the two-time-scale system can be approximated by the solutions of the slow and fast systems.

THEOREM 16-2 **(Tikhonov's result (see [142, p. 361])**
Consider the singularly perturbed system (Σ_ε) with $f(0,0) = 0$ and $g(0,0) = 0$ and suppose that $z = h(x)$ is an isolated root of $0 = g(x,y)$ satisfying $h(0) = 0$. Suppose the reduced problem (Σ_{red}) has a unique solution $(\overline{x}(t), \overline{z}(t) = h(x(t)))$ defined on the interval $[t_0, t_1]$ and $\|\overline{x}(t)\| < r$ for all $t \in [t_0,t_1]$. If the following conditions are satisfied

(a) the origin of the boundary-layer system (Π_{BLS}) is exponentially stable uniformly[1] *with respect to $x \in B_r$ (an open ball of radius r),*

[1] The origin is exponentially stable uniformly with respect to a parameter $\lambda \in \Lambda$ if there exist constants k, γ and ρ such that the solutions of $\dot{y} = g(y, \lambda)$ satisfy $\|y(t)\| \le k\|y_0\|e^{-\gamma t}$ for all $\|y(0)\| < \rho$ and $\lambda \in \Lambda$.

(b) *the vector functions f, g, h are continuously differentiable and* $\frac{\partial g(x,z)}{\partial z}$ *has continuous first partial derivatives with respect to x and z for all* $x \in B_r$ and $y \in B_\rho$,

then, there are positive constants μ *and* ε^* *such that for every initial condition* z_0 *of the fast variable z satisfying* $\|z_0 - h(x_0)\| \leq \mu$ *and* $0 < \varepsilon < \varepsilon^*$, *the two-time-scale system* (Σ_ε) *has a unique solution* $x(t,\varepsilon)$, $z(t,\varepsilon)$, *defined on the interval* $[t_0, t_1]$, *satisfying:*

$$\begin{aligned} x(t,\varepsilon) - \bar{x}(t) &= O(\varepsilon) \\ z(t,\varepsilon) - h(\bar{x}(t)) - \bar{y}\left(\frac{t}{\varepsilon}\right) &= O(\varepsilon) \end{aligned} \tag{16.12}$$

uniformly with respect to $t \in [t_0, t_1]$, *where* $\bar{y}\left(\frac{t}{\varepsilon}\right)$ *is the solution of* (Π_{BLS}) *starting in* $z_0 - h(x_0)$. *Moreover, given a* $t_b > t_0$, *there exists a number* $\varepsilon^{**} < \varepsilon^*$ *such that* $z(t,\varepsilon) - h(\bar{x}(t)) = O(\varepsilon)$ *uniformly for* $t \in [t_b, t_1]$, *whenever* $\varepsilon < \varepsilon^{**}$.

Tikhonov's theorem justifies the decomposition of the dynamics of the singularly perturbed system, in a finite interval of time, into the dynamics of the fast and slow systems. It establishes a set of conditions under which the difference of the solution $x(t, \varepsilon)$ of the slow variable of the singularly perturbed system and that of the reduced system $\bar{x}(t)$ is in the order of ε.

It is important to observe that the slow variable solution is well approximated by the reduced model for every time in the interval $[t_0, t_1]$. However, we cannot expect this from the fast variable. In the reduced model, we do not have freedom to choose the initial condition of the fast variable, since this initial condition must belong to Γ. So we cannot expect that the initial condition of the fast variable z, say z_0, will be close to the initial condition of the fast variable $h(x_0)$ in the reduced system on Γ. Consequently, we cannot expect that $h(\bar{x}(t))$ will be a good approximation for $z(t, \varepsilon)$ at least in a short interval of time $[t_0, t_b]$. Fortunately, the stability of the fast system guarantees that the approximation is good for times greater than a certain time t_b. That means the simplified model given by the system (Σ_0) or (Σ_{red}) is a good approximation for the original system (Σ_ε), provided that ε is small enough and that the transient behavior of the fast system is stable.

Tikhonov's theorem proves the decomposition of the dynamics of two-time-scale systems in a finite interval of time. However, if an additional assumption of exponential stability of the slow sytem is made, then the result is valid for all $t \geq t_0$. For more details on this subject, see [142].

Example 16-1 Consider the DC motor equivalent circuit of Figure 16.3. By applying Kirchoff's law for voltages, we obtain the following model for the electrical part of the machine:

$$e(t) + Ri(t) + L\frac{di(t)}{dt} = u(t) \tag{16.13}$$

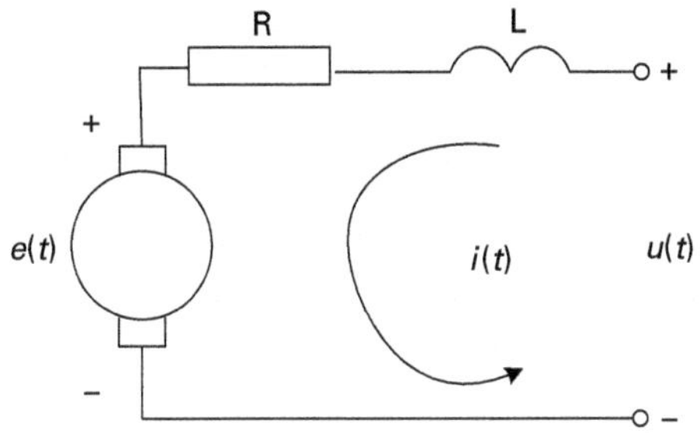

Figure 16.3 The electrical equivalent circuit of a DC motor.

with i as the armature current, R the armature resistance and L the armature inductance. The input u is the voltage applied to the armature terminals while e is the electromotive force. The mechanical model of the machine is obtained by applying the second law of Newton for rotation:

$$J\frac{d\omega}{dt} = T_{el} - T_m \tag{16.14}$$

with J as the inertia constant of the rotor, T_{el} the electrical torque and T_m the mechanical torque applied to the rotor. Variable ω is the angular rotor speed. Since the electromotive force is proportional to the rotor angular speed, the electrical torque is proportional to the armature current, i.e. $T_{el}=ki$, and $T_m=0$, and one obtains the following model for the DC motor:

$$\begin{aligned} J\frac{d\omega}{dt} &= ki \\ L\frac{di}{dt} &= -k\omega - Ri + u. \end{aligned} \tag{16.15}$$

Typically, the armature inductance L is very small if compared to J. So this is a particular example of systems in the form (Σ_ε) where $L=\varepsilon$ is a small number. In this example, $x=\omega$ is a slow variable while $z=i$ is a fast variable. Rewriting the system in these new variables, one obtains:

$$\begin{aligned} \dot{x} &= \frac{k}{J}z \\ \varepsilon\dot{z} &= -kx - Rz + u. \end{aligned} \tag{16.16}$$

Figure 16.4 shows the response of this system when a unitary step u of voltage is applied at $t = 1$ second to the machine for J=5, k=1, R=0.2 and L=0.01.

Since ε is a small number, the term di/dt is large, which means that the variable i varies very quickly until it approaches the equilibrium set $\Gamma = \{(\omega, i) \in R^2: -k\omega - Ri + u = 0\}$ for the fast variables. In this example, the set Γ is a straight line. So, trajectories of this system approach this set very fast and soon after the slow dynamic drives the system along the neighborhood of Γ. Figure 16.5 shows trajectories of this system approach the set Γ.

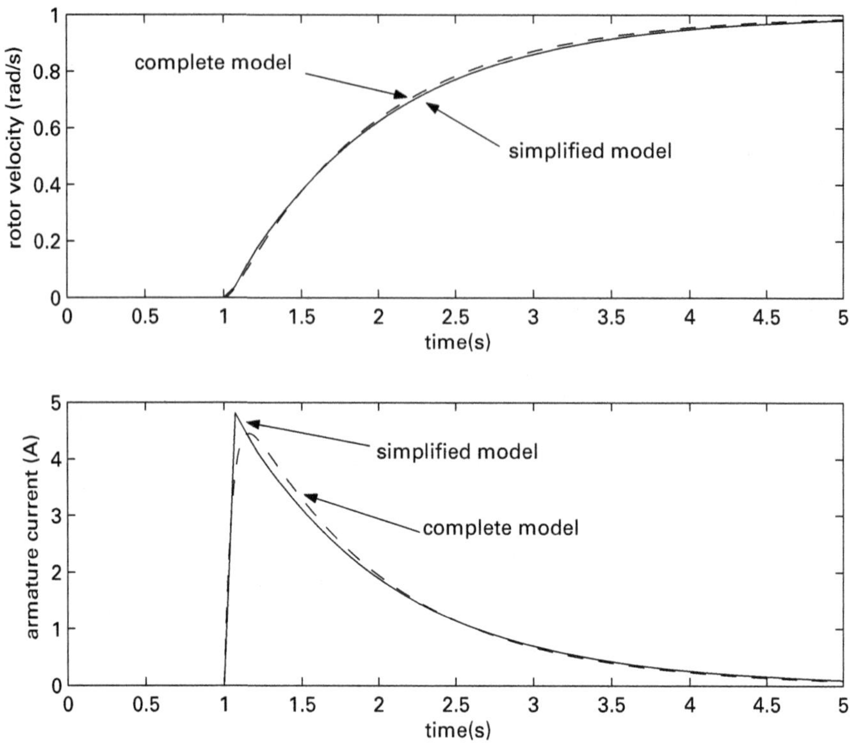

Figure 16.4 DC motor simulation, a unitary step of voltage is applied to the armature terminals at $t = 1$ second, $J = 5$, $k = 1$, $R = 0.2$ and $L = 0.01$.

To highlight the fast dynamic in the time domain, we define the new variable $\Delta i = i - i_0$, where $i_0(\omega)$ is the solution of equation $-k\omega - Ri + u = 0$. Figure 16.6 shows the time-domain behavior of Δi. Now it is clear that Δi is a fast variable since it reaches the steady state in about 0.5 seconds, while the variable ω takes around 4 seconds to settle down.

Time-scale features can be explored to justify the use of simplified models. In this particular example of a DC motor, it is common to neglect the inductance of the stator and to assume the electrical circuit is in a quasi-steady-state situation. This assumption is equivalent to making $\varepsilon = 0$ in the previous equation. By doing that, we find the following simplified model for the DC motor,

$$\begin{aligned} J\frac{d\omega}{dt} &= ki \\ 0 &= -k\omega - Ri + u \end{aligned} \tag{16.17}$$

which is equivalent to the following first-order model:

$$J\frac{d\omega}{dt} = \frac{k}{R}(-k\omega + u). \tag{16.18}$$

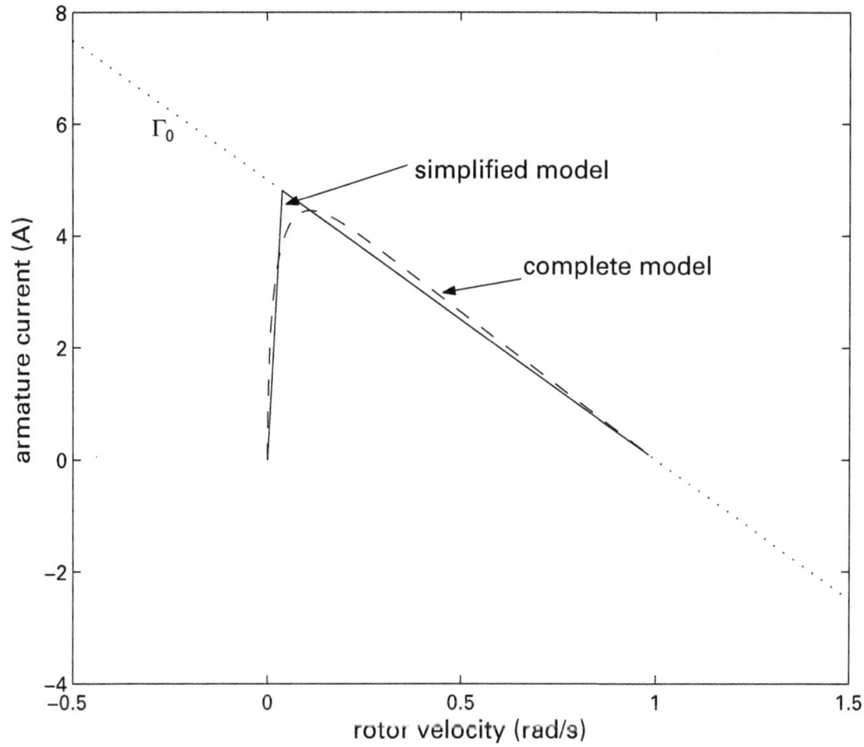

Figure 16.5 DC motor: set Γ and trajectories of the complete and simplified model. A unitary step of voltage is applied to the armature terminals at $t = 1$ second.

Figure 16.4 shows the solution of this simplified model. Solutions of both the complete model and the simplified model are very close to each other. So, if the fast dynamic is not a concern, then this model is very appropriate for analysis, with the added advantage of being simpler (it has dimension one, while the complete model has dimension two).

16.4 Stability of two-time-scale systems

Trajectories of singularly perturbed systems can be approximated, in a two-time-scale sense, by the solutions of the slow and fast systems. It is also possible to decompose a stability analysis of these systems into a stability analysis of the fast and slow systems [105,221]. In this section, the main result of decomposing the stability analysis of singularly perturbed systems derived by Saberi and Khalil [221] is presented.

THEOREM 16-3 **(Composite Lyapunov function)**
Consider the singularly perturbed system (Σ'_ε) *and let* $D_1 \subset R^n$ *and* $D_2 \subset R^m$ *be open connected sets. Assume that the following are satisfied.*

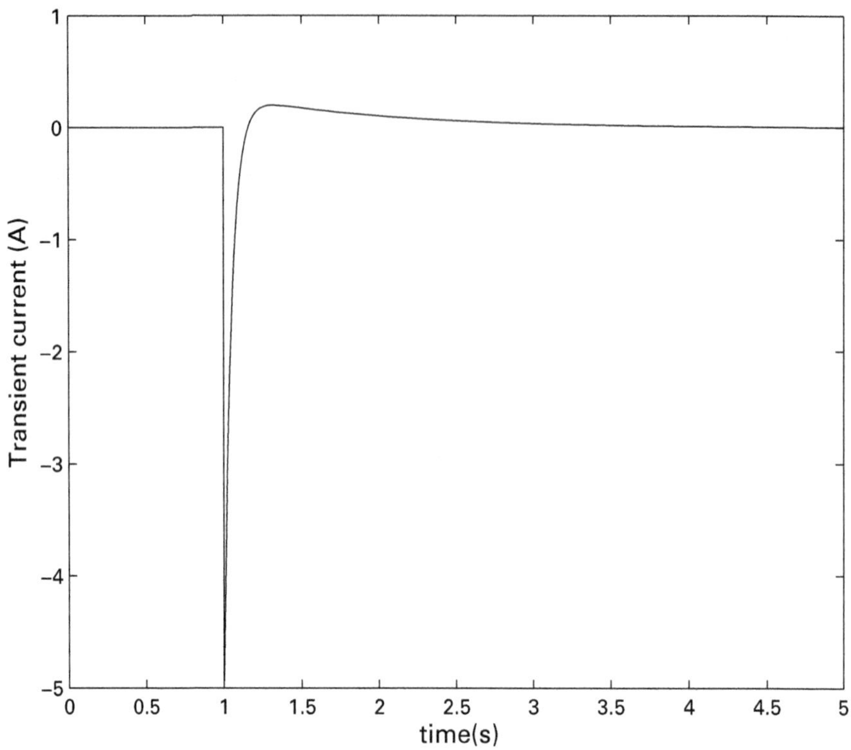

Figure 16.6 DC motor: the transient current Di. A unitary step of voltage is applied to the armature terminals at $t = 1$ second, $J = 5$, $k = 1$, $R = 0.2$ and $L = 0.01$.

(A1) Let $V(x)$ be a C^1-function such that

$$\frac{\partial V}{\partial x} f(x, h(x)) \leq -\alpha_1 \psi_1{}^2(x) \quad \forall x \in D_1 \tag{16.19}$$

where $\psi_1 : R^n \rightarrow R$ is a continuous function.

(A2) Let $W(x, y)$ be a C^1-function and consider the existence of two continuous functions $W_1(x)$ and $W_2(x)$ such that

$$W_1(y) \leq W(x, y) \leq W_2(y) \quad \forall (x, y) \in D_1 \times D_2 \tag{16.20}$$

and

$$\frac{\partial W}{\partial y} g(x, y + h(x)) \leq -\alpha_2 \psi_2^2(y) \quad \forall (x, y) \in D_1 \times D_2 \tag{16.21}$$

where $\psi_2 : R^m \rightarrow R$ is a continuous function.

(A3) There exists a positive real constant β_1 such that:

$$\frac{\partial V}{\partial x}[\, f(x, y + h(x)) - f(x, h(x))] \leq \beta_1 \psi_1(x) \psi_2(y) \quad \forall (x, y) \in D_1 \times D_2\,. \tag{16.22}$$

(A4) There exist positive real constants β_2 and γ such that:

$$\left[\frac{\partial W}{\partial x}+\frac{\partial W}{\partial y}\frac{\partial h}{\partial x}f(x,y+h(x))\right]\leq\beta_2\psi_1(x)\psi_2(y)+\gamma\psi_2^2(y)\quad\forall(x,y)\in D_1\times D_2. \tag{16.23}$$

Then, the following results hold.

(a) *There exists an upper bound* ε_d *such that, for* $0<\varepsilon<\varepsilon_d$, *every bounded trajectory in* $D_1\times D_2$ *converges to the largest invariant set contained in the set* $C:=\{(x,y)\in D_1\times D_2: \psi_1(x)=0,\ \psi_2(y)=0\}$.

(b) $\varepsilon_d=\dfrac{\alpha_1\alpha_2}{\alpha_1\gamma+\frac{1}{4d(1-d)}[(1-d\beta_1)+d\beta_2]^2}$, *where* $0<d<1$.

(c) *Moreover, if the origin is an equilibrium point of the singularly perturbed system,* $V(x)$ *is a Lyapunov function that proves stability for the reduced system, and* $W(x,y)$ *is a Lyapunov function that proves stability of the fast system uniformly with respect to the slow variable* $x\in D_1$, *then the composite function* $U(x,y)=(1-d)V(x)+dW(x,y)$ *is a Lyapunov function for the two-time-scale system for all* $0<d<1$.

This theorem provides sufficient conditions for decomposition of the stability analysis of a two-time-scale system into the stability analysis of a fast and a slow system. More precisely, if one finds a function $V(x)$ satisfying assumption (A1), ensuring the stability of the slow system, and another function $W(x,y)$ satisfying assumption (A2), ensuring the stability of the fast system uniformly with respect to $x\subset D_1$, one then can derive the stability of the original two-time-scale nonlinear system. However, this decomposition is not complete. Assumptions (A3) and (A4) are interconnected conditions that need to be checked.

The proof of this theorem exploits the fact that the derivative of the composite function $U(x,y)=(1-d)V(x)+dW(x,y)$, $0<d<1$, satisfies the following inequality:

$$\dot{U}(x,y)\leq-[\psi_1(x)\psi_2(y)]\Lambda\begin{bmatrix}\psi_1(x)\\ \psi_2(y)\end{bmatrix} \tag{16.24}$$

where

$$\Lambda=\begin{bmatrix}(1-d)\alpha_1 & -\frac{1}{2}[(1-d)\beta_1+d\beta_2]\\ -\frac{1}{2}[(1-d)\beta_1+d\beta_2] & d\left(\frac{\alpha_2}{\varepsilon}-\gamma\right)\end{bmatrix} \tag{16.25}$$

and Λ is positive definite for all $0<\varepsilon<\varepsilon_d$. More details can be found in [221] and [142]. The proof is accomplished using the invariance principle and the observation that the set where $\dot{U}=0$ is contained in the set C.

In assumptions (A1) and (A2), quadratic estimates of the derivatives for both $V(x)$ and $W(x,y)$ are required. These estimates are not important to ensuring the stability of slow and fast systems; however, they are crucial to the interconnected conditions. It is also important to highlight that there is freedom to choose the parameter d in the interval

(0,1). In general, there is a trade-off between maximizing ε_d and maximizing the stability region estimate. It is easy to check that the maximum

$$\varepsilon_d^* = \frac{\alpha_1\alpha_2}{\alpha_1\gamma + \beta_1\beta_2}$$

is reached for

$$d^* = \frac{\beta_1}{\beta_1 + \beta_2}.$$

16.5 Stability region of two-time-scale systems

A comprehensive theory for stability regions of several classes of nonlinear dynamical systems was developed in Chapters 4 to 9. In particular, a comprehensive theory for the stability region of continuous dynamical systems was derived in Chapter 4. Although, for each fixed $\varepsilon > 0$, the singularly perturbed system (Σ_ε) can be viewed as a regular continuous dynamical system, several problems are usually encountered when applying the results derived in Chapter 4 and Chapter 10 to characterize and estimate the stability region of two-time-scale dynamical systems. Moreover, little insight can be gained regarding how the stability region changes with the variation of the parameter ε.

The body of theory developed in Chapter 4, related to the characterization of the stability region, does not take into account the two-time-scale property of dynamical systems. There are numerous advantages in exploring two-time-scale properties. In this section, stability region theory for autonomous two-time-scale nonlinear dynamical systems will be developed. The relationship between the stability boundary of the two-time-scale system and the stability boundaries of the slow and fast systems will be explored. It will be shown that any compact subsets of the stability boundary of the two-time-scale system can be sufficiently approximated by subsets of the stability boundary and stability region of the slow and fast systems. We will then explore how these compact subsets can be used to provide accurate estimates of the relevant stability region of two-time-scale systems.

16.5.1 Stability regions

Consider the singularly perturbed system (Σ_ε) and let $\varphi_\varepsilon(t,x_0,z_0)$ denote the trajectory of (Σ_ε) starting at (x_0,z_0). Suppose (x_s,z_s) is an asymptotically stable equilibrium point (SEP) of (Σ_ε). The set

$$A_\varepsilon(x_s, z_s) = \{(x, z) \in R^n \times R^m : \varphi_\varepsilon(t, x, z) \rightarrow (x_s, z_s) \text{ as } t \rightarrow \infty\} \tag{16.26}$$

denoting the collection of all initial conditions of (Σ_ε), whose trajectories converge to (x_s,z_s) as $t \rightarrow \infty$, exists and is unique. This set is termed the *stability region* of the SEP (x_s,z_s). Our main interest is to study the behavior of the stability region $A_\varepsilon(x_s,z_s)$ as $\varepsilon \rightarrow 0$. In particular, we will establish a relationship between $A_\varepsilon(x_s,z_s)$ and the stability regions

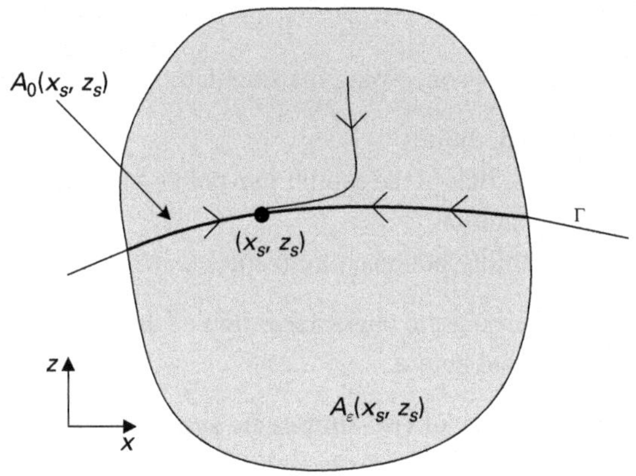

Figure 16.7 The stability region $A_0(x_s,z_s)$ of the slow system is an n-dimensional subset of set Γ, represented by the thick line. The shaded area illustrates the stability region $A_\varepsilon(x_s,z_s)$ of the two-time-scale system (Σ_ε), which is an $(n+m)$-dimensional subset of R^{n+m}.

of two simplified systems, namely the stability region of the slow system and the stability regions of the family of fast systems, for sufficiently small ε.

Suppose (x_s,z_s) is also an asymptotically stable equilibrium point of the slow system (Σ_0) and let $\varphi_0(t,x_0,z_0)$ denote the trajectory of (Σ_0) starting at (x_0,z_0). The stability region of the slow system is the set

$$A_0(x_s, z_s) = \{(x, z)\in\Gamma : \varphi_0(t, x, z)\rightarrow(x_s, z_s) \text{ as } t\rightarrow\infty\} \tag{16.27}$$

which is composed of all initial conditions for which the corresponding trajectories of the slow system converge to (x_s,z_s) as $t\rightarrow\infty$. Figure 16.7 illustrates the stability region of the slow system. The stability region of the slow system $A_0(x_s, z_s)$ is an n-dimensional subset of the constraint manifold G, while the stability region $A_\varepsilon(x_s,z_s)$ of the original system is an $(n+m)$-dimensional subset of R^{n+m}. As a consequence, the stability region of the slow system is not an approximation of the stability region of the system (Σ_ε) for small ε. However, by exploiting Tikhonov's results and imposing some conditions on the vector field, it can be shown that every point inside the stability region of the slow system lies inside the stability region of the singularly perturbed system (Σ_ε) for small ε.

Let $\Phi_\varepsilon(\tau,x_0,z_0)$ denote the trajectory of the fast system (Π_ε) starting at (x_0,z_0). It is evident that $\Phi_\varepsilon(\tau,x_0,z_0)= \varphi_\varepsilon(\varepsilon\tau,x_0,z_0)$. Let $\Phi_0(\tau,x_0,z_0)$ denote the trajectory of (Π_0) starting at (x_0,z_0) and $A_F(x^*, z^*) = \{z \in R^m:\Phi_0(\tau, x^*, z) \rightarrow (x^*, z^*) \text{ as } t\rightarrow \infty\}$ be the stability region of the asymptotically stable equilibrium point z^* of $(\Pi_F(x^*))$.

A relationship between the stability region of the two-time-scale system (Σ_ε) and the stability regions of its two simplified systems, the slow and fast systems, will be investigated. The stability boundaries of these systems will be respectively denoted by $\partial A_\varepsilon(x_s,z_s)$, $\partial A_0(x_s,z_s)$, and $\partial A_F(x,z)$. It will be shown that local approximations of the stability boundary of the two-time-scale system can be obtained in terms of the stability boundaries of the slow and fast systems.

16.5.2 Stability boundary characterization

Consider the following assumptions concerning the singularly perturbed system (Σ_ε).

(A1) All the equilibriums are hyperbolic.
(A2) The stable and unstable manifolds of the equilibrium points on the stability boundary satisfy the transversality condition.
(A3) Every trajectory on the stability boundary converges to an equilibrium point.

A fundamental theorem concerning the characterization of the stability boundary of two-time-scale systems is presented below.

THEOREM 16-4 **(Stability boundary of two-time-scale systems)**
For the two-time-scale system (Σ_ε) satisfying assumptions (A1) and (A3) for sufficiently small ε, let (x_i,z_i), $i=1,2,\ldots$ be the equilibrium points on the stability boundary $\partial A_\varepsilon(x_s, z_s)$ of the asymptotically stable equilibrium point (x_s,z_s), for a fixed sufficiently small ε, then the stability boundary can be characterized as

$$\partial A_\varepsilon(x_s, z_s) \subseteq \bigcup_i W^s_{(\Sigma_\varepsilon)}(x_i, z_i).$$

If in addition, system (Σ_ε) satisfies assumption (A2), then the stability boundary can be characterized as follows:

$$\partial A_\varepsilon(x_s, z_s) = \bigcup_i W^s_{(\Sigma_\varepsilon)}(x_i, z_i).$$

Proof The proof of Theorem 16-4 can be accomplished by observing, for each fixed small ε, that all assumptions of Theorem 4-10 of Chapter 4 are satisfied. Hence Theorem 16-4 follows.

This theorem gives a complete characterization of the stability boundary of two-time-scale systems for each fixed small ε. It asserts that, for each fixed small ε, the stability boundary is completely characterized as the union of the closure of stable manifolds of all equilibrium points on the stability boundary. The characterization is valid for each fixed small ε; however, it provides little information regarding the behavior of the stability boundary $\partial A_\varepsilon(x_s,z_s)$ as $\varepsilon\to 0$. We next show, by utilizing the two-time-scale properties, that the relevant stability region and the relevant stability boundary of the two-time-scale system (Σ_ε) can be sufficiently approximated (as $\varepsilon\to 0$) by the stability boundary and stability regions of the slow and fast systems.

16.5.3 Decomposition of the relevant stability boundary

We next explore the two-time-scale properties to develop a comprehensive theory of the stability region of two-time-scale systems. In particular, we will investigate how the stability region and stability boundary of slow and fast systems can provide an approximation to the relevant stability boundary of the two-time-scale system.

It is shown that, under some conditions on the vector field, every point of the stability region of the slow system (Σ_0) also belongs to the stability region of the singularly perturbed system (Σ_ε) for sufficiently small ε.

THEOREM 16-5 **(Stability boundaries of (Σ_0) and (Σ_ε)).**
Consider the two-time-scale system (Σ_ε) satisfying assumptions (A1) and (A3) for sufficiently small ε. Suppose (x_s,z_s) is a hyperbolic asymptotically stable equilibrium point of the slow system (Σ_0) on a stable component Γ_s and (x_0,z_0) is a point of the stability region $A_0(x_s,z_s)$ of the slow system (Σ_0). Then (x_0,z_0) lies in the stability region $A_\varepsilon(x_s,z_s)$ of the singularly perturbed system (Σ_ε) for sufficiently small ε.

Proof Let $(x_0, z_0) \in A_0(x_s, z_s)$. Since Γ_s is a stable component of Γ and the solution $\varphi_0(t, x_0, z_0)$ is bounded and exponentially converges to (x_s,z_s), one can ensure from Tikhonov's result for infinite intervals of time that for a given $\xi > 0$, there exists $\varepsilon^* > 0$ such that $\|\varphi_\varepsilon(t, x_0, z_0) - \varphi_0(t, x_0, z_0)\| < \xi$ for all $t>0$ and all $\varepsilon \in [0, \infty)$. That means the trajectory of the singularly perturbed system stays bounded and close to the equilibrium (x_s,z_s). Since (x_s,z_s) is also a hyperbolic asymptotically stable equilibrium of the singularly perturbed system (Σ_ε) for sufficiently small ε (see Theorem 7-1), assumption (A3) guarantees that $\varphi_\varepsilon(t, x_0, z_0) \to (x_s, z_s)$ as $t \to \infty$ for sufficiently small ε. In other words, $(x_0, z_0) \in A_\varepsilon(x_s, z_s)$ for sufficiently small ε and this theorem holds.

Type-one equilibrium points play a crucial role in the characterization of the relevant stability boundary of two-time-scale systems. We next establish the relationship between type-one equilibrium points that lie on the stability boundary of the two-time-scale system (Σ_ε) and those that lie on the stability boundaries of the slow (Σ_0) and fast systems ($\Pi_F(x)$).

THEOREM 16-6 **(Stability boundary and hyperbolic equilibrium points)**
Consider the two-time-scale system (Σ_ε) satisfying assumptions (A1) and (A3) for sufficiently small ε and the associated slow system (Σ_0) satisfying assumption (A1). Suppose (x_s,z_s) is a hyperbolic asymptotically stable equilibrium point of the slow system (Σ_0) on the stable component Γ_s and (x_u,z_u) is a hyperbolic equilibrium point of (Σ_0). Then there exists an $\varepsilon^* > 0$, such that the following results hold for all $\varepsilon \in (0, \varepsilon^*)$.

(a) (x_s,z_s) is a hyperbolic asymptotically stable equilibrium point of the two-time-scale system (Σ_ε).
(b) If (x_u,z_u) lies on the stability boundary $\partial A_0(x_s,z_s)$ of the slow system on Γ_s, *then (x_u,z_u) lies on the stability boundary $\partial A_\varepsilon(x_s,z_s)$ of the two-time-scale system (Σ_ε).*
(c) If (x_u,z_u) lies on the stability boundary $\partial A_F(x_u,z^*)$ of the fast system ($\Pi_F(x_u)$) and (x_u,z^*) lies in the stability region $A_0(x_s, z_s) \subset \Gamma_s$ of the slow system (Σ_0), then (x_u,z_u) lies on the stability boundary $\partial A_\varepsilon(x_s,z_s)$ of the two-time-scale system (Σ_ε).

Proof The proof of (a) is a consequence of Theorem 7-1. Proof of (b) follows from Theorem 7-3 by substituting the assumption of existence of an energy function by assumption (A3). To prove (c), we show the existence of points arbitrarily close to (x_u, z_u), such that trajectories of (Σ_ε) starting from these points tend to the asymptotically stable equilibrium point (x_s,z_s) as $t\to\infty$. By hypothesis, $(x_u, z_u) \in \partial A_F(x_u, z^*)$. Therefore, for any small number $r > 0$, the open ball $B_r(x_u,z_u)$ of radius r centered at (x_u,z_u) intersects the stability region of the fast system, that is, $B_r(x_u,z_u) \cap A_F(x_u,z^*) \neq \emptyset$, hence there must exist a point $(x_1, z_1) \in B_r(x_u, z_u)$ such that $\Phi_0(\tau,x_1,z_1)\to(x_u,z^*)$ as $\tau\to\infty$. For any arbitrarily small number $\rho>0$, there exists a time $T_1(\rho) > 0$ such that

$(\hat{x},\hat{z}) = \Phi_0(T_1, x_1, z_1) \in B_{\frac{\rho}{2}}(x_u, z^*)$. Applying the regular perturbation theory to the fast system, one can show the existence of $\varepsilon^{**}>0$ such that $\Phi_\varepsilon(T_1, x_1, z_1) \in B_{\frac{\rho}{2}}(\hat{x}, \hat{z})$ for all $\varepsilon \in (0, \varepsilon^{**})$. Therefore, from the triangle inequality, $\Phi_\varepsilon(T_1, x_1, z_1) \in B_\rho(x_u, z^*)$ for all $\varepsilon \in (0, \varepsilon^{**})$. On the other hand, we know that $(x_u, z^*) \in A_0(x_s, z_s)$, that is, $\varphi_0(t, x_u, z^*) \to (x_s, z_s)$ as $t \to \infty$. For an arbitrarily small number υ, there exists a time $T_2>0$ such that $\varphi_0(T_2, x_u, z^*) \in B_{\frac{\upsilon}{2}}(x_s, z_s)$. Since ρ can be chosen arbitrarily small, Tikhonov's theorem [142, p.434], for a finite interval of time, guarantees the existence of $\hat{\varepsilon}$ such that $\varphi_\varepsilon(T_2, \Phi(T_1, x_1, z_1)) \in B_\upsilon(x_s, z_s)$ for all $\varepsilon \in (0, \hat{\varepsilon})$.

A choice of sufficiently small υ and the exponential stability of (x_s, z_s) with respect to (Σ_0) guarantees, via Tikhonov's theorem for infinite intervals of times [142, pp.439–440], that $\varphi_\varepsilon(t, \varphi_\varepsilon(T_2, \Phi_\varepsilon(T_1, x_1, z_1)))$ is bounded for $t \geq 0$ and stays close to (x_s, z_s) for sufficiently small ε. Assumptions (A1) and (A2) imply $\varphi_\varepsilon(t, x_1, z_1) \to (x_s, z_s)$ as $t \to \infty$ for sufficiently small ε. This completes the proof.

Theorem 16-6 asserts that the task of checking whether a hyperbolic equilibrium point is on the stability boundary of the two-time-scale system can be decomposed into the tasks of checking whether the hyperbolic equilibrium point is on the stability boundary of the fast and/or slow systems. Result (b) offers a scheme to verify whether a hyperbolic equilibrium point on the stable component Γ_s lies on the stability boundary of a two-time-scale system by checking whether the same equilibrium is on the stability boundary of the slow system. On the other hand, result (c) provides a scheme to check whether a hyperbolic unstable equilibrium point on an unstable component Γ_u of Γ belongs to the stability boundary of (Σ_ε), for sufficiently small ε, depending on whether the equilibrium belongs to the stability boundary of the fast system. Figure 16.8 illustrates conclusion (c) of Theorem 16-6.

Example 16-2 This example illustrates the main implications of Theorem 16-6. The following two-time-scale nonlinear dynamical system appears in the literature of power systems [135]:

$$(\Sigma_\varepsilon)\begin{cases} \dot{\omega} = -\dfrac{1}{M_g} D_g \omega - \dfrac{1}{M_g}(B_{12}\theta \sin\delta - P_l) \\ \dot{\delta} = -\dfrac{1}{D_l}(B_{12}\theta \sin\delta - P_l) + \omega \\ \varepsilon\dot{\theta} = -\dfrac{1}{\theta}(Q_l - B_{12}\theta\cos\delta - B_{22}\theta^2). \end{cases} \tag{16.28}$$

The constraint set $\Gamma = \{(\omega, \delta, \theta): g(\delta, \theta) = 0\}$, shown in Figure 16.9, is a two-dimensional manifold composed of two components, the stable component Γ_s and the type-one component Γ_u, which are separated by the set of singular points NH. For $M_g = 20$, $D_g = 0.4$, $D_l = 50$, $P_l = 4$, $Q_l = 0.5$, $B_{12} = 10$, $B_{22} = -10$, the slow system (Σ_0) possesses a hyperbolic asymptotically SEP $(\omega_s, \delta_s, \theta_s) = (0, 0.52, 0.81)$ on Γ_s. As a

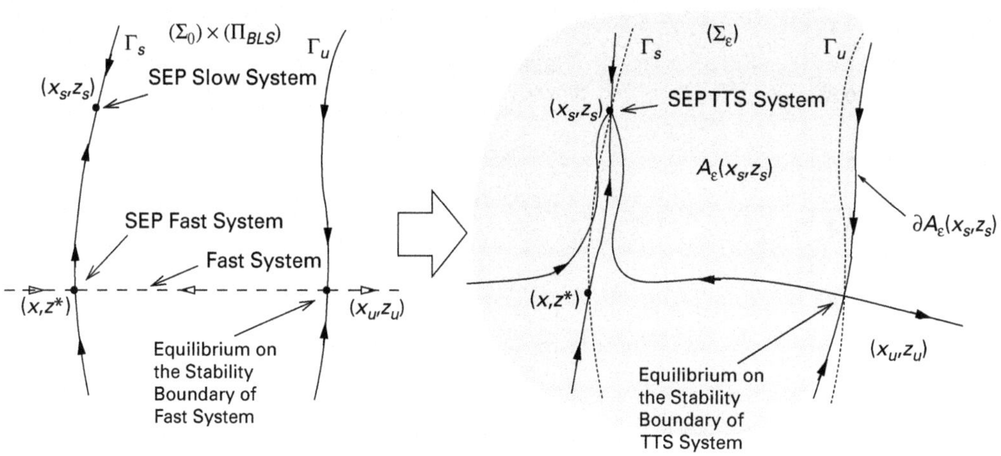

Figure 16.8 Illustration of conclusion (c) of Theorem 16-6. On the left side, (x_u,z^*) belongs to the stability region of the slow system and (x_u,z_u) lies on the stability boundary of the fast system; as a consequence, on the right side, the equilibrium (x_u,z_u) lies on the stability boundary of the two-time-scale system for sufficiently small ε.

consequence of result (a) of Theorem 16-6, $(\omega_s,\delta_s,\theta_s)$ is a hyperbolic asymptotically SEP of the two-time-scale system (Σ_ε) for sufficiently small ε.

The stability boundary of the slow system is composed of the stable manifold of a type-one UEP $(\omega_u,\delta_u,\theta_u) = (0, 0.93, 0.5)$, which lies on the stability boundary of the slow system, and a piece of singular points (see Figure 16.9(a)). Applying result (b) of Theorem 16-6, it follows that $(\omega_u,\delta_u,\theta_u)$ also lies on the stability boundary of the two-time-scale system (16.28) for sufficiently small ε. The result (b) is numerically confirmed in this example by checking the non-empty intersection between the stability region of the two-time-scale system and the unstable manifold of $(\omega_u,\delta_u,\theta_u)$.

Consider the same two-time-scale system (16.28) with the same parameters except for $P_l = 0.5$. The constraint manifold Γ does not vary with changes in P_l, however the equilibria do change. The slow system (Σ_0) still has a hyperbolic asymptotically SEP $(\omega_s,\delta_s,\theta_s) = (0,0.053,0.946)$ on Γ_s. Again, applying result (a) of Theorem 16-6, $(\omega_s,\delta_s,\theta_s)$ is a hyperbolic asymptotically SEP of the two-time-scale system (Σ_ε) for sufficiently small ε. The type-one UEP $(\omega_u,\delta_u,\theta_u) = (0, 0.73, 0.075)$ of the two-time-scale system for this new parameter does not belong to the stable component of Γ_s but to the type-one component Γ_u (see Figure 16.9(b)). The UEP of the two-time-scale system is a type-one UEP of the fast system which lies on the stability boundary of the SEP $(\omega_u,\delta_u,\theta_{sFAST}) = (0, 0.73, 0.67)$ of the fast system. The SEP of the fast system $(\omega_u,\delta_u,\theta_{sFAST})$ lies inside the stability region of the slow system (see Figure 16.9(b)) and, as a consequence of result (c) of Theorem 16-6, the UEP $(\omega_u,\delta_u,\theta_u)$ lies on the stability boundary of the two-time-scale system for sufficiently small ε. This result is also confirmed in the numerical example, observing a non-empty intersection between the stability region of the two-time-scale system and the unstable manifold of $(\omega_u,\delta_u,\theta_u)$.

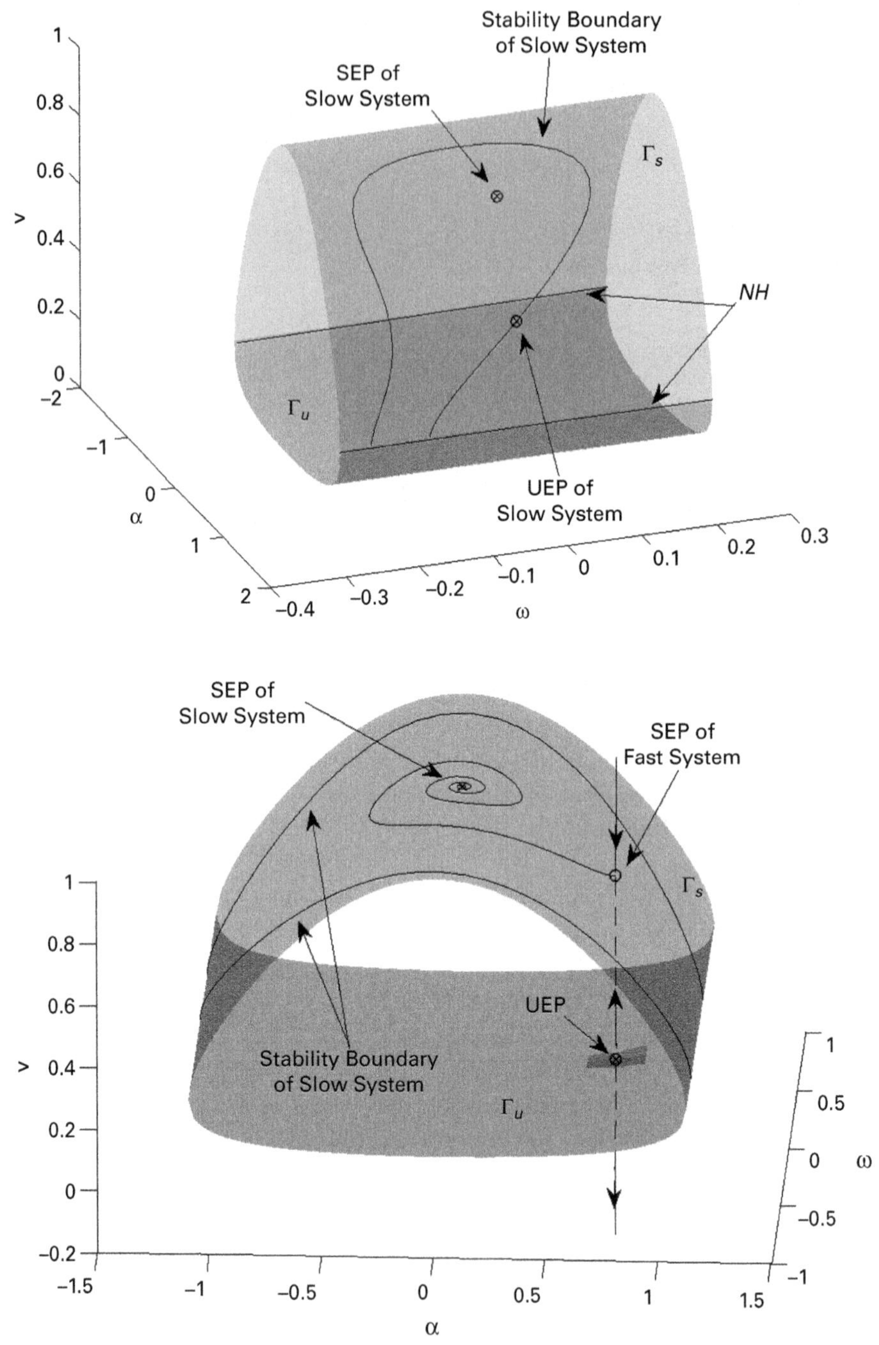

Figure 16.9 (a) The stability region of the slow system of Example 16-2 for $M_g = 20$, $D_g = 0.4$, $D_l = 50$, $P_l = 4$, $Q_l = 0.5$, $B_{12} = 10$, $B_{22} = -10$. The UEP on the stability boundary of the slow system on the stable component Γ_s of Γ also belongs to the stability boundary of the two-time-scale system for sufficiently small ε. (b) The stability region of the slow system of Example 16-2 for $M_g = 20$, $D_g = 0.4$, $D_l = 50$, $P_l = 0.5$, $Q_l = 0.5$, $B_{12} = 10$, $B_{22} = -10$. The type-one UEP on the stability boundary of the fast system also lies on the stability boundary of the two-time-scale system for sufficiently small ε.

We next establish a relationship between the stability boundary of the two-time-scale system and the stability boundary and stability region of fast and slow systems via two theorems. Theorem 16-7 below offers an approximation of the stability boundary of the two-time-scale system, for sufficiently small ε, by means of the union of stability regions of the fast system and the stability boundary of the slow system.

The stable and unstable manifolds of a hyperbolic equilibrium point (x^*,z^*) of (Σ_ε) will be respectively denoted $W^s_{(\Sigma_\varepsilon)}(x^*, z^*)$ and $W^u_{(\Sigma_\varepsilon)}(x^*, z^*)$. The subindex (Σ_ε) will be used to indicate that these are invariant manifolds of system (Σ_ε). We say a set A is η-close to a set B if $d(A,B)\leq \eta$, in which $d(A, B) = \sup_{x\in A} \inf_{y\in B} d(x, y)$.

THEOREM 16-7 **(Closeness of stability boundary of slow and two-time-scale systems)**

Consider system (Σ_ε) satisfying assumptions (A1)–(A3), for sufficiently small ε, and the associated slow system (Σ_0) satisfying assumption (A1). Let (x_s,z_s) be a hyperbolic asymptotically stable equilibrium point and (x_u,z_u) be a type-one equilibrium point lying on the stability boundary $\partial A_0(x_s,z_s)$ of the slow system (Σ_0). Let $S\subset W^s_{(\Sigma_0)}(x_u, z_u)$ be a connected compact subset of the stability boundary of the slow system (Σ_0) containing (x_u,z_u). Let N be a neighborhood of S in R^{n+m} and, for each $(\hat{x}, \hat{z})\in S$, let $F_{\hat{x}} = W^s_{(\Pi_F(\hat{x}))}(\hat{x}, \hat{z})\cap N$, the intersection of the stability region of the fast system $(\Pi_F(\hat{x}))$ with N. Then, given $\eta > 0$, there exists $\varepsilon^* > 0$ such that $\bigcup_{(\hat{x},\hat{z})\in S} F_{\hat{x}}$ is an $(n+m-1)$-dimensional set that is η-close to the stability boundary $\partial A_\varepsilon(x_s,z_s)\cap N$ of the two-time-scale system for all $\varepsilon \in (0, \varepsilon^*)$.

Proof Set S is a compact set composed of equilibrium points of the fast system (Π_0). Consider the extended system:

$$(\Pi_\varepsilon)\times 0\left\{\frac{dx}{d\tau} = \varepsilon f(x, z), \quad \frac{dz}{d\tau} = g(x, z), \quad \frac{d\varepsilon}{d\tau} = 0.\right.$$

For each $(\hat{x}, \hat{z})\in S$, let $E^s_{(\hat{x},\hat{z})}$, $E^u_{(\hat{x},\hat{z})}$ and $E^c_{(\hat{x},\hat{z})}$ denote the invariant subspaces of $R^{n+m}\times(-\varepsilon_0, \varepsilon_0)$ associated with the eigenvalues of the Jacobian of the system $(\Pi_\varepsilon)\times 0$ calculated at $((\hat{x}, \hat{z}), \varepsilon = 0)$. According to Theorem 9.1 of [93], there exists a center-stable manifold C^s for $(\Pi_\varepsilon)\times 0$ near S for sufficiently small ε. This center-stable manifold C^s satisfies the following properties: (i) $K\times\{\varepsilon=0\}\subset C^s$; (ii) C^s is locally invariant [93] under the flow $(\Pi_\varepsilon)\times 0$; (iii) C^s is tangent to $E_{(x,z)}{}^s\oplus E_{(x,z)}{}^c$ at $((\hat{x}, \hat{z}), \varepsilon = 0)$ for all $(\hat{x}, \hat{z})\in S$. Let $(\hat{x}, z)$ be a point lying in $F_{\hat{x}}$ and, as a consequence, lying in $W^s_{(\Pi_F(\hat{x}))}(\hat{x}, \hat{z})$. Given $\rho/2>0$, there exists $\tau>0$ such that $\rho/2 < \|\Phi_0(\tau, \hat{x}, z) - (\hat{x}, \hat{z})\| < \rho$.

Let P be the projection onto $E_{(x,z)}{}^s$. Define $q = P(\Phi_0(\tau, \hat{x}, z) - (\hat{x}, \hat{z}))$. According to [36], given $\rho>0$, there exists a unique solution $\varphi_\varepsilon(t)$, bounded for $t\geq 0$, such that $P(\varphi_\varepsilon(0) - (\hat{x}, \hat{z})) = q$ and $\|\varphi_\varepsilon(t) - \varphi_0(t, \hat{x}, \hat{z})\|\leq\rho$ for every $t\geq 0$ and ε sufficiently small. Since $\varphi_\varepsilon(t)$ is bounded for $t\geq 0$ and $\varphi_0(t, \hat{x}, \hat{z})\to(x_u, z_u)$ as $t\to\infty$, then $\varphi_\varepsilon(t)\to(x_u, z_u)$ as $t\to\infty$ as a consequence of assumptions (A1) and (A2). As a result, $\varphi_\varepsilon(0)\in C^s$ and $\varphi_\varepsilon(0)\in W^s_{(\Sigma_\varepsilon)}(x_u, z_u)$. Since $(\hat{x}, \hat{z})$ is a hyperbolic equilibrium point of the fast system $(\Pi_F(\hat{x}))$, the stable manifold $W^s_{(\Pi_F(\hat{x}))}(\hat{x}, \hat{z})$ is also tangent to $E_{(x,z)}{}^s$ at $(\hat{x}, \hat{z})$. Consequently, given a number $\alpha>0$, there exists $\rho>0$ such that

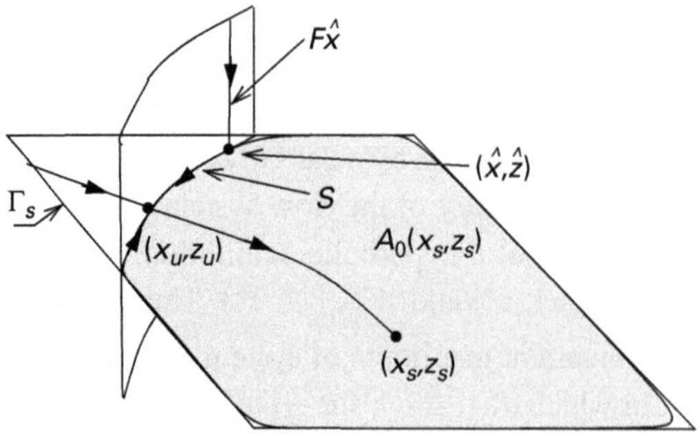

Figure 16.10 Illustration of Theorem 16-7. The type-one UEP (x_u,z_u) lies on the stability boundary of the slow system and, as a consequence, it lies on the stability boundary of the two-time-scale system for sufficiently small ε. The set S, a subset of the stability boundary of the slow system, and the union of sets $F_{\hat{x}}$ provide an approximation of the stability boundary of the two-time-scale system.

$\|(I-P)(x,z)-(\hat{x},\hat{z})\| < \alpha\|(x,z)-(\hat{x},\hat{z})\|$for all (x,z) lying on $W^s_{(\Pi_F(\hat{x}))}(\hat{x},\hat{z})$ or C^s $\|(\mathrm{x},\mathrm{z})-(\hat{x},\hat{z})\| < \rho$. As a consequence $\|\Phi_0(\tau,\hat{x},z)-\varphi_\varepsilon(0)\| = \|P(\Phi_0(\tau,\hat{x},z)-\varphi_\varepsilon(0)) - (I-P)\,(\Phi_0(\tau,\hat{x},z)-\varphi_\varepsilon(0))\| = \|(I-P)(\Phi_0(\tau,\hat{x},z)-\varphi_\varepsilon(0))\| \leq 2\alpha\rho$.

Finally, we apply the regular perturbation technique to the system (Π_ε). Given $\eta>0$ there exists $\gamma>0$ such that $\Phi_\varepsilon(-\tau,x,z)\in B_\eta(\hat{x},z)$ for every $(x,z)\in B_\gamma(\Phi_0(\tau,\hat{x},z))$ for sufficiently small ε. Choosing $\alpha<1/2$ and ρ such that $2\alpha\rho<\gamma$, one guarantees that $\varphi_\varepsilon(0)\in B_\gamma(\Phi_0(\tau,\hat{x},z))$. For this reason, there exists a point $p=\Phi_\varepsilon(-\tau,\varphi_\varepsilon(0))\in W^s_{(\Sigma_\varepsilon)}(x_u,z_u)$ that is η-close to $(\hat{x},z)$. By way of Theorem 16-6, $(x_u,z_u)\in\partial A_\varepsilon(x_s,z_s)$ for all $\varepsilon\in(0,\varepsilon^*)$, then $(\hat{x},z)$ is η-close to $\partial A_\varepsilon(x_s,z_s)$ for all $\varepsilon\in(0,\varepsilon^*)$. The proof is completed observing this is true for every $(\hat{x},z)$ in the compact set $\bigcup_{(\hat{x},\hat{z})\in S} F_{\hat{x}}$.

Example 16-3 Consider the two-time-scale system of Example 16-2. Figure 16.11 shows a subset of the stability region of the two-time-scale system. We can see that the points on the stability boundary of the slow system are close to the stability boundary of the two-time-scale system, confirming the results of Theorem 16-7.

The next theorem offers an approximation of the stability boundary of the two-time-scale system $\partial A_\varepsilon(x_s,z_s)$ of (Σ_ε), for sufficiently small ε, by means of the union of stability boundaries of the fast systems and the stability region of the slow system.

THEOREM 16-8 **(Closeness of the stability boundary of the fast subsystem and the two-time-scale system)**

Consider the two-time-scale system (Σ_ε) *satisfying assumptions (A1)–(A3), for sufficiently small* ε, *and the associated slow system* (Σ_0) *satisfying assumption (A1). Let* (x_s,z_s) *be a hyperbolic asymptotically stable equilibrium point and* (x_u,z_u) *a type-one equilibrium point on the type-one component* Γ_u *of* Γ *of the two-time-scale system* (Σ_ε) *lying on the stability boundary* $\partial A_\varepsilon(x_s,z_s)$ *of the two-time-scale system* (Σ_ε) *for sufficiently small* ε. *Let* $Q\subset A_{(\Sigma_0)}(x_u,z_u)$ *be a connected compact subset of dimension n of the*

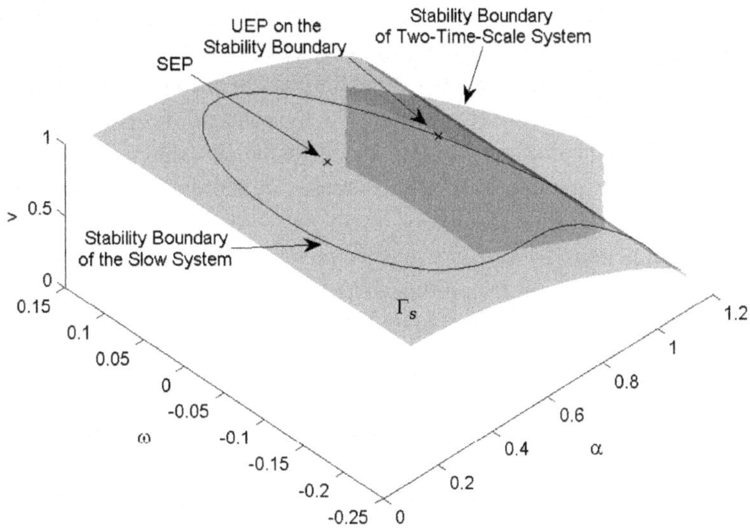

Figure 16.11 An illustration of the close relationship between the stability boundary of the slow system and that of the two-time-scale system under the stated conditions. In particular, the figure shows that a subset of the stability boundary of the slow system is close to a subset of the stability boundary of the two-time-scale system (16.28).

stability region $A_{(\Sigma_0)}(x_u, z_u)$ *of the slow system on* Γ_u. *Let* N *be a neighborhood of* Q *in* R^{n+m} *and for each* $(\hat{x}, \hat{z})\in Q$, *define the* $(m-1)$*-dimensional set* $G_{\hat{x}} = W_{(\Pi F(\hat{x}))}^{s}(\hat{x}, \hat{z})\cap N$. *Then, given* $\eta > 0$, *there exists* $\varepsilon^* > 0$ *such that* $\bigcup_{(\hat{x},\hat{z})\in Q} G_{\hat{x}}$ *is an* $(n+m-1)$*-dimensional set that is* η*-close to the stability boundary* $\partial A_\varepsilon(x_s,z_s)\cap N$ *of the two-time-scale system for all* $\varepsilon \in (0, \varepsilon^*)$.

Proof Using arguments very similar to those used in the proof of Theorem 16-7, it follows that every $(\hat{x}, \hat{z})$ in Q and $(\hat{x}, z)$ in $W^s_{(\Pi_F(\hat{x}))}(\hat{x}, \hat{z})$ are η-close to the stable manifold $W^s_{(\Sigma_\varepsilon)}(x_u, z_u)$ of the type-one equilibrium point (x_u, z_u) for sufficiently small ε. The type-one equilibrium point (x_u, z_u) is on the stability boundary $\partial A_\varepsilon(x_s,z_s)$ of the two-time-scale system (Σ_ε) for sufficiently small ε. As a consequence of Theorem 16-4 on stability boundary characterization, $(\hat{x}, \hat{z})$ and $(\hat{x}, z)$ are η-close to the stability boundary $\partial A_\varepsilon(x_s,z_s)$ of the two-time-scale system for sufficiently small ε. The proof is completed observing that this is true for every point in $\bigcup_{(\hat{x},\hat{z})\in Q} G_{\hat{x}}$.

16.6 Two-time-scale energy function theory

Energy functions give a powerful approach to analyzing the global dynamics of a fairly large class of nonlinear systems. The existence of an energy function is a sufficient condition for the non-existence of complex limit sets such as closed orbits (limit cycles), quasi-periodic solutions and chaos [54]. In addition, energy functions provide estimates of the stability region of an attractor of nonlinear systems.

In this section, we generalize the theory of energy functions, which was developed in Chapter 3 for continuous dynamical systems, to a class of two-time-scale systems. The theory developed in this section combines the concept of composite Lyapunov functions, which were introduced in Section 16.4, with the concept of energy function introduced in Chapter 3. More precisely, the idea of composition is explored to develop a family of composite energy functions that are composed of energy functions for the associated slow and fast systems.

Energy functions require the satisfaction of more conditions when compared to Lyapunov functions, but they provide sharper results about global dynamics. We can achieve global dynamic analysis of two-time-scale nonlinear systems by analysis of the slow system and of the fast system. Several advantages offered by this decomposition approach are illustrated. One implication of this generalization is that the existing body of the theory of stability regions can be extended to two-time-scale nonlinear systems which admit an energy function.

We develop a global result for a class of two-time-scale nonlinear systems, asserting that limit sets of two-time-scale systems admitting (two-time-scale) energy functions cannot contain complicated behavior such as closed orbits (limit cycles), quasi-periodic trajectories and chaotic motions. It is also shown that the composite energy function leads to improved estimates of stability regions and to deeper insight into system dynamics and characterizations of limit sets of two-time-scale nonlinear systems. Some numerical examples are used to illustrate the advantages of exploring time-scale properties of two-time-scale nonlinear systems.

16.6.1 Uniform energy function

In this section we will define the concept of uniform energy function to deal with systems that are dependent on a parameter.

Consider the nonlinear autonomous dynamical system:

$$\dot{x} = F(x) \tag{16.29}$$

where $F{:}R^n \rightarrow R^n$ is a C^1-function. Let $E = \{x \in R^n{:}\ F(x) = 0\}$ be the set of equilibrium points of (16.29) and $\phi(t, x_0)$ be the trajectory of (16.29) passing through the point x_0. According to the results of Chapter 3, a C^1-function $V{:}\ R^n \rightarrow R$ is an energy function for system (16.29) if it satisfies the following three conditions:

(C1) $\dot{V}(x) \leq 0$ for every $x \in R^n$;

(C2) if $x_0 \notin E$ then the set $\{t \in R : \dot{V}(\phi(t, x_0)) = 0\}$ has the measure 0 in R;

(C3) for each $x_0 \in R^n$, if $\{V(\phi(t, x_0))$ is bounded for $t \geq 0$, then $\phi(t, x_0)$ is bounded for $t \geq 0$.

In this section, however, we will need to replace condition (C2) with the following stronger condition:

(C2′) The vector field F is transversal to the set $C = \{x \in D : \dot{V}(x) = 0\}$ at every point of $C \setminus E$.

It is easy to see that condition (C2′) implies condition (C2).

Now consider the system:

$$\dot{y} = G(x, y) \tag{16.30}$$

where $G{:}R^n \times R^m \rightarrow R^m$ is a C^1-function and x is a parameter vector that belongs to some set $\Lambda \subset R^n$.

DEFINITION **(Uniform energy function)**

A C^1-function $W{:}R^n \times R^m \rightarrow R$ is a uniform energy function for system (16.30) if the following conditions are satisfied.

(C4) There exist continuous functions $W_1{:}R^m \rightarrow R$ and $W_2 : R^m \rightarrow R$ such that $W_1(y) \leq W(x, y) \leq W_2(y)$ for every $(x, y) \in \Lambda \times R^m$.

(C5) $\dot{W}(x, y) \leq 0$ for every $(x, y) \in \Lambda \times R^m$.

(C6) W is a proper function.

We are now in a position to present the following energy function theorem for two-time-scale nonlinear systems.

THEOREM 16-9 **(Composite energy function)**

Consider the two-time-scale system (Σ_ε) and suppose V is an energy function for the slow system (Σ_{red}) and W is a uniform energy function for the fast system (Π_F) satisfying assumptions (A1) to (A4) of Theorem 16-3. If the slow vector field $f(x,h(x))$ is transversal to $C_{slow} = \{x{:}\psi_1(x) = 0\} - E_{slow}$ and the fast vector field is transversal to $C_x = \{y{:}\psi_2(y) = 0\} - E_x$ for every x, then there exists ε^* such that, for $0 < \varepsilon < \varepsilon^*$, the composite function $U_d = (1 - d)V + dW$ is an energy function for system (Σ_ε) for every $d \in (0, 1)$.

Proof We must prove that U_d satisfies conditions (C1) to (C3) of an energy function for every $d \in (0, 1)$. From the proof of Theorem 16-3, we know that $\dot{U}_d \leq 0$ for every $d \in (0, 1)$ if ε is small enough. Therefore condition (C1) is satisfied. Let $(x(t), y(t)) = \phi_t(x_0, y_0)$ be the trajectory of (Σ_ε) passing through (x_0, y_0) and suppose for some $d^* \in (0, 1)$, $\{U_{d^*}(\phi_t(x_0, y_0))\}$ is bounded for $t \geq 0$. Since the composite function U_d is a uniformly continuous function of d, there exists an open interval (d_1, d_2) containing d^* such that $\{U_d(\phi_t(x_0, y_0))\}$ is bounded for $t \geq 0$. Suppose now that $\phi_t(x_0, y_0)$ is not bounded. According to assumption (C6), neither $\{V(x(t))\}$ nor $\{W(x(t), y(t))\}$ is bounded. This means that there exists a nondecreasing sequence of times $\{t_n\}$ such that both $\|V(x(t_n))\| \rightarrow \infty$ and $\|W(x(t_n), y(t_n))\| \rightarrow \infty$ as $n \rightarrow \infty$. Two situations are possible: either $t_n \rightarrow \infty$ or $t_n \rightarrow w < \infty$. In the first case, since $\dot{U}_d \leq 0$ and U_d is bounded for $t \geq 0$, we conclude that U_d converges to a number α_d as $n \rightarrow \infty$. In the second case, the continuity of U_d and the boundeness for $t \geq 0$ guarantee that U_d converges to a number $\alpha_d = \lim_{n\to\infty} U_d(t_n)$ as $n \rightarrow \infty$.

Suppose that $V(x(t_n)) \rightarrow +\infty$ as $n \rightarrow \infty$. This implies, since $U_d = (1-d)V + dW$ is bounded, that $W(x(t_n), y(t_n)) \rightarrow -\infty$. Now, if d' and d'' are chosen such that $d', d'' \in (d_1, d_2)$, then,

$$\begin{aligned} U_{d'} &= (1-d')V + d'(W) \to \alpha_{d'} \\ U_{d''} &= (1-d'')V + d''(W) \to \alpha_{d''} \end{aligned} \quad \text{as} \quad n \to \infty.$$

By subtracting these previous equations we conclude that $(d'' - d')(V - W) \to \alpha_{d'} - \alpha_{d''}$ as $n \to \infty$. But this is a contradiction since $V - W \to +\infty$.

Similar arguments can be used for the case where $V \to -\infty$. Therefore condition (C3) is satisfied. To prove condition (C2′), observe that according to Eq. (16.24), $\dot{U}_{d(x,y)} = 0$ if and only if both $\psi_1 = 0$ and $\psi_2 = 0$, where ψ_1 and ψ_2 are the continuous functions respectively defined in assumptions (A1) and (A2) of Theorem 16-3. By assumption, trajectories of the slow system intercept the set C_{slow} transversally. Now, using Tikhonov's results, we know that solutions close to the algebraic manifold Γ satisfy $x(t) - \bar{x}(t) = O(\varepsilon)$. Since transversality persists under small smooth perturbations, $x(t)$ intercepts the set $C \setminus E_{slow}$ transversally for small values of ε. Also, we know that if $(x_{eq}, y_{eq}) \in E$, then $x_{eq} \in E_{slow}$. Consequently, $(x(t), y(t))$ intercepts $C \setminus E$ transversally as required by condition (C2′). When the trajectory is far away from the set Γ, the result follows, with similar arguments, from the local approximation $y(t/\varepsilon) - \bar{y}(t/\varepsilon) = O(\varepsilon)$ and the transversal intersection of $\bar{y}$ with C_x.

Theorem 16-9 provides sufficient conditions allowing the decomposition of stability analysis of two-time-scale nonlinear systems with an energy function into the stability analysis of a fast and slow system. Hence, from a computational viewpoint, stability analysis of two-time-scale nonlinear systems with an energy function is reduced to the stability analysis of two smaller subsystems: the slow subsystem and the fast subsystem.

In general, trajectories of general nonlinear systems can be very complicated. The asymptotic behavior of trajectories can be quasi-periodic or chaotic, for example. If the underlying dynamical system has some special properties, then the system may admit only simple trajectories. For instance, every trajectory of two-time-scale systems having an energy function has only two modes of behavior: its trajectory either converges to an equilibrium point or goes to infinity (becomes unbounded) as time increases or decreases. This result is explained in the following theorem.

THEOREM 16-10 **(Global behavior of trajectories)**
Consider the two-time-scale system (Σ_ε) and suppose V is an energy function for the slow system (Σ_{red}) and W is a uniform energy function for the fast system (Π_{BLS}). If these two functions satisfy assumptions (A1) to (A4) of Theorem 16-3, then there exists ε^* such that, for $0 < \varepsilon < \varepsilon^*$, every bounded trajectory of the two-time-scale system converges to one of its equilibrium points.

Theorem 16-10 asserts that there does not exist any limit cycle (oscillation behavior) or bounded complicated behavior such as an almost periodic trajectory, chaotic motion, etc. in the system. In Theorem 16-10, we have shown that every trajectory of two-time-scale systems either converges to one of the equilibrium points or goes to infinity. In the next section, we will show a sharper result, asserting that every trajectory on the stability boundary must converge to one of the equilibrium points on the stability boundary as time increases.

16.6.2 Characterization of the stability boundary

One important application of Theorem 16-9 is that all stability region theory based on energy functions can now be applied to the decomposition of stability analysis of two-time-scale nonlinear systems. The following theorem provides a fundamental characterization and insight into the stability boundaries of two-time-scale systems.

THEOREM 16-11 **(Stability boundary characterization of two-time-scale systems)** *Consider the two-time-scale system (Σ_ε) and suppose all the equilibrium points of (Σ_0) are hyperbolic. Suppose V is an energy function for the slow system (Σ_{red}) and W is a uniform energy function for the fast system (Π_{BLS}), satisfying assumptions (A1) to (A4) of Theorem 16-3. Let x_s be an asymptotically stable equilibrium point on the stable component Γ_s of the slow system (Σ_{red}). If the slow vector field $f(x,h(x))$ is transversal to $C_{slow} = \{x:\psi_1(x) = 0\} - E_{slow}$ and the fast vector field is transversal to $C_x = \{y:\psi_2(y) = 0\} - E_x$ for every x, then there exists ε^* such that for $0 < \varepsilon < \varepsilon^*$, the stability boundary $\partial A_\varepsilon(x_s, z_s)$ of x_s is contained in the union of the stable manifolds of the unstable equilibrium points on the stability boundary. More precisely, if (x_i, z_i), $i = 1, 2, \ldots$ are the unstable equilibrium points on the stability boundary of system (Σ_ε), then $\partial A_\varepsilon(x_s, z_s) \subset \cup_i W_\varepsilon^s(x_i, z_i)$.*

Proof According to Theorem 16-9, the composite function $U_d = (1 - d)V + dW$ is an energy function for system (Σ_ε) with $d \in (0, 1)$ for sufficiently small ε. If (x^*, z^*) is a hyperbolic equilibrium point of (Σ_0), then (x^*, z^*) is a hyperbolic equilibrium point of (Σ_ε) for sufficiently small ε [288]. Since all equilibrium points of (Σ_0) are hyperbolic, then all equilibrium points of (Σ_ε) are hyperbolic for sufficiently small ε. The proof is accomplished as a consequence of Theorem 4-9 of Chapter 4.

The significance of this theorem is that it offers a way to completely characterize the stability boundary. In fact, Theorem 16-11 asserts that the stability boundary $\partial A_\varepsilon(x_s, z_s)$ of the two-time-scale system is contained in the stable manifolds of all the unstable equilibrium points on the stability boundary.

THEOREM 16-12 **(Unbounded stability region)**
Consider the two-time-scale system (Σ_ε) which has an energy function. If the equilibrium points on the stability boundary $\partial A_\varepsilon(x_s, z_s)$ are hyperbolic and have no source, then the stability region $\partial A_\varepsilon(x_s, z_s)$ is unbounded.

The significance of Theorem 16-12 is that the condition derived only requires checking the types of equilibrium points lying on the stability boundary, as opposed to checking every point in the state space. If there are no source-type equilibrium points lying on the stability boundary, then the corresponding stability region is unbounded.

16.6.3 Examples

Several numerical examples are used to show that we can obtain limited results on stability analysis and the structure of limit sets without taking into account time-scale properties. On the other hand, more accurate results on stability analysis and limit set structures can be obtained by exploring the time-scale properties.

Example 16-4 Consider the following singularly perturbed system:

$$\begin{aligned} \dot{x} &= x - x^3 + z \\ \varepsilon\dot{x} &= -x - z \end{aligned} \qquad (16.31)$$

The origin is the unique equilibrium point of this system. Without taking into account the two-time-scale property of this system, it is quite natural to try

$$V(x, z) = \frac{x^2 + \varepsilon z^2}{2}$$

as an energy function. It is easy to see that the derivative of V is given by

$$\dot{V} = -x^4 + x^2 - z^2.$$

Since $\dot{V}$ is greater than zero at points close to the origin, function V is not an energy function. However, we note that the set where the derivative is positive is bounded; hence, the extension of the invariance principle [218] can be applied to obtain an accurate estimate of the location of the limit sets. For $\varepsilon < 1$, the level set $\Omega = \{(x, z): V(x, z) \leq 1/2\}$ is the smallest level set that contains the region where the derivative of V is positive. Since V is radially unbounded, all trajectories enter Ω in finite time. As a consequence, all limit sets are located inside Ω. The set Ω as well as the phase portrait of (16.31) are depicted in Figure 16.12.

Although the proposed function ensures the existence of a bounded set Ω containing all the limit sets of system (16.31) for any ε, function V cannot guarantee, as the phase portrait suggests, that the origin is a global stable equilibrium point. We note that function V was given without consideration of the two-time-scale behavior.

Now, let us apply the ideas of fast–slow decomposition to construct energy functions for the slow and fast systems. The constraint manifold is composed of a single stable connected component given by $\Gamma_s = \{(x, z) \in R^2 : z = h(x) = -x\}$. This constraint manifold is depicted in Figure 16.12. The slow system is given by $\dot{x} = -x^3$. Defining the new variable $y := z - h(x) = z + x$ and the fast time $\tau = t/\varepsilon$, it is easy to see that the fast system is given by $\dfrac{dy}{d\tau} = -y$. Now, it is quite natural to choose $V(x) = x^4/_4$ as an energy function for the slow system and $W(y) = y^2/_2$ as an energy function for the fast system. Assumptions (A1)–(A4) of Theorem 16-3 are satisfied with $\psi_1(x) = |x|^3$, $\psi_2(y) = |y|$, $\alpha_1 = 1$, $\alpha_2 = 1$, $\beta_1 = 1$, $\beta_2 = 1$, and $\gamma = 1$. The transversality condition of Theorem 16-9 is also trivially satisfied. So, all conditions of Theorem 16-9 are satisfied and $U_d = (1 - d)V + dW$, $0 < d < 1$ is a family of energy functions for the original system if $\varepsilon < \varepsilon^*_d = 1/2$, where

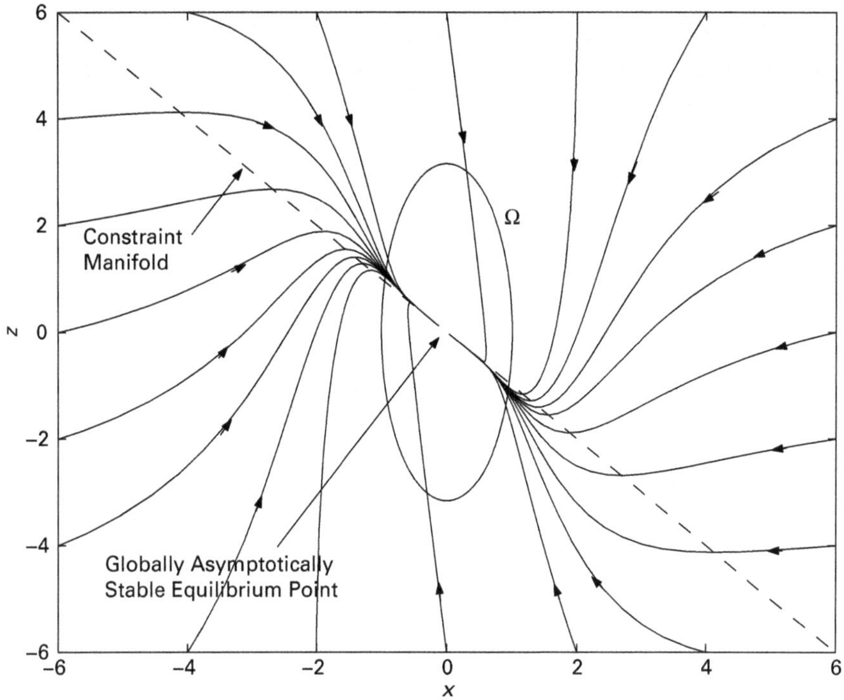

Figure 16.12 Phase portrait of the system of Example 16-4 for $\varepsilon = 0.1$. Every trajectory enters the set Ω in a finite amount of time. The set Ω is an estimate of the limit sets obtained using function $V = (x^2 + \varepsilon z^2)/2$.

ε^*_d is defined in Section 16.4. These energy functions allow us to conclude that the origin of system (16.31) is globally asymptotically stable if $\varepsilon < 1/2$. It is important to emphasize that fast–slow decomposition proves the global asymptotic stability of the origin for $\varepsilon < 1/2$.

The above procedure of fast–slow decomposition naturally leads us to construct a two-time-scale energy function, which can be used to assess the dynamical behavior of two-time-scale nonlinear systems in a global sense. We may, therefore, obtain improved results, such as improved characterization of limit sets.

Example 16-5 Consider the following singularly perturbed system:

$$\begin{aligned} \dot{x} &= 2x - x^3 + z \\ \varepsilon \dot{z} &= -x - z. \end{aligned} \tag{16.32}$$

The analytical results of Theorem 16-9 offer a family of energy functions for this example, leading to improved estimations of the stability region and an accurate characterization of its limit sets. This system contains only three equilibriums at (−1,1), (0,0) and (1,−1). The origin is unstable while the other points are stable. Again, it is quite natural to try

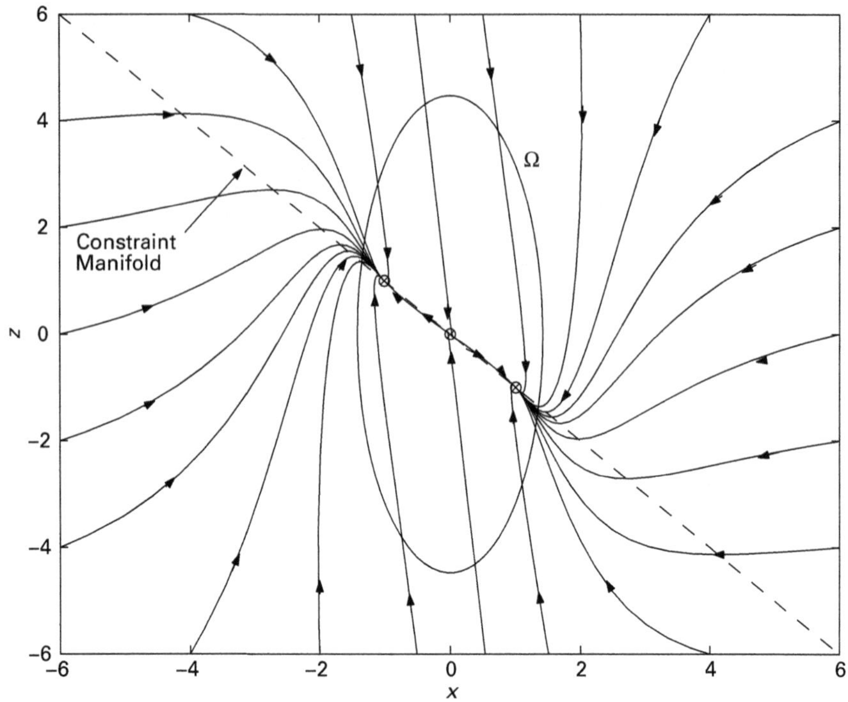

Figure 16.13 Phase portrait of the system of Example 16-5 for $\varepsilon = 0.1$. Every trajectory enters the set Ω in a finite amount of time. The set Ω is an estimate of the limit sets obtained using the function $V = (x^2 + \varepsilon z^2)/2$.

$$V(x, z) = \frac{x^2 + \varepsilon z^2}{2}$$

as a candidate energy function for the system. The derivative of V is given by

$$\dot{V} = -x^4 + 2x^2 - z^2$$

and so V is not an energy function for system (16.32). However, using arguments similar to those of the last example, we conclude that $\Omega = \{(x, z): V(x, z) \leq 1\}$ is the best estimate of the attractor that can be obtained with function V. The set Ω, as well as the phase portrait, is depicted in Figure 16.13. This estimation is uniform with respect to ε.

Now let us apply the idea of slow–fast decomposition to derive an energy function for this system. The slow system is given by $\dot{x} = x - x^3$ and the constraint manifold is given by $\Gamma_s = \{(x, z) \in R^2 : z = h(x) = -x\}$. Defining the new variable $y := z - h(x) = z + x$ and the fast time $\tau = t/\varepsilon$, it is easy to see that the fast system is given by $\dfrac{dy}{\partial\tau} = -y$. It is quite natural to choose $V(x) = x^4/_4 - x^2/_2 + 1/_4$ as an energy function for the slow system and $W(y) = y^2/_2$ as an energy function for the fast system. With these functions, the following estimates are obtained:

$$\begin{aligned}\frac{\partial V}{\partial x} f(x, h(x)) &\leq -\left|x^3 - x\right|^2 = -\psi_1^2(x) \\ \frac{\partial V}{\partial y} g(x, y) &\leq -y^2 = -\psi_2^2(x).\end{aligned} \tag{16.33}$$

It is straightforward to check that assumptions (A1)–(A4) of Theorem 16-3 and the conditions of Theorem 16-9 are satisfied. Therefore, $U = (1 - d)V + dW$, $0 < d < 1$ is a family of energy functions for system (16.32). These energy functions ensure that all the trajectories converge to the largest invariant set contained in the set $C = \{(x, y) \in D_1 \times D_2 : \psi_1(x) = 0, \psi_2(y) = 0\}$ that, in this case, is composed of the three equilibrium points, i.e. $\{(0, 0); (1, -1); (-1, 1)\}$.

Again, the composite energy function offers a deeper insight into the structure of limit sets. More precisely, the composite energy function allows us to conclude that the ω-limit sets are the three equilibrium points while the traditional approach just ensures that the ω-limit sets are contained in a bounded set.

The slow–fast decomposition also provides improved estimations of stability regions. Suppose one needs to estimate the stability region of the stable equilibrium point (−1,1). The traditional function V, although providing an estimate of the attractor, does not provide estimates of the stability region for the stable equilibrium points. One traditional alternative is to try a local quadratic Lyapunov function obtained from the linearized system in the neighborhood of the equilibrium point. Usually the stability region estimates obtained with these functions are very conservative compared to the exact stability region. The two-time-scale energy function provides more improved estimation results. Figure 16.14 shows an estimate for $\varepsilon = 0.1$ and $d = 0.02$ as well as the estimate obtained from a local quadratic Lyapunov function. It can be seen that the stability region estimate obtained with the two-time-scale energy function is much less conservative than that obtained via a quadratic function. Of course, the stability region estimate obtained with the two-time-scale energy function depends on the parameter $d \in (0, 1)$. Usually, there is a trade-off between maximizing the upper bound ε_d and the stability region estimate. In this example, d was determined using a trial and error procedure. Further research would be required to provide a guideline for optimizing the stability region estimate with respect to parameter d.

The exact stability boundary is also shown in Figure 16.14. The non-existence of a source indicates, according to Theorem 16-12, that the stability boundary is unbounded. More precisely, the boundary of the stability region is the stable manifold of the origin, corroborating the results of Theorem 16-11.

16.6.4 Applications to power systems

The analytical results derived in this chapter can be applied to two-time-scale systems that have energy functions. In this section, we illustrate one application to electrical power systems.

We consider the following simple power system model proposed in [221] and [231]:

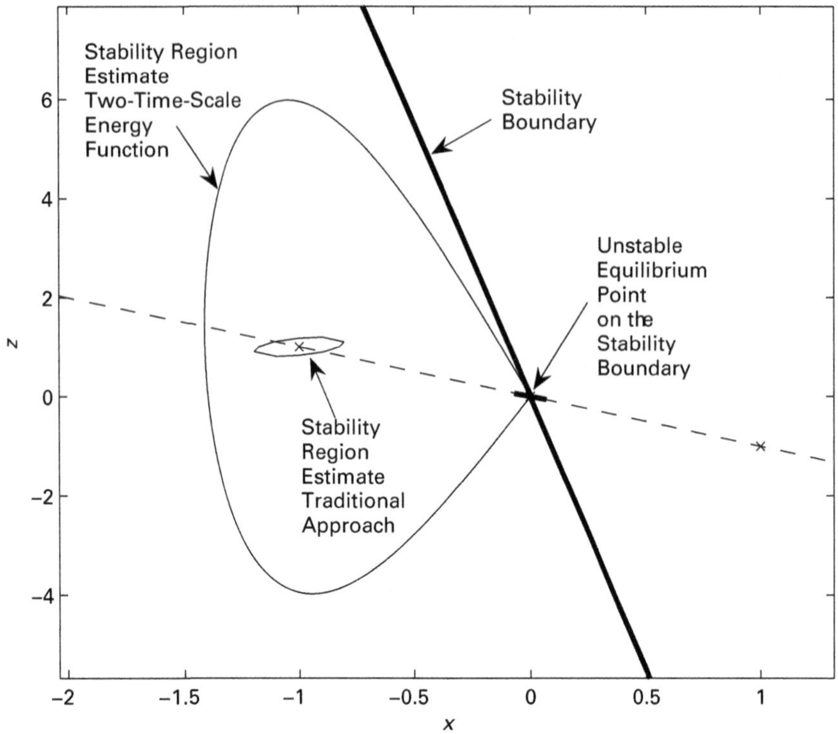

Figure 16.14 Stability region estimation of the stable equilibrium point (−1,1) of Example 16-5 with $\varepsilon = 0.1$. A more accurate estimate is obtained from the composite energy function U with $d = 0.02$ while a very conservative result is obtained using function $V = 0.275(x+1)2 - 0.225(x+1)(z-1) + 0.275(z-1)2$.

$$\begin{cases} \dot{E} = -\alpha E + b\cos\delta + E_{fd} \\ \varepsilon\dot{\delta} = \omega \\ \varepsilon\dot{\omega} = -\lambda\omega + P - cE\sin\delta. \end{cases} \tag{16.34}$$

It models a power system composed of one generator connected to an infinite bus through a transmission line. The task of estimating stability regions of power systems plays a key role in power system operation and planning. In this example, we compare three energy function approaches for estimating the stability region of the asymptotically stable equilibrium point (β,α,0). It will be shown that composite energy functions provide better estimates of the stability region compared to the conventional energy function.

Conventional approach

In power system literature, the following energy function is usually employed to estimate the stability region of (16.34):

$$V_{conv} = \varepsilon\frac{\omega^2}{2} - P\delta - cE\cos\delta + \frac{ca}{b}\frac{E^2}{2} - \frac{c}{b}E_{fd}E.$$

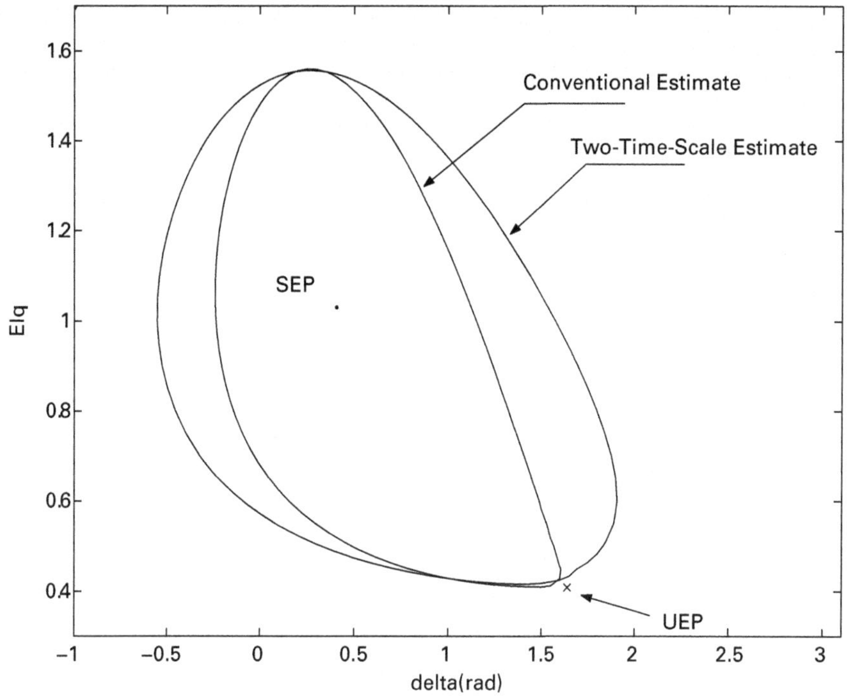

Figure 16.15 Intersection of the estimated stability region with the subspace $\{(E, \delta, \omega) \in R^3 : \omega = 0\}$. Comparison between the conventional and two-time-scale energy functions.

It is straightforward to show that

$$\dot{V} = -\lambda\omega^2 - \frac{c}{b}\dot{E}^2 \leq 0.$$

The optimal stability region estimate that can be obtained with this function is shown in Figure 16.15 for $\lambda = 4$, $P = 55.4$, $\varepsilon = 0.1515$, $a = 2.214$, $b = 1.214$, $c = 97.181$, and $E_{fd} = 1.22$. For these parameters, the asymptotically stable equilibrium point (SEP) is $(\beta,\alpha, 0) = (1.031, 0.4067, 0)$.

Two-time-scale energy function approach

Define the new variables $x := E - \beta$, $z_1 := \delta - \alpha$ and $z_2 := \omega$. In these new variables, the asymptoticaly stable equilibrium point $(\beta,\alpha, 0)$ is translated to the origin and system (16.34) assumes the form:

$$\begin{aligned}\dot{x} &= -ax + b[\cos((z_1 + \alpha) - \cos\alpha)]\\ \varepsilon\dot{z}_1 &= z_2\\ \varepsilon\dot{z}_2 &= -\lambda z_2 - c[(\beta + \mathrm{x})\sin(z_1 + \alpha) - \beta\sin\alpha].\end{aligned} \tag{16.35}$$

It is straightforward to obtain the slow subsystem:

$$\dot{x} = -ax + bN(x) \tag{16.36}$$

where $N(x)\colon = \cos(h_1(x) + \alpha) - \cos\alpha$,

$$h_1(x) = \sin^{-1}\left\{\frac{\beta \sin\alpha}{\beta + x}\right\} - \alpha,$$

and the fast subsystem

$$\begin{cases} \dfrac{dy_1}{d\tau} = y_2 \\ \dfrac{dy_2}{d\tau} = -\lambda y_2 - cM(x, y) \end{cases} \tag{16.37}$$

where $y_1\colon = z_1 - h_1(x)$, $y_2\colon = z_2$, $M(x, y) = (\beta + x)\sin(y_1 + h_1(x) + \alpha) - \beta \sin\alpha$.

We consider the following candidates for energy functions of the slow and fast systems, respectively,

$$V(x) = -\int_0^x [-a\sigma + bN(\sigma)]d\sigma$$

and

$$W(y_1, y_2, x) = [y_1\, y_2]\begin{bmatrix} 2 & \frac{1}{2} \\ \frac{1}{2} & \frac{\gamma}{2} \end{bmatrix}\begin{bmatrix} y_1 \\ y_2 \end{bmatrix} + \Upsilon c\int_0^{y_1} M(x, \sigma)d\sigma.$$

It can be shown that these candidates satisfy assumptions (A1)–(A4) of Theorem 16-3. As a consequence, the composite function $U = (1-d)V + dW$, $0 < d < 1$ is an energy function for system (16.35) for sufficiently small ε. The composite energy function U offers a better estimation of the stability region. The stability region estimate for $d = 0.01$ and $\gamma = 0.52$ is shown in Figure 16.15. It can be seen that the two-time-scale energy function provides an improved stability region estimate compared to the conventional energy function.

16.7 Concluding remarks

We have developed a comprehensive characterization of the stability boundary of two-time-scale dynamical systems and explored the two-time-scale property to derive its relationship with the stability boundaries and stability regions of slow and fast systems. The derived characterization shows that subsets of the stability region of the two-time-scale system can be sufficiently approximated by subsets of the stability boundary and stability region of slow and fast systems. It is also shown that type-one equilibrium points play an important role in the characterization of the stability boundary of two-time-scale systems. From a computational viewpoint, we have shown that the task of checking whether a type-one equilibrium point belongs to the stability boundary of the two-time-scale system can be decomposed into the simpler tasks of checking whether the same equilibrium is on the stability boundary and/or stability region of the slow and fast systems.

We have also developed generalizations of well-known stability results in the literature of singularly perturbed nonlinear systems. It was shown that global dynamic analysis of two-time-scale nonlinear systems can be decomposed into analysis of the slow and fast subsystems. One implication of this generalization is that the existing body theory of the stability region can be extended to two-time-scale nonlinear systems. We have derived composite energy functions and have shown that improved estimates of stability regions for two-time-scale nonlinear systems can be obtained. We have developed a global result for a class of two-time-scale nonlinear systems, asserting that limit sets cannot contain complicated behavior such as closed orbits (limit cycles), quasi-periodic trajectories and chaotic motions.

A complete characterization of the stability boundary of two-time-scale nonlinear systems admitting an energy function has been derived. These analytical results confirm our belief that by exploring the fast–slow properties of two-time-scale nonlinear systems, one can gain more insight into system dynamics, obtain improved estimations of stability regions, and derive more accurate characterizations of their limit sets.

16.8 Supplementary discussion and bibliographical notes

Most of the literature on two-time-scale systems is concerned with dynamics in the neighborhood of a stable component of the so-called algebraic manifold (or equilibrium manifold). Tikhonov explored the uniform stability of the fast system and some regularity of the vector field to derive conditions for ensuring the proximity between solutions of singularly perturbed systems and solutions of the slow and fast subsystems in the neighborhood of a stable component of the algebraic manifold [173]. This theory was further extended in [259,260]. Tikhonov's theorem was extended to differential inclusions in [261]. Stability analysis of a class of singularly perturbed systems was decomposed into the stability analysis of slow and fast systems using Lyapunov-based methods in [143]. Local exponential stability of equilibriums in the stable component of the algebraic manifold was studied in [103] using Tikhonov's approach and in [75] via Lyapunov theorems in terms of exponential stability of the fast and slow subsystems. Near asymptotic stability of the slow system was studied in [17] independently of the stability of the fast system.

The composite approach usually leads to less conservative estimates of stability and the stability region, when compared to traditional Lyapunov functions. The idea of composite Lyapunov functions was further developed in the design of composite controllers [146], including the design of robust composite controllers [37,74].

17 Stability regions for a class of non-hyperbolic dynamical systems: theory and estimation

17.1 Introduction

There is a rich history of the development of stability theory for nonlinear dynamical systems based on generic properties. Comparatively, there has traditionally been much less attention devoted to the analysis of non-hyperbolic dynamical systems because the dynamics of these systems are more complex and difficult to analyze mathematically.

In this chapter, we study the stability regions of a class of non-hyperbolic dynamical systems. In particular, we present a fairly comprehensive theory of the stability regions for the class of non-hyperbolic dynamical systems. Several necessary and sufficient conditions for an equilibrium manifold (the generalized concept of an equilibrium point) to lie on the stability boundary will be presented. A complete characterization of the stability boundary is derived; in particular, it is shown that the stability boundary equals the union of stable manifolds of equilibrium manifolds on the stability boundary. An effective scheme for estimating stability regions of stable equilibrium manifolds by using energy functions will be presented. On the application side, a systematic method for finding multiple separate feasible regions of the so-called *constrained satisfaction programming* problem can be developed.

17.2 Definitions and notation

We consider a class of *non-hyperbolic* dynamical systems described by systems of ordinary differential equations of the form

$$\frac{dx}{dt} = F(x) : \; = M(x)H(x), \; F : R^n \to R^n \tag{17.1}$$

where H: $R^n \to R^m$ and M: $R^n \to R^{n \times m}$, $m \leq n$, are assumed to be smooth to ensure the existence and uniqueness of solutions. The solution of (17.1) starting from x at $t = 0$ is called an *orbit* (or *trajectory*), denoted by $\phi(t, x)$: $R \to R^n$. Note that for the vector field F of (17.1), the vector field V_F: $R^n \to R^n$ defined by

$$\frac{dx}{dt} = V_F(x) = \frac{F(x)}{1 + \|F(x)\|} \tag{17.2}$$

is complete and both vector fields, F and V_F, are topologically equivalent. Hence we will assume, without loss of generality, that the vector field of (17.1) is complete (i.e. $\phi(t, x)$ is defined for all $t \in R$ for any $x \in R^n$). Let E_F: $= F^{-1}(0) = M^{-1}(0) \cup H^{-1}(0)$ be the zero sets of F. Then, by Sard's theorem, it is a *generic* property that $M^{-1}(0) \cap H^{-1}(0) = \varnothing$ and E_F is a smooth manifold where $H^{-1}(0)$: $= \{x \in R^n$: $H(x) = 0\}$ and $M^{-1}(0)$: $= \{x \in R^n$: $M(x) = 0\}$.

DEFINITION
A path-connected component of $F^{-1}(0)$ is called an equilibrium manifold *of system (17.1).*

DEFINITION
An equilibrium manifold Σ of system (17.1) is called stable *if, for each $\varepsilon > 0$, there is $\delta = \delta(\varepsilon) > 0$ such that*

$$x \in B_\delta\left(\sum\right) \Rightarrow \phi(t,x) \in B_\varepsilon\left(\sum\right) \forall t \in R^+$$

and is called asymptotically stable, *if it is stable and δ can be chosen such that*

$$x \in B_\delta\left(\sum\right) \Rightarrow \lim_{t \to \infty} \phi(t,x) \in \sum$$

where $B_\delta(\Sigma) = \{x \in R^n$: $\|x - y\| < \delta$, $\forall y \in \Sigma\}$. An equilibrium manifold Σ is called unstable *if it is not stable.*

DEFINITION
An l-dimensional equilibrium manifold Σ of (17.1) is called pseudo-hyperbolic *if for each $x \in \Sigma$, the Jacobian of $F(\cdot)$ at x, denoted by $J_F(x)$, has no eigenvalues with a zero real part on the normal space $N_x(\Sigma)$, the orthogonal complement of the tangent space $T_x(\Sigma)$, of Σ at x in R^n and there exists an $\varepsilon > 0$ such that $\phi_{-\infty}$: $B_\varepsilon(\Sigma) \rightarrow \Sigma$, defined by $\phi_{-\infty}(x) = lim_{t \to -\infty}\phi(t, x)$, is locally homeomorphic to a projection from R^n to R^l.*

DEFINITION
An l-dimensional pseudo-hyperbolic equilibrium manifold Σ is termed a type-*k* equilibrium manifold *if for each $x \in \Sigma$, the corresponding Jacobian $J_F(x)$ has exactly k eigenvalues with positive real part on $N_x(\Sigma)$.*

DEFINITION
A pseudo-hyperbolic equilibrium manifold Σ is called a (asymptotically) stable equilibrium manifold *if, for each $x \in \Sigma$, all the eigenvalues of its corresponding Jacobian on $N_x(\Sigma)$ have negative real parts.*

It is clear that if an asymptotically stable equilibrium manifold is pseudo-hyperbolic, then it is a stable equilibrium manifold. On the other hand, a pseudo-hyperbolic stable equilibrium manifold is also asymptotically stable. Hence, these two terms, stable equilibrium manifold and asymptotically stable equilibrium manifold, will be used interchangeably.

For a pseudo-hyperbolic equilibrium manifold Σ, we can decompose the normal space $N_x(\Sigma)$ of Σ at x uniquely as a direct sum of two subspaces and the tangent space $E_x^s \oplus E_x^u \oplus T_x(\Sigma)$ such that each subspace is invariant under the linear operator $DF(x)$. The eigenvalues of $DF(x)$ restricted to *the stable subspace* E_x^s have negative real parts, and the eigenvalues of $DF(x)$ restricted to *the unstable subspace* E_x^u have positive real parts. Let the dimension of E_x^u be k and the dimension of E_x^s be $n - l - k$. Then we can express such subspace as the following:

$$\begin{aligned} E_x^u &= \text{span}\{e_1, e_2, \ldots, e_k\} \\ E_x^s &= \text{span}\{e_{k+1}, e_{k+2}, \ldots, e_{n-1}\} \end{aligned}$$

where $e_1, e_2, \ldots, e_k$ are the k (generalized) eigenvectors whose eigenvalues have positive real parts, and $e_{k+1}, e_{k+2}, \ldots, e_{n-l}$ are the $n - l - k$ (generalized) eigenvectors whose eigenvalues have negative real parts.

The *stable and unstable manifolds* $W^s(\Sigma)$, $W^u(\Sigma)$ of the equilibrium manifold Σ are defined as follows:

$$\begin{aligned} W^s(\Sigma) &= \left\{x \in R^n : \lim_{t\to\infty} \phi(t,x) \in \Sigma\right\} \\ W^u(\Sigma) &= \left\{x \in R^n : \lim_{t\to\infty} \phi(t,x) \in \Sigma\right\}. \end{aligned}$$

If Σ is a type-k equilibrium manifold, then the dimensions of $W^u(\Sigma)$ and $W^s(\Sigma)$ are $l + k$ and $n - k - l$, respectively. The stable and unstable manifolds need not be embedded submanifolds of R^n since they may wind around in a complex manner approaching themselves arbitrarily closely [31,108,201]. Note that

$$\begin{aligned} W^s(\Sigma) &= \bigcup_{x\in\Sigma} W^s(x) \\ W^u(\Sigma) &= \bigcup_{x\in\Sigma} W^u(x). \end{aligned}$$

The idea of transversality is basic to the study of dynamical behaviors of nonlinear systems. If Σ_i and Σ_j are l-dimensional pseudo-hyperbolic equilibrium manifolds, we say that their intersection satisfies the *transversality condition* if either (i) they do not intersect at all, or (ii) $W^u(\Sigma_i) \cap W^s(\Sigma_j) \neq \varnothing$ implies

$$\dim\left(W^u(\Sigma_i) \cap W^s(\Sigma_j)\right) \geq 1 + 1$$

and

$$T_x(W^u(\Sigma_i)) \oplus T_x\left(W^s(\Sigma_j)\right) = R^n, \ \forall x \in W^u(\Sigma_i) \cap W^s(\Sigma_j).$$

For a stable equilibrium manifold, it can be shown that there exists a number $\varepsilon > 0$ such that $x \in B_\varepsilon(\Sigma) \Rightarrow \lim_{t\to\infty} \phi(t, x) \in \Sigma$. If such an ε is arbitrarily large, then Σ is called a *globally stable equilibrium manifold*; otherwise, there exists a stability region of the stable equilibrium manifold.

DEFINITION

The stability region of a stable equilibrium manifold Σ_s *is defined as*

$$A(\Sigma_s) := \left\{x \in R^n : \lim_{t\to\infty} \phi(t,x) \in \Sigma_s\right\}.$$

The boundary of the stability region $A(\Sigma_s)$ is called the *stability boundary* of Σ_s and will be denoted by $\partial A(\Sigma_s)$. The stability boundary is topologically an $(n-1)$-dimensional closed, invariant set.

From a topological point of view, the stability region $A(\Sigma_s)$ is an open, invariant, and connected set which is diffeomorphic to R^n. We will present a comprehensive theory of the stability region and the stability boundary for the class of *non-hyperbolic* dynamical systems (17.1). Several necessary and sufficient conditions for an equilibrium manifold lying on the stability boundary of non-hyperbolic dynamical systems (17.1) will be derived. The stability boundary will be completely characterized and shown to consist of the union of the stable manifolds of the unstable equilibrium manifolds on the stability boundary. An effective scheme for estimating the stability region of a stable equilibrium manifold by using an energy function will be presented.

17.3 Energy functions

This section introduces the notion of the energy functions and develops analytic results for non-hyperbolic dynamical systems (17.1). The analytical results derived in this section, which are similar to those derived for hyperbolic dynamical systems presented in Chapter 3, will be applied to characterize the stability boundary of a stable equilibrium manifold for the class of non-hyperbolic dynamical systems.

Energy functions related to non-hyperbolic dynamical systems (17.1) are defined as follows.

DEFINITION

A function $V: R^n \rightarrow R$ *is said to be an* energy function *for system (17.1) if the following three conditions (E1), (E2), (E3) or (E1), (E2), (E3′) are satisfied.*

(E1) $F(x) = 0$ *implies* $\nabla V(x) = 0$.

(E2) Along any nontrivial trajectory $\phi(t, x_0)$ (i.e. $x_0 \notin E_F$; *i.e. any trajectory that does not lie in the equilibrium manifold*)

$$\frac{d}{dt}V(\phi(t,x_0)) \leq 0$$

and the set

$$\{t \in R : \dot{V}(\phi(t,x_0)) = 0\}$$

has measure zero in R.

(E3) If $\{V(\phi(t, x_0)): t \geq 0\}$ *is bounded, then* $\{\phi(t, x_0): t \geq 0\}$ *is bounded.*

or

(E3′) For any given interval $[a, b]$ and for any closed subset S of $V^{-1}([a, b])$ which does not contain equilibrium manifolds,

$$\sup\left\{\frac{\nabla V(x)^T F(x)}{||F(x)||} : x \in S\right\} < 0.$$

Remarks

[1] Properties (E1) and (E2) imply that the energy function strictly decreases along any nontrivial trajectory. Property (E3) means that along any system trajectory, the function V is proper, i.e. the inverse image of any compact set is compact, but this V need not be proper in the entire state space.

[2] The property (E3′) is a non-traditional one, but is very useful from a practical point of view since a large class of non-hyperbolic dynamical systems satisfies this property.

Now consider the following set $S(k)$:

$$S(k) = \{x \in R^n : V(x) < k\}.$$

Generally speaking, the set $S(k)$ can be very complicated with several different components, i.e. the level set $S(k)$ can be decomposed into disjoint path-connected components, such as

$$S(k) = \underset{j=1}{\cup} S^j(k)$$

where $S^j(k)$ are disjoint path-connected components which are positively invariant sets.

We next show that in spite of the possibility that a constant energy surface may contain several disjoint connected components, there is an interesting relationship between the constant energy surface and the stability region. In this relationship, at most one connected component of the constant energy surface $S(r)$ has a non-empty intersection with the stability region $A(\Sigma_s)$ as shown in the following theorem.

THEOREM 17-1 **(Constant energy surface and stability region)**
Suppose that there exists an energy function for the class of non-hyperbolic dynamical systems (17.1). Let Σ be an equilibrium manifold and Σ_s be a stable equilibrium manifold. Then

(a) $V(x) =$ constant for all $x \in \Sigma$,
(b) the set $S(r)$ contains only one connected component, say, $S(r, \Sigma_s)$ which has a non-empty intersection with $A(\Sigma_s)$ if and only if $r > V(\Sigma_s)$.

Proof We consider any point $x_0 \in \Sigma$ and let $x_1 \in \Sigma$. Then by the path connectedness of the equilibrium manifold Σ, there exists a path $c: R \rightarrow \Sigma$ such that $c(0) = x_0$ and $c(1) = x_1$. Since $F(c(s)) = 0$ for all s, we have $\nabla V(c(s)) = 0$ by (E1) and so

$$\frac{d}{ds}V(c(s)) = \nabla V(c(s))^T \frac{d}{ds}c(s) = 0.$$

Hence $V(c(s)) =$ constant for all s, which implies that $V(x_0) = V(c(0)) = V(c(1)) = V(x_1)$. Therefore, $V(x) =$ constant for all $x \in \Sigma$. It is obvious from the definition of $S(r)$ that $S(r)$ has a non-empty intersection with $A(\Sigma_s)$ if and only if $r > V(\Sigma_s)$. To prove uniqueness, we will proceed with a contradictory argument. Suppose that there is another component $S^0(r)$ that has a non-empty intersection with $A(\Sigma_s)$. Let $y \in S^0(r) \cap A(\Sigma_s)$. Then $S^0(r)$ is an invariant set and the trajectory starting from y approaches Σ_s. Therefore, this is a contradiction, as $S(r, \Sigma_s)$ must also intersect $S^0(r)$. This completes the proof.

In general, trajectories of general nonlinear dynamical systems can be very complicated, unless the underlying dynamical systems have some special properties. One nice property of non-hyperbolic dynamical systems which have an energy function is that the asymptotic behavior of the system trajectories on the stability boundary consists solely of the equilibrium manifold. This property is a consequence of the next theorem.

THEOREM 17-2 **(Asymptotic behavior on the stability boundary)**
Suppose that there exists an energy function for system (17.1) and all the equilibrium manifolds are pseudo-hyperbolic. Then every trajectory on $\partial A(\Sigma_s)$ of system (17.1) is bounded and converges to one of the equilibrium manifolds.

Proof First we will show that every bounded trajectory converges to one of the equilibrium manifolds. Let $x_0 \in \partial A(\Sigma_s)$ and $\phi(t, x_0)$ be a bounded trajectory starting at $x_0 \in R^n$. Since

$$\frac{\partial}{\partial t} V(\phi(t, x_0)) < 0$$

$g(t) = V(\phi(t, x_0))$ is a non-increasing function of t. Since $\phi(t, x_0)$ is a bounded trajectory, $\phi(t, x_0)$ is complete and $\{ \phi(t, x_0): t \geq 0\}$ is non-empty and compact, so g is bounded from below because V is continuous. Hence $g(t)$ has a limit a as $t \to \infty$. Let $\omega(x_0)$ be the ω-limit set of x_0. Then for any $p \in \omega(x_0)$, there exists a sequence $\{t_n\}$ with $t_n \to \infty$ and $\phi(t, x_0) \to p$ as $n \to \infty$. By the continuity of V, $V(p) = \lim_{n \to \infty} V(\phi(t_n x_0)) = a$. Hence, $V(p) = a$, for all $p \in \omega(x_0)$. Since $\omega(x_0)$ is an invariant set,

$$\dot{V}(x) = 0 \text{ on } \omega(x_0)$$

or, equivalently, $F(\omega(x_0)) = 0$ by conditions (E1) and (E2). Since every bounded trajectory converges to its ω-limit set and $\phi(t, x_0)$ is bounded, $\phi(t, x_0)$ approaches $\omega(x_0)$ as $t \to \infty$. Hence it follows that every bounded trajectory of system (17.1) converges to one of the equilibrium manifolds.

Let $x_0 \in \partial A(\Sigma_s)$ and $\phi(t, x_0)$ be a trajectory starting at $x_0 \in R^n$. Since V is bounded from below by $V(\Sigma_s)$, the set $\{V(\phi(t, x_0)): t \geq 0\}$ is bounded. Hence, if condition (E3) holds, then $\phi(t, x_0)$ is bounded and converges to one of the equilibrium manifolds. Therefore we only need to prove that $\phi(t, x_0)$ converges to one of the equilibrium manifolds under the condition (E3′). This will be proved by contradiction. Suppose that the condition (E3′) holds and $\phi(t, x_0)$, $t \geq 0$ is an unbounded trajectory starting at $x_0 \in \partial A(\Sigma_s)$ which does not converge to one of the equilibrium manifolds. Since $\{V(\phi(t, x_0)): t \geq 0\}$ is bounded, we have $\phi(t, x_0) \in V^{-1}([a,b])$ for some $a, b \in R$ for all $t \geq 0$.

Note that since all of the equilibrium manifolds are isolated, the set $E_F \cap V^{-1}([a,b])$ consists of at most a finite number of equilibrium manifolds, say, $\Sigma_1, \ldots, \Sigma_m$. (If there are an infinite number of equilibrium manifolds, say, $\Sigma_1, \Sigma_2, \ldots$ in $V^{-1}([a,b])$, then we can choose $\varepsilon_i > 0, i = 1, 2, \ldots$ such that the $B_{\varepsilon i}(\Sigma_i)$ are pair-wise disjoint subsets of $V^{-1}([a,b])$ and

$$\sup_{x \in (V^{-1}([a,b]) \backslash B_{\varepsilon_i}(\Sigma_i))} \frac{\|\nabla V(x)^T F(x)\|}{\|F(x)\|} < \frac{1}{i}.$$

Since $K = V^{-1}([a,b]) \backslash \cup B_{\varepsilon_i}(\Sigma_i))$ is a closed subset of $V^{-1}([a,b])$ with $\sup\{\nabla V(x)^T F(x)/\|F(x)\| : x \in K\} = 0$, this leads to a contradiction.)

Now suppose that $\omega(x_0) \cap \Sigma_i = \varnothing$ for all $i = 1, \ldots, m$. Then, since all the equilibrium manifolds are isolated and $\phi(t, x_0)$ does not converge to one of the equilibrium manifolds, there exists an $\varepsilon_k < 0$, $k = 1, \ldots, m$ such that the $B_{\varepsilon_k}(\Sigma_k)$ are pair-wise disjoint and $\phi(t, x_0) \in S := (R^n \backslash \cup B_{\varepsilon_k}(\Sigma_k)) \cap V^{-1}([a,b])$ for all $t > t_0$. Since S is a closed subset of $V^{-1}([a,b])$ which does not contain equilibrium manifolds, there exists a $\delta > 0$ such that the following holds:

$$\nabla V(\phi(t, x_0))^T F(\phi(t, x_0)) < -\delta \|F(\phi(t, x_0))\|, \ \forall t > t_0.$$

Then we have

$$\begin{aligned}
V(\phi(t, x_0)) - V(\phi(t_0, x_0)) &= \int_{t_0}^{t} \dot{V}(\phi(\tau, x_0)) d\tau \\
&= \int_{t_0}^{t} \nabla V(\phi(\tau, x_0))^T F(\phi(\tau, x_0)) d\tau \\
&< -\delta \int_{t_0}^{t} \|F(\phi(\tau, x_0))\| d\tau \\
&< -\delta \left\| \int_{t_0}^{t} \frac{\mathrm{d}}{d\tau} \phi(\tau, x_0)) d\tau \right\| \\
&< -\delta \sum_i \|\phi(t, x_0) - \phi(t_0, x_0)\|
\end{aligned}$$

Since $\phi(t, x_0)$ is unbounded, the above inequality shows that

$$\lim_{t \to \infty} V(\phi(t, x_0)) = -\infty$$

which contradicts the fact that $V(\cdot)$ is bounded from below. Therefore at least one of the elements, say Σ_1, has a non-empty intersection with $\omega(x_0)$, i.e. $\omega(x_0) \cap \Sigma_1 \neq \varnothing$. Now choose an $\varepsilon > 0$ such that $B_{2_\varepsilon}(\Sigma_1) \cap E_F = \Sigma_1$. Since $\phi(t, x_0)$ is an unbounded trajectory starting at $x_0 \in R^n$ which does not converge, there exist strictly increasing sequences $\{t_i\}$, $\{s_i\}$ such that t_i is the ith time to enter $\partial B_\varepsilon(\Sigma_1)$ and s_i is the first time after t_i to enter $\partial B_{\varepsilon/2}(\Sigma_1)$ where $t_{i+1} > s_i > t_i$. If we choose a $\delta > 0$ such that $\nabla V(x)^T F(x) < \delta \|F(x)\|$ for all $x \in cl\,(B_\varepsilon(\Sigma_1) / B_{\varepsilon/2}(\Sigma_1))$, then we have,

$$
\begin{aligned}
V(\phi(t,x_0)) - V(\phi(0,x_0)) &= \int_{t_0}^{t} \dot{V}(\phi(\tau,x_0))dr \\
&< \sum_i \int_{t_i}^{s_i} \nabla V(\phi(\tau,x_0))^T \times F(\phi(\tau,x_0))d\tau \\
&< -\delta \sum_i \int_{t_i}^{s_i} \|F(\phi(\tau,x_0))\| d\tau < -\delta \sum_i \left\| \int_{t_i}^{s_i} \frac{\mathrm{d}}{d\tau}\phi(\tau,x_0)d\tau \right\| \\
&< -\delta \sum_i \|\phi(s_i,x_0) - \phi(t_i,x_0)\| \\
&= -\delta \sum_i \left(\frac{\varepsilon}{2}\right)
\end{aligned}
$$

This shows that $\lim_{t\to\infty} V(\phi(t, x_0)) = -\infty$ which contradicts the fact that $V(\cdot)$ is bounded from below. Therefore, $\phi(t, x_0)$ is bounded and converges to one of the equilibrium manifolds. This completes the proof.

17.4 Characterization of the stability boundary

This section develops a fairly comprehensive theory of the stability boundary for a class of non-hyperbolic dynamical systems. To this end, we will first derive necessary and sufficient conditions which an equilibrium manifold must possess to be on the stability boundary of a stable equilibrium manifold. We will then show that under fairly general assumptions, the stability boundary is the union of the stable manifolds of the equilibrium manifolds on the stability boundary.

We make the following four basic assumptions concerning a class of the vector field of (17.1) for which the stability boundary of a stable equilibrium manifold Σ_s can be completely characterized.

Assumptions

(N1) If an equilibrium manifold Σ has a non-empty intersection with $\partial A(\Sigma_s)$, then $\Sigma \subset \partial A(\Sigma_s)$.

(N2) All the equilibrium manifolds on $\partial A(\Sigma_s)$ are pseudo-hyperbolic, and have the same dimensions.

(N3) The stable and unstable manifolds of equilibrium manifolds on $\partial A(\Sigma_s)$ satisfy the transversality condition.

(N4) There exists an *energy function* $V: R^n \to R$ for system (17.1).

Remarks

[1] It can be shown that assumptions (N1)–(N3) are generic for the case $m = 1$ using the same techniques as in [200,201].

[2] Assumption (N4) rules out complicated dynamics, such as periodic solutions, horseshoes or homoclinic trajectories which can exist in the state space of system (17.1). It implies that the asymptotic behavior of the system trajectories on the

stability boundary consists solely of the equilibrium manifold as shown in Theorem 17-2. This assumption is not generic, but nonlinear dynamical systems satisfy this assumption if they admit a dissipating Lyapunov function.

Assumptions (N1)–(N4) are assumed to hold in the subsequent sections unless stated otherwise. Now we present two key theorems which characterize an equilibrium manifold lying on a stability boundary.

THEOREM 17-3 **(Characterization in terms of an unstable manifold)**
Let $\Sigma \neq \Sigma_s$ be an l-dimensional pseudo-hyperbolic, isolated equilibrium manifold of the non-hyperbolic dynamical system (17.1) and suppose that assumptions (N1)–(N4) hold. Then the following are equivalent.

(a) $W^u(\Sigma) \cap A(\Sigma_s) \neq \varnothing$.
(b) $W^u(y_0) \cap A(\Sigma_s) \neq \varnothing$ *for all* $y_0 \in \Sigma$.
(c) $\Sigma \subset \partial A(\Sigma_s)$.

Proof For the main proof, we need the following two results which characterize the equilibrium manifold on the stability boundary. The first lemma gives a consequence of the transversal intersection property and the second lemma is a generalized version of the λ-lemma [201].

LEMMA 17-4 *Let Σ_i, Σ_j be l-dimensional pseudo-hyperbolic equilibrium manifolds of (17.1). Suppose that $W^u(\Sigma_i) \cap W^s(\Sigma_j) \neq \varnothing$ and that the intersection is transversal. Then* $\dim W^u(\Sigma_i) > \dim W^u(\Sigma_j)$

Proof Let $\dim(W^u(\Sigma_i)) = l + u$ and $\dim(W^s(\Sigma_j)) = l + s$. Since $W^u(\Sigma_i) \cap W^s(\Sigma_j) \neq \varnothing$, from the transversality condition, we have

$$\dim(W^u(\Sigma_i) \cap W^s(\Sigma_j)) = (l+u) + (l+s) - n \geq 1 + 1$$

or

$$s \geq 1 - u - 1 + n.$$

Since $\dim W^s(\Sigma_j) + \dim W^u(\Sigma_j) = n + l$, we have

$$\dim(W^u(\Sigma_j)) = n - s \leq 1 + u - 1 < \dim W^u(\Sigma_i).$$

LEMMA 17-5 **(λ-lemma)**
Let Σ be an l-dimensional pseudo-hyperbolic equilibrium manifold of (17.1) with $\dim W^u(\Sigma) = l + u$. *Let D be an $(l + u)$-disk in $W^u(\Sigma)$. Let N be an $(l + u)$-disk having a point of transversal intersection with $W^s(\Sigma)$. Then D is contained in the closure of the set $\cap_{t \geq 0} \varnothing(t, N)$.*

We now return to the proof of Theorem 17-3. We first consider the case for $l = 0$. In this case, an equilibrium manifold becomes an equilibrium point and the theorem can be proved as in Chapter 4. For the rest of the proof, we will assume $l \geq 1$.

(a) $\Rightarrow$ (b): Since $W^u(\Sigma) = \cup_{y_0 \in \Sigma} W^u(y_0)$ where the $W^u(y_0)$ are relatively disjoint over $y_0 \in \Sigma$,

$$\bigcup_{y_0 \in \Sigma} (W^u(y_0) \cap A(\Sigma_s)) = W^u(\Sigma) \cap A(\Sigma_s) \neq \varnothing.$$

Hence $W^u(y_0) \cap A(\Sigma_s) \neq \varnothing$ for some $y_0 \in \Sigma$.

Next we define set S by

$$S = \{x \in \Sigma : W^u(x) \cap A(\Sigma_s) \neq \varnothing\}.$$

We will show that S is open in Σ. Since $\Sigma \in V^{-1}(c)$ for some $c \in R$, $W^u(\Sigma) \in V^{-1}((-\infty, c])$. Now we can choose a sufficiently small $\delta > 0$ such that $U = W^u(\Sigma) \cap V^{-1}((c-\delta, c])$ is a neighborhood of Σ in $W^u(\Sigma)$ and

$$\partial U = W^u(\Sigma) \cap V^{-1}(c\text{–}\delta) = \bigcup_{y \in \Sigma} \partial U_y$$

is a submanifold in $W^u(\Sigma)$ (which is called a *fundamental domain*) where $\partial U_y = W^u(y) \cap V^{-1}((c-\delta)$ is a compact set for each $y \in \Sigma$. Then we have

$$W^u(\Sigma)\backslash\Sigma = \bigcup_{y_0 \in \Sigma} (W^u(y_0)\backslash\{y_0\}) = \bigcup_{t \in R} \phi(t, \partial U)$$

or

$$W^u(y)\backslash\{y\} = \bigcup_{t \in R} \phi(t, \partial U_y) \quad \forall y \in \Sigma.$$

Since for each $y \in \Sigma$, the set $\{\phi(t, y): t \leq 0\}$ is bounded, we can define a continuous, surjective, open mapping $\phi_{-\infty}: U \to \Sigma$ which leaves Σ fixed. (Note that $\phi_{-\infty}$ is locally homeomorphic to a projection map from R^{l+u} to R^l, which is an open mapping.) Now let $x \in S$ be arbitrarily given. Then $\exists z \in W^u(x) \cap A(\Sigma_s) \cap V^{-1}((c-\delta, c])$. Since $A(\Sigma_s)$ is open, $\exists \varepsilon > 0$ such that $B_\varepsilon(z) \subset A(\Sigma_s)$. Let $N = U \cap B_\varepsilon(z)$. Then $\phi_{-\infty}(N)$ is open in Σ since N is open in U and $\phi_{-\infty}$ is an open mapping. Therefore, $W^u(y) \cap A(\Sigma_s) \neq \phi$ for all $y \in B_\gamma(x) \cap \Sigma$ for some $\gamma > 0$ and so S is open in Σ. The openness of $(\Sigma \setminus S)$ can be similarly proved. Since Σ is connected and $S \neq \varnothing$, we should have $\Sigma = S$. Therefore $W^u(x) \cap A(\Sigma_s) \neq \varnothing$ for all $x \in \Sigma$.

(b) ⇒ (c): Let $y_0 \in \Sigma$ be arbitrarily given. Choose a $y \in W^u(y_0) \cap A(\Sigma_s)$. Then

$$\lim_{t \to -\infty} \phi(t, y) = y_0.$$

But since $\overline{A(\Sigma_s)}$ is invariant and closed, we have

$$y_0 \in \overline{A(\Sigma_s)}.$$

Since y_0 cannot be in the stability region $A(\Sigma_s)$, we have

$$y_0 \in \partial A(\Sigma_s).$$

Therefore, $y_0 \in \partial A(\Sigma_s)$ for all $y_0 \in \Sigma$ and so $\Sigma \subset \partial A(\Sigma_s)$.

(c) ⇒ (a): Let $\Sigma \subset \partial A(\Sigma_s)$. Choose a sufficiently small $\delta > 0$ such that $U = W^u(\Sigma) \cap V^{-1}((c-\delta, c]))$ where $\Sigma \in V^{-1}(c)$ for some $c \in R$ as above. Let $N_\varepsilon(\partial U)$ be the ε-neighborhood of ∂U in R^n. Then $\cup_{t<0} \phi(t, N_\varepsilon(\partial U))$ contains a set of the form $O \setminus W^s$, where O is a neighborhood of Σ in R^n. Since $\Sigma \subset \partial A(\Sigma_s)$, $O \cap A(\Sigma_s) \neq \varnothing$ and $W^s(\Sigma) \cap A(\Sigma_s) = \varnothing$. Therefore, we have

$$\cup_{t<0}\phi(t, N_\varepsilon(\partial U)) \cap A(\Sigma_s) = (O \backslash W^s(\Sigma_s)) \cap A(\Sigma_s) \neq \varnothing.$$

This implies that $N_\varepsilon(\partial U) \cap \phi_t(A(\Sigma_s)) = \varnothing$ for some t. Since $A(\Sigma_s)$ is invariant, we have $N_\varepsilon(\partial U) \cap A(\Sigma_s) = \varnothing$. Since $\varepsilon > 0$ is arbitrary and ∂U is a closed set, ∂U contains at least a point of $\overline{A(\Sigma_s)}$. Therefore, there exists a point $y \in (W^u(\Sigma) \backslash \Sigma) \cap \overline{A(\Sigma_s)}$. If $y \in A(\Sigma_s)$, the proof is complete. If $y \in A(\Sigma_s)$, by assumption (N1) and Theorem 17-1, there exists an equilibrium manifold $\Sigma_1 \in A(\Sigma_s)$ such that $y \in (W^s(\Sigma_1) \backslash \Sigma_1)$. Let the type of Σ and Σ_1 be $(l + r)$, $(l + u)$, respectively. By assumption (N3), $W^u(\Sigma) \cap W^s(\Sigma_1) \neq \varnothing$ and the intersection is transverse. Hence, by assumption (N2) and Lemma 17-4, $r > u$. Now, consider two cases.

(i) If $r = 1$, then $u = 0$ and Σ_1 becomes a stable equilibrium manifold which is contradictory to the fact that no stable equilibrium manifold exists on the stability boundary. Consequently, $W^u(\Sigma) \cap A(\Sigma_s) \neq \varnothing$.
(ii) If $r > 1$, without loss of generality, we may assume inductively that $W^u(\Sigma_1) \cap A(\Sigma_s) \neq \varnothing$. Since $y \in W^u(\Sigma) \cap W^s(\Sigma_1)$ and the intersection is transverse, $W^s(\Sigma)$ contains an $(l + u)$-disk N centered at y, transverse to $W^s(\Sigma_1)$. Applying Lemma 17-4 for Σ_1, we have $\phi(t, N) \cap A(\Sigma_s) = \varnothing$ for some $t > 0$. Since $A(\Sigma_s)$ is invariant, this implies $N \cap A(\Sigma_s) = \varnothing$. Hence $W^u(\Sigma) \cap A(\Sigma_s) = \varnothing$. This completes the proof.

THEOREM 17-6 **(Characterization in terms of stable manifold)**
Let $\Sigma \neq \Sigma_s$ be a pseudo-hyperbolic equilibrium manifold of the non-hyperbolic dynamical system (17.1) and suppose Σ is not a source (i.e. $W^s(\Sigma) \setminus \Sigma \neq \varnothing$). Suppose that assumptions (N1)–(N4) hold. Then the following are equivalent.

(a) $W^s(\Sigma) \subset \partial A(\Sigma_s)$.
(b) $W^s(y_0) \subset \partial A(\Sigma_s)$ *for some* $y_0 \in \Sigma$.
(c) $\Sigma \subset \partial A(\Sigma_s)$.

Proof (a) ⇒ (b): Obvious since $W^s(y_0) \subset W^s(\Sigma) \subset \partial A(\Sigma_s)$.

(b) ⇒ (c): Suppose that $W^s(y_0) \subset \partial A(\Sigma_s)$ for some $y_0 \in \Sigma$. Since $y_0 \in W^s(y_0)$, $y_0 \in \partial A(\Sigma_s)$. Therefore, by assumption (N1), $\Sigma \subset \partial A(\Sigma_s)$.

(c) ⇒ (a): Suppose $\Sigma \subset \partial A(\Sigma_s)$. By Theorem 17-3, $W^u(\Sigma) \cap A(\Sigma_s) \neq \varnothing$. Let dim $W^u(\Sigma) = l + u$. Then $W^u(\Sigma) \cap A(\Sigma_s)$ contains an $(l + u)$-disk D. Let $y_0 \in W^s(\Sigma)$ be arbitrary. For any $\varepsilon > 0$, let N be an $(l + u)$-disk centered at y, transverse to $W^s(\Sigma)$. By Lemma 17-5 for Σ, $\exists t_0 > 0$ such that $\phi(t_0, N)$ is so close to D that $\phi(t_0, N)$ contains a point $p \in A(\Sigma_s)$. Thus $\phi(-t_0, p) \in N$. Since $A(\Sigma_s)$ is invariant, this implies $N \cap A(\Sigma_s) \neq \varnothing$. Letting $\varepsilon > 0$ proves $y_0 \in \overline{A(\Sigma_s)}$ which implies $W^s(\Sigma) \subset \overline{A(\Sigma_s)}$. Since $W^s(\Sigma)$ is disjoint from $A(\Sigma_s)$, it follows that $W^s(\Sigma) \cap \partial A(\Sigma_s)$. This completes the proof.

The next theorem presents a complete characterization of the stability boundary of the non-hyperbolic dynamical system (17.1) that satisfies assumptions (N1)–(N4). It asserts that the stability boundary is the union of the stable manifolds of all the unstable equilibrium manifolds on the stability boundary, which gives an explicit description of the geometrical and dynamical structure of the stability boundary.

THEOREM 17-7 **(Complete characterization of the stability boundary)**
Let Σ_s be a stable equilibrium manifold of the non-hyperbolic dynamical system (17.1) and suppose that assumptions (N1)–(N4) hold. Let $\{\Sigma_i: i = 1, 2, \ldots\}$ be the set of all equilibrium manifolds on $\partial A(\Sigma_s)$. Then

$$\partial A(\Sigma_s) = \bigcup_{\Sigma_i \in \partial A(\Sigma_s)} W^s(\Sigma_i).$$

Proof Let $\{\Sigma_i: i = 1, 2, \ldots\}$ be the set of all equilibrium manifolds on $\partial A(\Sigma_s)$. By Theorem 17-3, we have

$$\bigcup_i W^s(\Sigma_i) \subset \partial A(\Sigma_s).$$

Assumption (N3) and Theorem 17-2 imply

$$\partial A(\Sigma_s) \subset \bigcup_i W^s(\Sigma_i).$$

Therefore, we have

$$\partial A(\Sigma_s) = \bigcup_i W^s(\Sigma_i).$$

This completes the proof.

17.5 Optimal estimation of the stability region

The key task in the energy function approach to estimating the stability region of the stable equilibrium manifold is the determination of the critical level value. In this section, we present a scheme for optimally determining the critical level value of an energy function for estimating the stability region.

THEOREM 17-8 **(Optimal estimation of the stability region)**
Let Σ_s be a stable equilibrium manifold of the non-hyperbolic dynamical system (17.1) and $\{\Sigma_i: i = 1, 2, \ldots\}$ be the set of the equilibrium manifolds on $\partial A(\Sigma_s)$. Suppose there exists an energy function $V: R^n \to R$ for system (17.1) and let $S(r) = \{x \in R^n: V(x) < r\}$ be the level set of V and $S(r, \Sigma_s)$ be the connected component of $S(r)$ containing Σ_s. Then we have the following.

(a) The points with the minimal value of the energy function over the stability boundary $\partial A(\Sigma_s)$ exist and these points all lie in an equilibrium manifold.
(b) $c = \min_{x \in \partial A(\Sigma_s)} V(x) = \min_i V(\Sigma_i)$.
(c) $S(c, \Sigma_s) \subset A(\Sigma_s)$.
(d) $\{S(r, \Sigma_s) \setminus A(\Sigma_s)\} \neq \varnothing$ for any number $r > c$.

Proof We first state two facts needed in this proof.

Fact 1: The value of the energy function at a regular point is always greater than its value at its ω-limit set.

Fact 2: The ω-limit set of any trajectory on the stability boundary must be in one of the equilibrium manifolds on the stability boundary.

Based on the above two facts, the equality below follows:

$$c = \min_{x \in \partial A(\Sigma_s)} V(x) = \min_i V(\Sigma_i).$$

Let $a = V(\Sigma_s)$ and $b = V(\Sigma_1)$. Then $a < c \leq b$. By using the same argument in the proof of Theorem 17-1, it can be shown that the set $E_F \cap V^{-1}([a, b]) \cap \partial A(\Sigma_s)$ consists of at most a finite number of equilibrium manifolds, say, $\Sigma_{1_1}, \ldots, \Sigma_{i_r}$. Hence, it follows that

$$c = \min_k V(\Sigma_{i_k})$$

which implies that the equilibrium manifold with the minimal value of energy function over the stability boundary $\partial A(\Sigma_s)$ exists. Since $S(c, \Sigma_s) \subset \partial A(\Sigma_s) = \varnothing$, it follows that $S(c, \Sigma_s) \subset \overline{(A(\Sigma_s))}^c = \varnothing$ because otherwise $S(c, \Sigma_s)$ would be disconnected. Therefore, we have $S(c, \Sigma_s) \subset A(\Sigma_s)$. This also implies that $\{S(r, \Sigma_s) \cap A(\Sigma_s)^c\} \neq \varnothing$ for any number $r > c$, which completes the proof of part (b).

The above theorem asserts that the scheme of choosing $c = \min_i V(\Sigma_i)$ as the critical value to estimate the stability region $A(\Sigma_s)$ is *optimal* because the estimated stability region characterized by the corresponding energy function is the largest one within the entire stability region, as asserted by part (d) of the theorem.

Based on the above theoretical development, we present the following scheme to optimally estimate the stability region $A(\Sigma_s)$ of a non-hyperbolic system (17.1).

Scheme (Optimal estimation of the stability region $A(\Sigma_s)$ via an energy function $V(\cdot)$)

Stage I: Determining the critical level value of an energy function.

Step 1.1: Find all the equilibrium manifolds.

Step 1.2: Arrange those equilibrium manifolds whose corresponding values $V(\cdot)$ are greater than $V(\Sigma_s)$ in increasing order.

Step 1.3: Of these, identify the one with the lowest energy function value whose unstable manifold converges to the stable equilibrium manifold Σ_s. Let this be Σ.

Step 1.4: The value of the energy function at any point of Σ gives the critical level value of this energy function.

Stage II: Estimating the stability region $A(\Sigma_s)$.

Step 2.1: The connected component of $\{x \in R^n\colon V(x) < V(\Sigma)\}$ containing the stable equilibrium manifold Σ_s gives the optimal estimation of the stability region $A(\Sigma_s)$.

17.6 Illustrative examples

In this section, we present several examples for the purpose of illustrating some analytical results and the optimal estimation scheme developed in this chapter.

Example 17-1 This example was considered as a global optimization problem in [49,161]. Let h: $R^2 \to R$ be given by

$$h(x,y) = -\frac{332}{3}x^4 - 136y^2x^2 + \frac{76}{3}x^2 - 128x^8 + 192x^6 - 256y^4x^4$$

$$+192y^4x^2 + 192y^4x^2 - \frac{332}{3}y^4 + \frac{76}{3}y^2 - 128y^8 + 192y^6.$$

We define a non-hyperbolic dynamical system of the form

$$\begin{bmatrix} \dot{x} \\ \dot{y} \end{bmatrix} = -\nabla h(x,y)h(x,y).$$

If we define $V(x,y) = \frac{1}{2}h(x,y)^2$, then

$$\dot{V}(x,y) = -\|\nabla h(x,y)h(x,y)\|^2 \leq 0.$$

Therefore, $V(x, y)$ is an energy function for this system. There are six stable equilibrium manifolds $E_1, \ldots, E_6$ as shown in Figure 17.1(a). We are interested in the exact stability region, $A(E_1)$, of E_1, which is represented by the shaded area in Figure 17.1(a). Applying the computational scheme presented in the previous section we have the following.

Step 1: There are two zero-dimensional type-one equilibrium manifolds on the stability boundary; they are (0.80, 0.4198) and (0.4198, 0.80) which have the same energy function value 0.6021.

Step 2: The connected region characterized by the energy function $\{x \in R^n: V(x) < 0.6021\}$ containing the stable equilibrium manifold E_1 is used to estimate the stability region $A(E_1)$ as represented by the shaded area in Figure 17.1(b).

Example 17-2 Consider the following problem which is related to power system stabiliy [60]. Let h: $R^2 \to R$ be given by

$$h(x,y) = \sin(y) + 0.5\sin(-x+y) - 0.110 + 1.0\sin(x).$$

We define a non-hyperbolic dynamical system of the form

$$\begin{bmatrix} \dot{x} \\ \dot{y} \end{bmatrix} = -\nabla h(x,y)h(x,y).$$

Then, $V = \frac{1}{2}h(x,y)^2$ is an energy function for this system. There are infinitely many equilibrium manifolds as shown in Figure 17.1(c). We are interested in the stability region $A(E_1)$ which is represented by the shaded area in Figure 17.1(c). Now, applying the computational scheme presented in the previous section, we have the following.

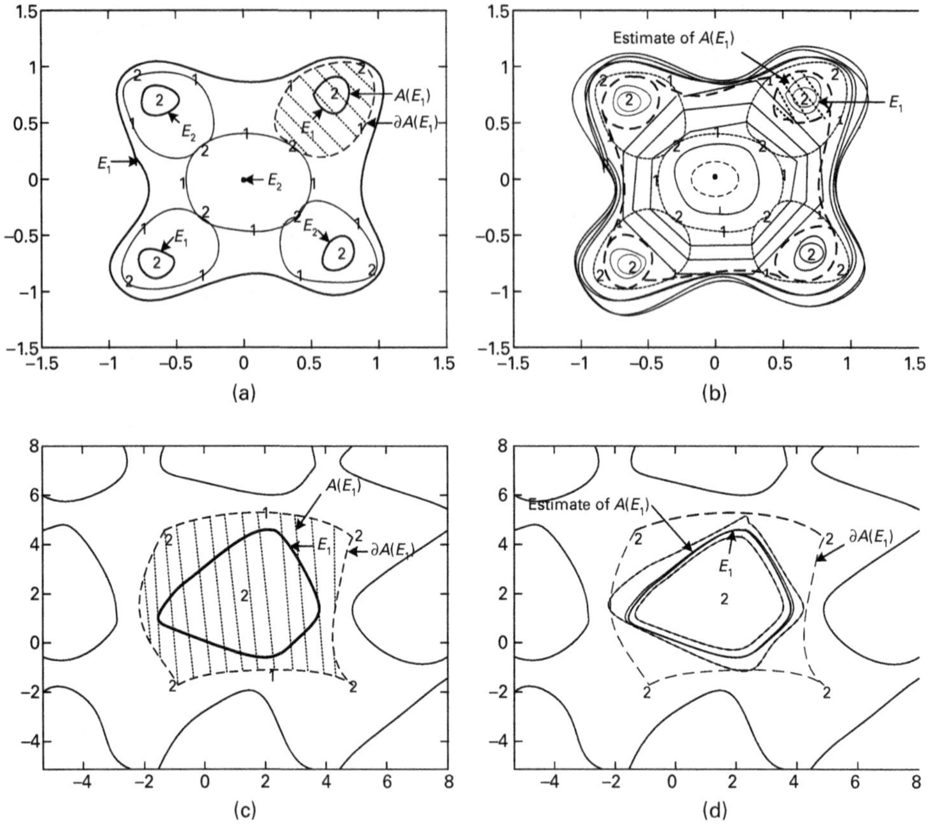

Figure 17.1 Figures (a) and (c) show the exact stability regions $A(E_1)$ of a stable equilibrium manifold E_1 in Example 17-1 and Example 17-2, respectively. The numbers 1, 2 represent the type of the equilibrium manifold. Figures (b) and (d) show the estimated stability region of E_1 obtained using the energy function of Example 17-1 and Example 17-2, respectively. Figure (d) shows the conservativeness in estimating the stability region using the energy function.

Step 1: There are four zero-dimensional type-one equilibrium manifolds on the stability boundary, say, (2.0695, 5.1786), (−2.0770, 1.0504), (2.0921, −1.1239), (4.1314, 1.0504). Of these, the type-one equilibrium manifold (2.0695, 5.1786) is the one with the lowest energy function value, 0.0118.

Step 2: The constant energy surface of $x \in R^n$: $V(x) < 0.0118\}$ containing the stable equilibrium E_1 is used to estimate the stability boundary of E_1 which is represented by the shaded area in Figure 17.1(d).

This example shows that the energy function approach can lead to more conservative estimation of stability regions when applied to stability equilibrium manifolds than when applied to stable equilibrium points.

17.7 Concluding remarks

This chapter has presented a fairly comprehensive theory of stability regions of stable equilibrium manifolds for a class of non-hyperbolic dynamical systems. A complete characterization of the stability boundary of such systems has been derived; in particular, it is shown, under three generic assumptions and the existence of an energy function, that the stability boundary equals the union of the stable manifolds of the unstable equilibrium manifolds on the stability boundary. An effective scheme to optimally estimate stability boundaries of stable equilibrium manifolds via level surfaces of energy functions has been presented. The estimated stability region based on this scheme is, however, only a subset of the stability region and is usually more conservative than that of a hyperbolic dynamical system. Further work toward reducing the conservativeness in estimating stability regions of non-hyperbolic dynamical systems is needed.

18 Optimal estimation of stability regions for a class of large-scale nonlinear dynamical systems

18.1 Introduction

In this chapter we study the closest UEP method for a class of large-scale second-order nonlinear dynamical systems from both theoretical and computational points of view. This method is developed by exploring the special structure of the second-order systems.

Consider a nonlinear dynamical system described by the following second-order vector differential equation:

$$M\ddot{x} + D\dot{x} + f(x) = 0 \tag{18.1}$$

where $M = \text{diag}\{M_1, M_2, \ldots, M_n\}$, $M_i > 0$, for all i, $D = [D_{ij}]$, $i, j = 1, 2, \ldots, n$. $D_{ii} > 0$, $D_{ij} = D_{ji}$ and $D_{ii} > \Sigma_{j \neq i}|D_{ij}|$, f: $R^n \rightarrow R^n$ is a bounded gradient vector with a bounded Jacobian. The zero vector $0 \in R^n$ is a regular value of $f(x)$.

Several physical systems can be modeled by (18.1). By exploring special structures of the class of second-order dynamical systems, an improved closest UEP method for estimating stability regions of large-scale systems will be presented. In addition, the robustness of the closest UEP with respect to changes in different parameters of a second-order dynamical system will be investigated. We consider changes in the matrices M and D (for example, due to error in modeling the system). This robustness property highlights the practical applicability of the closest UEP method to second-order dynamical systems.

This improved closest UEP method is based on exploration of a reduced model and will be shown to be optimal in the sense that the estimated stability region characterized by the corresponding energy function is the largest one within the entire stability region. Compared with the closest UEP method, this improved method provides computational advantages and yet yields optimally estimated stability regions.

The organization of this chapter is as follows. Section 18.2 is concerned with the class of nonlinear dynamical systems to be studied in this chapter. In Section 18.3 we examine several properties of the closest UEP, and we introduce another system which is a reduction of the second-order dynamical system. The relationship between the closest UEP of the reduced system and the closest UEP of the original system will explored in Section 18.4. Based on the results, an improved closest UEP method will be presented and applied to large-scale second-order dynamical systems in Section 18.5. A study of the robustness of the closest UEP with respect to both small and large variations in different parameters of a second-order dynamical system is conducted in Section 18.6. Examples are given to illustrate the robustness of the closest UEP and the improved closest UEP method.

18.2 Second-order nonlinear dynamical systems

For many physical system models, the matrix M represents masses or moments of inertia and D represents the damping matrix. Note that the damping term D_{ij} can be negative for $i \neq j$. Both matrices M and D are positive definite. Since $f(x)$ is continuously differentiable, by the Morse–Sard Theorem [2, p.69] almost every vector $x \in R^n$ is a regular value of $f(x)$. In particular, the zero vector is a regular value of $f(x)$. The analytical results to be developed in this chapter also hold if the matrix M is symmetric positive definite. Note that with the transformation $z = M^{1/2}x$, where $M^{1/2}$ is the unique positive definite square root of the matrix M, the system (18.1) is transformed into

$$\ddot{z} + (M^{1/2})^{-1} D(M^{1/2})^{-1}\dot{z} + (M^{1/2})^{-1} f\left((M^{1/2})^{-1}z\right) = 0,$$

which belongs to system (18.1).

Let us rewrite (18.1) in the state-space form:

$$\begin{aligned} \dot{x} &= y \\ M\dot{y} &= -Dy - f(x), \end{aligned} \tag{18.2}$$

We shall denote the system described by (18.2) as $d(M, D)$. Note that all the equilibrium points of (18.2) are of the form $E{:} = \{(x, 0){:}\ f(x) = 0,\ 0 \in R^n\}$. Let $p_s \in E$ be a stable equilibrium point of $d(M, D)$. We shall use the notation $A(p_s)$ to denote the stability region of p_s and the notation $\partial A(p_s)$ to denote the boundary of $A(p_s)$. The notation $B_r(\hat{p})$(respectively $B_r^0(\hat{p})$) is used to denote the closed (respectively open) ball centered at $\hat{p}$ with radius r, i.e. $B_r(\hat{p}) = \{p : |p - \hat{p}| \leq r\}$, $B_r^0(\hat{p}) = \{p : |p - \hat{p}| < r\}$ and $\partial B_r(\hat{p}) = \{p : |p - \hat{p}| = r\}$denotes the sphere of radius r.

First, we show that the inertia of the equilibrium points of $d(M, D)$ are independent of the matrices M and D, as long as they are positive definite. Note that the *inertia* of the equilibrium point $\hat{x}$ of $\dot{x} = g(x)$ is a triple integer $(n_u(\hat{x}), n_s(\hat{x}), n_c(\hat{x}))$ representing, respectively, the number of eigenvalues of $((\partial g) / (\partial x))(\hat{x})$ with positive, negative, and zero real part. We say the equilibrium point $\hat{x}$ is *hyperbolic* if $n_c(\hat{x}) = 0$. We say $\hat{x}$ is a type-k equilibrium point if $n_s(\hat{x}) = k$.

THEOREM 18-1 *All the equilibrium points of $d(M, D)$ are hyperbolic. Moreover, $(\hat{x}, 0)$ is a type-k equilibrium point of $d(M, D)$ if and only if $\hat{x}$ is a type-k equilibrium point of $\dot{x} = f(x)$.*

Proof We prove this result by use of the inertia theorem [3] which states that if H is a nonsingular, Hermitian matrix and A has no eigenvalues on the imaginary axis, then

$AH + HA^* \geq 0$ implies that the matrices A and H have the same number of eigenvalues with strictly positive (and negative) real part.

The Jacobian matrix of $d(M, D)$ at $(\hat{x}, 0)$ is

$$J(\hat{x}, 0) = \begin{bmatrix} 0 & I \\ -M^{-1}J_{\hat{x}} & -M^{-1}D \end{bmatrix} \tag{18.3}$$

where $J_{\hat{x}}$ is the Jacobian matrix of $f(x)$ at $\hat{x}$, M^{-1} is the inverse of M, which is also positive diagonal. Let λ be an eigenvalue of the Jacobian matrix $J(\hat{x}, 0)$ with the corresponding eigenvector denoted by $\overline{x} = [x_1, y_1]^T$, namely,

$$\begin{bmatrix} 0 & I \\ -M^{-1}J_{\hat{x}} & -M^{-1}D \end{bmatrix} \begin{bmatrix} x_1 \\ y_1 \end{bmatrix} = \lambda \begin{bmatrix} x_1 \\ y_1 \end{bmatrix}. \tag{18.4}$$

It is obvious from (18.3) that $J(\hat{x}, 0)$ is nonsingular, i.e. $\lambda \neq 0$. Next, we claim that λ cannot be purely imaginary.

Equation (18.4) can be rewritten as

$$y_1 = \lambda x_1 \tag{18.5}$$

$$-M^{-1}J_{\hat{x}}x_1 - M^{-1}Dy_1 = \lambda y_1 \tag{18.6}$$

Substituting (18.5) into (18.6) and multiplying both sides by $x_1^* M$ we get

$$\lambda^2 x_1^* M x_1 + \lambda^1 x_1^* D x_1 + x_1^* J_{\hat{x}} x_1 = 0. \tag{18.7}$$

Suppose λ is a pure imaginary number

$$\lambda_1 = ja.$$

Substituting (18.6) into (18.7) and letting $x_1 = z_1 + jz_2$ yields

$$-a^2(z_1^T M z_1 + z_2^T M z_2) + (z_1^T J_{\hat{x}} z_1 + z_2^T J_{\hat{x}} z_2) = 0$$
$$ja(z_1^T D z_1 + z_2^T D z_2) = 0.$$

This is impossible since M, D are positive definite matrices. Therefore, we conclude that the matrix $J(\hat{x}, 0)$ does not contain any eigenvalue with zero real part, i.e. the equilibrium points of $d(M, D)$ are hyperbolic. Next, we choose H to be the symmetric matrix

$$H = \begin{bmatrix} -J_{\hat{x}}^{-1} & 0 \\ 0 & -M^{-1} \end{bmatrix}. \tag{18.8}$$

It follows that

$$J(\hat{x}, 0)H + HJ(\hat{x}, 0)^T = 2\begin{bmatrix} 0 & 0 \\ 0 \quad M^{-1} & D \quad M^{-1} \end{bmatrix} \geq 0. \tag{18.9}$$

Applying the matrix inertia theorem, we conclude that the matrix $J(\hat{x}, 0)$ has the same number of eigenvalues with positive real part as $J_{\hat{x}}^{-1}$does. Since $J_{\hat{x}}^{-1}$ and $J_{\hat{x}}$ have the same number of eigenvalues with positive (negative) real part [4, p.266], the proof is complete.

The above proof also shows that all the eigenvalues of the Jacobian matrix (18.3) on the open right half-plane must be real. Note that this indicates that the Hopf bifurcation phenomenon cannot occur in system *d(M, D)*.

The following two results are derived from the proof of Theorem 18-1. Corollary 18-2 gives a complete characterization of the stable equilibrium point of *d(M, D)*. It states that instead of checking the system Jacobian matrix to determine the stability of the equilibrium points, we need only to check the Jacobian matrix of $f(\cdot)$ which is of lower dimension. Corollary 18-3 reveals a relationship between the dimension of a stable manifold $W^s(\hat{x}, 0)$ and an unstable manifold $W^u(\hat{x}, 0)$. We will use the notation dim W to denote the dimension of W.

COROLLARY 18-2 **(Characterization of the stable quilibrium point)**
Let $(\hat{x}, 0)$ *be an equilibrium point of d(M, D). Then,* $(\hat{x}, 0)$ *is a stable equilibrium point of d(M, D) if and only if* $-J_{\hat{x}}$ *is positive definite.*

COROLLARY 18-3 **(Relationship between dimensions of stable and unstable manifolds)**
Let $(\hat{x}, 0)$ *be an equilibrium point of d(M, D). Then* $W^s(\hat{x}, 0) \geq \dim W^u(\hat{x}, 0)$.

We next show that the asymptotic behavior of the trajectories of *d(M, D)* is very simple: every bounded trajectory converges to an equilibrium point.

THEOREM 18-4 **(Asymptotic behavior of bounded trajectory)**
Every trajectory $(x(t), y(t))$ *of system (18.2) with* $x(t)$ *being bounded converges to an equilibrium point.*

Proof First, we show the ω-limit set of every bounded trajectory $(x(t), y(t))$ consists of equilibrium points. Then, we show that every bounded trajectory converges to its ω-limit set. The following lemma is used in the proof.

LEMMA 18-5 *There exists a* C^1*-function* $V: R^{2n} \to R$ *for d(M, D) such that*

(a) $\dot{V}(x(t), y(t)) \leq 0$, if $(x(t), y(t)) \notin E$.
(b) *let* $(x(0), y(0)) \notin E$, *then the set* $\{t \in R: \dot{V}(x(t), y(t)) = 0\}$ *has measure zero in R.*

Proof We define $V: R^n \times R^n \to R$

$$V(x, y) = \frac{1}{2}\langle y, My\rangle + \int_0^x \langle f(u), du\rangle. \tag{18.10}$$

The derivative of $V(\cdot)$ along the trajectory of *d(M, D)* is

$$\begin{aligned}\dot{V}(x, y) &= \frac{\partial V}{\partial x}\dot{x} + \frac{\partial V}{\partial y}\dot{y} \\ &= -\langle y, Dy\rangle \leq 0\end{aligned} \tag{18.11}$$

Hence, part (a) holds. Suppose that part (b) is not true, then there exists an interval $T = (t_1, t_2)$ with $t_2 > t_1 \geq 0$ such that $\dot{V}(x(t), y(t)) = 0$ for $t \in T$. From (18.11), we have

$y(t) = 0$, $t \in T$. This implies that $y(t) = 0$ and $x(t) =$ constant, for $t \in (t_1, t_2)$. From (18.2), this, in turn, implies that $f(x(t)) = 0$. So, we have $(x(t), y(t)) \in E$ for $t \in (t_1, t_2)$. Since (18.2) is an autonomous dynamical system, it follows that $(x(t), y(t)) \in E$ for $t \in R$. This is contradictory to the fact that $(x(0), y(0)) \notin E$. Therefore, part (ii) is also true. The proof of this lemma is complete.

Lemma 18-5 asserts that the conditions (E1) and (E2) of energy functions are satisfied. By applying Theorem 3-1 of Chapter 3 and the fact that the equilibrium points of (18.2) are isolated (see Theorem 18-1), we conclude that every bounded trajectory of (18.2) converges to an equilibrium point. The proof of this theorem is complete.

One implication of Theorem 18-4 is that the behavior of the trajectories of the system $d(M, D)$ is very simple: every trajectory either converges to an equilibrium point or becomes unbounded, i.e. there does not exist any limit cycle or bounded complicated behavior such as almost periodic trajectory, chaotic motion, etc.

From now on, $V(\cdot)$ stands for the particular function (18.10). The following assumption (A) will be in effect in the rest of this section. We denote $E: = \{x: f(x) = 0\}$.

(A) There exists $\varepsilon > 0$ and $\delta > 0$ such that

$$\begin{aligned} &\text{dist}(B_\varepsilon(x_i), B_\varepsilon(x_j)) > \varepsilon, \quad \text{for all} \quad x_i, x_j \in \overline{E} \\ &|f(x)| > \delta, \quad \text{for} \quad x \notin \bigcup_{\overline{x} \in \overline{E}} B_\varepsilon(\overline{x}) \, . \end{aligned} \tag{18.12}$$

We note that if $f(x)$ is a periodic vector field, then assumption (A) is satisfied. In addition, if there exist $\delta_1 > 0$ and $\delta_2 > 0$ such that $|f(x)| > \delta_2$ for $x \notin B_{\delta_1}(0)$, then assumption (A) is satisfied.

We have shown in Theorem 18-4 the asymptotic behaviors of bounded trajectories of $d(M, D)$. This leads to the following question: under what condition is a given trajectory $(x(t), y(t))|_{t \geq 0}$ of $d(M, D)$ bounded? We will show in Theorem 18-6 below that if $V(x(t), y(t))$ is bounded below for $t \geq 0$, then the trajectory $(x(t), y(t))|_{t \geq 0}$ is bounded.

THEOREM 18-6 **(Asymptotic behavior of trajectory with bounded $V(.)$)**
Let $(x(t), y(t))$ be a trajectory of $d(M, D)$. If $V(x(t), y(t))|_{t \geq 0}$ is bounded below, then $(x(t), y(t))$ converges to an equilibrium point as $t \to \infty$.

Proof Because of Theorem 18-4, it suffices to show that if $V(x(t), y(t))|_{t \geq 0}$ is bounded below this implies that the trajectory $(x(t), y(t))$ is bounded for $t \geq 0$. This proof consists of two steps. In the first step we show that the component $y(t)$ of $(x(t), y(t))$ is bounded no matter what the initial state $(x(0), y(0))$ is. We then show in the second step that the component $x(t)$ of $(x(t), y(t))$ is bounded when $V(x(t), y(t))|_{t \geq 0}$ is bounded. We rewrite (18.2)

$$\dot{y} = -M^{-1}Dy - M^{-1}f(x). \tag{18.13}$$

It is easy to see that the matrix $M_1; = -M^{-1}D$ has all real eigenvalues in the open left half-plane. Integrating both sides of (18.13) gives

$$y(t) = e^{M_1 t}y(0) + \int_0^t e^{M_1(t-s)}M^{-1}f(x(s))ds$$

$$= \sum_{k=1}^{q}\sum_{l=0}^{m_k-1} t^l e^{\lambda_k t}P_{kl}(M_1)\lambda(0) + \int_0^t \sum_{k=1}^{q}\sum_{l=1}^{m_k-1}(t-s)l_e^{\lambda_k(t-s)} \cdot P_{kl}(M_1)M^{-1}f(x(s))ds$$

where $\lambda_1, \lambda_2, \ldots, \lambda_q$ are the distinct eigenvalues of M_1 and m_k is the multiplicity of λ_k and the matrix $P_{kl}(M_1)$ depends only on M_1 [11, pp.307–310]. In the sequel, |.| stands for the vector norm, and ||.|| stands for the induced matrix norm. Applying the triangular inequality,

$$|y(t)| \leq \sum_{k=1}^{q}\sum_{l=0}^{m_k-1}\left(t^l e^{\lambda_k t}\|P_{kl}(M_1)\||\lambda(0)| + t^l e^{\lambda_k t}\cdot \int_0^t e^{-\lambda_k s}\|P_{kl}(M_1)\|\|M^{-1}\||f(x(s))|ds\right)$$

$$\leq \sum_{k=1}^{q}\sum_{l=0}^{m_k-1}\left(t^l e^{-a_k t}\|P_{kl}(M_1)\||\lambda(0)| + c\frac{t^l}{a_k}e^{-a_k t}\|P_{kl}(M_1)\|\right) \tag{18.14}$$

where $\lambda_k = a_k$, $a_k > 0$, and c is a positive number. Since the function $t \to t^l e^{-a_k t}$ is bounded on $[0, \infty)$, it follows that $|y(t)|$ is bounded on $[0, \infty)$.

Next, since

$$\begin{aligned} \dddot{x} &= -M^{-1}D\ddot{x} - M^{-1}\frac{\partial f}{\partial x}\dot{x} \\ &= (M^{-1}D)^2\dot{x} + M^{-1}DM^{-1}f(x) - M^{-1}\frac{\partial f}{\partial x}\dot{x} \end{aligned} \tag{18.15}$$

so $\dddot{x}(t)$ is also bounded. For positive numbers t and h, the mean-value theorem gives

$$\dot{x}(t+h) - \dot{x}(t) - h\ddot{x}(t) = \frac{1}{2}h^2 q \tag{18.16}$$

where

$$q_i = \dddot{x}_i(t + s_i),\ 0 \leq s_i \leq h.$$

Multiplying both sides by M, and rearranging the terms we have

$$M\dot{x}(t+h) - [M - hD]\dot{x}(t) = -hf(x) + \frac{1}{2}h^2 Mq \tag{18.17}$$

Taking the norm on both sides, we get

$$\|M\||\dot{x}(t+h)| + \|M - hD\||\dot{x}(t)| \geq h|-f(x)| - \frac{1}{2}h^2\|M\||q|. \tag{18.18}$$

Let us choose

$$h < \min\left\{\frac{M_i}{D_{ii}},\ i = 1, \ldots, n\right\}. \tag{18.19}$$

Then, the following inequality holds:

$$\|M - hD\| < \|M\|.$$

Equation (18.18) can be rewritten as

$$m|\dot{x}(t + h)| + m|\dot{x}(t)| > h| - f(x)| - \frac{1}{2}h^2 ma \tag{18.20}$$

where $m = \|M\|$ and $a = |q|$. Because $y(t)$ is bounded, it follows from assumption (A) and (18.2) that the time required for the component $x(t)$ to leave a ball $B_\varepsilon(\hat{x}_i)$ and enter another ball $B_\varepsilon(\hat{x}_1)$ is bounded below, say by $\hat{c}$, where $\hat{x}_i$, $\hat{x}_1 \in \overline{E}$, $\hat{c} = (c_1\delta)/(ma)$, c_1 is a positive number. Without loss of generality, we assume that $x(t)$ passes through a sequence of balls$B_\varepsilon(\hat{x}_i)$. Let us define an increasing sequence $\{s'_i\}$, where s'_i is the time that $x(t)$ leaves the ball $B_\varepsilon(\hat{x}_i)$. In the same manner, we define another increasing sequence $\{t'_i\}$, where t'_{i-1} is the time that $x(t)$ enters the ball $B_\varepsilon(\hat{x}_i)$. Two situations are possible.

(1) $t'_i - s'_i < \hat{c}$; *this means that* $x(t)$ *is coming back to the same ball after traveling a distance smaller than* ε.
(2) $t'_i - s'_i \geq c$; *this indicates that either* $x(t)$ *leaves a ball and enters another ball or* $x(t)$ *is coming back to the same ball.*

Since we are interested in knowing whether $x(t)$ is bounded or not, we only need to consider case (2). Let us take two subsequences $\{s_i\} \subset \{s'_i\}$, $\{t_i\} \subset \{t'_i\}$, such that $t'_i - s'_i \geq c$; in other words, $t'_i - s'_i = n_i\hat{c} + a_i\hat{c}$, where $n_i \geq 1$ are integers and $0 \leq a_i < 1$. We also get

$$T_\varepsilon = \{t : t \in [s_i, t_i], i = 1, 2, \ldots\}$$

and let $L(T_\varepsilon)$ denote the Lebesgue measure of T_ε. Now, we want to show that $L(T_\varepsilon)$ is bounded.

Let us set

$$h = \frac{\delta}{ma}, \quad \text{if } c_1 \geq 2 \tag{18.21}$$

or

$$h = \frac{c_1\delta}{2ma}, \quad \text{if } 0 < c_1 \leq 2.$$

This choice guarantees that $h < \hat{c}$. From (18.20) and (18.21) we have the following relations at $t \in T_\varepsilon$:

$$|\dot{x}(t + h)| + |\dot{x}(t)| > h\left(\frac{\delta}{m} - \frac{1}{2}ah\right) = \frac{1}{2a}\left(\frac{\delta}{m}\right)^2, \quad \text{if } c_1 \geq 2$$

or

$$|\dot{x}(t + h)| + |\dot{x}(t)| = \frac{c_1(4 - c_1)}{8a}\left(\frac{\delta}{m}\right)^2, \quad \text{if } 0 < c_1 \leq 2. \tag{18.22}$$

Since $V(x(t), y(t))$ is bounded below, say, by $\hat{V}$, it follows that

$$\int_{T_\varepsilon} \langle y(t), Dy(t)\rangle \, dt < \int_0^\infty \langle y(t), Dy(t)\rangle \, dt < V(x(0), y(0)) - \hat{V}. \tag{18.23}$$

Because D is a positive symmetric matrix

$$\langle y(t), Dy(t)\rangle \geq \overline{d}|y(t)|_2 > \frac{\overline{d}}{c_2}|y(t)|_\infty = d|y(t)|_\infty$$

where $\overline{d}$ is the minimum eigenvalue of D and c_2 is a positive number. Consequently, the following inequality holds:

$$\begin{aligned}
\int_{T_\varepsilon} \langle y(t), Dy(t)\rangle\, dt &\geq d \int_{T_\varepsilon} |y(t)|^2\, dt \\
&= d\left[\int_{s_1}^{(s_1+h)} |y|^2\, dt + \int_{(s_1+h)}^{(s_1+2h)} |y|^2\, dt + \ldots + \int_{(s_1+2(n_1-2)h)}^{(s_1+2(n_1-1)h)} |y|^2\, dt\right] \\
&\quad + d\left[\int_{(s_1+(2n_1-1)h)}^{(s_1+2n_1h)} |y|^2\, dt + \int_{(s_1+2n_1h)}^{(s_1+2n_1h+a_1h)} |y|^2\, dt\right] \\
&\quad + d\left[\int_{(s_2}^{(s_2+h)} |y|^2\, dt + \int_{(s_2+h)}^{(s_1+2h)} |y|^2\, dt + \ldots + \int_{(s_2+2(n_2-2)h)}^{(s_2+2(n_2-1)h)} |y|^2\, dt\right. \\
&\quad + \int_{(s_2+(2n_2-1)h)}^{(s_2+2n_2h)} |y|^2\, dt + \int_{(s_2+2n_2h)}^{(s_2+2n_2h+a_2h)} |y|^2\, dt\Big] + \ldots .
\end{aligned} \tag{18.24}$$

Since for $\beta \in [s_i + 2jh, j = 1,2, \ldots, n_i, i= 1,2, \ldots]$ we have

$$\begin{aligned}
\int_{\beta}^{(\beta+h)} |y|^2\, dt + \int_{(\beta+h)}^{(\beta+2h)} |y|^2\, dt &= \int_{\beta}^{(\beta+h)} (|y(t)|^2 + |y(t+h)|^2 dt \\
&\geq \int_{\beta}^{(\beta+h)} \frac{(|y(t)|^2 + |y(t+h)|^2)}{2} dt, \quad \text{if } c_1 \geq 2 \\
&\geq \frac{h}{2}\left[\frac{1}{2a}\left(\frac{\delta}{m}\right)^2\right]^2
\end{aligned} \tag{18.25}$$

$$\text{or, } \geq \frac{h}{2}\left[\frac{c_1(4-c_1)}{8a}\left(\frac{\delta}{m}\right)^2\right]^2, \quad \text{if } 0 < c_1 \leq 2.$$

Thus (18.24) can be rewritten as

$$\int_{T_\varepsilon} \langle y(t), Dy(t)\rangle\, dt \geq \frac{d}{4}\left[\frac{1}{2a}\left(\frac{\delta}{m}\right)^2\right]^2 L(T_\varepsilon), \quad \text{if } c_1 \geq 2 \tag{18.26}$$

or

$$\int_{T_\varepsilon} \langle y(t), Dy(t)\rangle\, dt \geq \frac{d}{4}\left[\frac{c_1(4-c_1)}{8a}\left(\frac{\delta}{m}\right)^2\right]^2 L(T_\varepsilon), \quad \text{if } 0 < c_1 \leq 2.$$

Combining (18.23) and (18.26), it follows that

$$L(T_\varepsilon) \leq \frac{4}{d} \frac{[V(x(0), y(0)) - V]}{\left[\frac{1}{2a}\left(\frac{\delta}{m}\right)^2\right]^2}, \quad \text{if } c_1 \geq 2$$

or

$$L(T_\varepsilon) \leq \frac{4}{d} \frac{[V(x(0), y(0)) - V]}{\left[\frac{c_1(4 - c_1)}{8a}\left(\frac{\delta}{m}\right)^2\right]^2}, \quad \text{if } 0 < c_1 \leq 2.$$

From the above inequalities, it is clear that if $x(t)$ is not bounded, then $L(T_\varepsilon)$ cannot be bounded either. Since $L(T_\varepsilon)$ is bounded, it follows that $x(t)$is bounded and this proof is completed.

Theorem 18-6 enables us to characterize the stability boundary of the second-order system (18.2). Because of the existence of the energy function (18.10), the stability boundary of the system $d(M_g, D)$ can be characterized as follows.

THEOREM 18-7 **(Characterization of the stability boundary)**
Let (x_s, y_s) be a stable equilibrium point of system $d(M, D)$ and (x_i, y_i), $i = 1, \ldots$ be the equilibrium points on the stability boundary of (x_s, y_s), denoted by $\partial A(x_s, y_s)$. Then,

$$\partial A(x_s, y_s) \subseteq \cup_i W^s(x_i, y_i). \tag{18.27}$$

We next study a property of stability boundaries of the system $d(M, D)$.

THEOREM 18-8 **(Unboundedness of the stability region)**
Let $A(p_s)$ denote the stability region of a stable equilibrium p_s of $d(M, D)$, then $A(p_s)$ is unbounded.

Proof It suffices to show that $\partial A(p_s)$ is unbounded. By contradiction, suppose that $\partial A(p_s)$ is bounded. Then $\partial A(p_s)$ is compact because $\partial A(p_s)$ is also closed. Therefore, the maximum of $V(\cdot)$ over $\partial A(p_s)$ exists. Let $\hat{p}$ be the point with the maximum value of $V(\cdot)$ over $\partial A(p_s)$. Since the value of $V(\cdot)$ must decrease along any nontrivial trajectory, we conclude that $\hat{p}$ must be a source (type-$2n$ equilibrium point). This contradicts the fact that $d(M, D)$ does not contain a source. Thus this theorem is true.

18.3 Closest UEP

In this section, we study the closest UEP of the system $d(M, D)$ and the closest UEP of $d(I)$, see Section 18.4, which is related to the system $d(M, D)$. The issue concerning the existence and uniqueness of the closest UEP of the system $d(M, D)$ and that of system $d(I)$ have been addressed by the general theorem stated in Chapter 11. Dynamical and topological characterizations of the closest UEP are that the closest UEP is a type-one

UEP whose unstable manifold converges to the stable equilibrium point and whose stable manifold forms part of the stability boundary. These characterizations are useful in the computation of the closest UEP.

Recall that $\hat{x}$ is the *closest UEP* of a stable equilibrium point (SEP) x_s with respect to an energy function $V(\cdot)$ if the following condition is met:

$$V(\hat{x}) = \min_{x \in \partial A(x_S) \cap E} V(x).$$

In other words, the closest UEP is the UEP with lowest value of $V(\cdot)$ among all the UEPs on the stability boundary of the corresponding SEP x_s, rather than all the UEPs in the state-space. For ease of exposition, we sometimes say an equilibrium point $\hat{x}$ is a closest UEP without referring to its corresponding stable equilibrium point and energy function.

It has been shown in Chapter 10 that the closest UEP of a stable equilibrium point x_s of general dynamical systems admitting an energy function $V(\cdot)$ always exists. Moreover, this closest UEP is generically unique. To show the closest UEP is generically unique, it is sufficient to show that for any two equilibrium points of the system $d(M, D)$, say x_1 and x_2, generically $V(x_1) \neq V(x_2)$. It is constructive to see that the condition making $V(x_1) = V(x_2)$ is the existence of the solution of the following simultaneous equations:

$$\begin{aligned} V(x_1) &= V(x_2) \\ \frac{\partial V}{\partial x_i}(x_1) &= 0, \ 1 \le i \le 2n \\ \frac{\partial V}{\partial x_i}(x_2) &= 0, \ 1 \le i \le 2n. \end{aligned}$$

The difficulty is that there are $(4n+1)$ equations in the $4n$ unknowns (x_1, x_2). This amounts to saying that, generically, no two equilibrium points have the same value of $V(\cdot)$. A rigorous proof of the property that all critical values of a C^2-function $W(\cdot)$ are distinct, is generic in $C^2(R^{2n}, R)$ is shown in [248].

In Theorem 4-3 we have shown that, under the assumption of transversality, a UEP lies on a stability boundary if and only if its unstable manifold intersects with the stability region; in brief, $p_i \in \partial A(p_s)$ if and only if $W^u(p_i) \cap A(p_s) \neq \emptyset$. Although the transversality condition holds for "almost all" dynamical systems described by C^1 vector fields, verification of the transversality condition for a given dynamical system is not an easy task. However, it can be easily shown that an equilibrium point p_i of $d(M, D)$ on the stability boundary $\partial A(p_s)$ has the property that $W^u(p_i) \cap A(p_s) \neq \emptyset$ if it is a local minimizer of an energy function over the stability boundary. This property holds even without the assumption of the transversality condition. Note that we say that a point $\hat{p}$ on the stability boundary $\partial A(p_s)$ is a local minimizer of $V(\cdot)$ if there exists a neighborhood U of p such that $V(\hat{p}) = \min_{x \in U \cap \partial A(p_s)} V(x)$. Hence, without the transversality condition, the unstable manifold of a closest UEP converges to an SEP and the stable manifold of the closest UEP forms a subset of the stability boundary of the SEP.

18.4 A dimension-reduced system

Next, we shall study another dynamical system *d(I)* which is related to the system *d(M, D)*, but with a reduced state space. The motivation for investigating this reduced system is multiple. Among others, it provides an efficient way to optimally estimate stability regions of the second-order dynamical system. Indeed, it will be shown that there is a close relationship between the closest UEP of the reduced system *d(I)* and the closest UEP of system *d(M, D)*. Furthermore, several properties of the closest UEP of system *d(M, D)* and those of the reduced system *d(I)* will be explored.

Consider the following system *d(I)* which is reduced from system *d(M, D)*:

$$\dot{x} = -f(x). \tag{18.28}$$

The Jacobian matrix of system *d(I)* at an equilibrium point $\hat{x}$ is $J_{\hat{x}}$. Because of assumption (A) and the fact that $J_{\hat{x}}$ is symmetric, the equilibrium points of system *d(I)* are hyperbolic.

Consider the following function:

$$V_p(x) = \int_0^x \langle f(u), du \rangle. \tag{18.29}$$

To show $V_p(\cdot)$ is an energy function for *d(I)*, we first differentiate $V_p(\cdot)$ along the trajectory of (18.28)

$$\dot{V}_p(x) = -|f(x)|^2. \tag{18.30}$$

Second, it is observed that $\dot{V}_p(x) = 0$ if and only if x is an equilibrium point of *d(I)*. Lastly, we will show that the function $V_p(\cdot)$ also satisfies the condition (E3) of the energy function. Hence, $V_p(\cdot)$ is an energy function for system *d(I)*. Let $B_\varepsilon(\delta)$ be the ball with radius ε and center δ. Let $\hat{E}$ be the set of equilibrium points of *d(I)*. Choose two positive numbers ε, η such that the distance between two different balls is at least ε and $|f(x)| > \eta$ for all $x \notin \bigcup_{\delta \in \hat{E}} B_\varepsilon(\delta)$. Along the trajectory $x(t)$ of *d(I)* we set

$$T_\varepsilon := \left\{ t : x(t) \notin \bigcup_{\delta \in \hat{E}} B_\varepsilon(\delta) \right\} \tag{18.31}$$

and let $L(T_\varepsilon)$ denote the Lebsegue measure of T_ε. From (18.30) we notice that the function $V_p(\cdot)$ is strictly decreasing along its trajectory; hence, it follows that

$$\int_{T_\varepsilon} \langle \dot{x}, \dot{x} \rangle \, dt \le \int_0^\infty \langle \dot{x}, \dot{x} \rangle \, dt = V_p(x(0)) - V_p(x(\infty)) \tag{18.32}$$

but

$$\int_{T_\varepsilon} \langle \dot{x}, \dot{x} \rangle \, dt \geq \eta^2 L(T_\varepsilon) \tag{18.33}$$

Combining (18.32) and (18.33) gives

$$L(T_\varepsilon) \leq \frac{V_p(\delta(0)) - V_p(\delta(\infty))}{\eta^2}. \tag{18.34}$$

From (18.34) we see that $L(T_\varepsilon)$ is bounded if $V_p(\delta(\infty))$ is bounded. Note that the vector field of $d(I)$ is bounded for all $x \in R^n$. These two results imply that after some finite distance the trajectory $x(t)$ will remain in some ball $B_\varepsilon(\delta)$. Thus, we have shown that $V_p(x(t))$ being bounded implies $x(t)$ is bounded, and the function $V_p(\cdot)$ also satisfies the condition (E3) of the energy function.

With the existence of the energy function (18.29) for system $d(I)$, we are in a position to characterize the stability boundary of system $d(I)$.

PROPOSITION 18-9 **(Characterization of the stability boundary)**
Let x_s be a stable equilibrium point of system $d(I)$ and x_i, $i = 1, \ldots, n$, be the equilibrium points on the stability boundary of x_s, denoted by $\partial A(x_s)$. Then, the stability boundary $\partial A(x_s)$ is contained in the union of stable manifolds of x_i, $i = 1, \ldots, n$; i.e.

$$\partial A(x_s) \subseteq \bigcup_i W^s(x_i).$$

We next establish the relationship between the equilibrium points of the system $d(M, D)$ and the equilibrium points of system $d(I)$.

THEOREM 18-10 **(A static relationship)**
Let x_s be a stable equilibrium point and x be an equilibrium point of system $d(I)$, then

(a) x_s is a stable equilibrium point of the system $d(I)$ if and only if $(x_s, 0)$ is a stable equilibrium point of the system $d(M, D)$,
(b) x is a type-k equilibrium point of system $d(I)$ if and only if $(x, 0)$ is a type-k equilibrium point of system $d(M, D)$.

Proof We shall prove this result by applying the matrix inertia theorem, which states that if H is a nonsingular, Hermitian matrix and A has no eigenvalues on the imaginary axis, then $AH + HA^* \geq 0$ implies that the matrices A and H have the same number of eigenvalues with strictly positive and negative real parts.

The Jacobian matrix of $d(I)$ at x_S is J_{x_S} and the Jacobian matrix of $d(M, D)$ at $(x_s, 0)$ is

$$J(x_s, 0) = \begin{bmatrix} 0 & I \\ -M^{-1}J_{x_s} & -M^{-1}D \end{bmatrix}.$$

We choose matrix H to be the symmetric matrix

$$H = \begin{bmatrix} -J_{x_S}^{-1} & 0 \\ 0 & -M^{-1} \end{bmatrix}.$$

It follows that

$$J(x_s,\,0)H + HJ(x_s,\,0)^T = 2\begin{bmatrix}0 & & 0 & \\ 0 & M^{-1} & D & M^{-1}\end{bmatrix} \geq 0.$$

Applying the inertia theorem, we see that the matrix $J(x_s, 0)$ has the same number of eigenvalues with positive real part as the matrix J_{xs}^{-1} does. Since J_{xs}^{-1} and J_{xs} have the same number of eigenvalues with positive (negative) real parts, the matrix $J(x_s, 0)$ and the matrix J_{xs} have the same number of eigenvalues with positive real part, and we obtain the required result.

The closest UEP method can be applied to the system $d(I)$ as stated in the following. Let $\hat{x}$ be the closest UEP of the stable equilibrium point x_s of system $d(I)$ with respect to an energy function $V_p(\cdot)$. Let $S_p(r)$ denote the connected component of the set $\{\delta\text{: } V(\delta) < r\}$ containing x_s. Then, $S_p(r) \subset A(x_s)$ for $V(x_s) < r < V(\hat{x})$ and $S_p(r) \cap \partial A(x_s \neq \varnothing)$ for $r < V(\hat{x})$.

Next, we establish the relationship between the closest UEP of system $d(M, D)$ and the closest UEP of the reduced system $d(I)$, on which the improved closest UEP method is based.

THEOREM 18-11 **(Relationship between the closest UEPs)**
The equilibrium point $(\hat{x},\,0)$ is the closest UEP of the stable equilibrium point $(x_s, 0)$ of system $d(M, D)$ with respect to the energy function $V(\cdot)$ in (18.10) if and only if $\hat{x}$ is the closest UEP of the stable equilibrium point x_s of system $d(I)$ with respect to the energy function $V_p(\cdot)$ in (18.29).

Proof Let $S_c(r)$ denote the connected component of the set $\{(x, y)\text{: } V(x, y) < r\}$ containing $(x_s, 0)$ and let $S_p(r)$ denote the connected component of the set $\{\delta\text{: } V_p(x) < r\}$ containing x_s.

⇒ Observing $V(\cdot)$ in (18.10) and $V_p(\cdot)$ in (18.29), it follows that $V(x, y) = V_p(x) + V_k(y)$, here $V_k(y) = \frac{1}{2}y^T My$. It is easily verified that

$$S_c(V(\hat{x},\,0)) \cap \{(x,\,y) : y = 0,\, x \in R^n\} = S_p(V_p(\hat{x})). \tag{18.35}$$

Since $(\hat{x},\,0)$ is the closest UEP of $d(M, D)$, it follows from Theorem 18-1 and condition (E3) of the energy function that $S(V(\hat{x},\,0))$ is a bounded set containing no other equilibrium points but the stable equilibrium point $(x_s, 0)$. It is also noted that $(\hat{x},\,0)$ is an equilibrium point of $d(M, D)$ if and only if (x) is an equilibrium point of $d(I)$. Thus, by (18.35), $S_p(V_p(x))$ is a bounded set and contains no other equilibrium points but the stable equilibrium point x_s. On the other hand, condition (E1) of the energy function assures that $S_p(V_p(x))$ is a positive invariant set of system $d(I)$. Conditions (E1) and (E2) of the energy function imply that the ω-limit set of system $d(I)$ consists entirely of equilibrium points. With these two facts plus the fact that the ω-limit set of every bounded trajectory exists [124, p.156], we conclude that every trajectory in the set $S_p\,(V_p(\hat{x}))$ converges to x_s (the only ω-limit set in $S_p\,(V_p(\hat{x}))$. Hence, $S_p\,(V_p(\hat{x})) \subseteq A\,(x_s)$. By the continuity property of energy functions, $S_p\,(V_p(\hat{x})) \subseteq A\,(x_s)$ implies that $\hat{x}$ is in the closure of the stability region of x_s. But, $A(x_s)$ cannot contain any equilibrium points other than x_s. Thus, $\hat{x}$ must be lying on the stability boundary of x_s, and consequently possess the minimum value of energy function $V_p(\cdot)$ among all the equilibrium points on the stability boundary.

$\Leftarrow$ Suppose $\hat{x}$ is the closest UEP of $d(I)$. From condition (E3) of the energy function we have that the set $S_p\left(V_p(\hat{x})\right)$ is bounded and contains no other equilibrium points but x_s. In addition, the following is true:

$$\begin{aligned}\overline{S}_c(V(\hat{x},\ 0)) := &\text{ the connected component of}\\ &\{(x,y) : V(x_s,0) \le V(x,y) \le V(\hat{x},0)\}\\ &\text{containing}(x_s,0)\\ = &\{(x,y) : V(x_s,0) \le V_p(x) + V_k(y) \le V(\hat{x},0)\}\\ &\text{containing}(x_s,0)\\ \subseteq &\{(x,y) : x \in S_p(V_p(\hat{x})), V(x_s,0) - V_p(\hat{x})\\ &\le V_k(y) \le V_p(\hat{x}) + V(\hat{x},0)\}.\end{aligned} \qquad (18.36)$$

Since the map $V_k(y) := y^T My: R^n \to R$ is a proper map. (A map $f: X \to Y$ is called proper if the pre-image of every compact set in Y is compact in S.) Thus the set $\{y : V(x_s, 0) - V_p(\hat{x}) < V_k(y) < V_p(\hat{x}) + V(\hat{x}, 0)\}$ is bounded. Because the product of compact spaces is compact, from (18.36) we have that the set $S_c(V(\hat{x},\ 0))$ is also bounded. Since the only equilibrium point in the set $S_c(V_p(\hat{x}))$ is x_s, it follows from (18.35) that the set $S_c(V(\hat{x},\ 0))$ contains no other equilibrium point but $(x_s, 0)$. Since the ω-limit set of the system $d(M, D)$ consists entirely of equilibrium points and the ω-limit set of every bounded trajectory exists [124, p.156], we have that every trajectory in $S_c(V(\hat{x},\ 0))$ converges to $(x_s, 0)$. In other words, $S_c(V(\hat{x},\ 0)) \subset A(x_s, 0)$. Since equilibrium points cannot lie inside the stability region, $(\hat{x},\ 0)$ must be on the stability boundary of $(x_s, 0)$ and possess the minimum value of energy function $V(\cdot)$ among all the equilibrium points on the stability boundary. This completes the proof.

The significance of Theorem 18-11 is that it provides an approach resembling the model-reduction approach to find the closest UEP of the system $d(M, D)$. This theorem suggests that, in order to find the closest UEP with respect to the function $V(\cdot)$ of the system $d(M, D)$, we only need to find the closest UEP with respect to the function $V_p(\cdot)$ of system $d(I)$ which is of lower dimension than the original system $d(M, D)$; thus it offers a means to greatly reduce the required computational burden.

18.5 An improved closest UEP method

In this section we present an improved closest UEP method based on the theoretical results developed in the previous section. It will be shown that the improved closest UEP method is optimal in the sense that the estimated stability region characterized by the corresponding energy function is the largest one within the entire stability region.

Let $\hat{x}$ be the closest UEP of the stable equilibrium point x_s of the system $d(M, D)$ with respect to an energy function $V(\cdot)$. Let $S_c(r)$ denote the connected component of the set $\{x: V(x) < r\}$ containing x_s. According to the theoretical development presented in Chapter 10 for the closest UEP method, estimation of the stability region by the closest UEP method is optimal:

$$S_c(r) \subset A(x_s, 0) \text{ for } V(x_s, 0) < r < V(\hat{x}) \text{ and } S_c(r) \cap \partial A(x_s, 0) \neq \emptyset \text{ for } r > V(\hat{x})$$

This theoretical basis of optimal estimation leads to the following closest UEP method for estimating the stability region of system $d(M, D)$.

Closest UEP method

Step 1: Find all the type-one equilibrium points of $d(M, D)$.

Step 2: Order these type-one equilibrium points according to the values of their energy function $V(\cdot)$.

Step 3: Of these type-one equilibrium points, starting from the one with the lowest value of $V_p(\cdot)$ but greater than $V(x_s, 0)$, check whether it is on the stability boundary $\partial A(x_s, 0)$ by checking whether its unstable manifold, which is of dimension one, converges to the corresponding stable equilibrium point $(x_s, 0)$. The first one (lowest energy function) that lies on the stability boundary is the closest UEP, say $\hat{x}$.

Step 4: The connected component of $\{x : V(x) < V(\hat{x})\}$ containing $(x_s, 0)$ is the estimated stability region of $(x_s, 0)$.

This closest UEP method is different from the classical UEP method in one fundamental way: it only searches for those type-one UEPs on the stability boundary $\partial A(x_s, 0)$ rather than all types of UEPs in the state space. In other words, this closest UEP method considers not only the static behavior of the system equilibrium points, but also the dynamical behavior of the system: the type-one equilibrium point on the stability boundary $\partial A(x_s, 0)$ and its unstable manifold. In comparison, the classical closest UEP method only considers the static behavior of the system: the equilibrium points.

The approach we take for the development of an improved closest UEP method is to explore the relationship between the closest UEP of system $d(M, D)$ and the closest UEP of system $d(I)$. This approach is motivated by the fact that the value of the energy function $V_p(\cdot)$ in (18.29) at the equilibrium point (x) of system $d(I)$ is equal to the value of energy function $V(\cdot)$ in (18.10) at the equilibrium point $(x, 0)$ of system $d(M, D)$. Theorem 18-11 has shown that the closest UEP of the stable equilibrium point $(x_s, 0)$ of system $d(M, D)$ with respect to the energy function $V(\cdot)$ in (18.10) corresponds to the closest UEP of the stable equilibrium point (x_s) of system $d(I)$ with respect to the energy function $V_p(\cdot)$ in (18.29). This result leads to an improved closest UEP method.

We now present the following improved closest UEP method for estimating the stability region $A(x_s, y_s)$ of system $d(M, D)$.

Improved closest UEP method for estimating the stability region $A(x_s, y_s)$

Step 1: Compute all the type-one equilibrium points lying on the stability boundary $\partial A(x_s)$ of $d(I)$.

Step 2: Order these type-one equilibrium points according to the values of their energy function $V_p(\cdot)$.

Step 3: Identify the closest UEP by starting from the one with lowest value of $V_p(\cdot)$ but greater than $V_p(x_s)$ to check whether it lies on the stability boundary $\partial A(x_s)$. The first one (lowest energy function) that lies on the stability boundary is the closest UEP of system *d(I)*, say $\hat{x}$.

Step 4: The connected component of $\{(x, y) : V(x, y) < V_p(\hat{x})\}$ containing (x_s, y_s) is the estimated stability region of (x_s, y_s) of the system *d(M, D)*.

In order to illustrate this improved closest UEP method in a simple context, we consider the following example which closely represents a three-machine power system with machine number 3 as the reference machine.

Consider the following system *d(M,D)*:

$$\begin{aligned} \dot{x}_1 &= y_1 \\ \dot{y}_1 &= -\sin x_1 - 0.5\sin(x_1 - x_2) - 0.3y_1 + 0.01 \\ \dot{x}_2 &= y_2 \\ \dot{y}_2 &= -0.5\sin x_2 - 0.5\sin(x_2 - x_1) - 0.3y_2 + 0.05. \end{aligned} \tag{18.37}$$

The corresponding reduced system *d(I)* is

$$\begin{aligned} \dot{x}_1 &= -\sin x_1 - 0.5\sin(x_1 - x_2) + 0.01 \\ \dot{x}_2 &= -0.5\sin x_2 - 0.5\sin(x_2 - x_1) + 0.05. \end{aligned} \tag{18.38}$$

It is easy to show that the following function is an energy function for system (18.37)

$$V(x_1, x_2, y_1, y_2) = y_1^2 + y_2^2 - 2\cos x_1 - \cos x_2 - \cos(x_1 - x_2) - 0.02x_1 - 0.1x_2$$

and the following function is an energy function for system (18.38)

$$V_p(x_1, x_2) = -2\cos x_1 - \cos x_2 - \cos(x_1 - x_2) - 0.02x_1 - 0.1x_2.$$

The point $(x^s, y^s) := (x_1^s, y_1^s, x_2^s, y_2^s) = (0.02801, 0, 0.06403, 0)$ is a stable equilibrium point of the original system (18.37) while point $(x^s) = (x_1^s, x_2^s) = (0.02801, 0.06403)$ is a stable equilibrium point of the reduced system (18.38). Applying the improved closest UEP method to approximate the stability boundary $\partial A(x^s, y^s)$, we have the following.

Step 1: There are two type-one equilibrium points of the reduced system (18.38), see Table 18.1, within the region $\{(x_1, x_2) : x_1^s - \pi < x_1 < x_1^s + \pi, x_2^s - \pi < x_2 < x_2^s + \pi\}$.

Step 2: The ordered list is: (−3.03743, 0.33413) with $V_p(x_1, x_2) = 2.04547$, followed by (0.04667, 3.11489) with $V_p(x_1, x_2) = -0.31329$.

Table 18.1 Coordinates of two type-one equilibrium points on the stability boundary of the reduced system

Type-one equilibrium points	x_1	x_2	Energy function value $V(.)$
1	0.04667	3.11489	−0.31329
2	−3.03743	0.33413	2.04547

Step 3: The type-one equilibrium point (0.04667, 3.11489) is the closest UEP because its unstable manifold converges to the SEP (0.02801, 0.06403) of the reduced system and is the one with the lowest value of energy function among the two type-one UEPs on the stability boundary of the reduced system (18.38).

Step 4: The constant energy surface of $V(\cdot)$ with the level value −0.31329 containing the SEP (0.02801, 0, 0.06403, 0) is used to approximate the stability boundary ∂A (0.02801, 0, 0.06403, 0) as shown in Figure 18.1. Curve A in this figure is the intersection between the exact stability boundary $\partial A(0.02801, 0.06403)$ and the subspace $\{(x_1, y_1, x_2, y_2): y_1 = 0, y_2 = 0\}$. Curve B is the intersection between the approximated stability boundary by the improved close UEP method and the subspace.

To illustrate the optimality of the improved closest UEP method, we have found the estimated stability boundary with two different level values other than the value determined by the improved closest UEP method. As shown in Figure 18.2, the optimality of the improved reduced-state closest UEP method is verified.

18.6 Robustness of the closest UEP

The problem concerning the robustness of the closest UEP with respect to changes in different parameters of the system $d(M, D)$ is studied in this section. In particular, we

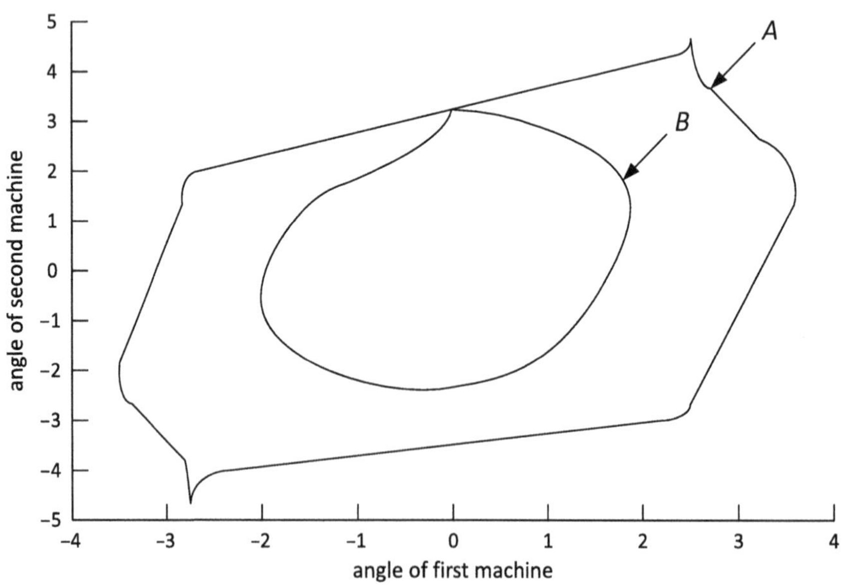

Figure 18.1 Curve A is the intersection between the exact stability boundary of system and the subspace $\{(x_1, y_1, x_2, y_2): y_1 = 0, y_2 = 0\}$. Curve B is the intersection between the estimated stability boundary determined by the improved closest UEP method and the subspace.

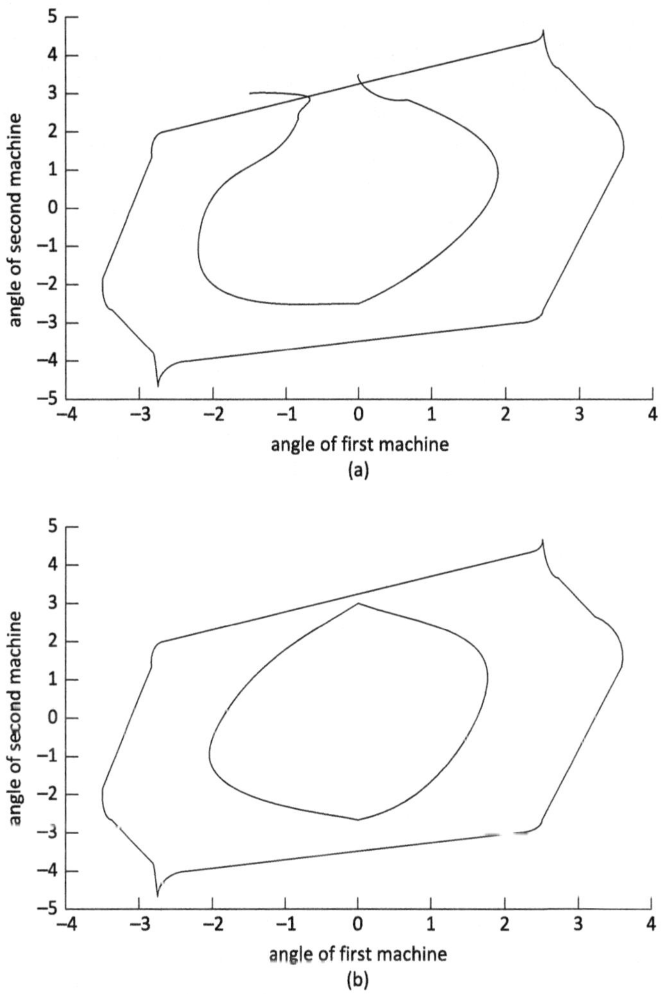

Figure 18.2 The estimated stability boundary with different level values other than the value determined by the presented closest UEP method. (a) The stability boundary estimated by the constant energy surface with level value −0.2. (b) The stability boundary estimated by the constant energy surface with level value −0.4. The figure shows the optimality of the improved reduced-state closest UEP method.

consider the changes in the matrices M and D (say, due to error in modeling the system). We will build the relation between the closest UEP of the system $d(M, D)$ and the closest UEP of the new system $d(\hat{M}, \hat{D})$. Theorem 18-12 below asserts that the closest UEP of system $d(M, D)$ is independent of the inertia matrix M and the damping matrix D.

THEOREM 18-12 **(Invariant property of the closest UEP)**
Let $\hat{p}$ be the closest UEP of a stable equilibrium point p_s of system $d(M, D)$ with respect to the energy function $V(\cdot)$ in (18.10). If the system $d(M, D)$ is changed into $d(\hat{M}, \hat{D})$, then

(a) $\hat{p}$ is on the stability boundary $\partial A(p_s)$ of $d(\hat{M}, \hat{D})$, and

(b) $\hat{p}$ is the closest UEP of the stable equilibrium point p_s of $d(\hat{M}, \hat{D})$ with respect to $\hat{V}(\cdot)$.

Proof From Theorem 18-11, we have that $\hat{X} := (\hat{x}, 0)$ is the closest UEP of the stable equilibrium point $(x_s, 0)$ of system $d(M, D)$ with respect to the energy function $V(\cdot)$ in (18.10) if and only if $\hat{x}$ is the closest UEP of the stable equilibrium point x_s of system $d(I)$ with respect to the energy function $V_p(\cdot)$ in (18.29). Application of Theorem 18-11 again shows that $\hat{x}$ is the closest UEP of the stable equilibrium point x_s of system $d(I)$ with respect to the energy function $V_p(\cdot)$ in (18.29) if and only if $\hat{p} = (\hat{x}, 0)$ is the closest UEP of the stable equilibrium point $(x_s, 0)$ of system $d(\hat{M}, \hat{D})$ with respect to the energy function $\hat{V}(\cdot)$. Combining these two facts we conclude this result.

In Theorem 18-12, we have illustrated another application of energy functions in increasing our understanding of the dynamical behavior of second-order dynamical systems. By choosing an appropriate energy function, one can show that a particular equilibrium point (i.e. the closest UEP) on the stability boundary of $d(M, D)$ remains on the stability boundary of new systems $d(\hat{M}, \hat{D})$ under large changes of both the inertia matrix M and damping factors D.

18.7 Conclusions

A closest UEP method tailored for large-scale second-order systems is presented in this chapter. From a theoretical point of view, it has been shown that the closest UEP of second-order systems corresponds to the closest UEP of the reduced-state systems associated with the original systems. In addition, the closest UEP has been shown to be independent of both the inertia matrix M and the damping matrix D of second-order systems. This invariant property is derived without the transversality condition.

From a computational point of view, an improved closest UEP method has been proposed based on the derived theoretical results. This improved closest UEP method uses a reduced model approach and has been shown to be optimal in the sense that the estimated stability region characterized by the corresponding energy function is the largest one within the entire stability region. Furthermore, it has been shown that the closest UEP is robust. This robustness property of the closest UEP heightens the practical applicability of the closest UEP method.

We have demonstrated another application of energy functions toward increasing our understanding of nonlinear dynamical behaviors of large-scale nonlinear systems. By using an appropriate energy function, we are able to show that the closest UEP of $d(M, D)$ remains on the stability boundary of a new system under large changes of both the machine inertia matrix and the damping matrix and, moreover, we have shown that, without assuming the transversality condition, the unstable manifold of the closest UEP always converges to the stable equilibrium point even if the system undergoes large variations of both the machine inertia matrix and damping matrix.

19 Bifurcations of stability regions

A complete characterization of the stability boundary was derived in Chapter 4 under the assumption that the model of the system is exact and that the parameters are precisely known. In practice, the model and the parameters are only approximations of the real values, and even if the model is exact, systems are subject to parameter variations. Because of both uncertainties and variation of parameters, it is natural to study the impact of parameter changes in the stability region and stability boundary.

In this chapter, we study the behavior of stability regions and the corresponding stability boundaries when a dynamical system is subject to parameter variations. Under certain conditions, only quantitative changes are observed on the stability region and stability boundary. More precisely, the critical elements, such as the equilibrium points, on the stability boundary change their location with the variation of parameters but they persist on the stability boundary, which is characterized by the perturbed version of the same stable manifolds or stable sets. The general shape of the stability region does not change and the perturbed stability region is, in a certain sense, close to the unperturbed one. However, qualitative changes in the shape of the stability region and in the characterization of the stability boundary may occur, when parameters vary through critical values.

19.1 Introduction

The study of qualitative changes of behaviors of dynamical systems due to parameter variations is called bifurcation theory. Under parameter variation, stability region bifurcations may occur. Stability region bifurcations are related to large changes of the stability region and stability boundary. These qualitative changes may significantly impact on the "size" of the stability region, which is a fundamental concern in many applications. Consequently, characterization of the stability boundary for a fixed parameter may not give a good approximation of the stability boundary if these parameters are perturbed from critical values.

Bifurcations of the stability region may be induced by either local bifurcations of invariant sets (in particular equilibrium points) on the stability boundary or global bifurcations. Bifurcations of the stability region induced by local bifurcations of equilibrium points are associated with the appearance of a non-hyperbolic equilibrium point on the stability boundary and violation of assumption (A1), which requires that every equilibrium point on the stability boundary is hyperbolic, see Chapter 4. Bifurcations of the stability region induced by global bifurcations are associated with the violation of assumption (A2), which requires the satisfaction of transversality conditions.

In this chapter, we will first develop conditions under which the stability region and stability boundary characterization persist with small variation of parameters. Then some particular cases of local bifurcations of equilibriums on the stability boundary will be addressed. More precisely, we will study bifurcations of the stability boundary induced by one type of generic local co-dimension one bifurcation, namely the *saddle-node bifurcation*. Global bifurcations are more difficult to analyze and they will not be considered in this chapter.

19.2 Persistence of the stability boundary characterization

In this section, we study the behavior of the stability region and stability boundary of a hyperbolic asymptotically stable equilibrium point under the influence of parameter variation. In particular, we will show, under the assumption of hyperbolicity of equilibrium points, that the characterization of the stability boundary derived in Chapter 4 persists under small variation of parameters.

Consider the nonlinear dynamical system

$$\dot{x} = f(x, \lambda) \tag{19.1}$$

with $x \in R^n$, depending on the parameter $\lambda \in R^m$. Each fixed λ defines the vector field $f_\lambda(\cdot) = f(\cdot, \lambda)$. Let $\phi_\lambda(t, x)$ denote the trajectory of $\dot{x} = f_\lambda(x)$ passing through x at time $t = 0$.

The implicit function theorem [235] guarantees that hyperbolic equilibrium points persist under small perturbation of the vector field. In other words, if $x^*_{\lambda_0}$ is a hyperbolic equilibrium point of system (19.1) for $\lambda = \lambda_0$, then there exist $\delta > 0$ and a neighborhood U of $x^*_{\lambda_0}$ such that system (19.1) possesses a unique hyperbolic equilibrium point x^*_λ in U for all $\lambda \in (\lambda_0 - \delta, \lambda_0 + \delta)$. Moreover, using the continuity of the eigenvalues with respect to the parameter λ, we can affirm that the perturbed equilibrium point x^*_λ has the same type of stability as $x^*_{\lambda_0}$. Consequently, hyperbolic equilibrium points do not suffer from bifurcations for sufficiently small variations of the parameter.

In particular, if $x^s_{\lambda_0}$ is a hyperbolic asymptotically stable equilibrium point of (19.1) for a fixed parameter $\lambda = \lambda_0$, then there exists $\delta > 0$ such that system (19.1) possesses a unique hyperbolic asymptotically stable equilibrium point x^s_λ in the neighborhood of $x^s_{\lambda_0}$ for all $\lambda \in (\lambda_0 - \delta, \lambda_0 + \delta)$. Therefore it makes sense to study the stability region of the perturbed stable equilibrium point and how it behaves under parameter variations.

The stability region of an asymptotically stable equilibrium point x^s_λ of system (19.1) for a fixed λ will be denoted:

$$A_\lambda(x^s_\lambda) = \{x \in R^n \,:\, \phi_\lambda(t, x) \to x^s_\lambda \text{ as } t \to \infty\}$$

and the corresponding stability boundary will be denoted $\partial A_\lambda(x^s_\lambda)$.

If the number of equilibrium points on the stability boundary $\partial A_{\lambda_0}(x^s_{\lambda_0})$ is finite and assumptions (A1)–(A3) of Chapter 4 are satisfied for every $\lambda \in (\lambda_0 - \alpha, \lambda_0 + \alpha)$, with $\alpha > 0$, then there exists $\varepsilon > 0$, with $\varepsilon \leq \alpha$, such that if $x^i_{\lambda_0}$, $i = 1, 2, \ldots, k$, belongs to the stability boundary $\partial A_{\lambda_0}(x^s_{\lambda_0})$, then the perturbed equilibrium point x^i_{λ}, $i = 1, 2, \ldots, k$, belongs to the stability boundary $\partial A_{\lambda}(x^s_{\lambda})$ of the perturbed system for all $\lambda \in (\lambda_0 - \varepsilon, \lambda_0 + \varepsilon)$ [50]. Consequently, $\cup_i W^s(x^i_{\lambda}) \subseteq \partial A_{\lambda}(x^s_{\lambda})$ for all $\lambda \in (\lambda_0 - \varepsilon, \lambda_0 + \varepsilon)$. In other words, the equilibrium points that were on the stability boundary continue to lie on the stability boundary of the perturbed system. If the estimate of the relevant stability boundary is of interest, then we can affirm that the relevant stability boundary of the perturbed system is the stable manifold of the perturbed controlling unstable equilibrium point.

For vector fields defined on compact manifolds, we conclude that $\partial A_{\lambda}(x^s_{\lambda}) = \cup_i W^s(x^i_{\lambda})$, i.e. the stability boundary $\partial A_{\lambda}(x^s_{\lambda})$ of the perturbed system contains the same number of equilibrium points of the stability boundary $\partial A_{\lambda_0}(x^s_{\lambda_0})$ as the unperturbed system and it is composed of the union of the stable manifolds of the perturbed equilibrium points x^i_{λ}, $i = 1, 2, \ldots, k$ for all $\lambda \in (\lambda_0 - \varepsilon, \lambda_0 + \varepsilon)$.

For vector fields in R^n, we have shown that if $x^i_{\lambda_0}$, $i = 1, 2, \ldots, k$, are the hyperbolic equilibrium points on the stability boundary of system (19.1) for $\lambda = \lambda_0$, then $\cup_i W^s(x^i_{\lambda}) \subseteq \partial A_{\lambda}(x^s_{\lambda})$ for every λ sufficiently close to λ_0. However, the inclusion $\cup_i W^s(x^i_{\lambda}) \supseteq \partial A_{\lambda}(x^s_{\lambda})$ is not always true. The reason for that is the possibility of the appearance of new hyperbolic equilibrium points at infinity. The next example illustrates the persistence of a hyperbolic equilibrium point on the stability boundary and the appearance of a new hyperbolic equilibrium point on the stability boundary as a result of parameter variation.

Example 19-1 Consider the following one-dimensional nonlinear system:

$$\dot{x} = \frac{-2x(x^2 + x) + 2x(x^2 - 1)}{(x^2 + 1)} + \lambda x. \tag{19.2}$$

For $\lambda = 0$, system (19.2) possesses two hyperbolic equilibrium points: $x^s_0 = 0$, which is an asymptotically stable equilibrium point, and $x^{u_1}_0 = -1$, which is a type-one hyperbolic unstable equilibrium point. The unstable equilibrium point $x^{u_1}_0$ lies on the stability boundary of the stable equilibrium point x^s_0 and the stability region $A_0(x^s_0)$ is the semi-straight line $(-1, \infty)$. Figure 19.1 illustrates the stability region of system (19.2) for $\lambda = 0$.

For $\lambda = 0.01$, the perturbed hyperbolic unstable equilibrium point $x^{u_1}_{0.01} \approx -0.98$ persists on the stability boundary and a new hyperbolic unstable equilibrium point $x^{u_2}_{0.01} \approx +6.04$ appears on the stability boundary of the stable equilibrium point $x^s_{0.01} = 0$. The stability region is bounded and equals the line segment $(x^{u_1}_{0.01}, x^{u_2}_{0.01})$. Figure 19.2 illustrates the stability region of system (19.2) for $\lambda = 0.01$.

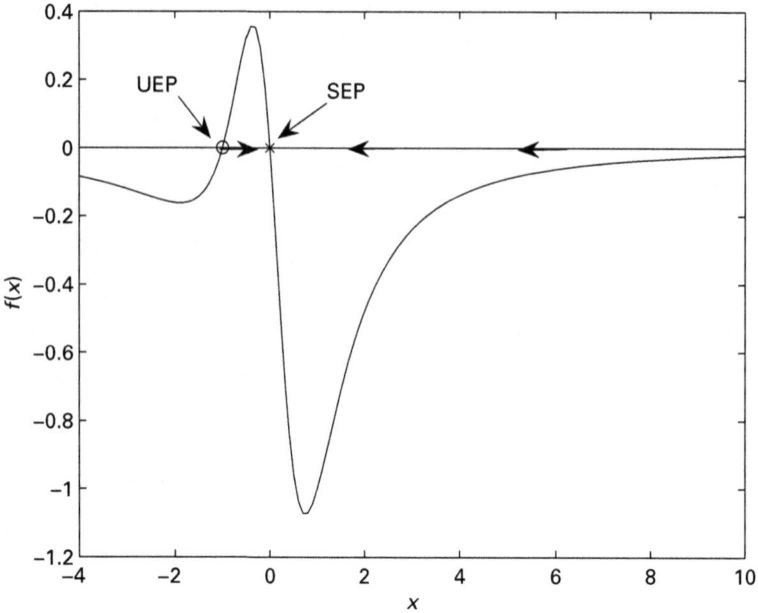

Figure 19.1 Stability region of system (19.2) for $\lambda = 0$. The stability boundary $\partial A_0(x_0^s)$ is composed of the unstable equilibrium point $x_0^{u_1} = -1$ and the stability region is the semi-straight line $(-1, \infty)$.

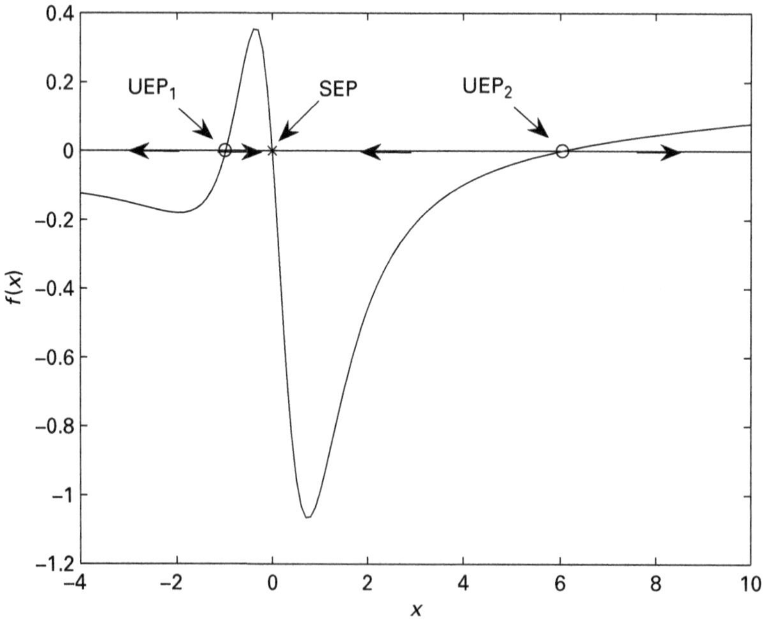

Figure 19.2 Stability region of system (19.2) for $\lambda = 0.01$. The stability boundary $\partial A_{0.01}(x_{0.01}^s)$ is composed of the unstable equilibrium points $x_{0.01}^{u_1}$ and $x_{0.01}^{u_2}$. The stability region is approximately the line segment $(-0.98, 6.04)$.

Under assumptions (A1)–(A3) and some control at infinity, the stability boundary does not suffer from drastic changes with small parameter variation. In this chapter, we study the behavior of the stability boundary under parameter variation for particular cases of violation of assumption (A1), that is, when certain types of non-hyperbolic equilibrium points lie on the stability boundary. In this case, drastic changes in the stability boundary and stability region may occur, with great impact on the size of the stability region.

19.3 Non-hyperbolic equilibrium points on the stability boundary

In Section 19.2, it was shown that hyperbolic equilibrium points persist on the stability boundary for sufficiently small variation of parameters. The only possibility of observing local bifurcations of equilibrium points on the stability boundary is with the presence of non-hyperbolic equilibrium points. Understanding the behaviors of these non-hyperbolic equilibrium points is fundamental to deriving a complete characterization of the stability boundary when parameters vary and assumption (A1) of Chapter 4 is violated.

We will study one particular type of non-hyperbolic equilibrium point, the saddle-node equilibrium point. We begin by studying behaviors of the dynamical system in a neighborhood of the equilibrium point, including the asymptotic behavior of solutions in the invariant local manifolds. Then a complete characterization of the stability boundary in the presence of this special type of non-hyperbolic equilibrium points is developed.

19.3.1 Saddle-node equilibrium points

The saddle-node equilibrium point is the simplest of the non-hyperbolic equilibrium points. In this case, the non-hyperbolicity is due to the existence of a simple null eigenvalue. Consider the following nonlinear dynamical system:

$$\dot{x} = f(x) \tag{19.3}$$

where $x \in R^n$ and f is a C^r-function.

DEFINITION [236] **(Saddle-node equilibrium point)**
A non-hyperbolic equilibrium point $p \in R^n$ of (19.3) is called a saddle-node equilibrium point if the following conditions are satisfied:

(a) $D_x f(p)$ has a unique simple null eigenvalue, with v as the right eigenvector and w the left eigenvector, and none of the other eigenvalues have real part equal to zero,
(b) $wD_x^2 f(p)(v, v) \neq 0$.

Condition (a) establishes the type of non-hyperbolicity, while condition (b) is a quadratic nondegeneracy condition of the vector field. Saddle-node equilibrium points can be classified in types according to the number of eigenvalues of $D_x f(p)$ with positive real part.

DEFINITION **(Saddle-node equilibrium type)**
A saddle-node equilibrium point p of (19.3), is called a type-k saddle-node equilibrium point if $D_xf(p)$ has k eigenvalues with positive real part and n−k−1 with negative real part.

If p is a saddle-node equilibrium point of (19.3), then there exist invariant local manifolds $W^s_{loc}(p)$, $W^{cs}_{loc}(p)$, $W^c_{loc}(p)$, $W^u_{loc}(p)$ and $W^{cu}_{loc}(p)$ of class C^r, tangent to E^s, $E^c \oplus E^s$, E^c, E^u and $E^c \oplus E^u$ at p, respectively [123]. These manifolds are respectively called stable, stable center, center, unstable and unstable center manifolds. The stable and unstable manifolds are unique, but the stable center, center and unstable center manifolds may not be.

If p is a saddle-node equilibrium point, then the following properties hold [236].

[1] If p is a type-zero saddle-node equilibrium point of (19.3), then we have the following.
   (i) The $(n-1)$-dimensional local stable manifold $W^s_{loc}(p)$ of p exists, is unique, and if $q \in W^s_{loc}(p)$ then $\phi(t, q) \to p$ as $t \to +\infty$.
   (ii) The unidimensional local center manifold $W^c_{loc}(p)$ of p can be split in two invariant submanifolds:

$$W^c_{loc}(p) = W^{c^-}_{loc}(p) \cup W^{c^+}_{loc}(p)$$

where $q \in W^{c^-}_{loc}(p)$ implies $\phi(t, q) \to p$ as $t \to +\infty$ and $q \in W^{c^+}_{loc}(p)$ implies $\phi(t, q) \to p$ as $t \to -\infty$. Moreover, $W^{c^+}_{loc}(p)$ is unique while $W^{c^-}_{loc}(p)$ is not.

Figure 19.3 illustrates the manifolds $W^{c^-}_{loc}(p)$, $W^{c^+}_{loc}(p)$ and $W^s_{loc}(p)$ for a type-zero saddle-node equilibrium point p on R^3.

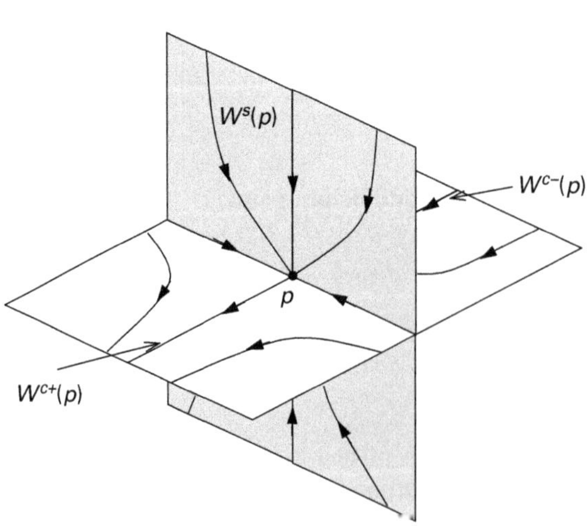

Figure 19.3 Manifolds $W^{c^-}_{loc}(p)$, $W^{c^+}_{loc}(p)$ and $W^s_{loc}(p)$ for a type-zero saddle-node equilibrium point p of system (19.3) on R^3.

[2] If p is a type-k saddle-node equilibrium point of (19.3), with $1 \leq k \leq n-2$, then we have the following.

(i) The k-dimensional local unstable manifold $W^u_{loc}(p)$ of p exists, is unique, and if $q \in W^u_{loc}(p)$ then $\phi(t, q) \rightarrow p$ as $t \rightarrow -\infty$.

(ii) The $(n{-}k{-}1)$-dimensional local stable manifold $W^s_{loc}(p)$ of p exists, is unique, and if $q \in W^s_{loc}(p)$ then $\phi(t, q) \rightarrow p$ as $t \rightarrow +\infty$.

(iii) The $(n{-}k)$-dimensional local stable center manifold $W^{cs}_{loc}(p)$ of p can be split in two invariant submanifolds:

$$W^{cs}_{loc}(p) = W^{cs^-}_{loc}(p) \cup W^{cs^+}_{loc}(p)$$

where $q \in W^{cs^+}_{loc}(p)$ implies $\phi(t, q) \rightarrow p$ as $t \rightarrow +\infty$. The local stable center manifold $W^s_{loc}(p)$ is contained in $W^{cs^-}_{loc}(p)$, moreover, $W^{cs^-}_{loc}(p)$ is unique while $W^{cs^+}_{loc}(p)$ is not.

(iv) The $(k+1)$-dimensional local unstable center manifold $W^{cu}_{loc}(p)$ of p can be split in two invariant submanifolds:

$$W^{cu}_{loc}(p) = W^{cu^-}_{loc}(p) \cup W^{cu^+}_{loc}(p)$$

where $q \in W^{cu^+}_{loc}(p)$ implies $\phi(t, q) \rightarrow p$ as $t \rightarrow -\infty$. The local unstable center manifold $W^u_{loc}(p)$ is contained in $W^{cu^+}_{loc}(p)$, moreover, $W^{cu^+}_{loc}(p)$ is unique while $W^{cu^-}_{loc}(p)$ is not.

Figure 19.4 illustrates the manifolds $W^{cu^-}_{loc}(p)$ and $W^{cu^+}_{loc}(p)$ for a type-1 saddle-node equilibrium point p on R^3.

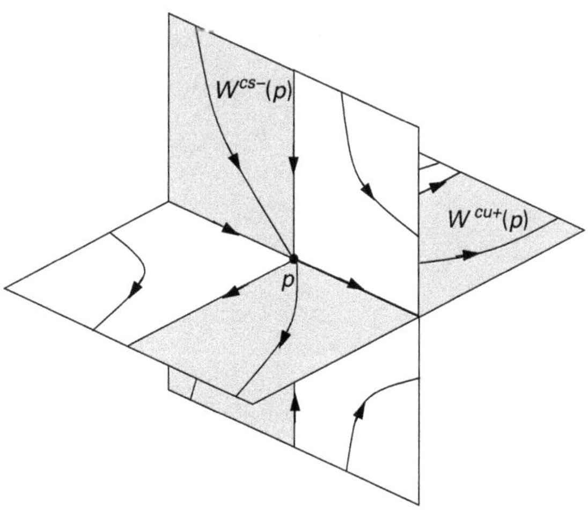

Figure 19.4 Manifolds $W^{cs^-}_{loc}(p)$ and $W^{cu^+}_{loc}(p)$ for a type-1 saddle-node equilibrium point p of system (19.3) on R^3.

[3] If p is a type-(n−1) saddle-node equilibrium point of (19.3), then we have the following.

(i) The (n−1)-dimensional local unstable manifold $W^u_{loc}(p)$ of p exists, is unique, and if $q \in W^u_{loc}(p)$ then $\phi(t, q) \to p$ as $t \to -\infty$.

(ii) The unidimensional local center manifold $W^c_{loc}(p)$ of p can be split in two invariant submanifolds:

$$W^c_{loc}(p) = W^{c^-}_{loc}(p) \cup W^{c^+}_{loc}(p)$$

where $q \in W^{c^-}_{loc}(p)$ implies $\phi(t, q) \to p$ as $t \to +\infty$ and $q \in W^{c^+}_{loc}(p)$ implies $\phi(t, q) \to p$ as $t \to -\infty$. Moreover, $W^{c^-}_{loc}(p)$ is unique while $W^{c^+}_{loc}(p)$ is not.

Figure 19.5 illustrates the manifolds $W^{c^-}_{loc}(p)$, $W^{c^+}_{loc}(p)$ and $W^u_{loc}(p)$ for a type-2 saddle-node equilibrium point p on R^3.

Although the stable and unstable manifolds of a hyperbolic equilibrium point are defined by extending the local manifolds through the flow, this technique cannot be applied to general non-hyperbolic equilibrium points. However, in the particular case of a saddle-node equilibrium point p, one still can define the global manifolds $W^s(p)$, $W^u(p)$, $W^{c^+}_{loc}(p)$, $W^{c^-}_{loc}(p)$, $W^{cs^-}_{loc}(p)$ and $W^{cu^+}(p)$ by extending the local manifolds $W^s_{loc}(p)$, $W^u_{loc}(p)$, $W^{c^+}_{loc}(p)$, $W^{c^-}_{loc}(p)$, $W^{cs^-}_{loc}(p)$ and $W^{cu^+}_{loc}(p)$ through the flow as follows:

$$W^s(p) = \bigcup_{t \le 0} \phi\left(t, W^s_{loc}(p)\right), \quad W^u(p) = \bigcup_{t \ge 0} \phi\left(t, W^u_{loc}(p)\right)$$

$$W^{cs^-}(p) = \bigcup_{t \le 0} \phi\left(t, W^{cs^-}_{loc}(p)\right), \quad W^{cu^+}(p) = \bigcup_{t \ge 0} \phi\left(t, W^{cu^+}_{loc}(p)\right)$$

$$W^{c^-}(p) = \bigcup_{t \le 0} \phi\left(t, W^{c^-}_{loc}(p)\right), \quad W^{c^+}(p) = \bigcup_{t \ge 0} \phi\left(t, W^{c^+}_{loc}(p)\right).$$

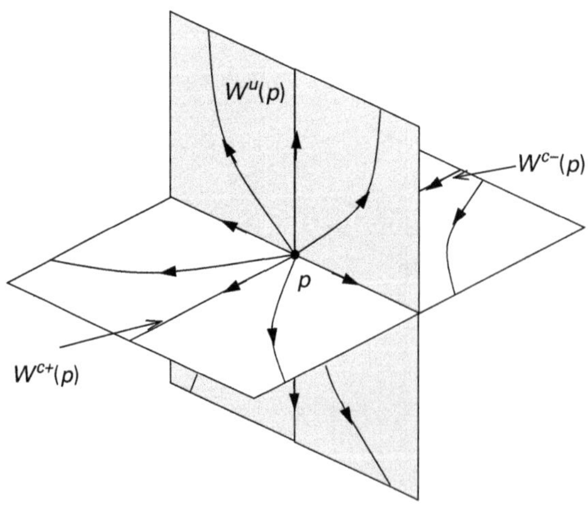

Figure 19.5 Manifolds $W^{c^-}_{loc}(p)$, $W^{c^+}_{loc}(p)$ and $W^u_{loc}(p)$ for a type-2 saddle-node equilibrium point p of system (19.3) on R^3.

This extension is justified by the invariance property and the asymptotic behavior of the local manifolds $W^s_{loc}(p)$, $W^u_{loc}(p)$, $W^{c^+}_{loc}(p)$, $W^{c^-}_{loc}(p)$, $W^{cs^-}_{loc}(p)$ and $W^{cu^+}_{loc}(p)$, see items [1], [2] and [3] above.

Example 19-2 (Saddle-node equilibrium point) Consider the following nonlinear dynamical system:

$$\begin{aligned} \dot{x} &= x^2 \\ \dot{y} &= y \\ \dot{z} &= z. \end{aligned} \tag{19.4}$$

The origin $p = (0, 0, 0)$ is the unique equilibrium point of this system. Linearizing the vector field at the origin, we conclude that the Jacobian matrix:

$$D_x f(p) = \begin{bmatrix} 0 & 0 & 0 \\ 0 & 1 & 0 \\ 0 & 0 & -1 \end{bmatrix}$$

possesses one null eigenvalue. Consequently, the first condition of the first definition is satisfied. The right and left eigenvectors associated with the null eigenvalue are respectively $v = (1, 0, 0)^T$ and $w = (1, 0, 0)$. Now we will check the second condition of the definition of a saddle-node equilibrium point. First we compute $D_x f(p)v$:

$$D_x f(p)v = \begin{bmatrix} 2x & 0 & 0 \\ 0 & 1 & 0 \\ 0 & 0 & -1 \end{bmatrix} \begin{bmatrix} 1 \\ 0 \\ 0 \end{bmatrix} = \begin{bmatrix} 2x \\ 0 \\ 0 \end{bmatrix}.$$

Then,

$$D_x[D_x f(p)v] = \begin{bmatrix} 2 & 0 & 0 \\ 0 & 0 & 0 \\ 0 & 0 & 0 \end{bmatrix}$$

and

$$[D_x[D_x f(p)v]]v = D_x^2 f(p)(v, v) = \begin{bmatrix} 2 \\ 0 \\ 0 \end{bmatrix}.$$

Finally,

$$w\left(D_x^2 f(p)(v, v)\right) = 2 \neq 0.$$

Therefore the second condition of the definition is also satisfied and the origin is a type-one saddle-node equilibrium point. The phase portrait of system (19.4) is depicted in Figure 19.6. The center stable and center unstable manifolds are highlighted in this figure.

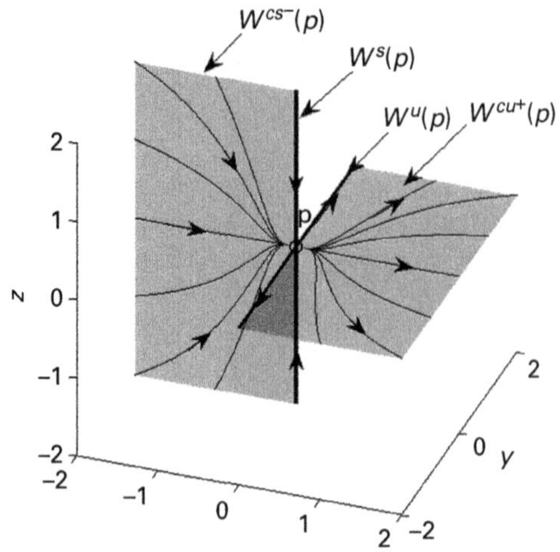

Figure 19.6 Phase portrait of system (19.4). The origin is a type-one saddle node equilibrium point. The component $W^{cs^-}(p)$ of the center stable manifold and the component $W^{cu^+}(p)$ of the center unstable manifold are highlighted in this figure.

19.3.2 Stability boundary characterization

In the presence of non-hyperbolic equilibrium points on the stability boundary, assumption (A1) is violated and Theorem 4.10 of Chapter 4 does not apply. In this section, a complete characterization of the stability boundary is developed when particular types of non-hyperbolic equilibrium points lie on the stability boundary $\partial A(x^s)$. The results of this section are a generalization of the results on stability boundary characterization developed in Chapter 4. In particular, a complete characterization of the stability boundary is developed when saddle-node equilibrium points are present on the stability boundary $\partial A(x^s)$.

Although the results of this section are restricted to saddle-node equilibrium points, a similar approach can be employed to deal with other types of non-hyperbolic equilibrium points, such as the Hopf equilibrium point. Our approach is the same as the one employed in Chapter 4. We start by studying a local characterization of the stability boundary by providing necessary and sufficient conditions for equilibrium points to lie on the stability boundary, and then a complete characterization of the stability boundary is derived.

The next theorem is complementary to Theorem 4-1. It offers necessary and sufficient conditions to characterize a saddle-node equilibrium point lying on the stability boundary in terms of the properties of its stable, unstable and center unstable manifolds.

THEOREM 19-1 **(Saddle-node equilibrium point on the stability boundary)**
Let p be a type-k saddle-node equilibrium point of the nonlinear dynamical system (19.3). Suppose also, the existence of an asymptotically stable equilibrium point x^s and let $A(x^s)$ be its stability region. Then the following holds:

$$(a)\ p \in \partial A(x^s) \Leftrightarrow \begin{cases} \left(W^{c^+}(p) - \{p\}\right) \cap \overline{A(x^s)} \neq \emptyset & if \quad k = 0 \\ \left(W^{cu^+}(p) - \{p\}\right) \cap \overline{A(x^s)} \neq \emptyset & if \quad 1 \leq k \leq n-1 \end{cases}$$

$$(b)\ p \in \partial A(x^s) \Leftrightarrow \left(W^s(p) - \{p\}\right) \cap \partial A(x^s) \neq \emptyset \quad if \quad 0 \leq k \leq n-2\,.$$

Theorem 19-1 offers necessary and sufficient conditions for a saddle-node equilibrium point lying on the stability boundary, without imposing any condition over the vector field. In the same spirit as Chapter 4, a further characterization of these non-hyperbolic equilibrium points on the stability boundary can be developed by imposing some assumptions on the vector field. These assumptions are more general than assumptions (A1)–(A3) of Chapter 4, leading to generalizations of the results of that chapter. Consider the following assumptions.

(A1′) All the equilibrium points on $\partial A(x^s)$ are hyperbolic, except possibly for p, a saddle-node equilibrium point.

(A2′) The following transversality conditions are satisfied.

- (i) The stable and unstable manifolds of hyperbolic equilibrium points on $\partial A(x^s)$ satisfy the transversality condition.
- (ii) The unstable manifolds of hyperbolic equilibrium points and the stable manifold $W^s(p)$ of the type-zero saddle-node equilibrium point p on $\partial A(x^s)$ satisfy the transversality condition.
- (iii) The unstable manifolds of hyperbolic equilibrium points and the stable component $W^{cs^-}(p)$ of the stable center manifold of the type-k saddle-node equilibrium point p, with $1 \leq k \leq n-2$, on $\partial A(x^s)$ satisfy the transversality condition.
- (iv) The unstable manifolds of hyperbolic equilibrium points and the stable component $W^{c^-}(p)$ of the center manifold of the type-(n−1) saddle-node equilibrium point p on $\partial A(x^s)$ satisfy the transversality condition.
- (v) The stable manifolds of hyperbolic equilibrium points and the ustable component $W^{cu^+}(p)$ of the unstable center manifold of the saddle-node equilibrium point p on $\partial A(x^s)$ satisfy the transversality condition.
- (vi) The stable manifold of equilibrium points and the unstable component $W^{c^+}(p)$ of the center manifold of the type-zero saddle-node equilibrium point p on $\partial A(x^s)$ satisfy the transversality condition.
- (vii) The stable component $\left(W^{cs^-}(p) - \{p\}\right)$ of the stable center manifold and the unstable component $\left(W^{cu^+}(p) - \{p\}\right)$ of the unstable center manifold of the type-k saddle-node equilibrium point p, with $1 \leq k \leq n-2$, on $\partial A(x^s)$ have empty intersection.
- (viii) The stable component $\left(W^{c^-}(p) - \{p\}\right)$ of the center manifold and the unstable component $\left(W^{cu^+}(p) - \{p\}\right)$ of the unstable center manifold of the type-(n−1) saddle-node equilibrium point p on $\partial A(x^s)$ have empty intersection.

Assumption (A1′) is weaker than (A1). Assumption (A1′) allows the presence of a non-hyperbolic equilibrium point p on the stability boundary. Assumptions (A1′) and (A2′) are generic properties of dynamical systems in the form of (19.3). Under assumptions (A1′), (A2′) and (A3), the next theorem offers necessary and sufficient conditions, which are sharper and more useful than conditions of Theorem 19-1, to characterize a hyperbolic or a saddle-node equilibrium point lying on the stability boundary.

THEOREM 19-2 **(Hyperbolic and saddle-node equilibrium points on the stability boundary)**
Let p be an equilibrium point of (19.3). Let x^s be an asymptotically stable equilibrium point and $A(x^s)$ be its stability region. If assumptions (A1′), (A2′) and (A3) are satisfied, then the following holds.

(a) If p is a hyperbolic equilibrium point, then:

$$(i)\ p \in \partial A(x^s) \Leftrightarrow W^u(p) \cap A(x^s) \neq \emptyset$$
$$(ii)\ p \in \partial A(x^s) \Leftrightarrow W^s(p) \subseteq \partial A(x^s).$$

(b) If p is a type-k saddle-node equilibrium point, then:

$$(i)\ p \in \partial A(x^s) \Leftrightarrow \begin{cases} W^{c^+}(p) \cap A(x^s) \neq \emptyset & \text{if} \quad k = 0 \\ W^{cu^+}(p) \cap A(x^s) \neq \emptyset & \text{if} \quad 1 \leq k \leq n-1 \end{cases}$$

$$(ii)\ p \in \partial A(x^s) \Leftrightarrow W^s(p) \subseteq \partial A(x^s) \quad \text{if} \quad n-k \geq 2.$$

Part (a) of Theorem 19-2 is the same as Theorem 4-3 of Chapter 4, showing that the conditions for a hyperbolic equilibrium point to lie on the stability boundary are not affected by the presence of non-hyperbolic equilibrium points on the stability boundary. For a type-k saddle-node equilibrium point, the result is similar to the case of a hyperbolic equilibrium point with the unstable center manifold playing the role of the unstable manifold of hyperbolic equilibrium points. Figure 19.7 illustrates a type-zero saddle-node equilibrium point p on the stability boundary. The component $W^{c^+}(p)$ of the center manifold of p intersects the stability region $A(x^s)$. Moreover, the stable manifold $W^s(p)$ lies on the stability boundary $\partial A(x^s)$.

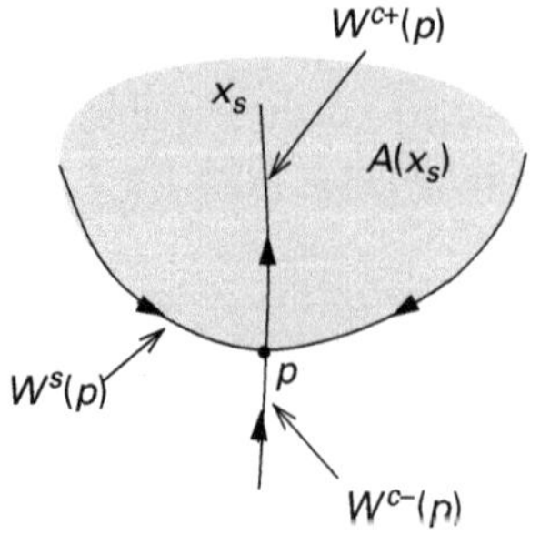

Figure 19.7 A type-zero saddle-node equilibrium point p on the stability boundary of the asymptotically stable equilibrium point x^s is characterized by its unstable center manifold and by its stable manifold.

The next theorem offers a complete characterization of the stability boundary of nonlinear autonomous dynamical systems in the presence of saddle-node equilibrium points on the stability boundary $\partial A(x^s)$.

THEOREM 19-3 **(Stability boundary characterization)**
Let x^s be an asymptotically stable equilibrium point of (19.3) and $A(x^s)$ be its stability region. Let x_i, $i = 1, 2, \ldots$, be the hyperbolic equilibrium points on $\partial A(x^s)$ and p be the only saddle-node equilibrium point on the stability boundary $\partial A(x^s)$. If assumptions (A1′), (A2′) and (A3) are satisfied, then the following holds.

(a) If p is a type-zero saddle-node equilibrium point, then:

$$\partial A(x^s) = \bigcup_i W^s(x^i) \cup W^s(p).$$

(b) If p is a type-k saddle-node equilibrium point with $1 \leq k \leq n-2$, then:

$$\bigcup_i W^s(x_i) \cup W^s(p) \subseteq \partial A(x^s) \subseteq \bigcup_i W^s(x_i) \cup W^{cs^-}(p).$$

If, in addition, the unstable manifold of the saddle-node equilibrium point p on the stability boundary $\partial A(x^s)$ intersects the stability region $A(x^s)$, then:

$$\partial A(x^s) = \bigcup_i W^s(x_i) \cup W^{cs^-}(p).$$

(c) If p is a type-$(n-1)$ saddle-node equilibrium point, then:

$$\bigcup_i W^s(x_i) \subseteq \partial A(x^s) \subseteq \bigcup_i W^s(x_i) \cup W^{c^-}(p).$$

If, in addition, the unstable manifold of the saddle-node equilibrium point p on the stability boundary $\partial A(x^s)$ intersects the stability region $A(x^s)$, then:

$$\partial A(x^s) = \bigcup_i W^s(x_i) \cup W^{c^-}(p).$$

Theorem 19-3 is a generalization of Theorem 4-11 of Chapter 4. It shows that the stability boundary is composed of the union of the stable manifolds of all equilibrium points on the stability boundary, including the stable manifold or the center stable manifold of the saddle-node equilibrium point on the stability boundary. The complete characterization of the stability boundary provided in Theorem 19-3 will be used in the next section to study stability region bifurcations. The next example illustrates this characterization.

Example 19-3 Consider the system of differential equations

$$\begin{aligned} \dot{x}_1 &= x_1^2 + x_2^2 - 1 \\ \dot{x}_2 &= x_1^2 - x_2 - 1 \end{aligned} \tag{19.5}$$

with $(x_1, x_2) \in R^2$. System (19.5) possesses three equilibrium points; they are $p = (0, -1)$, a type-zero saddle-node equilibrium point, $x_s = (-1, 0)$, a hyperbolic asymptotically stable equilibrium point, and $x^* = (1, 0)$, a type-one hyperbolic equilibrium point. Both the type-zero saddle-node equilibrium point p and the type-one hyperbolic equilibrium point x^* belong to the stability boundary of x_s. The stability boundary $\partial A(-1, 0)$ is formed, according to Theorem 19-3, of the union of the stable manifold of the type-one hyperbolic equilibrium point $(1, 0)$ and the stable manifold of the type-zero saddle-node equilibrium point $(0, -1)$, see Figure 19.8.

Figure 19.9 illustrates the importance of transversality conditions (A2) in the statement of Theorem 19-3. The illustration on the left of Figure 19.9 shows an example of a dynamical system in which the unstable component $W^{c^+}(p)$ of the central manifold of a type-zero saddle-node equilibrium point p has a nontransversal intersection with its stable manifold $W^s(p)$. Although p lies on the stability boundary, $W^{c^+}(p)$ does not intersect the stability region, the gray region in the figure. Moreover, the stable manifold of the type-zero saddle-node equilibrium point p is not contained on the stability boundary $\partial A(x^s)$. The illustration on the right of Figure 19.9 shows an example of a dynamical system in which the unstable component $W^{c^+}(p)$ of the central manifold of a type-zero saddle-node equilibrium point p has a nontransversal intersection with the stable manifold $W^s(x)$ of a type-one hyperbolic equilibrium point x^*. In this case, although both p and x^* lie on the stability boundary, $W^{c^+}(p)$ does not intersect the stability region, the gray region in the figure. Moreover, the stability boundary is not composed of the union of the stable manifolds of the equilibrium points p and x^*.

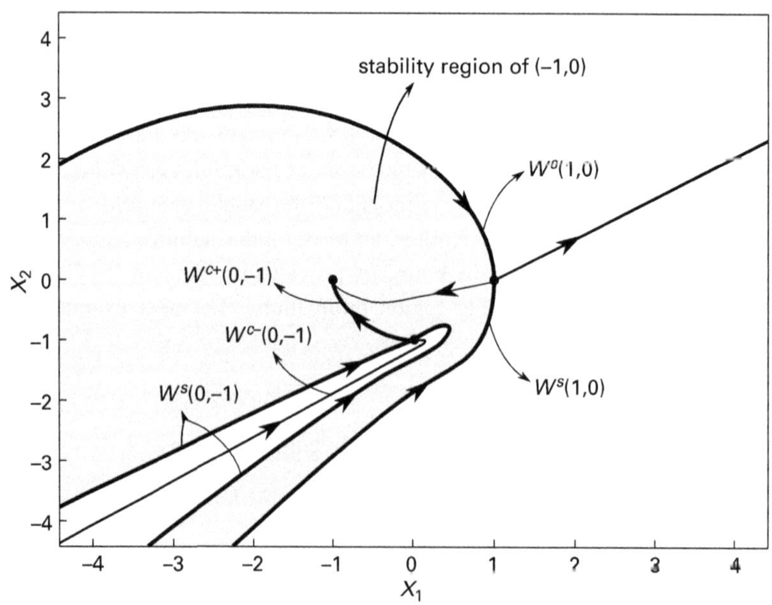

Figure 19.8 The phase portait of system (19.5).

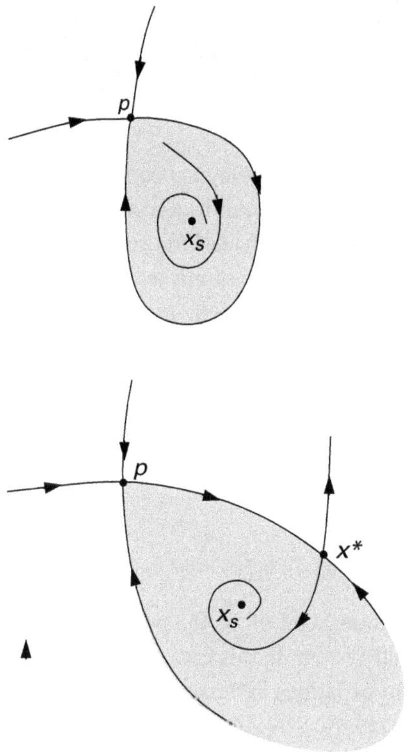

Figure 19.9 Examples of dynamical systems in which tranversality condition (A2′) is violated.

19.4 Overview of saddle-node bifurcation

In this section, a brief introduction to the theory of local bifurcation is presented. The focus is on the analysis of one generic co-dimension one bifurcation: the saddle-node bifurcation. For a more comprehensive treatment of bifurcations, see for example [114] and [153].

Intuitively speaking, two dynamical systems are equivalent if they have similar dynamical behaviors and phase portraits. For example, equivalent dynamical systems have the same number of equilibrium points and every corresponding equilibrium point possesses the same type of stability. Moreover, they have the same type and number of limit sets. The relationship between the corresponding limit sets will be preserved as well as the shapes of stability regions of attractors.

Two dynamical systems are equivalent if one can continuously transform the orbits of one system into the orbits of the other system. Consider the following two dynamical systems:

$$\dot{x} = f(x) \tag{19.6}$$

$$\dot{y} = g(y) \tag{19.7}$$

with $x,y \in R^n$. Formally, we consider the following definition of equivalence.

DEFINITION **(Topological equivalence)**

Dynamical systems (19.6) and (19.7) are topologically equivalent if there exists a homeomorphism h: $R^n \rightarrow R^n$ that maps orbits of the first system onto the orbits of the second system, preserving the orientation in time.

Suppose the homeomorphism h establishes a topological equivalence between dynamical systems (19.6) and (19.7). The point p is an equilibrium point of (19.6) if and only if $h(p)$ is an equilibrium point of (19.7). If $\omega(q)$ is the ω-limit set of a point q for system (19.6), then the image $h(\omega(q))$ of this set by h is the ω-limit set of $h(q)$ for system (19.7) [201]. Consequently, if $A(x^s)$ is the stability region of the asymptotically stable equilibrium point of system (19.6), then $h(A(x^s))$, the image of the stability region by h, is the stability region of the asymptotically stable equilibrium point $h(x^s)$ of system (19.7).

Consider the nonlinear dynamical system (19.1) for a fixed parameter $\lambda = \lambda_0$:

$$\dot{x} = f(x, \lambda_0) = f_{\lambda_0}(x) \tag{19.8}$$

and the perturbed system

$$\dot{x} = f(x, \lambda) = f_{\lambda}(x). \tag{19.9}$$

As the parameter varies in the neighborhood of λ_0, the perturbed system is usually topologically equivalent to the original system. In this case, we say that the vector field f_{λ_0} is structurally stable with respect to variations in the parameter λ. The next definition formally states this concept.

DEFINITION **(Structural stability)**

The vector field f_{λ_0} is structurally stable with respect to variations in the parameter λ if there exists $\varepsilon > 0$ such that the vector field f_{λ} is topologically equivalent to f_{λ_0} for all λ satisfying $\|\lambda - \lambda_0\| \leq \varepsilon$.

The notion of structural stability of this definition is weaker than the general notion of structural stability of dynamical systems. Usually, the notion of structural stability is given without specifying parameters. A metric in the space of all continuous vector fields is defined and a vector field is structurally stable if and only if it is topologically equivalent to every vector field in its neighborhood. For more details on the concept of structural stability, see [201].

The values of λ at which the vector field f_{λ} is not structurally stable with respect to variations of the parameter are called bifurcation values. At these values of λ, the system undergoes a qualitative change in its dynamical behavior. The theory of bifurcation studies the different scenarios at which these qualitative changes occur, classifies them in types via the equivalent classes of dynamical systems, and characterizes the dynamics of these systems in the neighborhood of these bifurcation values.

The set of bifurcation values divides the space of parameters into different regions. Inside each of these regions the vector fields are topologically equivalent. The boundaries of these regions are bifurcation surfaces or bifurcation boundaries. The system undergoes a bifurcation as parameters cross a bifurcation surface. The bifurcation surfaces divide the parameter space into regions induced by topological equivalence.

These surfaces are composed of several pieces. Every piece of the bifurcation surface is associated with a certain type of bifurcation. This stratification of the parameter space together with a phase portrait representation of each region is called a bifurcation diagram.

Example 19-4 Consider the following unidimensional dynamical system:

$$\dot{x} = x^2 + \lambda. \tag{19.10}$$

For $\lambda = 0$, the origin is the unique equilibrium point of system (19.10) and it is a type-zero saddle-node equilibrium point. For $\lambda < 0$, system (19.10) admits two equilibrium points: $x_1 = -\sqrt{-\lambda}$ and $x_2 = \sqrt{-\lambda}$. Both are hyperbolic equilibrium points, x_1 is asymptotically stable while x_2 is a type-one unstable hyperbolic equilibrium point. For $\lambda > 0$, system (19.10) possesses no equilibrium points.

In the neighborhood of $\lambda = 0$, the vector fields f_λ are not topologically equivalent to f_0. Consequently, $\lambda = 0$ is a bifurcation value. The qualitative changes in the dynamics of this system as a consequence of variation in λ are depicted in Figure 19.10.

The bifurcation value $\lambda = 0$ splits the space of parameters into two connected components. The semi-line $(-\infty, 0)$, in the space of parameters, is associated with a class of topologically equivalent dynamical systems. Each element of this class is a dynamical system that admits two hyperbolic equilibrium points, one stable and another of type one, and whose trajectories are similar to the diagram on the left of Figure 19.10. The dynamical systems associated with parameters in the semi-line $(0, \infty)$ also form a class of topologically equivalent systems.

This type of bifurcation, where two hyperbolic equilibrium points coalesce in a single non-hyperbolic saddle-node equilibrium point and disappear as a consequence of parameter variation, is called a saddle-node bifurcation. This bifurcation will be studied in more detail in the next section.

The co-dimension of a bifurcation gives an idea of how many parameters are necessary to model that bifurcation. The co-dimension of a bifurcation equals the co-dimension of its bifurcation surface. If, for example, we have three parameters and a

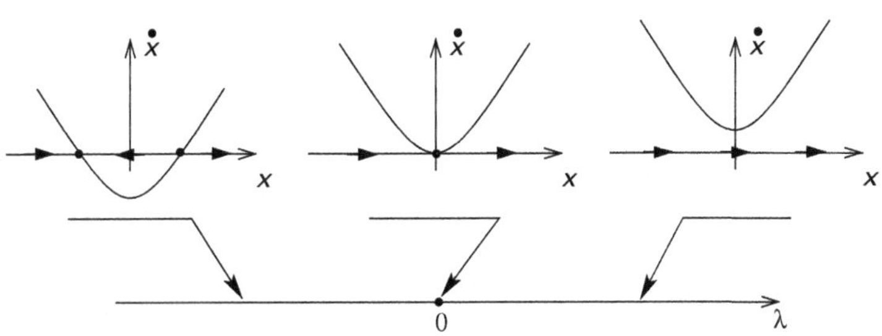

Figure 19.10 Bifurcation diagram of system (19.10).

bifurcation surface of dimension two, then the co-dimension of the bifurcation is one, that is, the difference between the number of parameters and the dimension of the bifurcation surface.

19.4.1 Saddle-node bifurcation

Consider the nonlinear dynamical system (19.1) and let $f: R^n \times R \to R^n$ be a vector field of class C^r, with $r \geq 2$.

DEFINITION **(Saddle-node equilibrium point)**
$(x_{\lambda_0}, \lambda_0) \in R^n \times R$ is a saddle-node bifurcation point of system (19.1) if $x_{\lambda_0} \in R^n$ is a non-hyperbolic equilibrium point of (19.1) for a fixed parameter $\lambda = \lambda_0$ and the following conditions are satisfied:

(SN1) $D_x f_{\lambda_0}(x_{\lambda_0})$ has a single simple eigenvalue equal to 0 with v as an eigenvector to the right and w to the left.

$$\text{(SN2)}\ w\left(\frac{\partial f_\lambda}{\partial \lambda}(x_{\lambda_0}, \lambda_0)\right) \neq 0.$$

$$\text{(SN3)}\ w\left(D_x^2 f_{\lambda_0}(x_{\lambda_0})(v, v)\right) \neq 0.$$

In other words, $(x_{\lambda_0}, \lambda_0) \in R^n \times R$ is a saddle-node bifurcation point of (19.1) if $(x_{\lambda_0}) \in R^n$ is a saddle-node equilibrium point of (19.1) for $\lambda = \lambda_0$ and the transversality condition (SN2) is satisfied. A saddle-node bifurcation point $(x_{\lambda_0}, \lambda_0)$ will be of type k if the non-hyperbolic equilibrium point x_{λ_0} is a type-k saddle-node equilibrium point. The parameter λ_0 will also be called a type-k saddle-node bifurcation value.

The next theorem studies the dynamical behavior of system (19.1) in the neighborhood of a saddle-node bifurcation point [236].

THEOREM 19-4 **(Saddle-node bifurcation)**
Let $(p_{\lambda_0}, \lambda_0)$ be a saddle-node bifurcation point of (19.1), for a fixed parameter $\lambda = \lambda_0$. Then there is a neighborhood N of p_{λ_0} and $\delta > 0$ such that depending on the signs of the expressions in (SN2) and (SN3), there is no equilibrium point in N when $\lambda \in (\lambda_0 - \delta, \lambda_0)$ $[\lambda \in (\lambda_0, \lambda_0 + \delta)]$ and two equilibrium points p_λ^k and p_λ^{k+1} in N for each $\lambda \in (\lambda_0, \lambda_0 + \delta)$ $[\lambda \in (\lambda_0 - \delta, \lambda_0)]$. The two equilibrium points on N are hyperbolic, more specifically p_λ^k is of type k and p_λ^{k+1} is of type $k + 1$, $k \in N$. Moreover, the stable manifold of the type-k equilibrium point and the unstable manifold of the type-(k+1) equilibrium point, intersect along a one-dimensional manifold.

In this chapter, we assume, without loss of generality, that the two equilibrium points exist in N for $\lambda \in (\lambda_0 - \delta, \lambda_0)$ and no equilibrium point exists in N for $\lambda \in (\lambda_0, \lambda_0 + \delta)$. Figure 19.11 illustrates Theorem 19-4 for a type-one saddle-node equilibrium point in R^3.

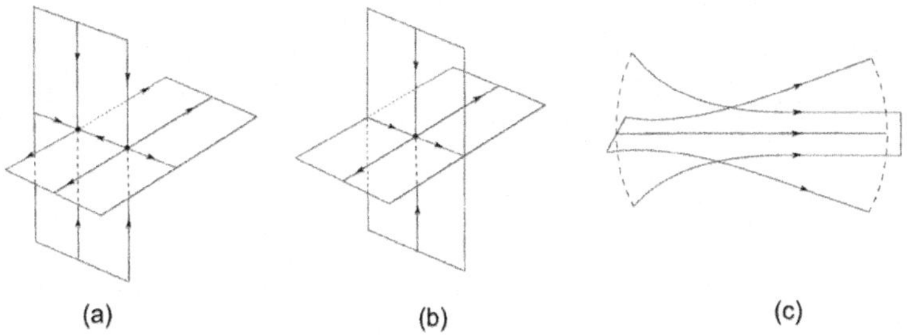

Figure 19.11 Phase portrait of system (19.1) in the neighborhood N of Theorem 19-4 for a type-1 saddle-node equilibrium point on R^3. In the neighborhood N, (a) system (19.1) has two hyperbolic equilibrium points, one type one and one type two for $\lambda < \lambda_0$. (b) System (19.1) possesses a unique equilibrium point, which is a type-one saddle-node equilibrium point, for $\lambda = \lambda_0$. (c) System (19.1) has no equilibrium point for $\lambda > \lambda_0$.

19.5 Stability region bifurcation

The characterizations of stability boundaries derived in Chapter 4 were developed under assumptions (A1)–(A3). Assumption (A1) admits that all the equilibrium points on the stability boundary are hyperbolic and assumption (A2) is a transversality condition. These assumptions are generic properties of nonlinear dynamical systems, that is, almost all dynamical systems satisfy them and, as a consequence, the existing characterizations of the stability boundary are valid and almost always assumption (A3) is satisfied.

Under parameter variation, bifurcations may occur on the stability boundary and assumption (A1) or (A2) may be violated at bifurcation points. Studying the characterization of the stability boundary at these bifurcation points is of fundamental importance to understanding how the stability region behaves under parameter variation. In this section, we study bifurcations of the stability boundary that are induced by local bifurcations of attracting sets and in particular by local bifurcation of equilibrium points.

We start this section with some examples of bifurcation of the stability boundary and then we present a complete characterization of the stability region and stability boundary when the system undergoes a type-zero saddle-node bifurcation on the stability boundary. Similar results can be derived for other types of bifurcations but they are beyond of the scope of this book.

19.5.1 Examples of stability region bifurcation

Two examples illustrating different types of bifurcation of the stability boundary are presented in this section. We start with an example in which a saddle-node bifurcation on the stability boundary leads to a drastic change in the size and shape of the stability region.

Example 19-5 Consider the following nonlinear dynamical system:

$$\begin{aligned}\dot{x}_1 &= x_1^2 + x_2^2 - 1\\ \dot{x}_2 &= x_1^2 - x_2 + \lambda\end{aligned} \tag{19.11}$$

with $(x_1, x_2) \in R^2$. The system of Example 19-3 is a particular case of this system for $\lambda = 1$. For $\lambda_0 = 1$, we have seen in Figure 19.8 that the stability boundary $\partial A_{\lambda_0}(-1, 0)$ is the union of the stable manifold of the type-one hyperbolic equilibrium point (1, 0) and the stable manifold of the type-zero saddle-node equilibrium point (0, −1). However, the saddle-node hyperbolic equilibrium point does not persist to parameter variations. For $\lambda = 1.02$, system (19.11) possesses four equilibrium points; they are $x_\lambda^s = (-0.99, -0.02)$, a hyperbolic asymptotically stable equilibrium point, $x_\lambda^* = (0.99, -0.02)$ a type-one hyperbolic equilibrium point, $y_\lambda^u = (-0.2, -0.97)$ a type-one hyperbolic equilibrium point and $y_\lambda^s = (0.2, -0.97)$ a type-zero hyperbolic equilibrium point. The equilibrium points y_λ^u and y_λ^s originated from the type-zero saddle-node equilibrium point in a type-zero saddle-node bifurcation. Moreover, $y_\lambda^u \in \partial A_\lambda(0.2, -0.97) \cap \partial A_\lambda(-0.99, 0.02)$ and $x_\lambda^* = (0.99, -0.02) \in \partial A_\lambda(-0.99, 0.02)$, see Figure 19.12. For $\lambda = -0.98$, system (19.11) possesses two equilibrium points; they are $x_\lambda^s = (-0.99, 0.01)$, a hyperbolic asymptotically stable equilibrium point and $x_\lambda^* = (0.99, 0.01)$, a type-one equilibrium point, which belongs to the stability boundary $\partial A_\lambda(-0.99, 0.01)$, see Figure 19.13.

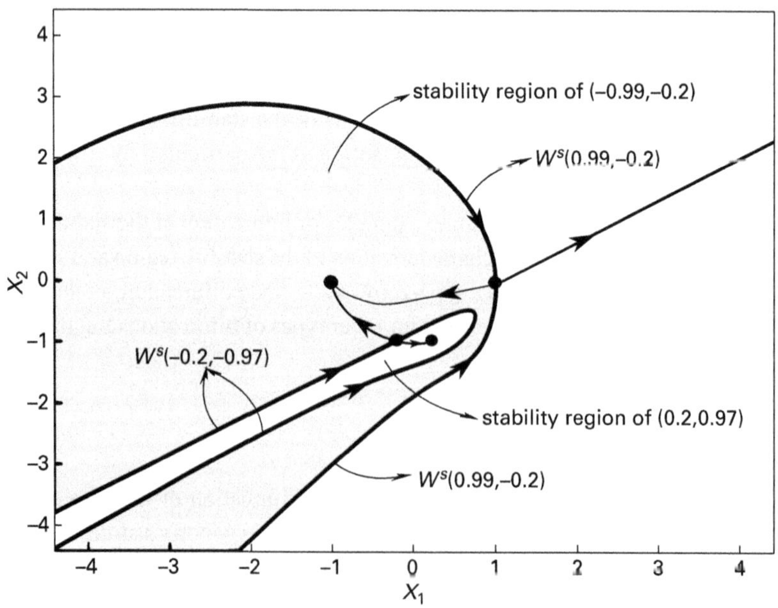

Figure 19.12 The phase portait of system (19.11) for $\lambda = -1.02$.

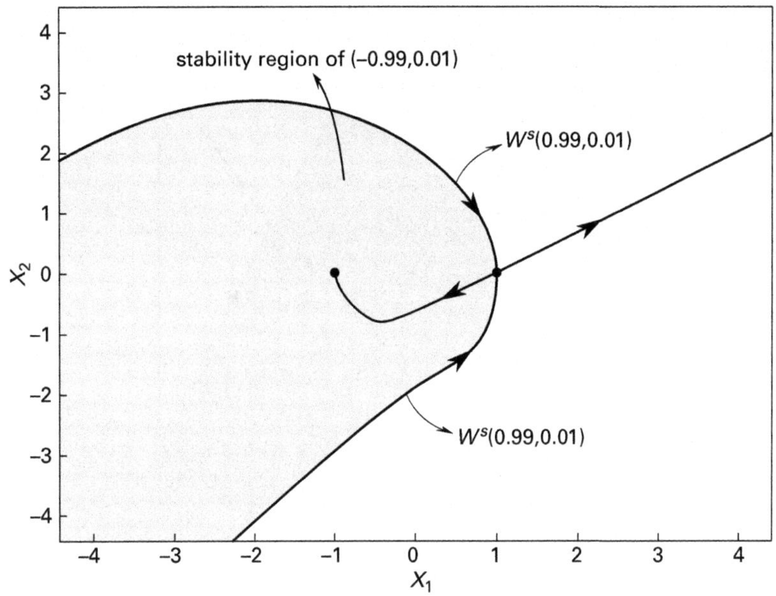

Figure 19.13 The phase portait of system (19.11) for $\lambda = -0.98$.

The next example illustrates a bifurcation of the stability boundary that is induced by a global bifurcation. More specifically, the appearance of a heteroclinic orbit breaks the transversality condition of assumption (A2).

Example 19-6 Consider the following nonlinear dynamical system:

$$\begin{aligned} \dot{x} &= y \\ \dot{y} &= 1 - 2\sin x + \lambda y \end{aligned} \tag{19.12}$$

which models a pendulum with a constant torque applied to the axis. This type of equation is also found in studies of power system stability and of Josephson junction models. The number λ is a positive parameter. This system possesses an infinite number of equilibrium points which are periodic repetitions of the following three equilibrium points: $x_s = \left(a\sin\frac{1}{2},\, 0\right)$, which is an asymptotically stable equilibrium point, $x_1 = \left(\pi - a\sin\frac{1}{2},\, 0\right)$ and $x_2 = \left(-\pi - a\sin\frac{1}{2},\, 0\right)$, which are type-one hyperbolic equilibrium points. These equilibrium points are invariant with respect to variation of the parameter λ. The phase portrait of system (19.12) for $\lambda = 0.57845$ is depicted in Figure 19.14. For this value of the parameter, there exists a saddle connection. More precisely, the unstable manifold of x_2 has a nontransversal intersection with the stable manifold of x_1. The stability region of the asymptotically stable equilibrium point is the gray area in the figure. Nontransversal intersections are not structurally stable, i.e. they can be easily broken by parameter variation. Figure 19.15 shows the phase

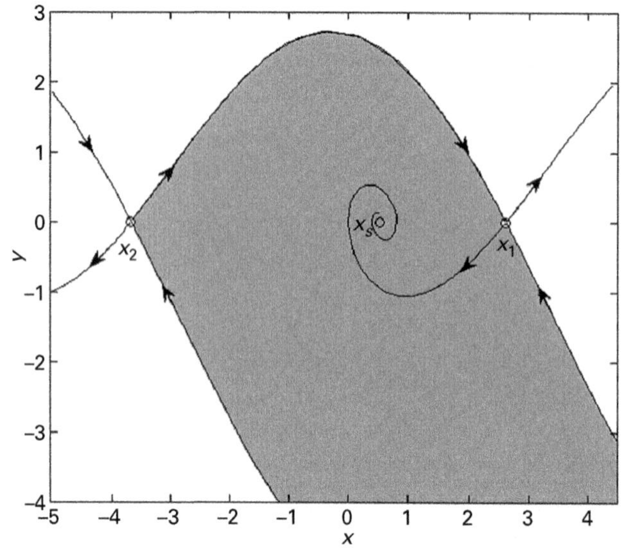

Figure 19.14 Phase portrait of system (19.12) for $\lambda = 0.57845$. The stable manifold of equilibrium point x_1 has a nontransversal intersection with the unstable manifold of the equilibrium point x_2.

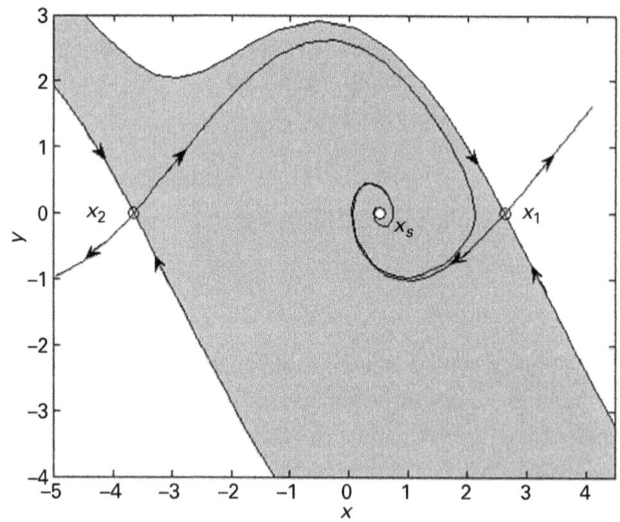

Figure 19.15 Phase portrait of system (19.12) for $\lambda = 0.65$. Both equilibrium points x_1 and x_2 lie on the stability boundary $\partial A(x_s)$.

portrait of the same system for $\lambda = 0.65$. In this case, both hyperbolic equilibrium points x_1 and x_2 lie on the stability boundary and the stability boundary is composed of the union of the stable manifold of these two equilibrium points. On the other hand, for $\lambda = 0.5$, only the equilibrium x_1 lies on the stability boundary, see Figure 19.16.

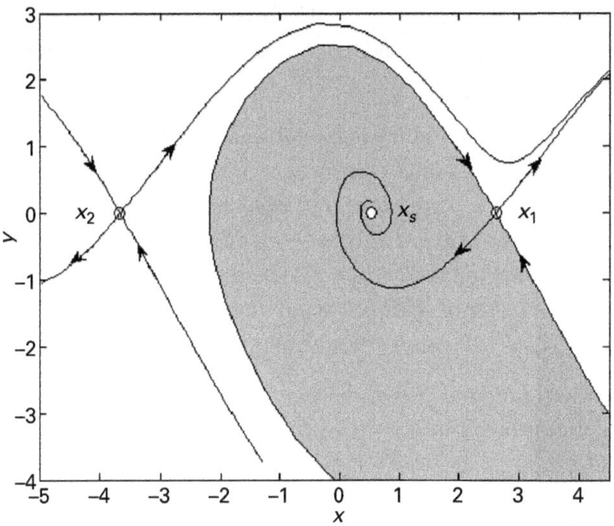

Figure 19.16 Phase portrait of system (19.12) for $\lambda = 0.5$. Only equilibrium point x_1 lies on the stability boundary $\partial A(x_s)$.

As the parameter λ passes through the bifurcation value $\lambda_0 = 0.57845$, the number of equilibrium points on the stability boundary as well as its form change. For values of λ smaller than λ_0, the stability boundary is composed of a single connected component, the stable manifold of x_1. For values of λ greater than λ_0, the stability boundary is composed of two connected components, the stable manifold of x_1 and the stable manifold of x_2.

19.5.2 Type-zero saddle-node bifurcations

In this section, we develop results that describe the behavior of the stability region and stability boundary in the neighborhood of a type-zero saddle-node bifurcation value. Similar results can be derived for other types of bifurcation. Nevertheless, we restrict ourselves to a detailed analysis of type-zero saddle-node bifurcations. These bifurcations are important in practice because they cause drastic changes in the "size" of the stability region.

Exploring the characterization of stability boundaries developed in this chapter, we study the behavior of the stability boundary in the neighborhood of a saddle-node bifurcation value. In particular, it is shown that stability regions bifurcate if type-zero saddle-node bifurcations occur on the stability boundary. A complete characterization of the bifurcation in the neighborhood of a type-zero saddle-node bifurcation value is presented in this section.

Hyperbolic equilibrium points that lie on the stability boundary at the type-zero bifurcation value λ_0 persist on the stability boundary under small variation of the parameters. In contrast, the non-hyperbolic type-zero saddle-node equilibrium point does not persist under parameter variation. The local behavior of the stability boundary in the neighborhood of a type-zero saddle-node equilibrium point is studied in the next theorem.

THEOREM 19-5 **(Stability boundary behavior near a type-zero saddle-node)**
Let x_{λ_0} be a type-zero saddle-node equilibrium point lying on the stability boundary $\partial A_{\lambda_0}(x^s_{\lambda_0})$ of the hyperbolic asymptotically stable equilibrium point $x^s_{\lambda_0}$ of (19.1) for $\lambda = \lambda_0$. If assumptions (A1)–(A3) are satisfied in an open interval containing λ_0, except at the type-zero saddle-node bifurcation value λ_0 where assumptions (A1′), (A2′) and (A3) are satisfied, and the number of equilibrium points on $\partial A_{\lambda_0}(x^s_{\lambda_0})$ is finite, then there is a neighborhood U of x_{λ_0}, $\beta' > 0$ and $\beta > 0$ such that the following holds.

(a) There are only two hyperbolic equilibrium points y^s_λ and y^u_λ on U of type zero and type one, respectively for all $\lambda \in (\lambda_0 - \beta', \lambda_0)$ and there is no equilibrium point on U for all $\lambda \in (\lambda_0, \lambda_0 + \beta')$. Moreover, the stable manifold of the type-zero equilibrium point and the unstable manifold of the type-one equilibrium point intersect at a one-dimension manifold.
(b) $y^u_\lambda \in \partial A_\lambda(x^s_\lambda) \cap \partial A_\lambda(y^s_\lambda)$ for all $\lambda \in (\lambda_0 - \beta, \lambda_0)$.
(c) $U \subset A_\lambda(x^s_\lambda)$ for all $\lambda \in (\lambda_0, \lambda_0 + \beta)$.

Theorem 19-5 studies the local behavior of the stability boundary in a neighborhood U of the type-zero saddle-node equilibrium point. For $\lambda < \lambda_0$ the type-one hyperbolic equilibrium point y^u_λ in U lies on the stability boundary of y^s_λ. As λ increases, a stable equilibrium point y^s_λ approaches the type-one equilibrium point y^u_λ. They coalesce, inside U at $\lambda = \lambda_0$ into a single type-zero saddle-node equilibrium point x_{λ_0}. At $\lambda = \lambda_0$, the type-zero saddle-node equilibrium point lies on the stability boundary of $x^s_{\lambda_0}$. As λ continues increasing, the equilibrium x_{λ_0} disappears and the neighborhood U of the saddle-node equilibrium point is now contained in the stability region of y^s_λ. Theorem 19-5 shows that the stability region and the stability boundary encounter drastic changes at $\lambda = \lambda_0$. The stability boundary intersects the neighborhood U for $\lambda \geq \lambda_0$ while U is totally contained in the stability region for $\lambda \geq \lambda_0$. The proof of Theorem 19-5 is presented in [9].

Let us now study the global behavior of the stability boundary under parameter variation. Given x_{λ_0} a saddle-node equilibrium point of type zero of (19.1), for a fixed parameter $\lambda = \lambda_0$, we define the following concept of weak stability region of x_{λ_0}.

DEFINITION **(Weak stability region)**
Let x_{λ_0} be a saddle-node equilibrium point of type zero of (19.1) for a fixed parameter $\lambda = \lambda_0$. The weak stability region *of x_{λ_0} is the set $S(x_{\lambda_0})$ of points $p \in R^n$ whose trajectories converge to x_{λ_0} as $t \to \infty$:*

$$S(x_{\lambda_0}) = \{p \in R^n : \phi_{\lambda_0}(t, p) \to x_{\lambda_0} \quad \text{as} \quad t \to \infty\}.$$

The weak stability region does not persist under small perturbations of the vector field f_{λ_0}, that is, both the type-zero saddle-node equilibrium point and $S(x_{\lambda_0})$ disappear with the variation of parameter λ. We explore the concept of weak stability region in the next theorem.

THEOREM 19-6 **(The inheritance of equilibrium points on the stability boundary)** *Let x_{λ_0} be a type-zero saddle-node equilibrium point lying on the stability boundary $\partial A_{\lambda_0}(x^s_{\lambda_0})$ of the hyperbolic asymptotically stable equilibrium point $x^s_{\lambda_0}$ of (19.1) for $\lambda = \lambda_0$. If assumptions (A1)–(A3) are satisfied in an open interval containing λ_0, except at the type-zero saddle-node bifurcation value λ_0 where assumptions (A1′), (A2′) and (A3) are satisfied, and the number of equilibrium points on $\partial A_{\lambda_0}(x^s_{\lambda_0})$ is finite, then there is a neighborhood U of x_{λ_0}, $\beta' > 0$, $\eta > 0$ and $\zeta > 0$ such that:*

(a) there are only two hyperbolic equilibrium points y^s_λ and y^u_λ on U of type zero and type one, respectively for all $\lambda \in (\lambda_0 - \beta', \lambda_0)$ and there is no equilibrium point on U for all $\lambda \in (\lambda_0, \lambda_0 + \beta')$.

Moreover, if $x^*_{\lambda_0}$ is a hyperbolic equilibrium point lying on the boundary of the weak stability region $S(x_{\lambda_0})$ for $\lambda = \lambda_0$, then:

(b) the perturbed equilibrium point $x^*_\lambda \in \partial A_\lambda(y^s_\lambda) \cup \partial A_\lambda(x^s_\lambda)$ for all $\lambda \in (\lambda_0 - \eta, \lambda_0)$,

(c) the perturbed equilibrium point $x^*_\lambda \in \partial A_\lambda(x^s_\lambda)$ for all $\lambda \in (\lambda_0, \lambda_0 + \eta)$.

See [9] for the proof of this theorem. Roughly speaking, Theorem 19-6 shows that the stability region of x^s_λ takes over the stability region of the stable equilibrium point y^s_λ that disappears in a saddle-node bifurcation at $\lambda = \lambda_0$. It shows that hyperbolic equilibrium points on the stability boundary of y^s_λ for $\lambda < \lambda_0$ are inherited by the stability boundary of x^s_λ as λ increases and passes through λ_0.

Using Theorem 19-5 and Theorem 19-6, we prove the following theorem regarding the complete characterization of the stability boundary in the neighborhood of a saddle-node bifurcation value.

THEOREM 19-7 **(Stability boundary characterization near a bifurcation value)** *Let x_{λ_0} be a type-zero saddle-node equilibrium point lying on the stability boundary $\partial A_{\lambda_0}(x^s_{\lambda_0})$ of the hyperbolic asymptotically stable equilibrium point $x^s_{\lambda_0}$ of system (19.1), for a fixed parameter $\lambda = \lambda_0$. Let hyperbolic equilibrium points $x^*_{\lambda^j_0}$, $j = 1, \ldots, m$ be on $\partial S(x_{\lambda_0})$. If assumptions (A1)–(A3) are satisfied in an open interval containing λ_0, except at the type-zero saddle-node bifurcation value λ_0, where assumptions (A1′), (A2′) and (A3) are satisfied, and the number of equilibrium points on $\partial A_{\lambda_0}(x^s_{\lambda_0})$ is finite. Then the following hold.*

(a) For $\lambda = \lambda_0$ we have

$$\partial A_{\lambda_0}(x^s_{\lambda_0}) = \bigcup_i W^s_{\lambda_0}(x_{\lambda^i_0}) \cup W^s_{\lambda_0}(x_{\lambda_0})$$

where $x_{\lambda^i_0}$, $i = 1, 2, \ldots, k$ are the hyperbolic equilibrium points on $\partial A_{\lambda_0}(x^s_{\lambda_0})$.

(b) There is $\varepsilon > 0$ such that, for all $\lambda \in (\lambda_0 - \varepsilon, \lambda_0)$,

$$\partial A_\lambda(x^s_\lambda) = \bigcup_i W^s_\lambda(x_{\lambda^i}) \cup W^s_\lambda(y^u_\lambda)$$

where x_{λ^i}, $i = 1, 2, \ldots, k$ are the perturbed hyperbolic equilibrium points on $\partial A_\lambda(x^s_\lambda)$ and y^u_λ is a type-one equilibrium point, which bifurcate from the type-zero saddle-node bifurcation, that also lies on $\partial A_\lambda(x^s_\lambda)$.

(c) There is $\varepsilon > 0$ *such that, for all* $\lambda \in (\lambda_0, \lambda_0 + \varepsilon)$,

$$\partial A_\lambda(x_\lambda^s) = \bigcup_i W_\lambda^s(x_{\lambda^i}) \bigcup_i W_\lambda^s(x_{\lambda^j}^*)$$

where x_{λ^i}, $i = 1, 2, \ldots, k$ *and* $x_{\lambda^j}^*$, $j = 1, \ldots, m$ *are the perturbed hyperbolic equilibrium points on* $\partial A_\lambda(x_\lambda^s)$.

We illustrate Theorem 19-5 and Theorem 19-6 in Figure 19.17, Figure 19.18 and Figure 19.19. Figure 19.17 shows stability regions of system (19.1), for $\lambda < \lambda_0$. There are two hyperbolic asymptotically stable equilibrium points x_λ^s and y_λ^s in this figure. The type-one equilibrium point y_λ^u belongs to the stability boundary of both asymptotically stable equilibrium points, while x_λ^* lies on the stability boundary of y_λ^s. As λ increases, the system undergoes a type-zero saddle-node bifurcation at $\lambda = \lambda_0$. Figure 19.18 depicts stability regions of system (19.1), for $\lambda = \lambda_0$. The equilibrium points y_λ^u and y_λ^s of Figure 19.17 are coalesced into a single equilibrium point x_{λ_0} in Figure 19.18. The equilibrium point x_{λ_0} is a type-zero saddle-node equilibrium point that belongs to the stability boundary of the asymptotically stable equilibrium point $x_{\lambda_0}^s$. The equilibrium

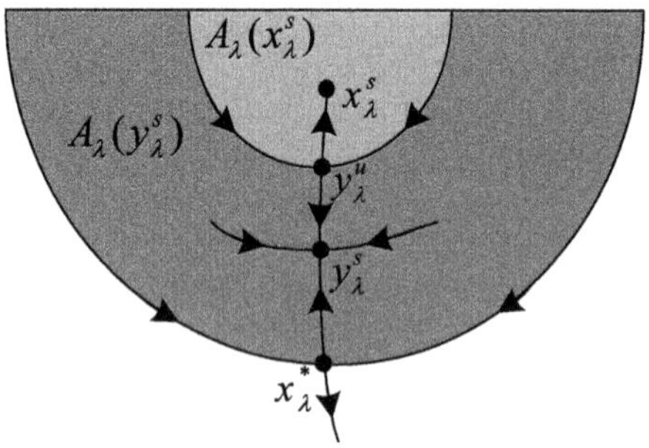

Figure 19.17 Stability regions of system (19.1) for $\lambda < \lambda_0$. The darkest shaded area is the stability region of the asymptotically stable equilibrium point y_λ^s while the lightest shaded area is the stability region of the asymptotically stable equilibrium point x_λ^s.

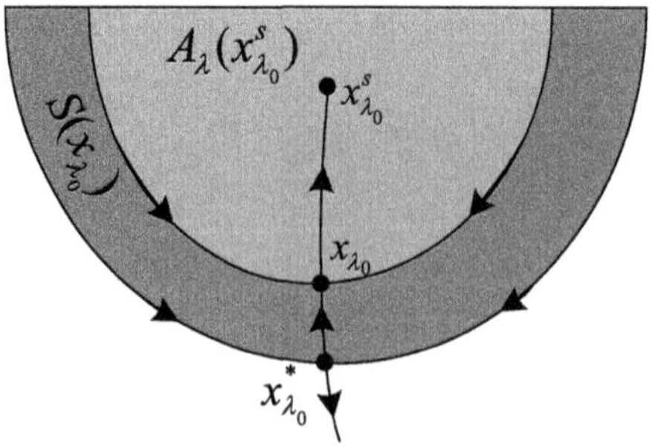

Figure 19.18 Stability regions of system (19.1) for $\lambda = \lambda_0$. The asymptotically stable equilibrium point y_λ^s coalesced with the type-one equilibrium point y_λ^u into a single type-zero saddle-node equilibrium point x_{λ_0}. The type-zero saddle-node equilibrium point x_{λ_0} lies on the stability boundary of $x_{\lambda_0}^s$ and the darkest shaded area is the weak stability region of x_{λ_0}.

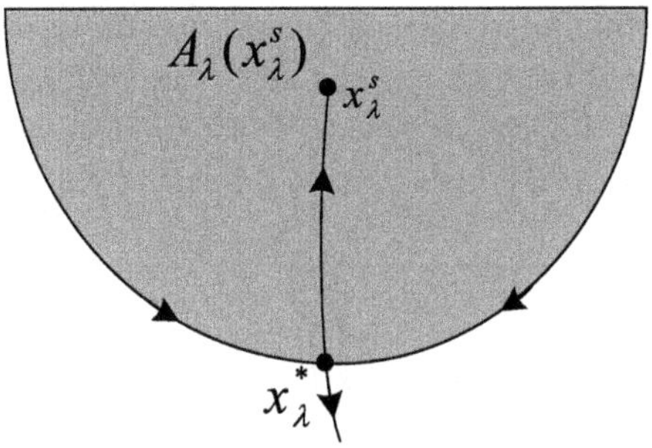

Figure 19.19 Stability regions of system (19.1) for $\lambda > \lambda_0$. The type-one hyperbolic equilibrium point x^*_λ was on the stability boundary of y^s_λ for $\lambda < \lambda_0$, and is now on the stability boundary of x^s_λ. The stability region of the equilibrium point x^s_λ takes over the stability region of y^s_λ.

point x^*_λ of Figure 19.17 persists. The equilibrium point $x^*_{\lambda_0}$, which was on the stability boundary of the asymptotically stable equilibrium point that had undergone a bifurcation, now lies on the boundary of the weak stability region of the type-zero saddle-node equilibrium point x_{λ_0}. As λ continues to increase, the type-zero saddle-node equilibrium point disappears and the perturbed type-one equilibrium point x^*_λ, which belonged to the stability boundary of y^s_λ for $\lambda < \lambda_0$, belongs now to the stability boundary of x^s_λ, Figure 19.19. In other words, the stability region of the perturbed equilibrium point x^s_λ "inherits" the stability region of the asymptotically stable equilibrium point that disappeared in the type-zero saddle-node bifurcation.

19.5.3 Example and applications

Example 19-7 Consider the system of differential equations

$$\begin{aligned} \dot{x}_1 &= x_1^4 - 1.25x_1^2 - x_2 + 0.25 \\ \dot{x}_2 &= -x_2 + \lambda \end{aligned} \tag{19.13}$$

with $(x_1, x_2) \in R^2$ and $\lambda \in R$. System (19.13) possesses for $\lambda_0 = 0.25$, three equilibrium points; they are $x_{\lambda_0} = (0, 0.25)$ a type-zero saddle-node equilibrium point, $x^s_{\lambda_0} = (-1.11, 0.25)$ a hyperbolic asymptotically stable equilibrium point and $x^*_{\lambda_0} = (1.11, 0.25)$ a type-one hyperbolic equilibrium point. The type-zero saddle-node equilibrium point belongs to the stability boundary of $x^s_{\lambda_0} = (-1.11, 0.25)$ and the type-one equilibrium point $x^*_{\lambda_0} = (1.11, 0.25)$ belongs to the boundary of the weak stability region of (0, 0.25). The stability boundary $\partial A_{\lambda_0}(-1.11, 0.25)$ is, according to Theorem 19-7, the stable manifold of the type-zero saddle-node equilibrium point (0, 0.25), see Figure 19.21. For $\lambda = 0.5$, system (19.13) possesses four equilibrium points; they are $x^*_{\lambda_0} = (-1.09, 0.2)$ a hyperbolic asymptotically stable equilibrium point, $x^*_{\lambda_0} = (1.09, 0.2)$ a type-one hyperbolic equilibrium point, $y^u_\lambda = (-0.2, 0.2)$ a type-one hyperbolic equilibrium point and $y^s_\lambda = (0.2, 0.2)$ a type-zero hyperbolic equilibrium point. The equilibrium points y_λ^u and y_λ^s originated from the type-zero saddle-node equilibrium point in a type-zero saddle-node bifurcation. Moreover, $y_\lambda^u \in \partial A_\lambda(-1.09, 0.2) \cap \partial A_\lambda(0.2, 0.2)$, according to Theorem 19-5, and $x_\lambda^* \in \partial A_\lambda(0.2, 0.2)$, confirming the results of Theorem 19-6, see Figure 19.20. For $\lambda = 0.3$,

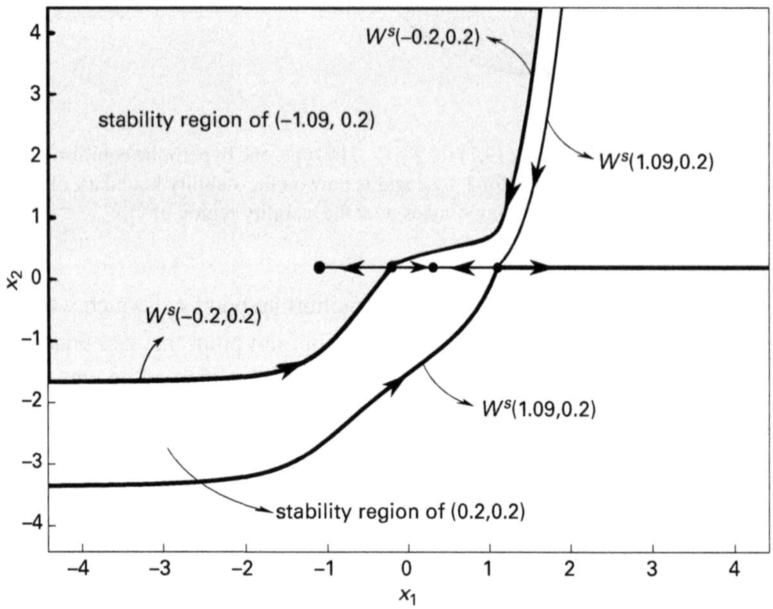

Figure 19.20 The phase portait of system (19.13) for $\lambda = 0.2$.

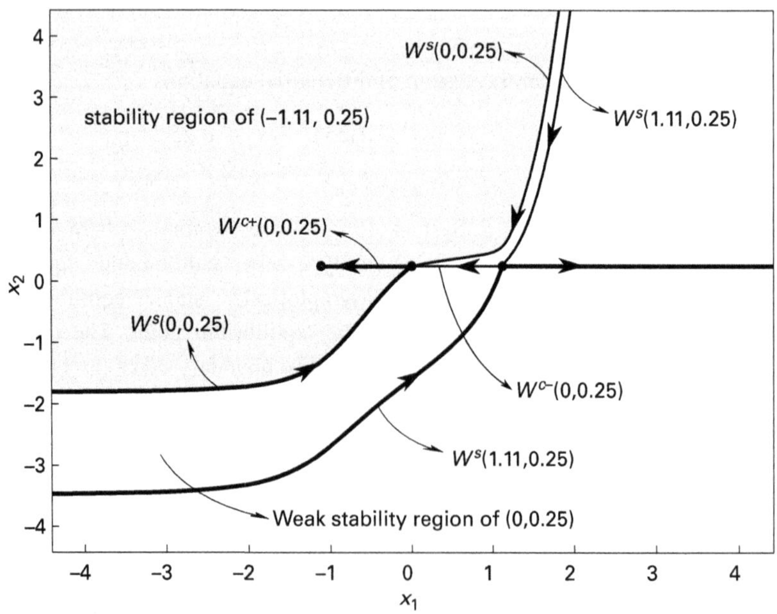

Figure 19.21 The phase portrait of system (19.13) for $\lambda_0 = 0.25$.

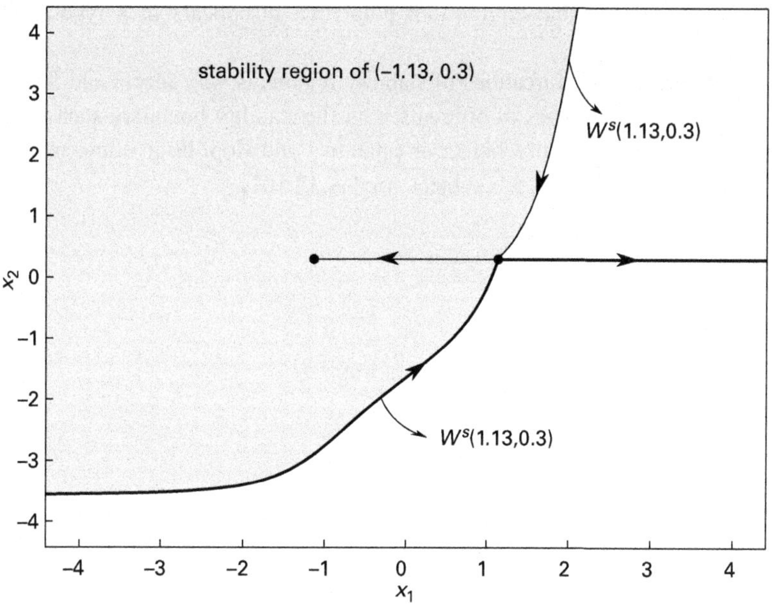

Figure 19.22 The phase portait of system (19.13) for $\lambda = 0.3$.

system (19.13) possesses two equilibrium points; they are $x_\lambda^s = (-1.13, 0.3)$ a hyperbolic asymptotically stable equilibrium point and $x_\lambda^* = (1.13, 0.3)$ a type-one equilibrium point, which belongs to the stability boundary $\partial A_\lambda(-1.13, 0.3)$ according to Theorem 19-6, see Figure 19.22.

19.6 Concluding remarks

In this chapter, we have studied the stability region of nonlinear dynamical systems under parameter variation. Bifurcations of the stability region can be induced by either local or global bifurcation of dynamical systems. Bifurcations of the stability region induced by saddle-node equilibrium points have been investigated. In particular, we have studied in detail the behavior of the stability boundary when a type-zero saddle-node bifurcation occurs on the stability boundary.

By studying the characterization of the stability boundary at the type-zero saddle-node bifurcation value λ_0, we have shown that the stability region encounters drastic changes as the parameter passes through λ_0. We have also shown that the stability region of the asymptotically stable equilibrium point that undergoes a type-zero saddle-node bifurcation is "inherited" by another asymptotically stable equilibrium point. In other words, the stability region of the asymptotically stable equilibrium point that persists is enlarged as the parameter passes through λ_0. It inherits the stability region of the asymptotically stable equilibrium point that disappears in a type-zero saddle-node bifurcation.

The theory of bifurcations of stability regions is very recent and future work in the analysis of other types of bifurcation on the stability boundary such as type-k saddle-node bifurcations with k higher or equal to 1 and Hopf bifurcations may prove fruitful. For further reading on this subject, see [11,12,102].

Part IV

Applications

20 Application of stability regions to direct stability analysis of large-scale electric power systems

20.1 Introduction

This chapter applies practical applications of the theory of relevant stability regions (presented in Chapter 8) and the optimal estimation scheme presented in Chapter 13 to direct stability analysis of large-scale electric power systems. The theoretical basis for direct stability analysis of power systems, without resort to numerical integration schemes, is the knowledge of stability regions: if an initial condition lies inside the stability region of a desired stable equilibrium point, then one can ensure, without performing numerical integrations, that the ensuing trajectory will converge to the desired stable equilibrium point.

Recent major blackouts in North America and Europe vividly demonstrated that power interruptions or blackouts can significantly impact the economy and are not acceptable to society. And yet, the ever increasing loading of transmission networks coupled with a steady increase in load demands have pushed the operating conditions of many power systems worldwide ever closer to their stability limits. This problem of reduced operating security margins is being further compounded by factors such as (i) the increasing number of bulk power interchange transactions and non-utility generators, and (ii) the trend toward installing higher output generators with lower inertia constants and higher short-circuit ratios. Under these conditions, it is now well recognized that any violation of power system dynamic security limits leads to far-reaching impacts on the entire power system.

By nature, a power system continually experiences two types of disturbances: *event disturbances* and *load variations*. Event disturbances (contingencies) include loss of generating units or transmission components (lines, transformers, substations) due to short-circuits caused by lightning, high winds, failures such as incorrect relay operations or insulation breakdown, sudden large load changes, or a combination of such events. Power systems are planned and operated to withstand the occurrence of certain disturbances. The North American Electric Reliability Council defines security as the ability to prevent cascading outages when the bulk power supply is subjected to severe disturbances [22]. The specific criteria which must be met are set by individual reliability councils. Each council establishes the types of disturbances which its system must withstand without cascading outages.

Power system stability analysis is concerned with the ability of a power system to reach an acceptable steady state (operating condition) following a disturbance. Stability analysis is one of the most important tasks in power system operations and planning. Today, stability analysis programs are being used by power system planning and operating engineers to simulate the response of the system to various credible disturbances. In these simulations, the dynamic behavior of a present or proposed power system is examined to determine whether stability has been maintained or lost. For operational purposes, power system stability analysis plays an important role in determining the system operating limits and operating guidelines. During the planning stage, power system stability analysis is performed to check relay settings, to set the parameters of control devices, or to assess the need for additional facilities and the locations at which to place additional control devices in order to enhance the system's static and dynamic security. Important conclusions and decisions about power system operations and planning are made based on the results of stability studies.

Transient stability problems, a class of power system stability problems, have been a major operating constraint in regions that rely on long-distance transfers of bulk power (e.g. in most parts of the Western Interconnection of the USA, Hydro Quebec, the interfaces between Ontario/New York area and the Manitoba/Minnesota area, in certain parts of China and Brazil). The trend now is that many parts of the various interconnected systems are becoming constrained by transient stability limitations. Hence, it is imperative to develop powerful tools to examine power system stability in a timely and accurate manner and to derive necessary control actions for both preventive control and enhancement control.

An on-line transient stability assessment (TSA) is an essential tool to avoid any violation of dynamic security limits. Indeed, with current power system operating environments, it is increasingly difficult for power system operators to generate all the operating limits for all possible operating conditions under a list of credible contingencies. Hence, it is imperative to develop a reliable and effective on-line TSA to obtain the operating security limits at or near real time in response to variability and uncontrollable renewable energies. In addition to this important function, power system transmission open access and restructuring further reinforce the need for the on-line TSA as it is the base upon which available transfer capability, dynamic congestion management problems and special protection systems can be effectively resolved. There are significant engineering and financial benefits expected from the on-line TSA.

From a computational viewpoint, the on-line TSA requires the handling of a mathematical model which is described by a large set of nonlinear differential equations in addition to a set of nonlinear algebraic equations. At present, stability analysis programs routinely used in utilities around the world are mostly based on step-by-step numerical integrations of power system stability models to simulate system dynamical behaviors. This practice of power system stability analysis based on the time-domain approach has a long history [22,95,151,224]. However, due to the nature of the time-domain approach, it has several disadvantages: (i) it requires intensive, time-consuming computation effort, therefore it has not been suitable for on-line application; (ii) it does not provide information as to how to derive preventive control when the system is deemed unstable and how to derive enhancement control when the system is deemed critically stable; and (iii) it does not provide information regarding the degree of stability (when the system is stable) and the degree of instability (when the system is unstable). This piece of information is valuable for both planning and operations.

An alternative approach to transient stability analysis employing energy functions, called *direct methods*, was originally proposed by Magnusson [172] in the late 1940s, and pursued in the 1950s by Aylett [18]. Direct methods have a long development history spanning seven decades. A comprehensive theoretical foundation for direct methods was developed in the 1980s [50]. Direct methods can determine transient stability without the time-consuming numerical integration of the (post-fault) power system [44,51,65]. In addition to speed, direct methods also provide a quantitative measure of the degree of system stability. This additional information makes direct methods very attractive when the relative stability of different network configuration plans must be compared or when system operating limits constrained by transient stability must be calculated quickly. Another advantage of direct methods is the ability to provide useful information regarding how to derive preventive control actions when the underlying power system is deemed unstable and how to derive enhancement control actions when the underlying power system is deemed critically stable.

After decades of research and developments in the energy-function-based direct methods and the time-domain simulation approach, it has become clear that the capabilities of direct methods and of the time-domain approach complement each other. The current direction of development is to include appropriate direct methods and time-domain simulation programs within the body of overall power system stability simulation programs [22,89,244]. For example, the direct method provides the advantages of fast computational speed and energy margins which make it a good complement to the traditional time-domain approach. The energy margin and its functional relations to certain power system parameters are an effective complement to develop tools such as preventive control schemes for credible contingencies which are unstable and to develop fast calculators for available transfer capability limited by transient stability. The direct method can also play an important role in the dynamic contingency screening for on-line transient stability assessment.

This chapter gives an overview of state-of-the-art direct methods for the on-line TSA. In particular, TEPCO-BCU methodology will be presented in detail. TEPCO-BCU has been evaluated on several practical power system models. The evaluation results reveal that theoretical work on the characterization of stability regions can be developed into a reliable tool for on-line transient stability assessment of practical power systems.

20.2 Stability problem and system model

The complete power system model for calculating the system dynamic response relative to a disturbance comprises a set of first-order differential equations

$$\dot{x} = f(x, y, u) \tag{20.1}$$

describing the internal dynamics of devices such as generators, their associated control systems, certain loads and other dynamically modeled components, and a set of algebraic equations

$$0 = g(x, y, u) \tag{20.2}$$

describing the electrical transmission system (the interconnections between the dynamic devices) and internal static behaviors of passive devices (such as static loads, shunt capacitors, fixed transformers and phase shifters). The differential equations (20.1) typically describe the dynamics of the speed and angle of generator rotors, flux behaviors in generators, the response of generator control systems such as excitation systems, voltage regulators, turbines, governors and boilers, the dynamics of equipment such as synchronous VAR compensators (SVCs), DC lines and their control systems, and the dynamics of dynamically modeled loads such as induction motors. The stated variables x typically include generator rotor angles, generator velocity deviations (speeds), mechanical powers, field voltages, power system stabilizer signals, various control system internal variables, and voltages and angles at load buses (if dynamic load models are employed at these buses). The algebraic equations (20.2) comprise the stator equations of each generator, the network equations of transmission networks and loads, and the equations defining the feedback stator quantities. The forcing functions u acting on the differential equations are terminal voltage magnitudes, generator electrical powers, and signals from boilers and automatic generation control systems. Some control system internal variables have upper bounds on their values due to their physical saturation effects. Let z be the vector of these constrained state variables; then the saturation effects can be expressed as

$$0 < z(t) \leq \bar{z}. \tag{20.3}$$

A detailed description of Eqs. (20.1)–(20.3) for each component can be found, for example, in [95,224]. For a 900-generator, 14,000-bus power system, the number of differential equations can easily reach as many as 20,000 while the number of nonlinear algebraic equations can easily reach as many as 32,000. The sets of differential equations (20.1) are usually loosely coupled.

To protect power systems from damage due to disturbances, protective relays are placed strategically throughout the power system to detect faults (disturbances) and to trigger the opening of circuit breakers necessary to isolate faults. These relays are designed to detect defective lines and apparatus or other power system conditions of an abnormal or dangerous nature and to initiate appropriate control circuit actions. Due to the action of these protective relays, a power system subject to an event disturbance can be viewed as going through changes in its network configuration in three stages: from the pre-fault, to the fault-on, and finally to the post-fault system. The pre-fault system is in a stable steady state; when an event disturbance occurs, the system then moves into the fault-on system before it is cleared by protective system operations. Stated more formally, in the pre-fault regime, the system is at a known stable equilibrium point (SEP), say (x_s^{pre}, y_s^{pre}). At some time t_0, the system undergoes a fault (an event disturbance), which results in a structural change in the system due to actions from relay and circuit breakers. Suppose the fault duration is confined to the time interval $[t_0, t_{cl}]$.

During this interval, the fault-on system is described by (for ease of exposition, the saturation effects expressed as $0 < z(t) \leq \bar{z}$ are neglected in the following):

$$\begin{aligned} \dot{x} &= f_F(x, y), \quad t_0 \leq t < t_{cl} \\ 0 &= g_F(x, y) \end{aligned} \tag{20.4}$$

where $x(t)$ is the vector of state variables of the system at time t. Sometimes, the fault-on system may involve more than one action from system relays and circuit breakers. In these cases, the fault-on systems are described by several sets of nonlinear equations:

$$\begin{aligned} \dot{x} &= f_F^1(x, y), \quad t_0 \leq t \leq t_{F,1} \\ 0 &= g_F^1(x, y) \\ \dot{x} &= f_F^2(x, y), \quad t_{F,1} \leq t \leq t_{F,2} \\ 0 &= g_F^2(x, y) \\ &\ldots \\ \dot{x} &= f_F^k(x, y), \quad t_{F,k} \leq t \leq t_{cl} \\ 0 &= g_F^k(x, y). \end{aligned} \tag{20.5}$$

The number of sets of equations equals the number of separate actions due to system relays and circuit breakers. Each set depicts the system dynamics due to one action from relays and circuit breakers. Suppose the fault is cleared at time t_{cl} and no additional protective actions occur after t_{cl}. The system, termed the post-fault system, is henceforth governed by post-fault dynamics described by

$$\begin{aligned} \dot{x} &= f_{PF}(x, y), \quad t_{cl} \leq t < \infty \\ 0 &= g_{PF}(x, y). \end{aligned} \tag{20.6}$$

The network configuration may or may not be the same as the pre-fault configuration in the post-fault system. We will use the notation $z(t_{cl}) = (x(t_{cl}), y(t_{cl}))$ to denote the fault-on state at switching time t_{cl}. The fault-on trajectory, post-fault trajectory due to an event disturbance corresponds respectively to the solution of Eq. (20.5) over the fault-on time period $t_0 \leq t < t_{cl}$ and to the solution of Eq. (20.6) over the post-fault time period $t_{cl} \leq t < t_{\infty}$.

The fundamental problem of power system stability due to a fault (i.e. event disturbance) can be roughly stated as follows: given a pre-fault SEP and a fault-on trajectory, will the post-fault trajectory settle down to an acceptable steady state?

A time-domain simulation program simulates the system trajectory starting from a pre-fault SEP, then moving to the fault-on trajectory and then the post-fault trajectory. If the post-fault trajectory settles down to an acceptable stable equilibrium point, then the post-fault system will be stable.

20.3 Direct methods for transient stability analysis

Of the several direct methods evolved in the last several decades, the current direct method, known as the controlling unstable equilibrium point (CUEP) method, is well recognized and widely accepted for its effectiveness. It uses an algorithmic procedure to determine, based on the energy function theory and controlling UEP, whether or not the post-fault trajectory will remain stable, without integrating the post-fault system model. The CUEP method assesses the stability property of the post-fault trajectory by comparing the system energy at the initial state of the post-fault trajectory with a critical energy value. The CUEP method not only avoids the time-consuming numerical integration of the post-fault system, but also provides a quantitative measure of the degree of system stability. Moreover, the CUEP method has a solid theoretical foundation which has been built upon the theory of stability regions, in particular, the theory of the relevant stability region.

Given a power system transient stability model with a specified fault-on system and a specified post-fault system, CUEP methods for transient stability analysis consist of four key steps.

Step 1: Construct an energy function for the post-fault power system.
Step 2: Compute the energy value at the initial point of the post-fault system.
Step 3: Compute the controlling UEP and use its energy function value as the critical energy for the fault-on trajectory.
Step 4: Perform direct stability assessments by comparing the energy computed at Step 2 with the critical energy computed at Step 3. If the former is smaller than the latter, then the post-fault trajectory will be stable. Otherwise, it may be unstable.

In Step 4, CUEP methods determine whether or not the post-fault trajectory is stable solely based on comparing the system energy (constructed in Step 1) at the initial point of post-fault system (computed in Step 2) with the critical energy (computed in Step 3). It is hence very important to correctly calculate the critical energy values.

The theoretical basis of CUEP methods for the direct stability assessment of a post-fault power system is the knowledge of a stability region; if the initial condition of the post-fault system lies inside the stability region of a desired post-fault stable equilibrium point, then one can ensure, without performing numerical integrations, that the ensuing post-fault trajectory will converge to the desired point.

We now formalize the CUEP method. We start with an example. Let us observe Figure 20.1 which depicts a fault-on trajectory $x_f(t)$ which moves toward the stability boundary $\partial A(x_s)$ and intersects it at the exit point x_e. The so-called critical clearing time (CCT) is the time difference between the pre-fault SEP and the exit point. If the fault is cleared before the fault-on trajectory reaches the exit point, say at $x(t_{cl}) = x_{cr}$, then the fault-clearing point must lie inside the stability region of the post-fault SEP. Hence, the post-fault trajectory starting from the fault-clearing point must converge to the post-fault SEP x_s and the post-fault trajectory is stable (see Figure 20.1).

Now the issue is how to determine whether or not the fault-clearing point $x(t_{cl})$ lies inside the stability region of the post-fault SEP without performing a time-domain simulation or without knowledge of the exit point. The controlling UEP method addresses this issue as follows. Instead of computing the exit point, the controlling UEP method approximates the relevant stability boundary by the constant energy surface passing through the controlling UEP. The method replaces the task of detecting the exit point with the task of detecting the intersection between the fault-on trajectory and

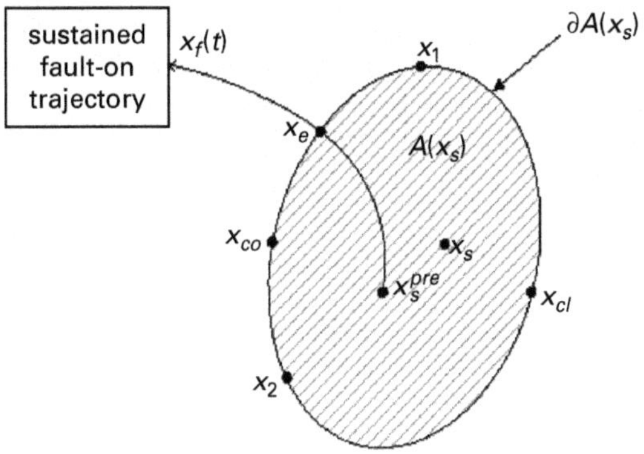

Figure 20.1 If the fault is cleared before the fault-on trajectory reaches the exit point, say at $x(t_{cl})$, then the fault-clearing point must lie inside the stability region of the post-fault SEP. Hence, the post-fault trajectory starting from the fault-clearing point must converge to the post-fault SEP x_s.

the constant energy surface passing through the controlling UEP. The latter task only involves a comparison between the energy at the initial state of the post-fault trajectory and the energy at the controlling UEP and is much easier than the former task.

We explain Step 1 and Step 4 in the next two sections.

20.4 Constructing energy functions

It can be shown that an analytical expression of energy functions does not exist for general lossy network-preserving transient stability models [47]. Consequently, numerical energy functions are employed. We present procedures to derive numerical energy functions for structure-preserving transient stability models. A majority of the existing network-preserving models can be rewritten as a set of general differential-algebraic equations of the following compact form [68]:

$$\begin{aligned}
0 &= -\frac{\partial U}{\partial u}(u, w, x, y) + g_1(u, w, x, y) \\
0 &= -\frac{\partial U}{\partial w}(u, w, x, y) + g_2(u, w, x, y) \\
T\dot{x} &= -\frac{\partial U}{\partial x}(u, w, x, y) + g_3(u, w, x, y) \\
\dot{y} &= z \\
M\dot{z} &= -Dz - \frac{\partial U}{\partial y}(u, w, x, y) + g_4(u, w, x, y)
\end{aligned} \tag{20.7}$$

where $u \in R^l$ and $w \in R^k$ are instantaneous variables while $x \in R^n$, $y \in R^m$ and $z \in R^m$ are state variables. T is a positive definite matrix, M and D are diagonal positive definite matrices. Here differential equations describe generator and/or load dynamics while algebraic equations express the power flow equations at each bus. Functions $g_1(u, w, x, y)$, $g_2(u, w, x, y)$, $g_3(u, w, x, y)$ and $g_4(u, w, x, y)$ are vectors representing the effects of the transfer conductance in the network Y-bus matrix. With the aid of the singularly perturbed systems, the compact representation of the network-preserving model becomes:

$$\begin{aligned}
\varepsilon_1 \dot{u} &= -\frac{\partial U}{\partial u}(u, w, x, y) + g_1(u, w, x, y) \\
\varepsilon_2 \dot{w} &= -\frac{\partial U}{\partial w}(u, w, x, y) + g_2(u, w, x, y) \\
T\dot{x} &= -\frac{\partial U}{\partial x}(u, w, x, y) + g_3(u, w, x, y) \\
\dot{y} &= z \\
M\dot{z} &= -Dz - \frac{\partial U}{\partial y}(u, w, x, y) + g_4(u, w, x, y)
\end{aligned} \tag{20.8}$$

where ε_1 and ε_2 are sufficiently small positive numbers. For the compact representation of the singularly perturbed network-preserving power system model (20.8) without the transfer conductance, we consider the following function $W: R^{k+l+2m+n} \to R$

$$W(u, w, x, y, z) = K(z) + U(u, w, x, y) = \frac{1}{2} z^T M z + U(u, w, x, y). \tag{20.9}$$

Suppose that along every nontrivial trajectory of system (20.8) with a bounded value of $W(u, w, x, y, z)$, the vector $(u(t), w(t), x(t))$ is also bounded for $t \in R^+$. Then $W(u, w, x, y, z)$ is an energy function for system (20.8).

A numerical network-preserving energy function $W_{num}(u, w, x, y)$ can be constructed by combining an analytic energy function $W_{ana}(u, w, x, y, z) = K(z) + U(u, w, x, y)$ and a path dependent potential energy $U_{path}(u, w, x, y)$, i.e.

$$\begin{aligned}
W_{num}(u, w, x, y, z) &= W_{ana}(u, w, x, y, z) + U_{path}(u, w, x, y) \\
&= K(z) + U(u, w, x, y) + U_{path}(u, w, x, y) \\
&= K(z) + U_{num}(u, w, x, y).
\end{aligned} \tag{20.10}$$

Details for the derivation of a numerical energy function for general power system stability models can be found in [45,68,212] and references therein.

20.5 The controlling UEP method

The fundamental problem of power system transient stability analysis is concerned with whether or not, given a pre-fault SEP and a fault-on trajectory, the post-fault trajectory will converge to a post-fault SEP at which all the engineering and operational constraints are satisfied. In the context of direct methods for transient stability analysis, we note that given a point in the state space (say, the initial point of the post-fault system), it is generally difficult to determine which connected component of a level set contains the point simply by comparing the energy at the given point and the energy of the level set. This is due to the fact that a level set usually contains several disjoint connected components and these components are not easy to differentiate based on an energy function value.

Fortunately, in the context of CUEP methods, this difficulty can be circumvented since CUEP methods compute the relevant pieces of information regarding (i) a pre-fault stable equilibrium point, (ii) a fault-on trajectory and (iii) a post-fault stable equilibrium point. These pieces of information are sufficient to identify the connected component of a level set that contains the initial point of the post-fault system. We next discuss the controlling UEP method for direct stability analysis.

Given a power system model with a pre-fault SEP x_s^{pre} a fault-on trajectory $x_f(t)$, and a post-fault (transient) stability system S^{post} with a post-fault SEP x_s^{post}, suppose there exists an energy function for the post-fault system S^{post} and x_s^{pre} lies inside the stability region of x_s^{post}. We next review the definition of the controlling UEP of Chapter 8.

DEFINITION

The controlling UEP *with respect to the fault-on trajectory $x_f(t)$ is the UEP of the post-fault system whose stable manifold contains the exit point of $x_f(t)$ (i.e. the controlling UEP is the first UEP whose stable manifold intersects the fault-on trajectory $x_f(t)$ at the exit point).*

This definition is motivated by the facts that a sustained fault-on trajectory must exit the stability boundary of a post-fault system and that the exit point (i.e. the point from which a given fault-on trajectory exits the stability boundary of a post-fault system) of the fault-on trajectory must lie on the stable manifold of a UEP on the stability boundary of the post-fault system. This UEP is the controlling UEP of the fault-on trajectory. Note that the existence and uniqueness of the controlling UEP with respect to a fault-on trajectory are assured and that the controlling UEP is independent of the energy function used in the direct stability assessment. With the formal definition of the controlling UEP, we are in a position to formalize the controlling UEP method.

20.5.1 The controlling UEP method

The controlling UEP method for direct stability analysis of large-scale power systems proceeds as follows.

Step 1: Construct a numerical energy function for the post-fault system, say $V(\cdot)$.

Step 2: Compute the initial point of the post-fault system, say $x_f(t_{cl})$, and calculate its energy function value, say $V_f = V(x_f(t_{cl}))$.

Step 3: Determine the critical energy

Step 3.1: Find the controlling UEP, say x_{co}, for a given fault-on trajectory $x_f(t)$.

Step 3.2: The critical energy, V_{cr}, is the value of energy function $V(\cdot)$ at the controlling UEP, i.e. $V_{cr} = V(x_{co})$.

Step 4: Direct stability assessment: if $V_f < V_{cr}$, then the post-fault trajectory is stable; otherwise, it can be unstable.

The controlling UEP method yields an approximation of the relevant part of the stability boundary of the post-fault system to which the fault-on trajectory is heading. It uses the (connected) constant energy surface passing through the controlling UEP to

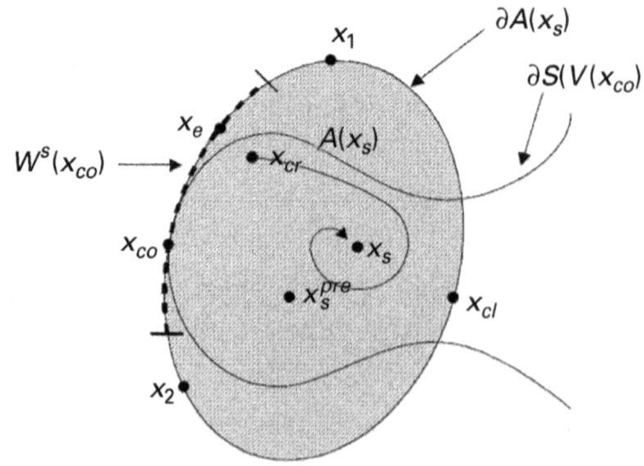

Figure 20.2 The controlling UEP method approximates the relevant stability boundary, the stable manifold of the controlling UEP $W^s(x_{co})$, by the constant energy surface, $\partial S(V(x_{co}))$, passing through the controlling UEP.

approximate the relevant stability boundary. If the fault is cleared before the fault-on trajectory reaches the constant energy surface containing the controlling UEP, then the post-fault trajectory will converge to the SEP and is stable (see Figure 20.2). To check when the fault-on trajectory passes through the constant energy surface containing the controlling UEP, one can monitor the value $V(x_f(t))$ along the fault-on trajectory $x_f(t)$ and identify when $V(x_f(t)) = V(x_{co})$ (see Figure 20.2).

20.5.2 Challenges in computing the controlling UEP

The task of finding the (exact) controlling UEP of a given fault for general power system models is very difficult. This difficulty comes in part from the following complexities.

(1) The controlling UEP is a particular UEP in a large-degree state-space.
(2) The controlling UEP is the first UEP whose stable manifold is hit by the fault-on trajectory (at the exit point).
(3) The task of computing the exit point is very involved; it usually requires a time domain approach.
(4) The task of computing the controlling UEP is complicated further by the size and the shape of its convergence region.

The ability to compute the controlling UEP (CUEP) is vital in direct stability analysis. The computational challenges and complexities of computing the controlling UEP described above serve to explain why the many existing methods fail to compute the CUEP. It is hence fruitful to develop tailored methods for computing the controlling UEP by exploring special properties as well as some physical and mathematical characteristics of the underlying power system transient stability model. In the next section, we will present such a method for computing the controlling UEP: the boundary of stability-region-based controlling unstable equilibrium point method (BCU method). The BCU method is thus far the only method which can reliably compute the CUEP.

20.6 The BCU method and theoretical basis

The BCU method is a systematic method for finding the controlling UEP [46,64]. The method has also been given other names such as the exit point method [212,224] and the hybrid method [95]. The BCU method has been evaluated in a large-scale power system and it has been compared favorably with other methods in terms of its reliability and the required computational effort [190,212]. The BCU method has been studied by several researchers, see for example, [4,86,167,197,198]. Descriptions of the BCU method can be found in books such as [45,95,151,199,224]. The theoretical foundation of the BCU method has been established in [46,50,51]. The BCU method and BCU classifiers have several practical applications. For example, a demonstration of the capability of the BCU method for on-line transient stability assessments using real-time data was held at two utilities, the Ontario Hydro Company and the Northern States Power Company [143,190]. The BCU method was implemented into EPRI TSA software which was integrated into an EMS installed at the control center of Northern States Power Company [85]. A TSA system, composed of the BCU classifiers, the BCU method and a time-domain simulation engine, was developed and integrated into the Ranger EMS system [58]. The TSA system has been installed and commissioned, as part of an EMS system, at several energy control centers. The BCU method has been applied to fast derivation of power transfer limits [251] and to real power rescheduling to increase dynamic security [152]. The BCU method has been improved, expanded and extended into the integrated package TEPCO-BCU [66,242,243].

We next present the BCU method from two viewpoints: numerical aspects and theoretical aspects. In developing a BCU method for a given power system stability model, an associated artificial, reduced-state model must be defined. The reason behind the reduced-state model is the following. It is very difficult, if not impossible, to directly compute the CUEP of the original power system stability model without resorting to the time-domain simulation approach. Furthermore, it is also very challenging to compute the CUEP of the original stability model using a time-domain simulation approach. This may serve to explain why the many existing methods fail to compute the CUEP.

To explain the reduced-state model, we consider the following generic network-preserving transient stability model,

$$\begin{aligned}
0 &= -\frac{\partial U}{\partial u}(u, w, x, y) + g_1(u, w, x, y) \\
0 &= -\frac{\partial U}{\partial w}(u, w, x, y) + g_2(u, w, x, y) \\
T\dot{x} &= -\frac{\partial U}{\partial x}(u, w, x, y) + g_3(u, w, x, y) \\
\dot{y} &= z \\
M\dot{z} &= -Dz - \frac{\partial U}{\partial y}(u, w, x, y) + g_4(u, w, x, y)
\end{aligned} \tag{20.11}$$

where $U(u, w, x, y)$ is a scalar function. It has been shown that the above canonical equations can represent existing transient stability models. In the context of the BCU method, the above model is termed the original model. Regarding the original model (20.11), we choose the following differential-algebraic system as the reduced-state model associated with the original model (20.11):

$$\begin{aligned} 0 &= -\frac{\partial U}{\partial u}(u, w, x, y) + g_1(u, w, x, y) \\ 0 &= -\frac{\partial U}{\partial w}(u, w, x, y) + g_2(u, w, x, y) \\ T\dot{x} &= -\frac{\partial U}{\partial x}(u, w, x, y) + g_3(u, w, x, y) \\ \dot{y} &= -\frac{\partial U}{\partial y}(u, w, x, y) + g_4(u, w, x, y). \end{aligned} \tag{20.12}$$

There are several close relationships between the reduced-state model (20.12) and the original model (20.11). The fundamental ideas behind the BCU method can be explained as follows. Given a power system stability model (which admits an energy function), say the original model (20.11), the BCU method first explores the special properties of the underlying model with the aim of defining a reduced-state model, say the model described in (20.12), such that the following static as well as dynamic relationships are met.

Static relations

(S1) The locations of equilibrium points of the reduced-state model (20.12) correspond to the locations of equilibrium points of the original model (20.11).

(S2) The types of equilibrium points of the reduced-state model are the same as those of the original model.

Dynamical relations

(D1) There exists an energy function for the reduced-state model (20.12).

(D2) An equilibrium point, say $(\hat{u}, \hat{w}, \hat{x}, \hat{y})$, is on the stability boundary of the reduced-state model (20.12) if and only if the equilibrium point $(\hat{u}, \hat{w}, \hat{x}, \hat{y}, 0)$ is on the stability boundary of the original model (20.11).

(D3) It is computationally feasible to detect the point at which the projected fault-on trajectory $(u(t), w(t), x(t), y(t))$ intersects the stability boundary $\partial A(u_s, w_s, x_s, y_s)$ of the post-fault reduced-state model (20.12) without resorting to an iterative time-domain procedure.

The dynamic relationship (D3) plays an important role in the development of the BCU method to circumvent the difficulty of applying an iterative time-domain procedure to compute the exit point of the original model. The BCU method computes the controlling UEP of the reduced-state model (20.12) by exploring the stability boundary and the energy function of the reduced-state model (20.12). The BCU method then uses the controlling UEP of the reduced-state model (20.12) to find the controlling UEP of the original model (20.11).

A conceptual BCU method

Step 1: Simulate the fault-on trajectory $(u(t), w(t), x(t), y(t), z(t))$ of the network-preserving model (20.11), detect the exit point (u^*, w^*, x^*, y^*) at which the projected fault-on trajectory $(u(t), w(t), x(t), y(t))$ exits the stability boundary of the post-fault reduced-state model (20.12).

Step 2: Use the exit point (u^*, w^*, x^*, y^*), detected in Step 1, as the initial condition and integrate the post-fault reduced-state model to an equilibrium point. Let the solution be $(u_{co}, w_{co}, x_{co}, y_{co})$.

Step 3: The controlling UEP with respect to the fault-on trajectory of the original network-preserving model (20.11) is $(u_{co}, w_{co}, x_{co}, y_{co}, 0)$. The energy function at $(u_{co}, w_{co}, x_{co}, y_{co}, 0)$ is the critical energy for the fault-on trajectory $(u(t), w(t), x(t), y(t), z(t))$.

Step 1 and Step 2 of the conceptual BCU method compute the controlling UEP of the reduced-state system. In Step 2, starting from the exit point (u^*, w^*, x^*, y^*), the reduced-state trajectory will converge to an equilibrium point, which is the CUEP relative to the reduced-state trajectory. The stable manifold of the reduced-state controlling UEP $(u_{co}, w_{co}, x_{co}, y_{co})$ contains the exit point (u^*, w^*, x^*, y^*). Step 3 relates the controlling UEP of the reduced-state system (with respect to the projected fault-on trajectory) to the controlling UEP of the original system with respect to the original fault-on trajectory.

20.6.1 Theoretical basis

Analytical results asserting that, under certain conditions, the original model (20.11) and the artificial, reduced-state model (20.12) satisfy static relationships (S1) and (S2) as well as dynamic relationships (D1) and (D2) are shown below. Detailed analysis of these relationships can be found in [16,50].

We propose the following seven steps to establish the static properties (S1) and (S2) and the dynamic property (D2) between the original model (20.11) and the reduced-state model (20.12).

Step 1: Determine the static as well as dynamic relationship between the reduced-state system (20.12) and the following singularly perturbed system:

$$\begin{aligned}
\varepsilon_1 \dot{u} &= -\frac{\partial}{\partial u} U(u, w, x, y) + g_1(u, w, x, y) \\
\varepsilon_2 \dot{w} &= -\frac{\partial}{\partial w} U(u, w, x, y) + g_2(u, w, x, y) \\
T\dot{x} &= -\frac{\partial}{\partial x} U(u, w, x, y) + g_3(u, w, x, y) \\
\dot{y} &= -\frac{\partial}{\partial y} U(u, w, x, y) + g_4(u, w, x, y).
\end{aligned} \tag{20.13}$$

Step 2: Determine the static as well as dynamic relationship between (20.13) and the following nonlinear dynamical system:

$$\begin{aligned}
\varepsilon_1 \dot{u} &= -\frac{\partial}{\partial u} U(u, w, x, y) \\
\varepsilon_2 \dot{w} &= -\frac{\partial}{\partial w} U(u, w, x, y) \\
T\dot{x} &= -\frac{\partial}{\partial x} U(u, w, x, y) \\
\dot{y} &= -\frac{\partial}{\partial y} U(u, w, x, y).
\end{aligned} \tag{20.14}$$

Step 3: Determine the static as well as dynamic relationship between (20.14) and the following system:

$$\begin{aligned}
\varepsilon_1 \dot{u} &= -\frac{\partial}{\partial u} U(u, w, x, y) \\
\varepsilon_2 \dot{w} &= -\frac{\partial}{\partial w} U(u, w, x, y) \\
T\dot{x} &= -\frac{\partial}{\partial x} U(u, w, x, y) \\
\dot{y} &= -\frac{\partial}{\partial y} U(u, w, x, y). \\
M\dot{z} &= -Dz
\end{aligned} \tag{20.15}$$

Step 4: Determine the static as well as dynamic relationship between (20.15) and the following one-parameter dynamical system $d(\lambda)$:

$$\begin{aligned}
\varepsilon_1 \dot{u} &= -\frac{\partial}{\partial u} U(u, w, x, y) \\
\varepsilon_2 \dot{w} &= -\frac{\partial}{\partial w} U(u, w, x, y) \\
T\dot{x} &= -\frac{\partial}{\partial x} U(u, w, x, y) \\
\dot{y} &= (1-\lambda)z - \lambda \frac{\partial}{\partial y} U(u, w, x, y) \\
M\dot{z} &= -Dz - (1-\lambda)\frac{\partial}{\partial y} U(u, w, x, y).
\end{aligned} \tag{20.16}$$

Step 5: Determine the static as well as dynamic relationship between (20.16) and the following nonlinear dynamical system:

$$\begin{aligned}
\varepsilon_1 \dot{u} &= -\frac{\partial}{\partial u} U(u, w, x, y) \\
\varepsilon_2 \dot{w} &= -\frac{\partial}{\partial w} U(u, w, x, y) \\
T\dot{x} &= -\frac{\partial}{\partial x} U(u, w, x, y) \\
\dot{y} &= z \\
M\dot{z} &= -Dz - \frac{\partial}{\partial y} U(u, w, x, y).
\end{aligned} \tag{20.17}$$

Step 6: Determine the static as well as dynamic relationship between (20.17) and the following intermediate system:

$$\begin{aligned}
\varepsilon_1 \dot{u} &= -\frac{\partial}{\partial u} U(u, w, x, y) + g_1(u, w, x, y) \\
\varepsilon_2 \dot{w} &= -\frac{\partial}{\partial w} U(u, w, x, y) + g_2(u, w, x, y) \\
T\dot{x} &= -\frac{\partial}{\partial x} U(u, w, x, y) + g_3(u, w, x, y) \\
\dot{y} &= z \\
M\dot{z} &= -Dz - \frac{\partial}{\partial y} U(u, w, x, y) + g_4(u, w, x, y).
\end{aligned} \tag{20.18}$$

Step 7: Determine the static as well as dynamic relationship between (20.18) and the original system (20.11).

We then apply the overall seven-step procedure to show that the original system (20.11) and the reduced-state system (20.12) satisfy the static properties (S1) and (S2) as well as the dynamic property (D2). These seven steps and their relationships are summarized in Figure 20.3.

In the following, we develop the properties stated in each step.

Step 1 and Step 7

In these two steps, the singular perturbation technique is applied to show that, for sufficiently small values of ε_1, ε_2, the following properties hold.

Static property: $(\bar{u}, \bar{w}, \bar{x}, \bar{y})$ is a type-k equilibrium point of model (20.12) if and only if $(\bar{u}, \bar{w}, \bar{x}, \bar{y})$ is a type-k equilibrium point of model (20.13). In particular, (u_s, w_s, x_s, y_s) is an SEP of model (20.12) if and only if (u_s, w_s, x_s, y_s) is an SEP of model (20.13).

Static property: $(\bar{u}, \bar{w}, \bar{x}, \bar{y}, 0)$ is a type-k equilibrium point of model (20.11) if and only if $(\bar{u}, \bar{w}, \bar{x}, \bar{y}, 0)$ is a type-k equilibrium point of model (20.18). In particular, $(u_s, w_s, x_s, y_s, 0)$ is an SEP of model (20.11) if and only if $(u_s, w_s, x_s, y_s, 0)$ is an SEP of model (20.18).

Dynamic property: (u_i, w_i, x_i, y_i) is an equilibrium point on the stability boundary $\partial A(u_s, w_s, x_s, y_s)$ of model (20.12) if and only if the equilibrium point (u_i, w_i, x_i, y_i) is on the stability boundary $\partial A(u_s, w_s, x_s, y_s)$ of model (20.13).

Dynamic property: $(u_i, w_i, x_i, y_i, 0)$ is an equilibrium point on the stability boundary $\partial A(u_s, w_s, x_s, y_s, 0)$ of model (20.17) if and only if the equilibrium point $(u_i, w_i, x_i, y_i, 0)$ is on the stability boundary $\partial A(u_s, w_s, x_s, y_s, 0)$ of model (20.18).

Step 3

In this step, the following properties can be shown.

Static property: $(\bar{u}, \bar{w}, \bar{x}, \bar{y})$ is a type-k equilibrium point of model (20.14) if and only if $(\bar{u}, \bar{w}, \bar{x}, \bar{y}, 0)$ is a type-k equilibrium point of model (20.15), where $0 \in R^m$ and m is an appropriate positive integer.

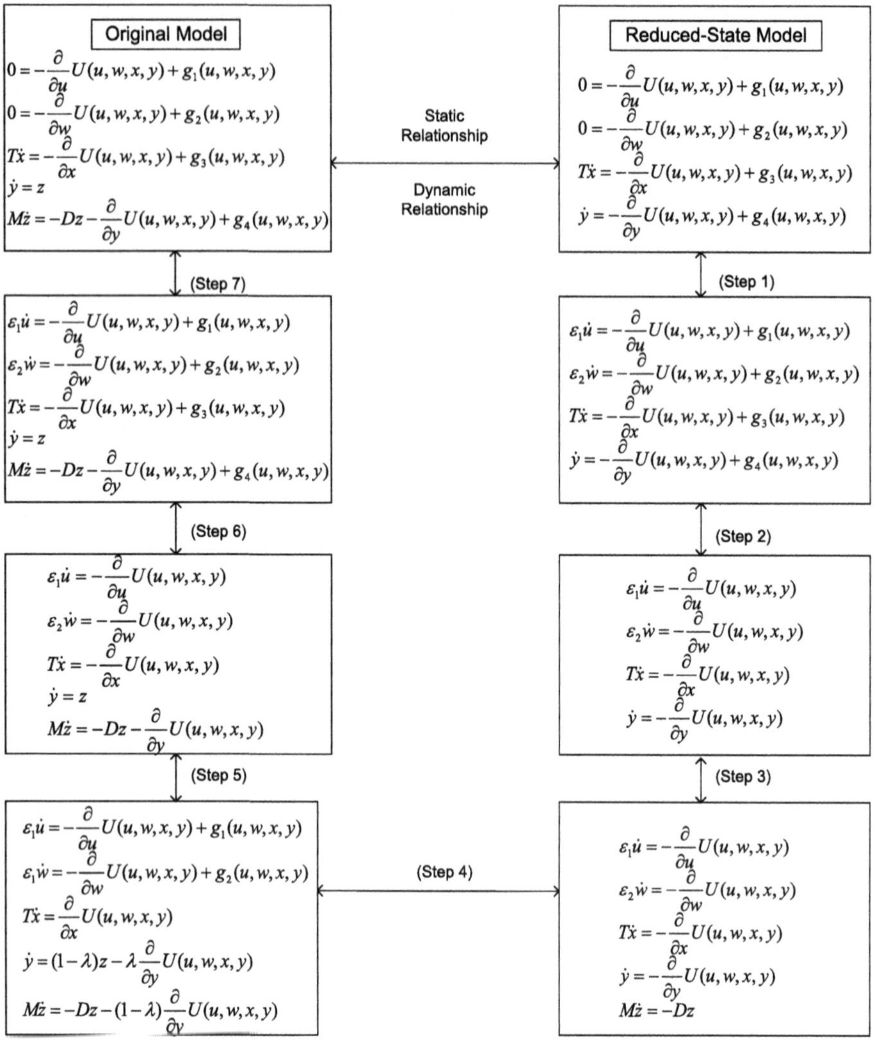

Figure 20.3 A seven-step procedure to show the static properties (S1) and (S2) as well as dynamic properties (D1) and (D2) between the original model (20.11) and the reduced-state model (20.12).

Dynamic property: $(\bar{u}, \bar{w}, \bar{x}, \bar{y})$ is an equilibrium point on the stability boundary $\partial A(u_s, w_s, x_s, y_s, 0)$ of model (20.14) if and only if $(\bar{u}, \bar{w}, \bar{x}, \bar{y}, 0)$ is an equilibrium point on the stability boundary $\partial A(u_s, w_s, x_s, y_s, 0)$ of model (20.15).

Step 4 and Step 5

In these two steps, the following properties can be shown.

Static property: $(\bar{u}, \bar{w}, \bar{x}, \bar{y}, 0)$ is a type-k equilibrium point of model (20.17) if and only if $(\bar{u}, \bar{w}, \bar{x}, \bar{y}, 0)$ is a type-k equilibrium point of model (20.15). In particular,

$(u_s, w_s, x_s, y_s, 0)$ is an SEP of model (20.17) if and only if $(u_s, w_s, x_s, y_s, 0)$ is an SEP of model (20.15).

Dynamic property: $(u_i, w_i, x_i, y_i, 0)$ is an equilibrium point on the stability boundary $\partial A(u_s, w_s, x_s, y_s, 0)$ of model (20.17) if and only if the equilibrium point $(u_i, w_i, x_i, y_i, 0)$ is on the stability boundary $\partial A(u_s, w_s, x_s, y_s, 0)$ of model (20.15).

Step 2 and Step 6

In these two steps, it is shown that if the transfer conductances $g_i(x)$ are sufficiently small, then the following properties hold.

Static property: $(\bar{u}, \bar{w}, \bar{x}, \bar{y})$ is a type-k equilibrium point of model (20.13) if and only if $(\bar{u}, \bar{w}, \bar{x}, \bar{y})$ is a type-k equilibrium point of model (20.14). In particular, (u_s, w_s, x_s, y_s) is an SEP of model (20.13) if and only if (u_s, w_s, x_s, y_s) is an SEP of model (20.14).

Static property: $(\bar{u}, \bar{w}, \bar{x}, \bar{y}, 0)$ is a type-k equilibrium point of model (20.17) if and only if $(\bar{u}, \bar{w}, \bar{x}, \bar{y}, 0)$ is a type-k equilibrium point of model (20.18). In particular, $(u_s, w_s, x_s, y_s, 0)$ is an SEP of model (20.17) if and only if $(u_s, w_s, x_s, y_s, 0)$ is an SEP of model (20.18).

Dynamic property: (u_i, w_i, x_i, y_i) is an equilibrium point on the stability boundary $\partial A(u_s, w_s, x_s, y_s)$ of model (20.13) if and only if the equilibrium point (u_i, w_i, x_i, y_i) is the stability boundary $\partial A(u_s, w_s, x_s, y_s)$ of model (20.14).

Dynamic property: $(u_i, w_i, x_i, y_i, 0)$ is an equilibrium point on the stability boundary $\partial A(u_s, w_s, x_s, y_s, 0)$ of model (20.17) if and only if the equilibrium point $(u_i, w_i, x_i, y_i, 0)$ is on the stability boundary $\partial A(u_s, w_s, x_s, y_s, 0)$ of model (20.18).

By combining the relationships derived in Steps 1 through 7, the following analytical results regarding the static as well as dynamic relationships between the original system (20.11) and the reduced-state model (20.12) are established.

THEOREM 20-1 **(Static relationship)**
Let (u_s, w_s, x_s, y_s) be a stable equilibrium point on a stable component Γ_s of the reduced-state model (20.12). If the following conditions are satisfied

(a) *zero is a regular value of* $\dfrac{\partial^4 U(u_i, w_i, x_i, y_i)}{\partial u \partial w \partial x \partial y}$ *for all the UEP (u_i, w_i, x_i, y_i), $i = 1, 2, \ldots, k$ on the stable component Γ_s,*

(b) *the transfer conductance of the reduced-state model (20.12) is sufficiently small,*

(c) *the stable and unstable equilibrium points of the original system (20.11) and those of the reduced-state model (20.12) that are of interest lie on a stable component of the constraint manifold,*

then the following static relation holds: $(\hat{u}, \hat{w}, \hat{x}, \hat{y})$ is a type-k equilibrium point of the reduced-state model (20.12) if and only if $(\hat{u}, \hat{w}, \hat{x}, \hat{y}, 0)$ is a type-k equilibrium point of the original system (20.11).

Theorem 20-1 asserts that, under the stated conditions, the static properties (S1) and (S2) between the original system (20.11) and the reduced-state model (20.12) hold.

Regarding dynamic property (D1), it can be shown that there exists a numerical energy function for the reduced-state model (20.12). More specifically, it can be shown that for any compact set S of the state-space of model (20.12), there is a positive number α such that, if the transfer conductance of the model satisfies $|G| < \alpha$, then there is an energy function defined on this compact set S. Hence, the dynamic property (D1) is satisfied.

To examine the dynamic property (D2), the following main result holds based on the analytical results derived in Step 1 through Step 7. Indeed, Theorem 20-2 asserts that, under the stated conditions, the dynamic property (D2) is satisfied. Furthermore, the stability boundaries of both the original model and the state-reduced model are completely characterized and they contain the "same" equilibrium points.

THEOREM 20-2 **(Dynamic relationship)**
Let (u_s, w_s, x_s, y_s) be a stable equilibrium point of the reduced-state model (20.12). If the following conditions are satisfied,

(a) zero is a regular value of $\dfrac{\partial^4 U(u_i, w_i, x_i, y_i)}{\partial u \partial w \partial x \partial y}$ for all the UEP (u_i, w_i, x_i, y_i), $i = 1, 2, \ldots, k$ on the stability boundary $\partial A(u_s, w_s, x_s, y_s)$,
(b) the transfer conductance of the reduced-state model (20.12) is sufficiently small,
(c) the stable and unstable equilibrium points of the original system (20.11) and those of the reduced-state model (20.12) that are of interest lie on a stable component of the constraint manifold,
(d) all the intersections of the stable and unstable manifolds of the equilibrium points on the stability boundary $\partial A(u_s, w_s, x_s, y_s, (0)$ of the one-parameterized model $d(\lambda)$ (20.16) satisfy the transversality condition for $\lambda \in [0, 1]$,

then the following results hold.

(1) The equilibrium point (u_i, w_i, x_i, y_i) is on the stability boundary $\partial A(u_s, w_s, x_s, y_s)$ of model (20.12) if and only if the equilibrium point $(u_i, w_i, x_i, y_i, 0)$ is on the stability boundary $\partial A(u_s, w_s, x_s, y_s, 0)$ of model (20.11).
(2) The stability boundary $\partial A(u_s, w_s, x_s, y_s)$ of the reduced-state model (20.12) is the union of the stable manifold of all the equilibrium points (u_i, w_i, x_i, y_i), $i = 1, 2, \ldots,$ on the stability boundary $\partial A(u_s, w_s, x_s, y_s)$; i.e. $\partial A(u_s, w_s, x_s, y_s) = \cup W^s(u_i, w_i, x_i, y_i)$.
(3) The stability boundary $\partial A(u_s, w_s, x_s, y_s, 0)$ of the original model (20.11) is the union of the stable manifold of all the equilibrium points $(u_i, w_i, x_i, y_i, 0)$, i = $1, 2, \ldots,$ on the stability boundary $\partial A(u_s, w_s, x_s, y_s, 0)$; i.e. $\partial A(u_s, w_s, x_s, y_s, 0) = \cup W^s(u_i, w_i, x_i, y_i, 0)$.

20.6.2 Dynamic property (D3)

We present a numerical scheme for efficiently detecting the intersection point at which the projected fault-on trajectory $(u_f(t), w_f(t), x_f(t), y_f(t))$ intersects with the stability boundary $\partial A_0(u_s, w_s, x_s, y_s)$ of the post-fault reduced-state model without resorting to detailed time-domain numerical simulation. The existence of such a numerical scheme ensures that the reduced-state model (20.12) satisfies the dynamic property (D3). Recall that a numerical energy function for the original system can be expressed as follows, which is a summation of a numerical potential energy and a kinetic energy:

$$\begin{aligned} W_{num}(u, w, x, y, z) &= W_{ana}(u, w, x, y, z) + U_{path}(u, w, x, y) \\ &= \frac{1}{2} z^T M z + U(u, w, x, y) + U_{path}(u, w, x, y) \\ &= \frac{1}{2} z^T M z + U_{num}(u, w, x, y). \end{aligned} \tag{20.19}$$

The numerical scheme proceeds as follows.

Step 1: Along the fault-on trajectory $(u_f(t), w_f(t), x_f(t), y_f(t), z_f(t))$, detect the point, say $(u*, w*, x*, y*)$ at which the projected fault-on trajectory $(u_f(t), w_f(t), x_f(t), y_f(t))$ reaches the first local maximum (along the fault-on trajectory) of the numerical potential energy $U_{num}(u, w, x, y)$ of (20.19).

Step 2: The point $(u*, w*, x*, y*)$ is a simulated exit point, which is an approximated point for the intersection point between the stability boundary of the (post-fault) reduced-state model (20.12) and the projected fault-on trajectory $(u_f(t), w_f(t), x_f(t), y_f(t))$.

Another numerical scheme for detecting the exit point of the reduced-state stability boundary is the following.

Step 1: Along the fault-on trajectory $(u_f(t), w_f(t), x_f(t), y_f(t), z_f(t))$, compute the dot product of the following two vectors: (i) the fault-on generator speed vector, and (ii) the post-fault power mismatch vector at each integration step. When the sign of the dot product changes from positive to negative, the exit point is detected.

The above numerical schemes can efficiently detect the exit point and do not require detailed time-domain simulations of the post-fault system. Hence, the dynamical property (D3) is satisfied.

20.7 Numerical BCU method

There are several possible ways to numerically implement the (conceptual) BCU method presented in the previous section for network-preserving power system models. A numerical implementation of this method along with several necessary numerical procedures are presented in this section.

A numerical BCU method

Step 1: Construct a numerical energy function for the post-fault system of a generic stability model, say the following function (from (20.19)) which is composed of a kinetic energy $K(z)$ and a potential energy $U_{num}(.,.,., .)$,

$$\begin{aligned}W_{num}(u, w, x, y, z) &= W_{ana}(u, w, x, y, z) + U_{path}(u, w, x, y)\\ &= \frac{1}{2}z^T Mz + U(u, w, x, y) + U_{path}(u, w, x, y)\\ &= \frac{1}{2}z^T Mz + U_{num}(u, w, x, y).\end{aligned}$$

Step 2: Integrate the fault-on system of the original model (20.11) to obtain the (sustained) fault-on trajectory $(u(t), w(t), x(t), y(t), z(t))$ until the point $(u*, w*, x*, y*)$ at which the projected trajectory $(u(t), w(t), x(t), y(t))$ reaches its first local maximum of the numerical potential energy function $U_{num}(.,.,.,.)$.

Step 3: Apply a stability-boundary-following procedure (see below) starting from the point $(u*, w*, x*, y*)$ until the point at which the (one-dimensional) local minimum of the following norm of the post-fault, reduced-state system is reached:

$$\left\|\frac{\partial U}{\partial u}(u, w, x, y) + g_1(u, w, x, y)\right\| + \left\|\frac{\partial U}{\partial w}(u, w, x, y) + g_2(u, w, x, y)\right\| + \left\|\frac{\partial U}{\partial u}(u, w, x, y) + g_3(u, w, x, y)\right\| + \left\|\frac{\partial U}{\partial w}(u, w, x, y) + g_4(u, w, x, y)\right\|.$$

Let the local minimum of the above norm be at the point $(u*_0, w*_0, x*_0, y*_0)$.

Step 4: Use the point $(u_0^*, w_0^*, x_0^*, y_0^*)$ as the initial guess and solve the following set of (vector) nonlinear algebraic equations:

$$\begin{aligned}&\left\|\frac{\partial U}{\partial u}(u, w, x, y) + g_1(u, w, x, y)\right\|\\ &+ \left\|\frac{\partial U}{\partial w}(u, w, x, y) + g_2(u, w, x, y)\right\|\\ &+ \left\|\frac{\partial U}{\partial x}(u, w, x, y) + g_3(u, w, x, y)\right\|\\ &+ \left\|\frac{\partial U}{\partial y}(u, w, x, y) + g_4(u, w, x, y)\right\| = 0.\end{aligned}$$

Let the solution be $(u_{co}, w_{co}, x_{co}, y_{co})$.

Step 5: The controlling UEP with respect to the fault-on trajectory $(u(t), w(t), x(t), y(t))$ of the original system is $(u_{co}, w_{co}, x_{co}, y_{co}, 0)$.

To reliably compute the CUEP, it is essential to move along the stable manifold of the CUEP starting from the exit point. The following scheme is designed to serve this purpose.

A stability-boundary-following scheme

Step 1: Integrate the post-fault reduced-state model for four or five time-step points and term the last point the current point.

Step 2. Check whether the post-fault reduced-state trajectory reaches a relative local minimum of the vector field of the reduced-state model. If yes, then the point with a local minimum of the vector field is the termed MGP and the scheme stops (output the MGP); otherwise, go to the next step.

Step 3. If the number of adjustments in the stability-boundary-following procedure is greater than a threshold, then stop the procedure and go back to the exit point detection procedure to compute an improved exit point and go to Step 1; otherwise go to the next step.

Step 4: Draw a ray connecting the current point on the post-fault reduced-state trajectory with the SEP of the post-fault reduced-state system.

Step 5: Start from the current point on the post-fault reduced-state trajectory and move along the ray to find the point on the ray that has the first local maximum value of the energy function for the post-fault original system. Replace the current point by the point found in this step and go to Step 1.

The theoretical basis for the above stability-boundary-following procedure is the structure of the stability boundary of the reduced-state system, which is composed of the stable manifold of the UEPs lying on the stability boundary. Steps 4 and 5 are based on the computational scheme developed to implement the detection of the exit point. Hence, these two steps serve to ensure the stability-boundary-following procedure moves closely along the stability boundary of the post-fault reduced-state system. Since the controlling UEP is of type one, it has the nice feature of being a "relative SEP" within the stability boundary.

20.8 Applications to a large-scale test system

The numerical network-preserving BCU method has been evaluated on several power systems, modeled by a network-preserving classical generator with static nonlinear load representations. The static nonlinear load model is a combination of constant power, constant impedance and constant current. The evaluation results for the IEEE 50-generator, 145-bus test system [132] and a 202-generator, 1293-bus system [63] are presented in this section.

Table 20.1 displays the estimated CCT of the IEEE 50-generator, 145-bus test system with different locations of three-phase faults using the time-domain method and the network-preserving BCU method. The load model used in this numerical study is a combination of 20% constant power, 20% constant current and 60% constant impedance. The CCTs estimated by the time-domain method are used as a benchmark. Table 20.1 is explained as follows. The second line of Table 20.1 states that a three-phase fault occurs at bus 7 and the post-fault system is the system with the transmission line between buses 6 and 7 tripped due to the openings of circuit breakers at both ends of the line. The CCT estimated by the network-preserving BCU method is 0.097 second while the CCT estimated by the time-domain simulation method is 0.103 second. The relative error is −5.8%, meaning that the CCT estimated by the network-preserving BCU is 5.8% conservative, as compared with the exact CCT. It should be pointed out that in these simulation results, the network-preserving BCU method consistently gives slightly conservative results in estimating the CCTs. The range of conservativeness is between 1.1% and 11.8%. This is consistent with

Table 20.1 Simulation results of the network-preserving BCU method on the 50-generator, 145-bus IEEE test system modeled by the network-preserving model with nonlinear static load (20% constant power, 20% constant current and 60% constant impedance)

Faulted bus	Opened line	CCT estimated by BCU method (second)	CCT estimated by time-domain method (second)	Relative error (%)
7	7–6	0.097	0.103	−5.8
59	59–72	0.208	0.222	−6.3
73	73–74	0.190	0.215	−11.8
112	112–69	0.235	0.248	−5.2
66	66–69	0.156	0.168	−7.1
115	115–116	0.288	0.292	−1.3
110	110–72	0.245	0.260	−5.7
101	101–73	0.232	0.248	−6.4
91	91–75	0.187	0.189	−1.1
6	6–1	0.153	0.170	−10.0
12	12–14	0.163	0.173	−5.8
6	6–10	0.162	0.177	−9.4
66	66–111	0.157	0.172	−9.7
106	106–74	0.170	0.186	−9.6
69	69–32	0.186	0.202	−7.9
69	69–112	0.110	0.118	−6.7
105	105–73	0.191	0.211	−9.4
73	73–75	0.194	0.210	−7.6
67	67–65	0.230	0.231	−0.4
59	59–103	0.221	0.223	−0.9
12	12–14, 12–14	0.156	0.167	−6.5
105	105–73, 105–73	0.110	0.118	−6.7
66	66–8, 66–8	0.167	0.174	−4.0
66	66–111, 66–111, 66–111	0.070	0.080	−12.5
73	73–26, 73–72, 73–82, 73–101	0.192	0.212	−9.4
73	73–69, 73–75, 73–96, 73–109	0.182	0.190	−4.2

the analytical results of the controlling UEP method that the critical energy value based on the controlling UEP gives accurate and yet slightly conservative stability assessments, despite the fact that numerical energy functions were used in these simulations. These simulation results also support the claim that the network-preserving BCU method computes the correct controlling UEPs of these contingencies.

Table 20.2 summarizes the simulation results of the 202-generator, 1293-bus system. The simulation results are the estimated CCTs with different locations of three-phase faults using the exact time-domain method and the BCU method with the following two load models: (1) constant impedance loads and (2) a nonlinear static load, which is a combination of 20% constant power, 20% constant current and 60% constant impedance. The CCTs estimated by the time-domain method are used as a benchmark. Table 20.2 is explained as follows. The last line of Table 20.2 states that a three-phase fault occurs at bus 360 and the post-fault system is the system with the transmission line between buses 360 and 362 tripped due to the openings of circuit breakers at both ends of the line. For the constant impedance load, the CCT estimated by the network-preserving BCU method is 0.272 second while the CCT estimated by the time-domain simulation method is 0.262 second. The relative error is −3.6%, meaning the CCT estimated by the network-preserving BCU is 3.6% conservative, as compared with the exact CCT.

Table 20.2 Simulation results of the network-preserving BCU method on the 202-generator, 1293-bus IEEE test system modeled by the network-preserving model with (1) constant impedance load, (2) nonlinear static load model (20% constant power, 20% constant current and 60% constant impedance)

		Load model (1)			Load model (2)		
Faulted bus	Opened line	CCT estimated by time-domain method (second)	CCT estimated by BCU method (second)	Relative error (%)	CCT estimated by time-domain method (second)	CCT estimated by BCU method (second)	Relative error (%)
77	77–124	0.325	0.320	−1.6	0.325	0.320	−1.6
74	74–76	0.343	0.313	−9.7	0.336	0.305	−9.2
75	75–577	0.210	0.212	−0.9	0.210	0.212	−0.9
136	136–103	0.262	0.260	−0.7	0.262	0.260	−0.7
248	248–74	0.165	0.160	−3.1	0.165	0.160	−3.1
360	360–345	0.273	0.262	−4.1	0.271	0.261	−3.6
559	559–548	0.215	0.197	−9.3	0.212	0.190	−11.3
634	634–569	0.212	0.197	−7.1	0.205	0.188	−9.2
661	661–669	0.123	0.103	−16.3	0.123	0.103	−16.3
702	702–1376	0.233	0.224	−3.8	0.230	0.222	−3.4
221	221–223	0.220	0.214	−3.2	0.216	0.209	−3.2
175	175–172	0.276	0.269	−3.5	0.272	0.262	−3.6
198	198–230	0.156	0.145	−7.1	0.155	0.145	−6.4
245	245–246	0.230	0.224	−2.6	0.228	0.220	−3.5
319	319–332	0.230	0.198	−13.9	0.229	0.198	−13.5
360	360–362	0.272	0.262	−3.6	0.272	0.261	−4.1

For the constant ZIP load, the CCT estimated by the network-preserving BCU method is 0.261 second while the CCT estimated by the time-domain simulation method is 0.272 second. The relative error is −4.1%, meaning the CCT estimated by the network-preserving BCU is 4.1% conservative, as compared with the exact CCT. It should be pointed out that in these simulation results, the network-preserving BCU method consistently gives slightly conservative results in estimating the CCTs. The range of conservativeness is between 0.7% and 16.3% for both the constant impedance load and the constant ZIP load. This is consistent with the analytical results of the controlling UEP method that the critical energy value based on the controlling UEP gives accurate and yet slightly conservative stability assessments, despite the fact that numerical energy functions were used in these simulations. The network-preserving BCU method consistently gives slightly conservative results in estimating the CCTs, which is in compliance with the controlling UEP method despite the fact that numerical energy functions were used in these simulations. Moreover, the performance of the network-preserving BCU method is very consistent when applied to different static load models.

To reduce the conservativeness of the BCU method in stability assessment, the constructive approach presented in Chapter 13 is useful. The next section is devoted to this topic.

20.9 Reducing the conservativeness in stability assessment

In this section, we present two schemes for constructing a sequence of functions from a given energy function $W(\cdot)$ with the aim of reducing the conservativeness in the stability assessment. We show that these functions are also energy functions, and that the level sets defined by these functions passing through the controlling UEP form a monotonically increasing sequence, thus reducing the conservativeness, and yet remain a conservative estimate for the relevant stability region.

Given an energy function $W(\cdot)$ for the nonlinear system (20.11), we construct a sequence of energy functions through the following first-order expansion scheme:

$$
\begin{aligned}
V_1(x) &= V\left(x + d_1 f(x)\right) \\
V_2(x) &= V_1\left(x + d_2 f(x)\right) \\
&\cdots \\
V_n(x) &= V_{n-1}\left(x + d_n f(x)\right)
\end{aligned}
$$

where $f(x)$ is the vector field of (20.11) and d_i, $i = 1, 2, \ldots, n$ are positive numbers. We show that the sequence of functions can be applied to estimate the relevant stability region. In addition, we show that the level sets defined by these energy functions form a sequence of sets with strictly monotonic increasing size and yet each set is a conservative estimate of the relevant stability region, hence it offers a larger (hence less conservative) estimate of the relevant stability region.

THEOREM 20-3 **(Energy-like function)**
Let $W(\cdot)$: $R^n \rightarrow R$ be an energy-like function for the nonlinear system (20.11) and K be a compact set, containing no other equilibrium points, in the state space of (20.11). Then, there exists a $\hat{d} > 0$ such that, for $d < \hat{d}$, the function $V_1(x) = V(x + df(x))$ is also an energy-like function on the compact set K for the nonlinear system.

Next, we show that the sizes of the sets which are the intersections between the level sets of the functions $V_i(\cdot)$ and any compact set are monotonically increasing.

THEOREM 20-4 **(Expansion)**
Let $W(\cdot)$: $R^n \rightarrow R$ be an energy function for the nonlinear system (20.11) and K be a compact set, containing no equilibrium points, in the state space of (20.11). Suppose that the set $S_W(c)$: = $\{x$: $W(x) \leq c$ and $x \in K\}$ is non-empty for some constant c. Then, there exists a $\tilde{d} > 0$ such that for the set characterized by $S_{W_1}(c) :=$ $\{x : W_1(x) \leq c \text{ and } x \in K\}$, where $W_1(x) = W(x + df(x))$ and $d < \tilde{d}$, the following is true:

$$S_W(c) \subset S_{W_1}(c).$$

We present another scheme for constructing a sequence of functions from a given energy function $W(\cdot)$ for the nonlinear system (20.11). We show that these functions are also energy functions for the nonlinear system (20.11), and that the level sets defined by these functions form a monotonically increasing sequence.

Given an energy-like function $V(\cdot)$ for the nonlinear system (20.11), we construct a sequence of energy functions through the following second-order expansion scheme

$$\begin{aligned} W_1(x) &= W\left(x + \frac{d}{2}\left(f\left(x + df(x)\right) + f(x)\right)\right) \\ W_2(x) &= W_1\left(x + \frac{d}{2}\left(f\left(x + df(x)\right) + f(x)\right)\right) \\ &\ldots \\ W_n(x) &= W_{n-1}\left(x + \frac{d}{2}\left(f\left(x + df(x)\right) + f(x)\right)\right). \end{aligned}$$

We first show that every function $W_i(\cdot)$, $i = 1, 2, \ldots$ is an energy function on a compact set for system (20.11), and then show that the sequence formed by the levels sets defined by the sequence of functions is monotonically increasing.

THEOREM 20-5 **(Energy-like function)**
Let $W(\cdot)$: $R^n \rightarrow R$ be an energy function for the nonlinear system (20.11) and K be a compact set, containing no equilibrium points, in the state space of (20.11). Then, there exists a $\hat{d} > 0$ such that, for $d < \hat{d}$, the function $W_1(x) = W(x + (d/2)(f(x + df(x)) + f(x)))$ is also an energy function on the compact set K for the nonlinear system (20.11).

THEOREM 20-6 **(Expansion)**
Let $W(\cdot)$: $R^n \rightarrow R$ be an energy function for the nonlinear system (20.11) and K be a compact set, containing no other equilibrium points, in the state space of (20.11). Suppose that the set $S_W(c)$: = $\{x$: $W(x) \geq c$ and $x \in K\}$ is non-empty for some

constant c. Then, there exists a $\hat{d} > 0$ *such that for the set characterized by* $S_{W_1}(x) := \{x : W_1(x) \geq c \text{ and } x \in K\}$, *where* $W_1(x) = W(x + d(f(x + df(x)) + f(x)))$ *and* $d < \hat{d}$, *the following is true:*

$$S_W(c) \subset S_{W_1}(c).$$

The constructive methodology for estimating the relevant stability region starts with a given energy function such as $W(\cdot)$ and proceeds as follows.

Step A: Compute the critical energy. Determine the critical level value of $V(\cdot)$ via the BCU method. Let $\hat{x}$ be the controlling UEP relative to a given fault-on trajectory.

Step B: Estimate the relevant stability region. Estimate the relevant stability region via the function $W(\cdot)$. The connected component of $\{x : W(x) < W(\hat{x})\}$ containing the stable equilibrium point x_s gives the relevant stability region.

Step C: Modify the energy function. Modify the energy function $W(\cdot)$. Replace $W(x)$ by either $W(x + df(x))$ or $W(x + (d/2)(f(x + df(x)) + f(x)))$, $d > 0$ and go to Step B.

This constructive method yields a sequence of estimated relevant stability regions. In addition, for a finite number of iterations, there exists a positive number $\hat{d}$ such that, for $d < \hat{d}$, this sequence of estimated relevant stability regions is a strictly increasing sequence and yet gives conservative estimates for the relevant stability region.

20.10 Concluding remarks

On-line transient stability assessment (TSA) is an essential tool for obtaining the operating security limits at or near real time. In addition to this important function, the current trend of power system transmission open access and restructuring worldwide further reinforces the need for on-line TSA as it is the base upon which available transfer capability, dynamic congestion management problems and special protection systems can be effectively resolved. There are significant engineering and financial benefits expected from on-line TSA.

This chapter has presented an overview of state-of-the-art direct methods and effective computational methods useful for on-line TSA. The current direction of development is to include the CUEP method, the BCU method and time-domain simulation programs within the body of overall power system stability simulation programs. TEPCO-BCU has been developed under this direction by integrating the BCU method, BCU classifiers, and BCU guide time domain method. The current version of TEPCO-BCU is able to perform exact stability assessment and accurate energy margin computation of each contingency of large-scale power systems such as a 12,000-bus test system with a list of 3000 contingencies.

It is emphasized that the reliability of the TEPCO-BCU in performing dynamic contingency screening is built on the theory of the stability region and the high-yield drop-out ratio is built on the theory of relevant stability region and the CUEP method.

It is indeed due to the reliability and efficiency capability of TEPCO-BCU that, in conjunction with some relevant functions, it can lead to several practical applications. These relevant functions include the energy function method, the controlling UEP coordinates and their sensitivities with respect to parameters or control actions. One such application is the development of a dynamic security constrained optimal power flow method. A preliminary dynamic security constrained OPF algorithm was developed based on the TEPCO-BCU engine [241].

It should be pointed out that in these simulations the BCU method consistently gives slightly conservative estimates of CCTs. These estimates are in compliance with the analytical results of the controlling UEP method, confirming that the critical energy value based on the controlling UEP should give slightly conservative stability assessments if exact energy functions are used. The simulation results also reveal that the CCTs estimated by the other methods can be either overestimates or underestimates. Overestimating the CCT is undesirable because it could lead to classifying an unstable case as stable.

We have presented two schemes for constructing a sequence of functions from a given energy function $W(\cdot)$ with the aim of reducing the conservativeness in stability assessment by the controlling UEP method. We show that these functions are also energy functions, and that the level sets defined by these functions passing through the controlling UEP form a monotonically increasing sequence, thus reducing the conservativeness in stability assessment, and yet still giving a conservative estimate for the relevant stability region.

21 Stability-region-based methods for multiple optimal solutions of nonlinear programming

21.1 Introduction

Optimization technology has practical applications in almost every branch of science, economics and engineering. Indeed, a large variety of quantitative issues such as decision, design, operation, planning and scheduling can be perceived and modeled as either continuous or discrete nonlinear optimization problems. Typically, the overall performance (or measure) of a system can be described by a multivariate function, called the *objective function*. According to this generic description, one seeks the best solution in the solution space which satisfies all stated feasibility constraints and minimizes (or maximizes) the value of an objective function. The vector, if it exists, is termed the *global optimal solution*.

For most practical applications, the underlying objective functions are often nonlinear and depend on a large number of variables. This makes the task of searching the solution space for the global optimal solution very challenging. The primary challenge is that, in addition to the high dimension of the solution space, there are many local optimal solutions in the solution space. A local optimal solution is optimal in a local region of the solution space, but not in the global solution space. The global optimal solution is the best local optimal solution and yet both the global optimal solution and local optimal solutions share the same local properties. In general, the number of local optimal solutions is unknown and it can be quite large. Furthermore, the values of an objective function at local optimal solutions and at the global optimal solution may differ significantly. Hence, there is a strong motivation for developing effective methods for finding the global optimal solution.

One popular method for solving nonlinear optimization problems is to use a local improvement search which can be described as follows: start from an initial vector and search for a better solution in its neighborhood; if an improved solution is found, repeat the search procedure using the new solution as the initial solution and continue; otherwise, the search procedure is terminated. Local improvement search methods usually become trapped at local optimal solutions and are unable to escape from them. In fact, the great majority of existing nonlinear optimization methods usually produce local optimal solutions but not the global optimal solution.

The drawback of local improvement search methods has motivated the development of more sophisticated local search methods designed to find better solutions by introducing some mechanisms that allow the search process to escape from local optimal solutions. The underlying "escape" mechanisms employ certain search strategies which accept a cost-deteriorating neighborhood to make an escape from a local optimal solution possible. These sophisticated local search methods, also termed modern heuristics, include simulated annealing, genetic algorithms, Tabu searches, evolutionary programming and particle swarm operator methods. However, these sophisticated local search methods require intensive computational effort and usually cannot find the globally optimal solution.

Significant efforts have been made to develop hybrid search methods which combine a stochastic method and an iterative local search method. For instance, a class of hybrid methods is formed by the combination of genetic algorithms and local gradient methods (or sensitivity-based methods). However, these hybrid methods still suffer from several difficulties in computational effort and the quality of the optimal solutions found.

To overcome the difficulties encountered by the majority of the existing optimization methods, the following two important and challenging issues in the course of searching for multiple high-quality optimal solutions need to be fully addressed.

(C1) How to effectively move (escape) from a local optimal solution and move on toward another local optimal solution.
(C2) How to avoid revisiting local optimal solutions which are already known.

In the past, significant effort has been directed towards attempting to address these two issues, but without much success. Issue (C1) is difficult to solve and the existing methods all encounter this difficulty. Issue (C2), related to computational efficiency during the course of the search, is also difficult to solve and again, the majority of existing methods encounter this difficulty. Issue (C2) is a common problem, which degrades the performance of many existing methods which search for the global optimal solution. From the computational viewpoint, it is important to avoid revisiting the same local optimal solution in order to maintain a high level of efficiency.

The purpose of this chapter is to present a stability-region-based methodology, which is a new paradigm for solving nonlinear optimization problems. This new methodology has several distinguishing features to be described in the next section and can fully address the two challenges issues (C1) and (C2).

21.2 Trust-Tech methodologies

We will present in this section a new systematic method, which deterministically and effectively computes a large set of local optimal solutions, if not all the local optimal solutions, of nonlinear optimization problems. This methodology is based on the following transformations.

(i) The transformation of a local optimal solution of a nonlinear optimization problem into a stable equilibrium point of a continuous nonlinear dynamical system.
(ii) The transformation of the search space of a nonlinear optimization problem into the union of the closure of stability regions of stable equilibrium points.

Hence, the optimization problem (i.e. the problem of finding local optimal solutions) is transformed into a problem of finding stable equilibrium points, and it will become clear that the stability regions of stable equilibrium points play an important role in finding these local optimal solutions.

This methodology is termed Trust-Tech which stands for Transformation Under Stability-reTaining Equilibria Characterization. One distinguishing feature is that it systematically searches the set of local optimal solutions in a tier-by-tier manner; starting from a local optimal solution, finds the nearby first-tier local optimal solutions, then the second-tier local optimal solutions, and so on.

Trust-Tech-based methods are dynamical methods for efficiently obtaining a set of local optimal solutions of general nonlinear optimization problems. Without loss of generality, we explain the Trust-Tech framework for solving the following unconstrained nonlinear programming problem:

$$\min_{x \in R^n} C(x) \tag{21.1}$$

where $C: R^n \rightarrow R$ is a function bounded below and possesses only finite local optimal solutions. The focus is to efficiently compute all or multiple local optimal solutions of $C(x)$. To this end, instead of directly solving the unconstrained optimization problem, we construct the corresponding dynamical system and explore its trajectories:

$$\dot{x}(t) = -\nabla C(x) \tag{21.2}$$

where $x \in R^n$. This is a nonlinear, gradient dynamical system. It will be shown that each local optimal solution of problem (21.1) corresponds to a stable equilibrium point of the gradient system (21.2). This transformation allows us to locate each local optimal solution of problem (21.1) by locating each stable equilibrium point of the gradient system (21.2).

We next derive several geometrical and topological properties of the gradient system (21.2). These properties will be useful in the development of a systematic method for obtaining the set of local optimal solutions.

THEOREM 21-1 **(Completely stable)**
If there exist an ε and δ such that $\|\nabla C(x)\| > \varepsilon$ unless $x \in B_\delta(\tilde{x})$, where $\tilde{x}$ is an equilibrium point, then the gradient system (21.2) is completely stable and $C(x)$ is an associated Lyapunov function.

Proof This proof consists of two parts. The first part shows that every bounded trajectory will converge to one of the equilibrium points while the second part shows that every trajectory is bounded. Hence, every trajectory converges to an equilibrium point and the system is completely stable. For the first part, let $\phi(t, x)$ denote the bounded trajectory starting at x. Consider the time derivative $\frac{d}{dt} C\left(\phi(t, x)\right)$ along the trajectory. It is clear that

$$\frac{d}{dt} C\left(\phi(t, x)\right) = -(\nabla (C\left(\phi(t, x)\right)^T \nabla C\left(\phi(t, x)\right) \leq 0.$$

Moreover, $\frac{d}{dt} C\left(\phi(t, x)\right) = 0$ if and only if x is an equilibrium point. Therefore, $C(x)$ is a Lyapunov function of the gradient system (21.2) and the ω-limit point of any bounded trajectory consists of equilibrium points only, i.e. any bounded trajectory will approach one of the equilibrium points.

Since $C(x)$ is bounded below, we need to show that every trajectory $\phi(t, x)$ is also bounded. Suppose $\phi(t, x)$ is not bounded, we will show that this leads to a contradiction. Two possible cases are considered:

(a) After $t > T^*$, $\phi(t, x) \notin \bigcup_{\hat{x} \in E} B_\delta(\hat{x})$. Since $\|\nabla C(x)\| > \varepsilon$ for $x \notin \bigcup_{\hat{x} \in E} B_\delta(\hat{x})$, it follows that

$$C\left(\phi(t, x)\right) - C\left(\phi(T^*, x)\right) = \int_{T^*}^{t} \dot{C}\left(\phi(t, x)\right) dt \leq -k \int_{T^*}^{t} \|\nabla C(x)\|^2 \, dt \leq -k\varepsilon^2 (t - T^*)$$

where k is a positive constant. As $t \to \infty$, the right hand side of the above inequality approaches $-\infty$, which contradicts the fact that $C(\phi(t, x))$ is bounded from below.

(b) There is an infinite sequence $\hat{x}_j \in E$, the set of equilibrium point of (21.2) and a strictly increasing sequence p_j such that $\phi(p_j, x) \in B_\delta(\hat{x}_j)$. Let us define two strictly increasing sequences t_i and s_i in which the sequence t_i is the time instance that $\phi(t, x)$ enters the ball $B_\delta(\hat{x}_i)$ and s_i is the time instance that $\phi(t, x)$ leaves the ball $B_\delta(\hat{x}_i)$, where $t_{i+1} > s_i > t_i$. Fix an integer $m > 0$; then for $t \geq t_{m+1}$ we have

$$\begin{aligned} C\left(\phi(t, x)\right) - C\left(\phi(0, x)\right) &= \int_0^t \dot{C}\left(\phi(t, x)\right) dt \\ &< \sum_{i=1}^{m} \int_{s_i}^{t_{i+1}} \dot{C}\left(\phi(0, x)\right) dt < -k \sum_{i=1}^{m} \int_{s_i}^{t_i+1} \|\nabla C(x)\|^2 dt \\ &< -k \sum_{i=1}^{m} \int_{s_i}^{t_{i+1}} \varepsilon^2 dt = -k\varepsilon^2 \sum_{i=1}^{m} (t_{i+1} - s_i) \end{aligned}$$

where k is a positive constant. If $m \to \infty$, then the right hand side of the above inequality approaches $-\infty$, which contradicts the fact that $C(\phi(t, x))$ is bounded. Therefore, any trajectory $\phi(t, x)$ is bounded. This completes this proof.

For a completely stable system, every trajectory converges to one of its equilibrium points. Hence, the state space is the union of the closure of the stability regions; see for example Figure 21.1. In other words, every trajectory converges to one of its stable equilibrium points or to one of its unstable equilibrium points. From a numerical simulation viewpoint, every trajectory literally converges to one of its stable equilibrium points.

We next present a complete characterization of the stability boundary of the gradient system (21.2). This characterization is expressed in terms of the stable manifolds of the equilibrium points lying on the stability boundary.

PROPOSITION 21-2 **(Characterization of the stability boundary)**
Suppose that all the equilibrium points of the gradient system (21.2) are hyperbolic. Let x_i, $i = 1, 2, \ldots$ be the equilibrium points on the stability boundary $\partial A(x_s)$ of a stable

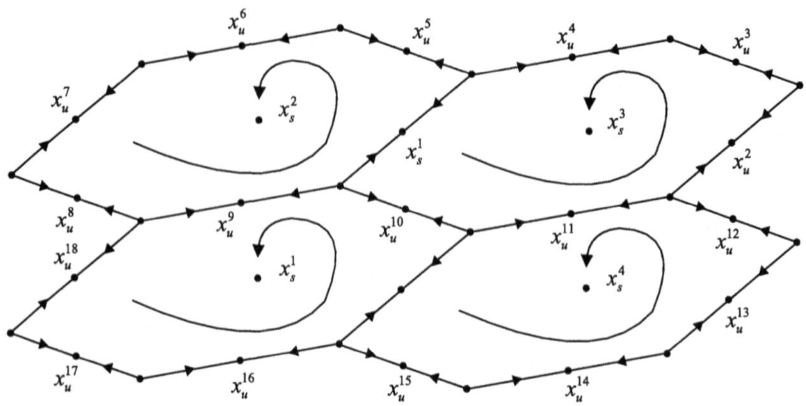

Figure 21.1 For a completely stable system, every trajectory converges to one of its equilibrium points. In other words, every trajectory converges to one of its stable equilibrium points or to one of its unstable equilibrium points.

equilibrium point, say x_s. Then, the stability boundary is contained in the union of the stable manifolds of the equilibrium points on the stability boundary; in other words,

$$\partial A(x_s) \subseteq \bigcup_{x_i \in \partial A} W^s(x_i).$$

THEOREM 21-3 **(Equilibrium points and local optimal solutions)**
If $\bar{x}$ is a hyperbolic equilibrium point of the gradient system (21.2), then $\bar{x}$ is a stable equilibrium point of system (21.2) if and only if $C(x)$ has an isolated minimum of the optimization problem (21.2) at $\bar{x}$.

Proof From Theorem 21-1, $C(x)$ is a Lyapunov function for the gradient system (21.2), since

(i) every trajectory within the stability region $A(x_s)$ converges to x_s, and
(ii) the function value $C(x)$ decreases along every nontrivial trajectory.

It is clear that x_s is an SEP of (21.2) if and only if there exists a neighborhood N such that $C(x)$ reaches the local minimum at $x = x_s$. If not, there is another point, say $\bar{x}$, that lies in a neighborhood and has a lower value of $C(x)$. Since the trajectory starting from $\bar{x}$ must converge to x_s, according to (i) and (ii), the value of $C(x)$ at $\bar{x}$ must be higher than that at x_s, which is a contradiction. This completes the proof.

Theorem 21-3 characterizes the relationship between the optimal solutions of the unconstrained optimization problem (21.1) and the stable equilibrium points of its corresponding dynamical system (21.2). Hence, if x_s is a stable equilibrium point of (21.2), then it is a local optimal solution of the unconstrained optimization problem (21.1). Conversely, if x_s is a local optimal solution of (21.1), then it is a stable equilibrium point of the gradient system (21.2). Our efforts are hence focused on developing effective algorithms to locate stable equilibrium points of (21.2) based on several topological and geometrical properties of the stability boundary.

21.3 Decomposition points and the quasi-stability boundary

Recall the concept of stability region $A(x_s)$ of the equilibrium point x_s:

$$A(x_s) = \{x \in R^n : \lim_{t \to \infty} \phi(t, x) = x_s\}.$$

Let $\overline{A(x_s)}$ represent the closure of $A(x_s)$. It was shown in Chapter 6 that the quasi-stability region eliminates the complex portion of the stability boundary lying inside the interior of $\overline{A(x_s)}$ and is useful for practical applications. The following results are derived from the theory of the quasi-stability boundary in Chapter 6.

THEOREM 21-4 **(Equilibrium points on the quasi-stability boundary)**
An equilibrium point x_d is on the quasi-stability boundary $\partial A_q(x_s)$ of the stable equilibrium point x_s if $x_d \in \partial A(x_s)$ and $W^u(x_d) \cap \left(\overline{A}(x_s)\right)^c \neq \emptyset$.

THEOREM 21-5 **(Types of equilibrium points on the boundary)**
If $\tilde{x}$ is a type-k equilibrium point, $k > 1$ on the quasi-stability boundary $\partial A_q(x_s)$, then there exists a type-one equilibrium point $x_1 \in \partial A_q(x_s)$ such that $\tilde{x} \in \overline{W^s(x_1)}$.

Motivated by the above two analytical results, we shall call a type-one equilibrium point lying on a quasi-stability boundary $\partial A_q(x_s)$ a *decomposition point (with respect to x_s)*. Decomposition points and their stable manifolds will be used to characterize the structure of a quasi-stability boundary. The searching procedure of Trust-Tech will be based on such a characterization.

THEOREM 21-6 **(Another characterization of $\partial A_q(x_s)$)**
Let σ_i, $i = 1, 2, \ldots$ be the decomposition points on the quasi-stability boundary $\partial A_q(x_s)$ of the stable equilibrium point x_s of the gradient system (21.2). Then,

$$\partial A_q(x_s) \subseteq \bigcup_{\sigma_i \in \partial A_q} \overline{W^s(\sigma_i)}$$

where $\partial A_q(x_s)$ is the quasi-stability boundary of x_s, σ_i is the decomposition point on the quasi-stability boundary of x_s and $\overline{W^s(\sigma_i)}$ represents the closure of the stable manifold of the decomposition point σ_i.

We will next explore some dynamic information for the decomposition points to develop a systematic search method for obtaining surrounding multiple local optimal solutions. To this end, we show that decomposition points can serve as a bridge linking two stable equilibrium points.

THEOREM 21-7 **(Decomposition point and stable equilibrium points)**
Suppose that every equilibrium point of the dynamical system (21.2) is hyperbolic and its stable and unstable manifolds satisfy the transversality condition. If x_s^1 is a stable equilibrium point of the dynamical system (21.2) and x_d is a decomposition point on its quasi-stability boundary, then there exists another stable equilibrium point x_s^2 such that x_d is a also a decomposition point of x_s^2.

Proof Since x_d is a decomposition point, its unstable manifold is of one dimension. Since x_d is on the quasi-stability boundary $\partial A_q(x_s)$, according to Theorem 21-4, $x_d \in \partial A(x_s)$ and $W^u(x_d) \cap \left(\overline{A}(x_s)\right)^c \neq \emptyset$. The complete stability of the dynamical system (21.2) shown in Theorem 21-1 asserts that the trajectories on the unstable manifold of x_d must converge to an equilibrium point. The transversality condition between the unstable manifold of x_d and the stable manifold of the equilibrium point ensures that the equilibrium point must be a stable equilibrium point. The proof is completed.

We note that the unstable manifold $W^u(x_d)$ is a one-dimensional manifold. Removing x_d from the unstable manifold leaves two invariant manifolds composed of only a single trajectory. These two trajectories must converge to two stable equilibrium points. These characterizations can be explored to devise a mechanism to escape from one stable equilibrium point to the adjacent stable equilibrium points. The basic idea is described as follows. The entire state space is decomposed into the closure of all the stability regions. Two adjacent stability regions are separated by the intersection of their stability boundaries which is the stable manifold of a decomposition point. To identify the adjacent stable equilibrium point from the initial stable equilibrium point, it suffices to devise a mechanism to go cross the stability boundary from one stability region to reach the adjacent stability region.

21.4 Trust-Tech-based decomposition point method

We present a Trust-Tech-based decomposition point (DP) method for, starting from a local optimal solution, finding another local optimal solution. Suppose that a local optimal solution, say x_s, has been found by an optimization method, such as an interior point method, a gradient-based method or a sequential quadratic programming method. According to Theorem 21-3, x_s is a stable equilibrium point of the gradient system (21.2). We seek to find another local optimal solution starting from a known one.

Given an unconstrained optimization problem (21.1) and a local optimal solution, say x_s^1, we have the following method.

Step 0: Construct the corresponding gradient system (21.2).

Step 1: Construct a path moving away from the local optimal solution x_s^1 (which is a stable equilibrium point of the gradient system (21.2)) and toward the stability boundary $\partial A(x_s^1)$ of x_s.

Step 2: Identify the exit point at which the constructed path intersects the stability boundary $\partial A(x_s^1)$.

Step 3: If the exit point exists, say x_e, then another stable equilibrium point (i.e. another local optimal solution) exists. (According to Theorem 21-6, this exit point must lie on the stable manifold of a dynamic decomposition point.)

Step 4: Starting from the exit point x_e and integrating the gradient system (21.2) results in a trajectory which moves along the stable manifold of the decomposition point (DP) until it reaches a point whose norm of the vector field of the gradient system (21.2) is zero (or close to zero). (As asserted by Theorem 21-6, this trajectory will converge to the DP.)

Step 5: Apply a solver such as the Newton method or the interior point method, starting from the point obtained at Step 4, to compute the dynamic decomposition point which separates the initial stable equilibrium point x_s and the corresponding stable equilibrium point.

Step 6: Starting from the found decomposition point, generate a point, which is a vector lying inside the stability region of the corresponding stable equilibrium point, as stated in Step 5.

Step 7: Starting from the point generated at Step 6 and integrating the gradient system (21.2) results in a trajectory that will converge to the corresponding stable equilibrium point (i.e. another local optimal solution), as asserted by Theorem 21-7. From some point along the trajectory, apply a local solver for fast convergence to the local optimal solution.

We note that Step 4 involves a trajectory moving along the stable manifold of a decomposition point while the computation of the DP is performed at Step 5. Theorem 21-6 provides a theoretical basis for Steps 4 and 5 while Theorem 21-7 provides a theoretical basis for Steps 6 and 7.

We next describe the numerical implementation of the key steps of the Trust-Tech DP method. The task of identifying the exit point in Step 2 plays an important role in the overall Trust-Tech-based methodology. A numerical implementation is presented below.

Step 1: Construct a path starting from a stable equilibrium point.

The path can be constructed by using heuristic schemes or by following the direction of one of the eigenvectors of the Jacobian matrix of the gradient system at the stable equilibrium point x_s.

Step 2: Identify the exit point.

Starting from a given initial stable equilibrium point x_s (i.e. a local optimal solution), we follow a specified search direction $\overline{d}$ and progressively sample a sequence of points $\{x_i\}$ with a small distance ε. By monitoring the value of the objective function $C(x)$ starting from the local optimal solution, it is clear that the value along the sequence will initially increase, since the sequence starts from a local optimal solution. The values along the sequence will continue to increase until the sequence reaches a point at which the value of the objective function $C(x)$ decreases. This point is termed an approximated exit point x_{exit}. This approximated exit point should be very close to the exact exit point, which is the intersection between the stability boundary and the curve connecting the sequence (see Figure 21.2).

Step 3: Carry out the stability-boundary following procedure and dynamic decomposition point computation.

This step computes the decomposition point by moving along its stable manifold. The exit point serves as the initial point of a trajectory lying on the stability boundary for

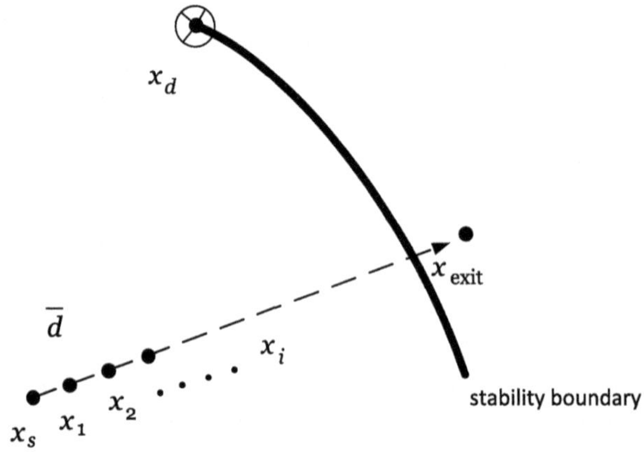

Figure 21.2 This exit point x_{exit} should be very close to the exact exit point.

locating the decomposition point. Due to the digital simulation of the trajectory, this stability-boundary following procedure constructs a sequence of points, starting from the exit point, moving along the stability boundary of x_s, and leading to a point which is close to the corresponding dynamic decomposition point. This stability-boundary following procedure will generate a sequence of points $\{z_i\}$ which is close to the stability boundary. Since each point in the sequence $\{z_i\}$ is very close to the stability boundary, the behavior of the sequence $\{z_i\}$ will approximately follow the behavior of a trajectory located on the stability boundary. Due to the structure of the stability boundary, where every trajectory on the stability boundary converges to a unique decomposition point, the point in the sequence $\{z_i\}$ having the minimal value of the norm of the vector field is the closest point to the decomposition point x_d. This point will be the output of the stability-boundary following procedure.

The point at the end of this procedure is termed the minimum gradient point (MGP). This point is, among all the points constructed by the procedure, the (first) minimum value of the norm of the vector field of the gradient system. From a geometric viewpoint, the MGP is close to the desired dynamic decomposition point. Using the MGP as an initial guess and applying a robust nonlinear algebraic solver, one can obtain the dynamic decomposition point x_d.

We next highlight some important features of the exit point detection procedure and the stability-boundary following procedure.

(i) All points in the sequence $\{x_i\}$ are constructed to identify the corresponding exit point moving away from the initial stable equilibrium point x_s and toward the stability boundary.
(ii) The values of the objective function of the sequence $\{x_i\}$ and the structure of the stability boundary are explored to identify the exit point of the stability boundary along the search path $\overline{d}$.
(iii) All points of the sequence $\{z_i\}$ are constructed to stay very close to the stability boundary (or more precisely, very close to the stable manifold of the decomposition point x_d).

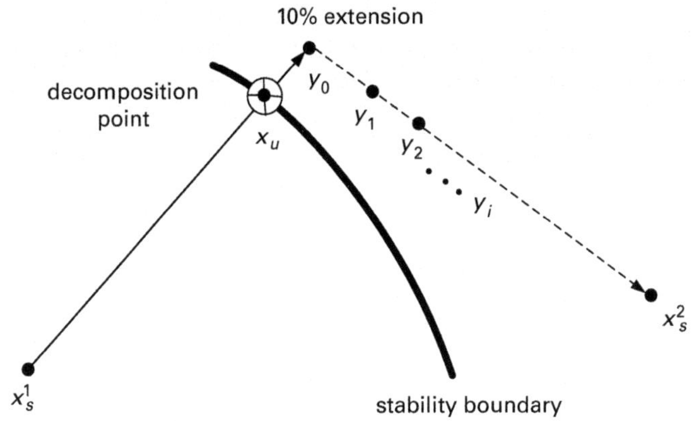

Figure 21.3 Locating one point lying inside the stability region of the nearby stable equilibrium point (i.e. another local optimal solution).

(iv) An L_2 norm or its equivalent of the sequence $\{z_i\}$ serves as a good index to measure the closeness of each point in the sequence $\{z_i\}$ and x_d.

Step 4: Locate a point lying in the stability region of another SEP.

Now, with information of a known local optimal solution x_s^1, and a DDP x_d located on the stability boundary, this step seeks to find a point lying inside the stability region of another stable equilibrium point (i.e. another local optimal solution). The theoretical basis of this step is Theorem 21-7. It asserts that the unstable manifold of the DDP x_d connects the initial stable equilibrium point x_s^1 and another SEP, say x_s^2.

We present two methods to numerically implement this step. The first method is to move outwards along the direction $\overrightarrow{x_s^1 x_d}$, say a step-size of 10%, to obtain a point, say y_0 (i.e. $y_0 = \overrightarrow{x_s^1 x_d} + 0,1 \,. \overrightarrow{x_s^1 x_d}$). It is clear that point y_0 locates inside the stability region of x_s^2 provided the step-size is not large, see Figure 21.3. We use point y_0 as an initial condition to numerically integrate the dynamical system (21.2). This numerical integration will generate a trajectory converging to the stable equilibrium point x_s^2. Hence, another local optimal solution is located.

The second method explores the fact that the DDP x_d is of type one. The dimension of the unstable manifold of x_d is hence one dimensional. The corresponding unstable eigenvector is tangent to the unstable manifold. This method uses the unstable eigenvector which points outwards along the direction $\overrightarrow{x_s^1 x_d}$ to locate a point lying in the stability region of another SEP.

One can monitor the ensuing trajectory until it reaches a point close to the targeted stable equilibrium point x_s^2 (i.e. the norm of the vector field is close to zero). One can stop the numerical integration and resort to a local optimization solver. The ending point of the numerical integration serves as an initial condition of the solver to locate another local optimal solution x_s^2. The efficiency of searching for another local optimal solution can be further improved by using a hybrid of the Trust-Tech DP method and an efficient local optimizer.

The Trust-Tech-based DDP method requires the stability-boundary following procedure, which involves numerical integration, functional evaluation (for closeness to the stability boundary) and a nonlinear algebraic solver (for locating the dynamic decomposition point). This method can avoid finding the same local optimal solution starting from a known local optimal solution. The theoretical basis of this feature is explained as follows.

From Theorem 21-7, it follows that the unstable manifold of the decomposition point x_d connects the initial stable equilibrium point x_s^1 and the targeted stable equilibrium point x_s^2. It is obvious that different search directions from the initial stable equilibrium point x_s^1 will lead to different exit points on the stability boundary of x_s^1. If multiple different exit points move along the same stable manifold of a DDP, then one locates the same decomposition point from these multiple different exit points. It follows from Theorem 21-7 that the same stable equilibrium point (i.e. the same local optimal solution) will be found if one applies Step 5 and Step 6. Therefore, the repetition of finding the same local optimal solution from two different search directions can be avoided. Hence, the challenging issue (C2) is addressed.

The Trust-Tech DP method can be modified for the sake of speed into the following Trust-Tech exit point method as follows.

Given an unconstrained optimization problem (21.1) and a local optimal solution, say x_s^1, we have the following method.

Step 0: Construct the corresponding gradient system (21.2).

Step 1: Construct a path moving away from the local optimal solution x_s^1 (which is a stable equilibrium point of the gradient system (21.2)) and toward the stability boundary $\partial A(x_s^1)$ of x_s.

Step 2: Identify the exit point at which the constructed path intersects the stability boundary $\partial A(x_s^1)$.

Step 3: If the exit point exists, say x_e, then another stable equilibrium point (i.e. another local optimal solution) exists.

Step 4: Starting from the exit point x_e, generate a point which is a vector lying inside the stability region of the corresponding stable equilibrium point, as stated in Step 3.

Step 5: Starting from the point at generated Step 4 and integrating the gradient system (21.2) results in a trajectory, which will converge to the corresponding stable equilibrium point (i.e. another local optimal solution), as asserted by Theorem 21-7. From some point along the trajectory, apply a local solver for fast convergence to the local optimal solution.

21.5 Numerical studies

We illustrate the Trust-Tech DP method on two test systems. The first optimization problem is a polynomial function as follows.

Table 21.1 Equilibrium points of the gradient system, their objective function values and their types

Number of EP	$x = (x_1, x_2)$		$f(x)$	Type of EP
1	−1.7476	−0.8738	1.7918	0
2	−1.0705	−0.5353	5.2642	1
3	**0.0000**	**0.0000**	**0.0000**	**0**
4	1.0705	0.5353	5.2642	1
5	1.7476	0.8738	1.7918	0

Note: the highlighted equilibrium point is the global optimal solution.

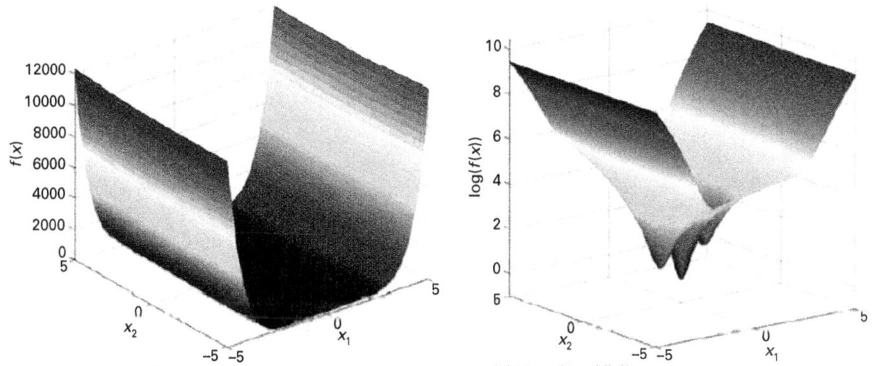

(a) Plot of $f(x)$ (b) Plot of $\log(f(x))$

Figure 21.4 Plot of the objective surface. It is seen that this objective function possesses a flat basin, within which all the local optimal solutions are located.

$$
\begin{aligned}
\min_{x \in R^2} f(x) &= 12x_1^2 - 6,\ 3x_1^4 + x_1^6 - 6x_1x_2 + 6x_2^2 \\
&\quad - 5 \le x_1, x_2 \le 5.
\end{aligned} \tag{21.3}
$$

The objective surface for this problem is shown in Figure 21.4(a). This objective function possesses a flat basin, within which all the local optimal solutions are located. For a better view of the details, a plot of the objective logarithm is shown in Figure 21.4(b).

To apply the Trust-Tech methodology to solve this optimization problem, the corresponding gradient dynamical system is constructed as follows:

$$
\begin{aligned}
\dot{x}_1 &= -\frac{\partial f}{\partial x_1} = -24x_1 + 25,\ 2x_1^3 - 6x_1^5 + 6x_2 \\
\dot{x}_2 &= -\frac{\partial f}{\partial x_2} = 6x_1 - 12x_2.
\end{aligned} \tag{21.4}
$$

There are five equilibrium points in the gradient system, which are listed in Table 21.1. Of these equilibrium points, two are of type one and three are type zero (i.e. SEP), which

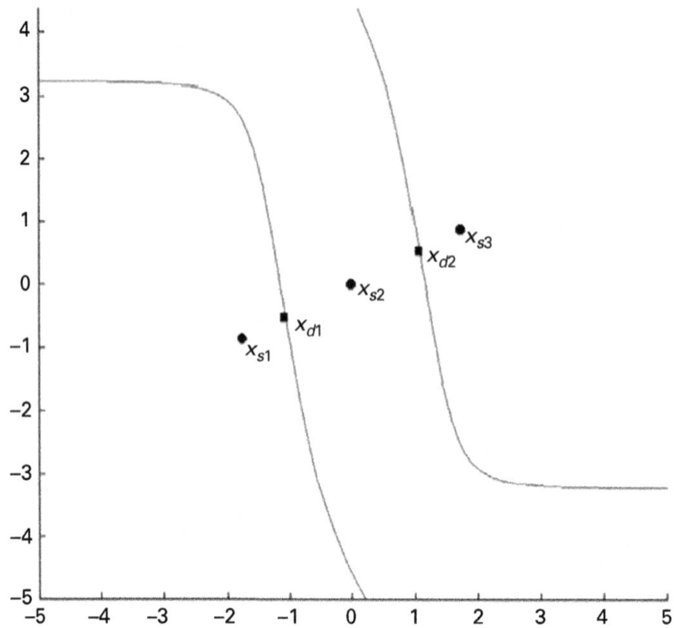

Figure 21.5 Equilibrium points (three SEPs and two type-one UEPs) of the negative gradient system (21.4).

correspond to the three local optimal solutions of the original optimization problem (21.3). Recall that a type-one equilibrium point is an unstable equilibrium point whose corresponding Jacobian matrix has only one positive eigenvalue.

A relationship among the three stable equilibrium points, their stability boundaries, and type-one unstable equilibrium points of the associated gradient system (21.4) is revealed in Figure 21.5. In this figure, stable equilibrium points are expressed with a subscript s, type-one equilibrium points are expressed with a subscript d. The curves represent the stability boundaries of the three SEPs of the gradient system.

The following dynamic properties are observed from this figure.

- There are three stable equilibrium points x_{s1}, x_{s2}, x_{s3} (in other words, there are three local optimal solutions) while there are two DPs (i.e. two type-one equilibrium points) x_{d1}, x_{d2}.
- The stability boundary of x_{s1} equals the stable manifold of x_{d1} while the stability boundary of x_{s3} equals the stable manifold of x_{d2}.
- The stability boundary of x_{s2} equals the union of the stable manifold of x_{d1} and of the stable manifold of x_{d2}.

The computational procedure of the Trust-Tech exit point method is described as follows. First, a local optimizer such as the limited-memory Broyden–Fletcher–Goldfarb–Shanno (L-BFGS) method is used as the local solver. The initial starting point is selected at (1.0, 1.0). The eigenvectors at each stable equilibrium point and their reverse directions are generated in order to achieve the following:

Table 21.2 Equilibrium points of interest involved in the Trust-Tech solution process

Number	Search direction	Exit point	Initial point	Local optimal solution	Tier
1	–	–	(1.000, 1.000)	(1.7476, 0.8738)	0
2	(−0.985, −0.172)	(1.036, 0.749)	(1.000, 0.743)	**(0.0000, 0.0000)**	1
3	(−0.985, −0.172)	(−1.124, −0.197)	(−1.180, −0.207)	(−1.7476, −0.8738)	2

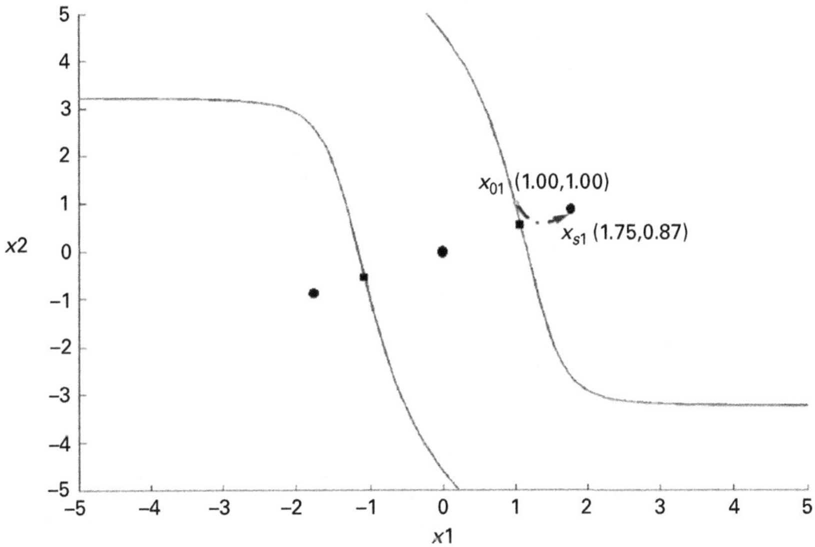

Figure 21.6 Starting from the specified initial point x_{01} = (1.0, 1.0), apply the L-BFGS method and the first local optimal solution x_{s1} = (1.7476, 0.8738) is found.

- escape from a local optimal solution;
- search for the next-tier stable equilibrium points (i.e. other local optimal solutions).

The Trust-Tech search process is described as follows.

Tier-zero local optimal solution

Starting from the specified initial point x_{01} = (1.0, 1.0), apply the L-BFGS method and the first local optimal solution x_{s1} = (1.7476, 0.8738) is found, as illustrated in Figure 21.6.

Procedure for finding the tier-one local optimal solution

Starting from the local optimal solution x_{s1}, the Trust-Tech methodology seeks to find the corresponding exit points near the stability boundary along some specified directions. Of these directions, an exit point (1.036, 0.749) is located via the search vector

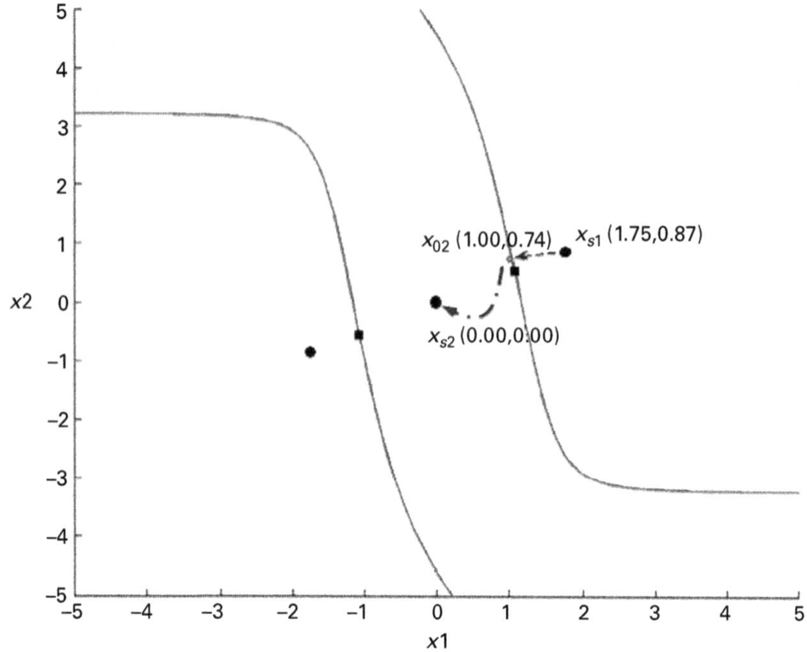

Figure 21.7 Starting from x_{s1}, the Trust-Tech methodology detects the corresponding exit points near the stability boundary along specified directions. Of these directions, an exit point (1.036, 0.749) is detected along the direction (−0.985, −0.172). A new initial point x_{02} = (1.000, 0.743) is generated using the exit point. The L-BFGS method is applied to this initial point and a tier-one local optimal solution x_{s2} = (0.0000, 0.0000) is found.

(−0.985, −0.172). A new initial point x_{02} = (1.000, 0.743) is generated using the exit point. Starting from the newly generated point, we apply the L-BFGS method to find a tier-one local optimal solution x_{s2} = (0.0000, 0.0000), as illustrated in Figure 21.7.

Procedure for finding the tier-two local optimal solution

Starting from x_{s2}, the Trust-Tech methodology detects an exit point (−1.124, −0.197) near the stability boundary along the search direction (−0.985, −0.172). A new initial point x_{03} = (−1.180, −0.207) is generated using the found exit point. The L-BFGS method is applied to this initial point and finds a tier-two local optimal solution $x_{s\text{-}3}$ = (−1.7476, −0.8738), as illustrated in Figure 21.8.

In summary, the Trust-Tech method has found a total of three local optimal solutions; of these, the solution (0.0, 0.0) is the global optimal solution.

The second optimization problem is described as follows:

$$\min_{x \in R^2} f(x) = 4x_1^2 - 2,\ 1x_1^4 + \frac{x_1^6}{3} + x_1 x_2 - 4x_2^2 + 4x_2^4 \qquad (21.5)$$
$$-5 \leq x_1,\ x_2 \leq 5.$$

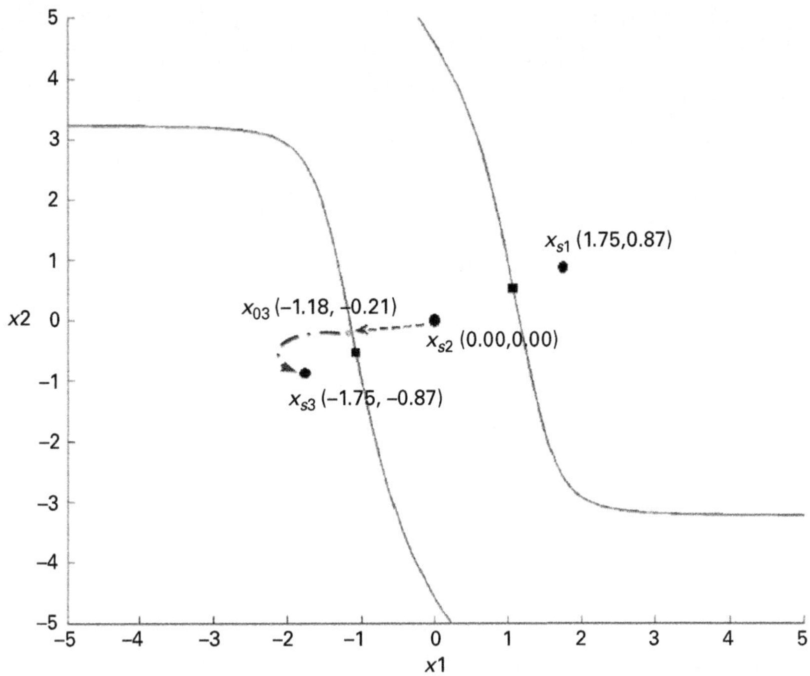

Figure 21.8 Starting from x_{s2}, the Trust-Tech methodology detects an exit point (−1.124, −0.197) near the stability boundary along the search direction (−0.985, −0.172). A new initial point x_{03} = (−1.180, −0.207) is generated using the found exit point. The L-BFGS method is applied starting from this initial point and finds a tier-two local optimal solution x_{s3} = (−1.7476, −0.8738).

The objective surface for this problem, shown in Figure 21.9(a), possesses a flat basin, within which all the local optimal solutions are located. For a better view of the details, a plot of the objective logarithm is shown in Figure 21.9(b).

To solve this global optimization problem using Trust-Tech method, a gradient dynamical system is constructed as follows:

$$\begin{aligned} \dot{x}_1 &= -\frac{\partial f}{\partial x_1} = -8x_1 + 8,\, 4x_1^3 - 2x_1^5 - x_2 \\ \dot{x}_2 &= -\frac{\partial f}{\partial x_2} = -x_1 + 8x_2 - 16x_2^3. \end{aligned} \tag{21.6}$$

There are at least 15 equilibrium points in the gradient system, which are listed in Table 21.3. Of these equilibrium points, seven are of type one and two are of type two. The remaining six equilibrium points are stable (type zero), and correspond to the six local optimal solutions of the original optimization problem.

A dynamic relationship between the equilibrium points and the stability boundaries of the gradient system is illustrated in Figure 21.10. In this figure, stable equilibrium points have subscript s, type-one equilibrium points have subscript d, while type-two equilibrium points have subscript u. The curves in the figure represent the stability boundaries of the gradient system. These UEPs all lie on the stability boundaries of SEPs,

Table 21.3 Equilibrium points of the gradient system, their objective function values and their type

Number of EP	$x = (x_1, x_2)$		$f(x)$	Type of EP
1	−1.6071	−0.5687	2.1043	0
2	−1.6381	−0.2287	2.2294	1
3	−1.7036	0.7961	−0.2155	0
4	−1.2961	−0.6051	2.2295	1
5	−1.2302	−0.1623	2.4963	2
6	**−0.0898**	**0.7127**	**−1.0316**	**0**
7	−1.1092	0.7683	0.5437	1
8	1.6071	0.5687	2.1043	0
9	**0.0898**	**−0.7127**	**−1.0316**	**0**
10	1.7036	−0.7961	−0.2155	0
11	0.0000	0.0000	0.0000	1
12	1.1092	−0.7683	0.5437	1
13	1.2961	0.6051	2.2295	1
14	1.6381	0.2287	2.2294	1
15	1.2302	0.1623	2.4963	2

Note: the highlighted equilibrium points are the global optimal solutions.

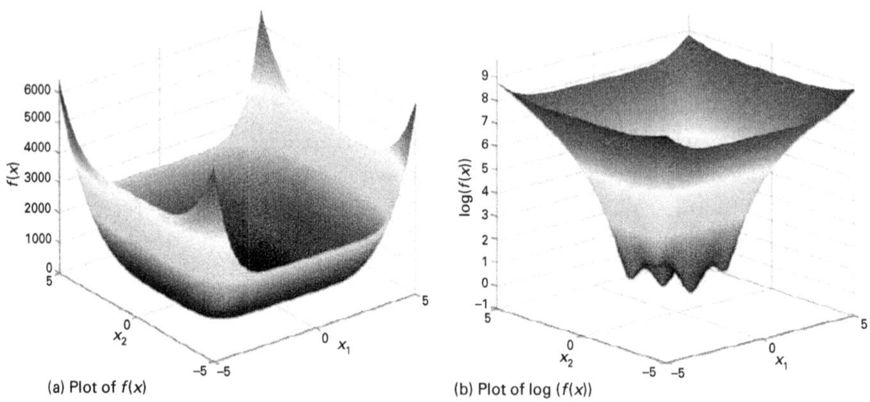

Figure 21.9 Plots of the objective surface. It is seen that this objective function possesses a flat basin, within which all the local optimal solutions are located.

confirming that the gradient system is completely stable. The following dynamic properties are observed from this figure.

- There are six stable equilibrium points $x_{s1}, x_{s2}, x_{s3}, x_{s4}, x_{s5}, x_{s6}$ (in other words, there are six local optimal solutions) while there are seven DPs (i.e. seven type-one equilibrium points) $x_{d1}, x_{d2}, x_{d3}, x_{d4}, x_{d5}, x_{d6}, x_{d7}$.

Table 21.4 Equilibrium points of interest involved in the Trust-Tech solution process

Number	Search direction	Exit point	Initial point	Local optimal solution	Tier
1	–	–	(0.000, 0.000)		
2	(0.918, −0.396)	(−1.249, −0.538)	(−1.311, −0.565)	(−1.6071, −0.5686)	1
3	(−0.929, 0.369)	(−0.993, 0. 394)	(−1.042, 0.414)	**(−0.0898, 0.7127)**	1
4	(−0.918, 0.396)	(−1.230, −0.484)	(−1.211, 0.216)	**(0.0898, −0.7127)**	2
5	(−0.040, 0.999)	(−1.620, −0.227)	(−1.622, −0.210)	(−1.7036, −0.7961)	2
6	(0.806, 0.592)	(1.246, −0.138)	(1.304, 0.180)	(1.7036, −0.7961)	3
7	(−0.040, −0.999)	(1.662, 0.231)	(1.660, 0.282)	(1.6071, 0.5687)	4

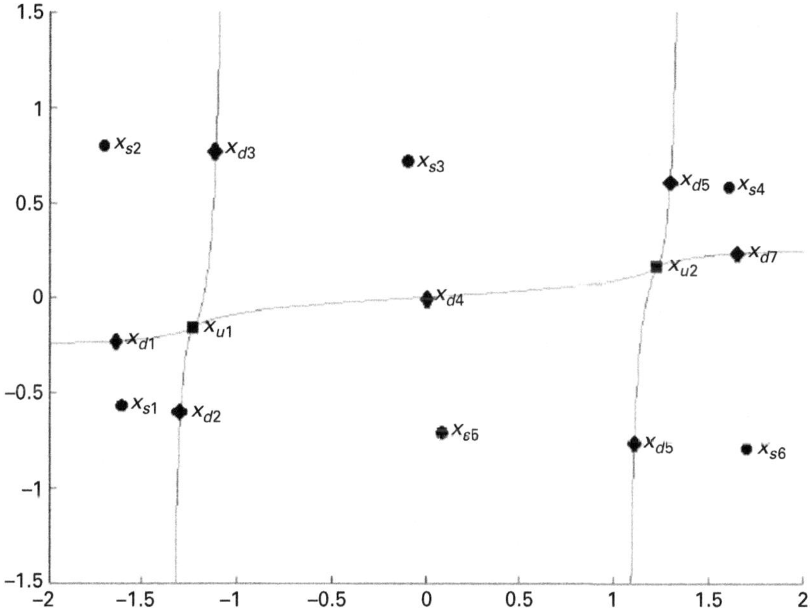

Figure 21.10 Locations of SEPs (i.e. the local optimal solutions) and their stability boundaries, of type-one and type-two equilibrium points of the gradient system. These UEPs all lie on the stability boundaries of SEPs.

- The stability boundary of x_{s1} equals the union of the stable manifold of x_{d1}, the stable manifold of x_{d2} and the stable manifold of x_{u1}.
- The stability boundary of x_{s2} equals the union of the stable manifold of x_{d1}, the stable manifold of x_{d3} and the stable manifold of x_{u1}.
- The stability boundary of x_{s3} equals the union of the stable manifold of x_{d3}, the stable manifold of x_{d4}, the stable manifold of x_{d5}, the stable manifold of x_{u1}, and the stable manifold of x_{u2}.

The computational procedure of the Trust-Tech method is described as follows. First, a local optimizer such as the limited-memory Broyden–Fletcher–Goldfarb–Shanno (L-BFGS) method is used as the local solver. The initial starting point is selected at (0.0, 0.0). The eigenvectors at each stable equilibrium point (i.e. local optimal solution) and their reverse directions are generated in order to achieve the following:

- escape from a local optimal solution;
- search for the next-tier stable equilibrium points (i.e. next-tier local optimal solutions).

The Trust-Tech solution procedure is described as follows.

Type-zero local optimal solution

Apply the L-BFGS method to the initial point $x_{01} = (0.0, 0.0)$, and a solution x_{d4} is found. This solution is identical to the initial point since the gradient of the objective function vanishes at this point. Note that x_{d4} is not a local optimal solution. In fact, it is just a type-one equilibrium point of the gradient system.

Procedure for finding tier-one local optimal solutions

Starting from x_{d4}, the Trust-Tech methodology seeks to find the corresponding exit points near the stability boundary along specified directions. Of these directions, two exit points are detected in two directions. The first exit point (−1.249, −0.538) is found along the search direction (0.918, −0.396), and the second exit point (−0.993, 0.394) is found along the search direction (−0.929, 0.369). Two new initial points $x_{02} = (-1.311, -0.565)$ and $x_{03} = (-1.042, 0.414)$ are generated using these two exit points, respectively. The L-BFGS method is then applied to these two initial points and two tier-one local optimal solutions are found, which are $x_{s2} = (-1.6071, -0.5686)$ and $x_{s3} = (-0.0898, 0.7127)$, respectively.

Procedure for finding tier-two local optimal solutions

Starting from x_{s1}, the Trust-Tech methodology detects two exit points (−1.230, −0.484) and (−1.620, −0.227) near the stability boundary along the search directions (−0.918, 0.396) and (−0.040, 0.999), respectively. New initial points $x_{04} = (-1.211, 0.216)$ and $x_{05} = (-1.622, -0.210)$ are generated using the found exit points. The L-BFGS method is then applied to these initial points and it finds two tier-two local optimal solutions $x_{s5} = (0.0898, -0.7127)$ and $x_{s2} = (-1.7036, -0.7961)$, respectively. Starting from x_{s3}, the Trust-Tech methodology finds finds no new local optimal solutions along the specified search directions.

Procedure for finding a tier-three local optimal solution

Starting from x_{s5}, the Trust-Tech methodology finds an exit point (1.246, −0.138) near the stability boundary along the search direction (0.806, 0.592). A new initial point

x_{06} = (1.304, 0.180) is generated via the newly found exit point. The L-BFGS method is then applied to this initial point and it finds a tier-three local optimal solution $x_{s\text{-}6}$ = (1.7036, −0.7961).

Procedure for finding a tier-four local optimal solution

Starting from x_{s6}, the Trust-Tech methodology finds an exit point (1.662, 0.231) near the stability boundary along the search direction (−0.040, −0.999). A new initial point x_{07} = (1.660, 0.282) is generated via the newly found exit point. The L-BFGS method is then applied to this initial point and it finds a tier-four local optimal solution $x_{s\text{-}4}$ = (1.6071, 0.5687).

For this test problem, the Trust-Tech method has found a total of six local optimal solutions; of these, the solutions (−0.0898, 0.7127), (0.0898, – 0.7127) are the global optimal solutions.

21.6 Concluding remarks

We have presented a stability-region-based methodology, termed the Trust-Tech methodology to systematically compute a set of, if not all of, the local optimal solutions of unconstrained (continuous) nonlinear optimization problems. The stability-region-based methodology has a mechanism to efficiently move away from one local optimal solution and move toward another local optimal solution. The mechanism utilizes the system trajectories of nonlinear dynamical systems and local optimization solvers, to systematically and effectively compute a set of local optimal solutions. The Trust Tech methodology employs knowledge of the stability boundary, of the stability region and of system trajectories to perform the following:

- identify the existence of another local optimal solution,
- detect the stability boundary of another SEP (i.e. another local optimal solution),
- compute multiple local optimal solutions by following system trajectories,
- find all the local optimal solutions in a tier-by-tier manner, and
- identify the non-existence of local optimal solution along a specified search direction.

From a theoretical viewpoint, the Trust-Tech-based method can find all the local optimal solutions. However, from a computational viewpoint, it requires the computation of a complete set of decomposition points that lies on the quasi-stability boundary of a local optimal solution so that all the next-tier local optimal solutions can be computed.

The Trust-Tech methodology addresses the following two challenges commonly faced by existing nonlinear optimization solvers.

(C1) How to effectively move (escape) from a local optimal solution and move on toward another local optimal solution.

(C2) How to avoid revisiting local optimal solutions which are already known.

Theorem 21-7 asserts that the unstable manifold of a decomposition point links one stable equilibrium point x_s^1 and another stable equilibrium point x_s^2. The Trust-Tech method finds a decomposition point and then another local optimal solution via the DP. Hence, the challenge (C1) is fully addressed. If one finds the same decomposition point from different exit points due to different selected search directions, it follows from Theorem 21-7 that the same stable equilibrium point (i.e. the same local optimal solution) will be found. Therefore, finding the same local optimal solution repeatedly from two different search directions can be avoided. Hence, challenge (C2) is addressed.

22 Perspectives and future directions

The theory of stability regions has made significant progress in the last three decades and reached a level of maturity at which practical applications are emerging. Indeed, recent stability-region-based applications include direct methods for fast stability analysis and control of power systems and Trust-Tech methods for solving high-dimensional nonlinear optimization problems. In this book, we have presented a comprehensive development of the theory of stability regions and of numerical methods for estimating stability regions of large-scale nonlinear systems. These developments can facilitate developing practical applications of stability regions.

Despite the significant advances in the area of stability regions and the maturity of the field, it is our belief that further development of stability region theory and numerical methods for estimating stability regions of large-scale nonlinear systems is needed. In fact, it is a new world in which we have just started our journey of exploration. In this chapter, some potential areas that can lead to further developments of theory of stability regions and of numerical methods for estimating stability regions will be presented. The extension of the theory of stability regions to other classes of dynamical systems and its potential applications will be described. Challenges in developing these extensions will be highlighted. The feasibility of exploring special features of some classes of dynamical systems for developing more powerful results on the stability boundary characterization as well as more efficient estimation methods will also be discussed.

22.1 Extension of the theory of stability regions

The theory of the stability boundary and methods to estimate stability regions were presented in this book for several classes of nonlinear dynamical systems, including the following:

- continuous nonlinear dynamical systems (Chapters 4, 6, 10, 15),
- discrete-time nonlinear dynamical systems (Chapters 9, 12),
- complex continuous nonlinear dynamical systems (Chapters 5, 11),
- constrained continuous nonlinear dynamical systems (Chapters 7, 10),
- two-time-scale nonlinear dynamical systems (Chapter 16),
- non-hyperbolic nonlinear dynamical systems (Chapter 17).

Complete characterizations of the stability boundary were developed for these classes of systems, and numerical methods for optimally estimating stability regions of nonlinear systems of these classes were also developed. The complete characterizations of stability boundaries of various classes of dynamical systems developed thus far have been expressed in terms of stable manifolds of unstable limit sets on the stability boundary. More specifically, for the classes of dynamical systems admitting energy functions, the stability boundary is contained in the union of the stable manifolds (or stable sets) of the equilibrium points (or invariant sets) on the stability boundary. Under the assumption of transversality, the stability boundary is equal to the union of the stable manifolds (or stable sets) of the equilibrium points (or invariant sets) on the stability boundary. We believe that this theory of the stability boundary can be extended to other classes of dynamical systems. For example, extending this theory to certain classes of stochastic dynamical systems, which model important applications in several areas such as control and economics, may result in fruitful results. Extending the theory of the stability region for hybrid dynamical systems, including the large class of switched nonlinear systems, is another important topic and this extension can lead to various applications. For instance, for controlling bipedal robots, the stability region of a periodic walking movement gives a measurement of how robust the robot is to perturbations [229].

The development of the theory of stability regions for some classes of functional differential equations, including the class of delayed nonlinear dynamical systems, is another challenging and relevant extension. It is well recognized that many nonlinear physical systems are subject to time delays. The feedback control signals of control systems that are implemented in networks are subject to delays in the communication channel. Delays are sometimes used for generating desired behavior or modeling the behavior of physical or biological systems. For many of these applications, knowledge of the stability region is essential. Some effort has been directed to determining stability regions of nonlinear discrete systems subject to time delays, see for example [281].

Extending the theory of the stability region to other classes of dynamical systems can be very challenging. We next describe several issues to be addressed in developing the theory of stability regions for the following class of nonlinear dynamical systems:

(i) systems described by retarded functional differential equations and
(ii) systems described by discrete-state systems.

22.1.1 Delayed functional differential equations

Consider the following class of autonomous retarded functional differential equations:

$$\dot{x} = f(x_t), \quad t \geq 0. \tag{22.1}$$

The vector field f in this class is a continuous function defined on the vector space of continuous functions $C([-h, 0], R^n)$, defined on $[-h, 0]$, into R^n and endowed with the norm $\|\phi\|_{C([-h,0],R^n)} = \sup_{\theta \in [-h,0]} \|\phi(\theta)\|_{R^n}$. Let $x{:}[-h, \alpha] \to R^n$, $\alpha > 0$, be a continuous function. For each $t \in [0, \alpha]$, the function x_t is one element of $C([-h, 0], R^n)$, defined as

$x_t(\theta) = x(t+\theta), \theta \in [-h, 0]$. The element x_t can be understood as a segment of the graph of $x(s)$, which is obtained by letting s vary from $t-h$ to t. For a $\phi \in C([-h, 0], R^n)$, consider the following initial value problem:

$$x_0 = \phi, \quad \phi \in C([-h, 0], R^n). \tag{22.2}$$

DEFINITION

A solution of the initial value problem (22.1)–(22.2) is a function $x(t)$ defined and continuous on an interval $[-h, 0]$, $\alpha > 0$, such that (22.2) holds and (22.1) is satisfied for all $t \in [0, \alpha]$.

If for each $f \in C([-h, 0], R^n)$ the initial value problem (22.1)–(22.2) has a unique solution $x(t, \phi)$ defined for all $t \geq 0$, then we will denote by $\pi^+(\phi)$ the positive orbit through ϕ, which is defined as $\pi(\phi) = \{x_t(\phi), t \geq 0\}$. The constant function $\varphi \in C([-h, 0], R^n)$ is an equilibrium point of the nonlinear retarded functional differential equation (22.1) if $f(\varphi) = 0$. The stability region $A(\phi_s)$ of an asymptotically stable equilibrium point $\phi_s \in C([-h, 0], R^n)$ is the set:

$$A(\phi_s) = \{\psi \in C([-h, 0], R^n) : x_t(\psi) \to \phi_s \quad \text{as} \quad t \to \infty\}. \tag{22.3}$$

The trajectories of the functional differential equation (22.1) are defined on the infinite-dimensional space $C([-h, 0], R^n)$. Consequently, the stability region is also a subset of an infinite-dimensional vector space. It is known that in such spaces, boundedness of solutions does not guarantee pre-compactness of solutions, as it does in a finite-dimension state space such as R^n. The pre-compactness of bounded trajectories is a key property which was explored in the development of the stability region of nonlinear continuous and discrete-time dynamical systems. The pre-compactness guarantees, for example, that the limit set of these solutions is not empty. In order to overcome this difficulty, certain conditions on the vector field f must be imposed to guarantee that solutions of (22.1) belong to a compact set. For a discussion on these conditions, see [113, 211].

Natural questions arise as we view the stability region as a subset of an infinite-dimensional space. Is the stability region diffeomorphic to the space $C([-h, 0], R^n)$? Is the co-dimension of the stability boundary equal to one? Can we show that, under certain conditions, the stability boundary is composed of the union of stable manifolds of invariant sets? And if we can, do these stable manifolds have infinite dimension? If they do, then the approaches used to derive the characterizations in finite-dimensional state spaces may not be applicable to infinite-dimensional state space. New approaches to derive the characterizations are needed.

The λ-lemma, for example, was extensively employed in the development of a complete characterization of the stability boundary of continuous as well as of discrete dynamical systems in R^n. Although the λ-lemma exists for functional differential equations, applying this technique to systems defined on infinite-dimensional state space encounters challenges. If the stable manifold of an equilibrium point on the stability boundary has infinite dimension, then the λ-lemma cannot be applied to obtain an approximation of the manifold by means of the flow of a disk (of infinite dimension) transversal to the unstable manifold.

22.1.2 Discrete-state systems

The development of a comprehensive theory of the stability region for discrete-state systems is another important and challenging area. Discrete-state systems are useful for modeling many practical systems. For example, they find applications in computer networks and manufacturing systems. The state space of discrete-state systems is discrete and their trajectories jump from one of these states to another. The concepts of stability, attractors and stability region for this class of systems were discussed, for example, in [28]. The task of developing a complete characterization of the stability region for this class of systems, however, is very challenging. Defining the stability boundary of an attracting set of discrete-state systems is still under development. It is likely that the tools needed to develop the theory of the stability region for this class of systems are different from those employed in this book to develop complete characterizations of stability regions.

22.2 Exploring special system structures

By exploring the special structure of the study system model, improved and more powerful results can be obtained. Some developments along this direction were presented in Chapter 18, in which system structures of second-order systems were explored to obtain more powerful results on the structure of the stability boundary and to develop more efficient methods for estimating stability regions. By exploring the relationship of second-order systems with reduced-state gradient systems, it was discovered that these two systems share the "same" closest UEP. This discovery leads to the development of a very effective method for computing the closest UEP and for estimating stability regions of large-scale second-order systems. In addition, it was shown that the stability region of vector second-order systems is always unbounded.

Another example of exploring system structures to achieve fruitful results presented in this book is the BCU method presented in Chapter 20. It is well recognized that the ability to compute the controlling UEP is vital to direct stability analysis of electric power systems. However, the task of computing the exact controlling UEP relative to a given fault is very difficult. The great computational challenges and complexities of computing the controlling UEP serve to explain why the great majority of existing methods fail. It is fruitful to develop tailored methods for computing the controlling UEP by exploring special properties as well as some physical and mathematical insights of the underlying power system transient stability model. The boundary of stability-region-based controlling unstable equilibrium point method (BCU method) [46,51,65,66] was developed along this line.

The BCU method is a systematic method for computing the controlling UEP of large-scale power systems. It first explores the special structure of the underlying model so as to define an artificial, reduced-state model that captures all the equilibrium points on the stability boundary of the original model. It then computes the controlling UEP of the original model by computing the controlling UEP of the reduced-state model, which is much easier to compute than that of the original model. Given a power system stability model with certain properties, one can develop the corresponding version of the BCU method by exploring the underlying model structure for direct stability analysis of the stability model.

In Chapter 16, the special structure of a class of two-time-scale systems was also explored to derive powerful results on characterizing the stability boundary. More specifically, the stability boundary characterization was decomposed into the characterization of the stability boundary and stability region of two simpler systems: the slow and the fast subsystems. This decomposition brings numerical advantages in speed and robustness for the development of numerical tools to estimate the stability region by decomposing the difficult problem of searching for the closest or controlling unstable equilibrium point into an easier problem of computing the corresponding unstable equilibrium points of the slow and fast subsystems.

In this section, we discuss the feasibility of exploring special structures of two classes of dynamical systems to further develop the theory of stability regions. More specifically, we discuss the possibilities and challenges of exploring the special structure of

(i) three-time-scale systems and
(ii) interconnected nonlinear systems.

22.2.1 Three-time-scale systems

The theory of stability regions developed in Chapter 16 can be extended to multi-time-scale nonlinear systems. In this section, we highlight challenges in extending the theory of stability regions to a class of three-time-scale systems.

A three-time-scale nonlinear system contains fast (state) variables, regular (medium) variables and slow variables. The following nonlinear dynamical system possesses x as a slow variable, y as a regular variable and z as a fast variable:

$$\begin{aligned} \dot{x} &= \varepsilon_1 f(x, y, z) \\ \dot{y} &= g(x, y, z) \\ \varepsilon_2 \dot{z} &= h(x, y, z) \end{aligned} \tag{22.4}$$

where ε_1, ε_2 are small positive scalars. The three-time-scale nonlinear systems can be approximated by the following three simpler systems for different time scales:

$$\text{(Fast system)} \begin{cases} \dfrac{dx}{d\tau} = 0 \\ \dfrac{dy}{d\tau} = 0 \\ \dfrac{dz}{du} = h(x, y, z) \end{cases}$$

$$\text{(Medium system)} \begin{cases} \dot{x} = 0 \\ \dot{y} = g(x, y, z) \\ 0 = h(x, y, z) \end{cases}$$

$$(\text{Slow system}) \begin{cases} \dfrac{dx}{du} = f(x, y, z) \\ 0 = g(x, y, z) \\ 0 = h(x, y, z) \end{cases} . \tag{22.5}$$

1. The fast system whose solution is a good approximation to the solution of the system for the fast time scale.
2. The medium system whose solution is a good approximation to the system for the mid-range time scale.
3. The slow system whose solution approximates the system solution for the slow time scale.

The solutions of these simpler systems can be used to approximate the solution of the original system for different time scales. Stability assessment of the original three-time-scale system can be made based on the stability properties of the three simpler systems. To this end, two singular-perturbation time-scale analyses can be applied to analyze the stability of the three-time-scale systems. This approach decomposes the task of stability analysis of a three-time-scale system into the tasks of stability analysis of two two-time-scale systems. By applying the results developed in Chapter 16 to these two simpler two-time-scale systems, it is expected that results similar to those obtained for two-time-scale systems can be derived for three-time-scale systems. For example, it is expected that the problem of checking whether or not an equilibrium point lies on the stability boundary of a three-time-scale system can be divided into the following three problems.

Problem 1: Checking whether or not the unstable equilibrium point lies on the stability boundary of a stable equilibrium point of a fast system.

Problem 2: Checking whether or not the stable equilibrium point of the fast system lies inside the stability region of a stable equilibrium point of a medium system.

Problem 3: Checking whether or not the stable equilibrium point of the medium system is inside the stability region of a stable equilibrium point of a slow system.

More generally, it can be expected that the relevant part of the stability region of a three-time-scale system can be sufficiently approximated (as $\varepsilon \to 0$) by the stability boundary and stability regions of the slow, medium and fast systems.

22.2.2 Interconnected nonlinear systems

Another area for development of the theory of the stability region is for the class of nonlinear interconnected systems, which is a particular class of continuous dynamical systems. Interconnected systems model a great variety of nonlinear systems, including power system models for stability analysis, complex systems, networked systems, economic systems and the behavior of populations. In the era of Big Data applications, the interconnected system approach may emerge as a powerful tool to deal with some issues with Big Data applications.

One common technique to analyze interconnected systems is to explore their special structure in order to decompose an interconnected system into a number of lower order free subsystems. The aim of the decomposition is to analyze the overall system using only information obtained from analysis of each free subsystem. The decomposition can be a natural physical decomposition or a mathematical decomposition, depending on the problem under study. Details regarding the interconnected system approach can be found in the books by Michel and Miller [180], by Šiljak [232], or in survey papers by Michel *et al.* [181] and Vidyasagar [266].

Consider a nonlinear dynamical system S described by the differential equation

$$(S) : \dot{x} = f(x) \tag{22.6}$$

where $x \in R^n$ and the function $f: R^n \to R^n$ is sufficiently smooth that (22.6) has a unique solution $x(t, x_0)$ over the interval $(0, \infty)$ corresponding to each initial condition x_0. It is frequently possible to view the system S as a nonlinear interconnection of several lower order subsystems

$$(S_i) : \dot{x}_i = g_i(x_i) + h_i(x_1, x_2, \ldots, x_l), \ i = 1, 2, \ldots, l \tag{22.7}$$

where $x_i \in R^{n_i}$, $g_i : R^{n_i} \to R^{n_i}$, $h_i : R^{n_1} \times R^{n_2} \times \cdots \times R^{n_l} \to R^{n_i}$, and $n = \sum_{i=1}^{l} n_i$.

Denote the system state vector by $x^T = [x_1^T, x_2^T, \ldots, x_l^T]$, the free subsystem vector field by the following

$$g(x)^T = [g_1(x_1)^T, g_2(x_2)^T, \ldots, g_l(x_l)^T],$$

and the interconnected vector field by the following

$$h(x)^T = [h_1(x_1, x_2, \ldots, x_l)^T, \ldots, h_l(x_1, x_2, \ldots, x_l)^T].$$

With these definitions of vectors, we can represent (22.6) equivalently as

$$\dot{x} = g(x) + h(x). \tag{22.8}$$

Thus the system S can be viewed as a nonlinear interconnection of the following (lower order) subsystems:

$$(S_i^0) : \dot{x}_i = g_i(x_i). \tag{22.9}$$

The system S is called an *interconnected system* or a *composite system* or a *large-scale system* and S_i^0 an *isolated system* or a *free subsystem of the interconnected system S.* Without loss of generality, it will be assumed throughout this chapter that $0 \in R^n$ is an equilibrium point for the interconnected system (22.8), and $0_i \in R^{n_i}$ is an equilibrium point for the free subsystem i (22.9).

The *stability region* $A(0)$ of the asymptotically stable equilibrium point 0 of the interconnected system S is the collection of the state defined as

$$A(0) = \{x^0 \in R^n : \lim_{t\to\infty} x(t, x^0) = 0\}.$$

For the *i*th subsystem S_i^0, the *stability region* $A(0_i)$ of the asymptotically stable equilibrium point 0_i is the collection of the state defined as

$$A(0_i) = \{x_i^0 \in R^{n_i} : \lim_{t\to\infty} x_i(t, x_i^0) = 0_i\}.$$

Basically, there are two approaches to estimating the stability regions of interconnected systems, namely, the scalar Lyapunov function approach and the vector Lyapunov function approach. In the scalar Lyapunov function approach, the stability region is estimated via a Lyapunov function, which is a function of the Lyapunov function for each free subsystem. In the vector Lyapunov function approach, an estimated stability region of the so-called comparison system is found first, based on which the stability region of the overall system is then determined. The vector Lyapunov function approach provides an estimated stability region of each subsystem independent of other subsystems, a feature not shared by the scalar Lyapunov function approach. However, the scalar Lyapunov function approach is more general in terms of assumptions imposed on the interconnected systems.

Both the scalar Lyapunov function approach and the vector Lyapunov function approach, however, give very conservative results when estimating the stability regions. This undesirable fact is partly due to the nature of the Lyapunov function approach and partly to the characteristic of the decomposition-aggregation technique used in analyzing the interconnected systems. Attempts to overcome the conservativeness in estimating the stability regions by the Lyapunov function approach were made by several researchers (see, for example, [38,136,182,222,273]).

To improve the conservativeness in estimating the stability region, the expansion schemes presented in Chapter 13 can be extended to interconnected systems. These expansion schemes are to be incorporated into a constructive method to estimate the stability region of the interconnected system (22.8). The constructive scalar Lyapunov function method starts with the following two components:

(1) a given function $V_i^0(x_i)$ for each free subsystem S_i^0, $i = 1, 2, \ldots, l$ satisfying certain conditions;
(2) a scalar c_i associated with $V_i^0(x_i)$ for S_i^0, $i = 1, 2, \ldots, l$, such that the set $D_{S_i^0} = \{x_i \in B_i(r_i) : V_i^0(x_i) < c_i\}$ is an estimated stability region of $x_i = 0$ for S_i^0.

Constructive scalar Lyapunov function method

Step 1: Given $V_i^0(x_i), c_i, i = 1, 2, \ldots, l,$ *for each free subsystem*, find α_i, $i = 1, 2, \ldots, l$ using an optimization method.

$$\text{Let } c = \min\{\alpha_i c_i\}. \tag{22.10}$$

Step 2: The set $\{x \in R^n : V^0(x) = \sum_{i=1}^{l} \alpha_i V_i^0(x_i) < c\}$ is the estimated stability region for $A(0)$.

Step 3: Modify the function $V_i^0(x_i)$ by $V_i^0(x_i) \leftarrow V_i^0(x_i + \hat{\lambda}_i g_i(x_i))$ or $V_i^0(x_i) \leftarrow V_i^0(x_i + (\overleftarrow{\lambda}_i/2)(g_i(x_i + \lambda_i g_i(x_i)) + g_i(x_i)))$, where $\hat{\lambda}_i$, $\overleftarrow{\lambda}_i$ are positive real numbers, and go to Step 2.

It can be shown that the above constructive method can yield a sequence of estimated stability regions which is a strictly increasing sequence, and each of these stability regions is contained in the entire stability region $A(0)$. The above constructive method can be extended to develop a constructive vector Lyapunov function method. This method would yield a sequence of estimated stability regions which is a strictly increasing sequence, with each stability region contained in the entire stability region $A(0)$. The analysis of the constructive vector Lyapunov function method parallels the analysis of the constructive scalar Lyapunov function method. Readers are encouraged to work out the details.

22.3 Deeper in the theory of the stability boundary

The development of the theory of the stability boundary has not been completed. In this book we have addressed some very challenging issues that are quite new and still under investigation. In Chapter 5, for example, we showed that complex behavior of dynamics on the stability boundary gives rise to very complex structures of the stability boundary, which are far from being fully understood. In Chapter 19, a glimpse of the enormous issue of bifurcations of stability regions and stability boundaries triggered by local or global bifurcations of vector fields was presented. Several routes can be followed to further develop the theory of the stability region. In this section we discuss two of these possibilities:

(i) types and number of UEPs on the stability boundary and
(ii) robustness of the stability region to parameter variations.

22.3.1 Types and number of UEPs on the stability boundary

This book presents the development of a comprehensive theory of stability regions and a complete characterization of the stability regions of a class of nonlinear dynamical systems described by the following (autonomous) nonlinear dynamical system

$$\dot{x} = f(x), \; x \in R^n. \tag{22.11}$$

We assume the function (i.e. the vector field) $f: R^n \to R^n$ satisfies a sufficient condition for the existence and uniqueness of the solution and also satisfies assumptions (A1)–(A3). The stability boundary of this class of nonlinear dynamical systems is characterized by the union of stable manifolds of equilibrium points and limit cycles on the boundary. Nevertheless, the characterization provides little information regarding the existence of certain types of equilibrium points on the stability boundary. Chapters 3 and 4 have shown the existence of type-one equilibrium points on stability boundaries and the existence of source points (i.e. type-n equilibrium points) on bounded stability boundaries.

However, it is still unclear whether or not every type of equilibrium point can lie on the stability boundary. In other words, whether or not type-one equilibrium points, type-two equilibrium points, type-three equilibrium points, ..., type-n equilibrium points exist on the stability boundary. We note that a type-k equilibrium point refers to a hyperbolic UEP at which the Jacobian matrix has k eigenvalues with positive real part and $(n-k)$ eigenvalues with negative real part. Furthermore, what is the number of each type of UEP that can lie on the stability boundary?

It has been shown that every isolated invariant set admits an index pair (in Conley index theory) such that the topology quotient has the homotopy type of a finite polyhedron, and the homotopy type of the quotient is independent of the choice of the index pair [73]. For an estimate of the number of critical points of nonlinear algebraic equations, the Morse inequality shows that the alternating sum of the Betti numbers is less than or equal to that of the numbers of critical points of index γ of a Morse function on a manifold [191,194]. For the critical points of electric power systems, it was shown that there are exactly $\frac{(2n-2)!}{(n-1)!(n-1)!}$ complex solutions to the power flow equations, for a completely interconnected lossless power network with n P–V nodes [20]. For an electric power system with low power injection, the number of equilibrium points on the hyperplane $\{(\delta, \omega) \in R\}$ contained in a stability boundary, has been estimated under the "general-position" hypothesis and the property that the dynamics commute with the identification map [169]. In addition, lower and upper bounds have been derived for the number of type-k equilibrium points on the stability boundary of a family of spatially periodic dynamical systems [61], which repeat the values in regular intervals for every variable. However, they provided little information on the existence of type-k equilibrium points. We next present insight into this issue.

Two stability regions A_1, A_2 are said to be neighboring to each other if their closures have non-empty intersection, i.e. $\overline{A_1} \cap \overline{A_2} \neq \emptyset$. Meanwhile, the corresponding stable equilibrium points are said to be neighboring to each other. Given two different stable equilibrium points, x_s^1, x_s^2, we observe that $A(x_s^1) \cap A(x_s^2) = \emptyset$ and the following relation:

$$
\begin{aligned}
\overline{A(x_s^1)} \cap \overline{A(x_s^2)} &= \partial A(x_s^1) \cap \partial A(x_s^2) \\
\overline{A(x_s^1)} \cap \overline{A(x_s^2)} &= \partial A_q(x_s^1) \cap \partial A_q(x_s^2).
\end{aligned}
$$

This implies that two stability regions are neighboring to each other if and only if the corresponding quasi-stability boundaries have non-empty intersection, a property that will be used later on. We will show the existence of all types of equilibrium points lying on the stability boundary under the following additional assumption.

(A4) The stability regions are uniformly bounded, and any bounded set only intersects finitely many stability regions. Moreover, two neighboring stability regions always share a common boundary of dimension $(n-1)$.

Again, we are exploring a special structure of the dynamical system to derive specialized results on stability boundary characterization. The boundedness assumption in (A4) ensures that all the stability regions are uniformly bounded, while the condition that the shared boundary is of dimension $(n - 1)$ in (A4) implies that the two neighboring stable equilibrium points are connected by the unstable manifold of a type-one equilibrium point. Indeed, assumptions (A1)–(A4) imply that the state space can be covered by the closures of the stability regions. The uniform-boundedness in (A4) asserts that there must be $(n + 1)$ stability regions that share a point in common on their boundaries.

By the complete characterization of stability regions of nonlinear dynamical systems satisfying assumptions (A1)–(A3), the common boundary of neighboring stability regions must be a union of stable manifolds of the shared unstable equilibrium points. The dimension theory further shows that any k-tuple stability regions having non-empty intersection on their boundaries, must contain a type-k equilibrium point.

By counting the number of such k-tuples including a given stability region as an element, we can finally estimate the number of type-k equilibrium points on the stability boundary as stated in the following [61].

THEOREM 22-1 **(Existence of and lower bound for type-k equilibriums)**
Consider the nonlinear dynamical system (22.9) satisfying assumptions (A1)–(A4). Then, for each stability boundary, the following results hold:

- *each type of unstable equilibrium point exists; in other words, there exist type-k equilibrium points, where $1 \leq k \leq n$, and n is the dimension of the state space,*
- *the number of type-k equilibrium points is at least $\frac{n!}{k!(n-k)!}$.*

The bound on the number of equilibrium points can be further improved by exploring the system structure of (22.9). We next give an illustration by showing that if the system possesses the spatial-periodicity property as stated in the following assumption, then better bounds can be derived. Recall that the nonlinear dynamical system (22.9), $\dot{x} = f(x) = \left(f_1(x), f_2(x), \ldots, f_n(x)\right)^T$ is called spatially periodic if there exist n positive constants p_i for $1 \leq i \leq n$ such that $f_j(x) = f_j(x + p_i e_i)$ for $x \in R^n$ and $1 \leq j \leq n$, where $e_i \in R^n$ denote the vector with 1 in the ith coordinate and zeros elsewhere. In addition, an n-tuple $\overline{p} = (\overline{p_1}, \overline{p_2}, \ldots, \overline{p_n})$ is called the spatial period if each $\overline{p_i}$ is the minimum number of p_i such that $f_j(x) = f_i(x + p_i e_i)$ for $x \in R^n$ and $1 \leq j \leq n$.

(A5) The system (22.9) is spatially periodic, with $p_i = 2\pi$. Moreover, at most one stable equilibrium point exists in each region of the form for all $x = (x_1, x_2, \ldots, x_n) \in R^n$

$$[x_1, x_1 + 2\pi] \times [x_2, x_2 + 2\pi] \times \ldots\ldots \times [x_n, x_n + 2\pi]$$

THEOREM 22-2 **(Lower bound for spatially periodic systems)**
Consider the nonlinear dynamical system (22.9) satisfying assumptions (A1)–(A5). Then, for each stability boundary, the following results hold:

- *each type of unstable equilibrium point exists; in other words, there exist type-k equilibrium points, where* $1 \leq k \leq n$*, and n is the dimension of the state space,*
- *the number of type-k equilibrium points is at least* $(k+1)\frac{n!}{k!(n-k)!}$.

THEOREM 22-3 **(Upper bounds for spatially periodic systems)**
Consider the nonlinear dynamical system (22.9) satisfying assumptions (A1)–(A5) and additionally suppose the system vector field can be converted into polynomials by a trigonometric transformation. Then, for each stability boundary, the following results hold:

- *each type of unstable equilibrium point exists; in other words, there exist type-k equilibrium points, where* $1 \leq k \leq n$*, and n is the dimension of the state space,*
- *the number of type-k equilibrium points is at least* $(k+1)\frac{n!}{k!(n-k)!}$,
- *the number of type-k equilibrium points is at most* $(k+1) \times 2n$ *multiplied by the degrees of the transformed polynomials.*

22.3.2 Robustness of stability regions

The complete characterizations of stability boundaries developed in this book were derived under the assumption that the model of the dynamic nonlinear system is exact and the associated parameters are precisely known. In practice, the model and the parameters are only approximations of the real values, and even if the model is exact, physical systems are subject to parameter variations. Due to both uncertainties and variation of parameters, it is natural to investigate the impact of parameter variations on the stability region and the stability boundary.

Chapter 19 is dedicated to the study of the impacts of parameter variation on the stability region and stability boundary. It is shown that bifurcations of the stability region may be induced by either local bifurcations of invariant sets (in particular of equilibrium points) on the stability boundary or global bifurcations. Bifurcations of the stability region induced by a particular type of local bifurcation of equilibrium points on the stability boundary, the so-called saddle-node bifurcations, are analyzed in detail in Chapter 19. For this type of bifurcation, it has been observed that drastic changes in the size of the stability region may occur. The theory of bifurcations of stability regions is very recent and future work in this area includes analysis of other types of local bifurcation on the stability boundary. Some work has already been done to characterize the stability region and the stability boundary near a supercritical Hopf bifurcation [102]. The Hopf equilibrium point is a non-hyperbolic equilibrium point due to the presence of a simple pair of conjugate eigenvalues in the imaginary axis. Hopf equilibrium points can also be classified in terms of the asymptotic behavior in the central manifold. The asymptotic behavior of a Hopf equilibrium point on the central manifold depends on the nonlinear terms of the vector field. This behavior can be predicted by the Lyapunov coefficients [153]. Trajectories in the central manifold of a supercritical Hopf equilibrium point tend to the equilibrium as time tends to infinity, while trajectories in the central manifold of a subcritical Hopf equilibrium point are repelled from the equilibrium point, i.e. they tend to the equilibrium point as time tends to minus infinity. Since

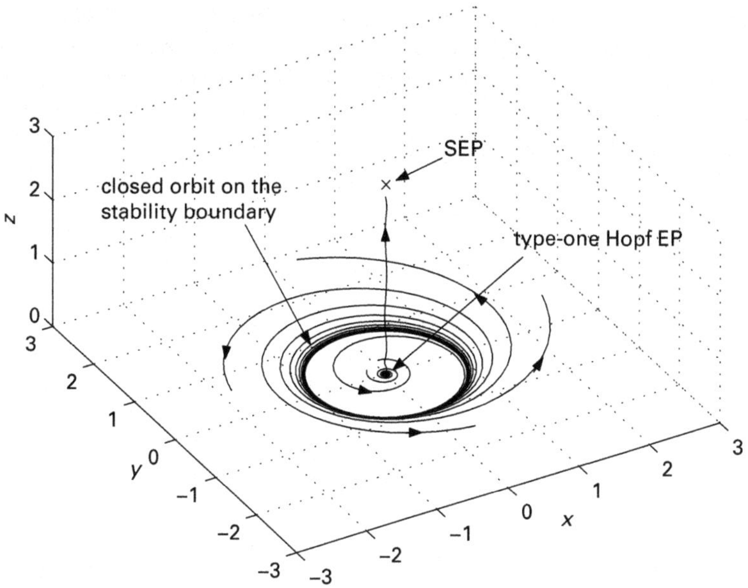

Figure 22.1 The origin is a supercritical type-one Hopf equilibrium point. This Hopf equilibrium point lies on the stability boundary of the stable equilibrium point (0,0,3).

the eigenvalues associated with the central manifold are on the imaginary axis, the attraction or repulsion is not exponencial in this manifold. The next example illustrates a Hopf equilibrium point.

Consider the following nonlinear dynamical system:

$$\begin{aligned}\dot{x}_1 &= -x_2 - x_1[(x_1^2 + x_2^2)^{1/2} - (1 - x_3)(x_1^2 + x_2^2)] \\ \dot{x}_2 &= x_1 - x_2[(x_1^2 + x_2^2)^{1/2} - (1 - x_3)(x_1^2 + x_2^2)] \\ \dot{x}_3 &= -\frac{1}{10}x_3(x_3 - 3)\end{aligned} \tag{22.12}$$

which possesses two equilibrium points, the origin and the point (0,0,3). These equilibrium points and the dynamics of the system are depicted in Figure 22.1. Trajectories in the central manifold tend to the equilibrium point as time tends to infinity, consequently this is a supercritical type-one Hopf equilibrium point.

The Hopf equilibrium point is on the stability boundary of the asymptotically stable equilibrium point. Beside the origin, a closed orbit lies on the stability boundary. A complete characterization of the stability boundary admitting the existence of supercritical Hopf equilibrium points was derived in [102].

The Hopf equilibrium point persists under perturbation of the vector field, but its stability property does not persist. The pair of conjugate eigenvalues associated with the Hopf bifurcation crosses the imaginary axis as a consequence of perturbation on the vector field. In a supercritical Hopf bifurcation, the stability property of the equilibrium point on the central manifold is lost and a closed orbit, stable on the central manifold, is born. Some natural questions arise from this observation. Does the equilibrium point persist on the stability boundary? Does the new closed orbit lie on the stability boundary? Does the size of the stability region drastically change? These and other questions are still open areas for research. Other types of bifurcations on the stability region also lead to an unexplored world of investigation. The analysis of bifurcations of the stability region due to global bifurcations is a completely open and difficult area of research. For further reading on this subject, see [9,10,11,12,102].

22.4 Advanced numerical methods

There are four types of steady state in nonlinear dynamical systems: equilibrium points, periodic trajectories, quasi-periodic trajectories and chaos. At present, the former two can be reliably computed, while tools for computing the latter two are yet to be developed. This numerical challenge poses a difficulty for characterizing these complex steady states on the stability boundary. Moreover, the complete characterization of the stability boundary of SEPs is expressed via the stable manifolds of equilibrium points and closed orbits. For low-dimensional manifolds, it is relatively easy to compute one-dimensional stable manifolds of type-($n-1$) equilibrium points and to compute the one-dimensional unstable manifolds of type-one equilibrium points. It is, however, very challenging to compute two-dimensional (or higher-dimensional) stable and unstable manifolds. From a computational viewpoint, developing advanced numerical methods to calculate stable manifolds leading to numerical characterization of the stability boundary is a challenge.

There have been several methods proposed to approximate the stability boundaries of nonlinear dynamical systems. It is our belief that the energy function approach described in this book is the best one for approximating the stability boundaries of high-dimensional nonlinear systems. To this end, in Chapter 15 we carried out a critical evaluation of three methods for approximating the relevant stability boundary of continuous dynamical systems: the energy function-based CUEP method, the hyperplane-based CUEP method and the quadratic-based CUEP method. It was shown that the energy function-based controlling UEP method excels in the following regard: accuracy in approximating the relevant stability boundary, consistency in providing a conservative approximation of the relevant stability boundary and the least computational burden.

It was observed in Chapter 5 that complex invariant sets, such as chaotic trajectories and closed orbits, can exist on the stability boundary. Although a characterization of the stability boundary has been derived for a very general class of continuous dynamical systems in terms of the stable sets of invariant sets on the stability boundary, several challenges still exist for estimating the stability regions of these systems. These challenges can be explained from both theoretical and computational viewpoints.

From a theoretical point of view, a deeper understanding of the nature of invariant sets on the stability boundary, as related to the characterization of the stability boundary, is desired. From a computational viewpoint, the development of advanced numerical methods to calculate complex invariant sets such as closed orbits and chaotic trajectories lying on the stability boundary can shed some light on the form and nature of the stability boundary and help devise more effective methods for estimating stability regions.

Stable and unstable manifolds play an important role in the characterization of the stability boundary. They are global objects that cannot generally be expressed in closed analytical form. One hence needs to resort to numerical methods to compute these objects. Advanced analytical and numerical methods to compute these manifolds not only provide deeper insight into the form of stable and unstable manifolds but also help devise effective methods for estimating stability regions. Several methods are available in the literature to calculate invariant manifolds, see for example [150]. These methods, however, are computationally intensive and they currently can only compute the invariant manifolds of low-dimensional dynamical systems. Computing the invariant manifolds of high-dimensional dynamical systems is of great importance. The parametrization method seems to be promising for this computation [33]. The recent development of a numerical method based on one-dimensional and two-dimensional characteristic invariant manifolds (CIMs) for evaluating the mismatches between the exact stability boundary and an approximated stability boundary [187] is interesting. This numerical method provides a means to measure the errors between an approximated stability boundary and the exact stability boundary. For a given threshold value, the method also computes a credible region which is defined as the neighborhood of a type-one equilibrium point lying on a stability boundary in which the mismatch between an approximated stability boundary and the exact stability boundary is below the threshold.

22.5 Conclusions

The theory of stability regions and numerical methods for estimating the stability regions of several classes of dynamical systems have been comprehensively presented in this book. The maturity of this theory and the development of effective numerical methods is leading to important applications in the areas of direct stability analysis of power systems and of nonlinear unconstrained as well as constrained optimization. Reaching this level of maturity of theory and applications has required an enormous effort of research in several fields of mathematics, physics and engineering. This chapter has illustrated possibilities in the developments of both the theory of stability regions and estimation methods for several classes of nonlinear dynamical systems. These developments come with potential fruitful results in applications. Following any of these routes of further developments will not be easy. Despite this challenge, we hope this book has paved a way to make these future journeys more attractive and rewarding for researchers and engineers. Please enjoy your journey!

Bibliography

[1] Available at www.ee.washington.edu/research/pstca/

[2] R. Abraham, J. Robbin, *Transversal Mappings and Flows*, Benjamin, New York, 1967

[3] A. Ailon, R. Segev, S. Arogeti, "A simple velocity-free controller for attitude regulation of a spacecraft with delayed feedback", *IEEE Transactions on Automatic Control*, v.**49**, n.1, pp.125–130, Jan 2004

[4] L. F. C. Alberto, *Transient Stability Analysis: Studies of the BCU method; Damping Estimation Approach for Absolute Stability in SMIB Systems* (In Portuguese), Escola de Eng. de São Carlos – Universidade de São Paulo, 1997

[5] L. F. C. Alberto, T. R. Calliero, A. Martins, N. G. Bretas, "An invariance principle for nonlinear discrete autonomous dynamical systems", *IEEE Transactions on Automatic Control*, v.**52**, n.4, pp.692–697, Apr 2007

[6] L. F. C. Alberto, H. D. Chiang, "Controlling unstable equilibrium point theory for stability assessment of two-time scale power system models", *IEEE Power and Energy Society General Meeting, Pittsburgh, PA*, pp.1–9, 2008

[7] L. F. C. Alberto, H. D. Chiang, "Uniform approach for stability analysis of fast subsystem of two-time-scale nonlinear systems", *International Journal of Bifurcation and Chaos*, v.**17**, n.11, pp.4195–4203, 2007

[8] L. F. C. Alberto, H. D. Chiang, "Characterization of stability region for general autonomous nonlinear dynamical systems", *IEEE Transactions on Automatic Control*, v.**57**, pp.1564 – 1569, 2012

[9] F. M. Amaral, L. F. C. Alberto, "Stability region bifurcations of nonlinear autonomous dynamical systems: type-zero saddle-node bifurcations", *International Journal of Robust and Nonlinear Control*, v.**21**, n.6, pp.591–612, 2011

[10] F. M. Amaral, L. F. C. Alberto, "Type-zero saddle-node bifurcations and stability region estimation of nonlinear autonomous dynamical systems", *International Journal of Bifurcation and Chaos in Applied Science and Engineering*, v.**22**, n.1, p.1250020 (16 pages), 2012

[11] F. M. Amaral, L. F. C. Alberto, "Stability boundary characterization of nonlinear autonomous dynamical systems in the presence of saddle-node equilibrium points", *TEMA Tendências em Matemática Aplicada e Computacional*, v.**13**, pp.143–154, 2012

[12] F. M. Amaral, L. F. C. Alberto, N. G. Bretas, "Stability boundary characterization of nonlinear autonomous dynamical systems in the presence of a type-zero saddle-node equilibrium point", *TEMA Tendências em Matemática Aplicada e Computacional*, v.**11**, pp.111–120, 2010

[13] F. Amato, C. Cosentino, A. Merola, "On the region of attraction of nonlinear quadratic systems", *Automatica*, v.**43**, pp.2119–2123, 2007

[14] B. Anderson, J. Keller, "Discretization techniques in control systems", *Control and Dynamic Systems*, v.**66**, pp.47–112, 1994

[15] V. Arnold, *Geometrical Methods in the Theory of Ordinary Differential Equations*, Springer-Verlag, New York, 1977

[16] K. L. Arrow, F. H. Hahn, *General Competitive Analysis*, Holden Day, San Francisco, CA, 1971

[17] Z. Artstein, "Stability in the presence of singular perturbation", *Nonlinear Analysis*, v.**34**, pp.817–827, 1998

[18] P. D. Aylett, "The energy integral-criterion of transient stability limits of power systems", *Proceedings of the IEEE*, v.**105**, n.8, pp.527–536, Sept 1958

[19] S. M. Baer, B. T. Li, H. L. Smith, "Multiple limit cycles in the standard model of three species competition for three essential resources", *Journal of Mathematical Biology*, v.**52**, n.6, pp.745–760, Jun 2006

[20] J. Baillieul, C. I. Byrnes, "Geometric critical point analysis of lossless power system models", *IEEE Transactions on Circuits and Systems*, v.**29**, pp.724–737, 1982

[21] J. Baker, "An algorithm for the location of transition states", *Journal of Computational Chemistry*, v.7, n.4, pp.385–395, 1986

[22] N. Balu, T. Bertram, A. Bose, V. Brandwajn, G. Cauley, D. Curtice, A. Fouad, L. Fink, M. G. Lauby, B. Wollenberg, J. N. Wrubel, "On-line power system security analysis" (Invited paper), *Proceedings of the IEEE*, v.**80**, n.2, pp.262–280, Feb 1992

[23] A. R. Bergen, D. J. Hill, "A structure preserving model for power system stability analysis", *IEEE Transactions on Power Apparatus and Systems*, v.**100**, pp.25–35, 1981

[24] N. P. Bhatia, G. P. Szego, *Stability Theory of Dynamical Systems*, Springer-Verlag, 1970

[25] F. Blanchini, "Set invariance in control", *Automatica*, v.**35**, pp.1747–1767, 1999

[26] P. Bondi, G. Casalino, L. Gambardella, "On the iterative learning control theory for robotic manipulators", *IEEE Journal of Robotics and Automation*, v.**4**, n.1, pp.14–22, Feb 1998

[27] S. Boyd, L. El Ghaoui, E. Feron, V. Balakrishnan, *Linear Matrix Inequalities in System and Control Theory*, SIAM Studies in Applied Mathematics, v.**15**, SIAM, Philadelphia, PA, 1994

[28] Y. Brave, M. Heymann, "On stabilization of discrete-event processes", *International Journal on Control*, v.**51**, n.5, pp.1101–1117, 1990

[29] N. G. Bretas, L. F. C. Alberto, "Lyapunov function for power systems with transfer conductances: an extension of the invariance principle", *IEEE Transactions on Power Systems*, v.**18**, n.2, pp.769–777, May 2003

[30] R. W. Brockett, Asymptotic stability and feedback stabilization. In *Differential Geometric Control Theory*, R. W. Brockett, R. S. Millmann, and H. J. Sussmann, Eds., Progress in Mathematics, Birkhauser, Boston, MA, 1983

[31] I. U. Bronstein, A. K. Ya, *Smooth Invariant Manifolds and Normal Forms*, World Scientific, Singapore, 1994

[32] R. R. Burridge, A. A. Rizzi, D. E. Koditschek, "Sequential composition of dynamically dexterous robot behaviors", *International Journal of Robotics Research*, v.**18**, n.6, pp.534–555, Jun 1999

[33] X. Cabré, E. Fontich, R. de la Llave, "The parametrization method for invariant manifolds III. Overview and applications", *Journal of Differential Equations*, v.**218**, n.2, pp.444–515, 2005

[34] D. Cai, "Multiple equilibria and bifurcations in an economic growth model with endogenous carrying capacity", *International Journal of Bifurcation and Chaos*, v.**20**, n.11, pp.3461–3472, 2010

[35] V. Chadalavada, V. Vittal, G. C. Ejebe, *et al.*, "An on-line contingency filtering scheme for dynamic security assessment", *IEEE Transactions on Power Systems*, v.**12**, n.1, pp.153–161, Feb 1997

[36] K. W. Chang, "Two problems in singular perturbations of differential equations", *Journal of the Australian Mathematical Society*, pp.33–50, 1969

[37] Y. Chen, J. Chen, "Robust composite control for singularly perturbed systems with time-varying uncertainties", *Journal of Dynamic Systems Measurement and Control – Transactions of the ASME*, v.**117**, n.4, pp.445–452, Dec 1995

[38] Y. K. Chen, R. Schinzinger, "Lyapunov stability of multimachine power systems using decomposition-aggregation method", *Proc. IEEE -PES Winter Meeting*, paper A 80036–4, 1980

[39] G. Chesi, "Estimating the domain of attraction for non-polynomial systems via lmi optimizations", *Automatica*, v.**45**, n.6, pp.1536–1541, 2009

[40] G. Chesi, *Domain of Attraction, Analysis and Control via SOS Programming*, Lecture Notes in Control and Information Sciences, v.**415**, Springer-Verlag, London, 2011

[41] G. Chesi, A. Garulli, A. Tesi, A. Vicino, "LMI-based computation of optimal quadratic Lyapunov functions for odd polynomial systems", *International Journal of Robust and Nonlinear Control*, v.**15**, pp.35–49, 2005

[42] G. Chesi, A. Tesi, A. Vicino, "Computing optimal quadratic Lyapunov functions for polynomial nonlinear systems via LMIs", *IFAC 15th Triennial World Congress*, Barcelona, Spain, 2002

[43] C. J. Chiang, A. G. Stefanopoulou, M. Jankoviv, "Nonlinear observer-based control of load transitions in homogeneous charge compression ignition engines", *IEEE Transactions on Control Systems Technology*, v.**15**, n.3, pp.438–448, May 2007

[44] H. D. Chiang, "Analytical results on the direct methods for power system transient stability analysis", *Control and Dynamic Systems: Advances in Theory and Application*, v.**43**, pp.275–334, Academic Press, New York, 1991

[45] H. D. Chiang, *Direct Methods for Stability Analysis of Electrical Power Systems: Theoretical Foundation, BCU Methodologies and Application*, John Wiley & Sons, 2011

[46] H. D. Chiang, The BCU method for direct stability analysis of electric power systems: pp. theory and applications, *Systems Control Theory for Power Systems*, IMA Volumes in Mathematics and Its Applications, v.**64**, pp.39–94, Springer-Verlag, New York, 1995

[47] H. D. Chiang, "Study of the existence of energy functions for power-systems with losses", *IEEE Transactions on Circuits and Systems*, v.**36**, n.11, pp.1423–1429, Nov 1989

[48] H. D. Chiang, J. H. Chen, C. K. Reddy, Trust-Tech-based global optimization methodology for nonlinear programming. In *Lectures on Global Optimization*, P. M. Pardalos and T. F. Coleman, Eds., American Mathematical Society, 2009

[49] H. D. Chiang, C.C. Chu, "A systematic search method for obtaining multiple local optimal solutions of nonlinear programming problems", *IEEE Transactions on Circuits and Systems: I*, v.**43**, n.2, pp.99–109, 1996

[50] H. D. Chiang, C. C. Chu. "Theoretical foundation of the BCU method for direct stability analysis of network-reduction power system models with small transfer conductances", *IEEE Transactions on Circuits and Systems: I*, v.**42**, n.5, pp.252–265, May 1995

[51] H. D. Chiang, C. C. Chu, G. Cauley, "Direct stability analysis of electric power systems using energy functions: theory, applications and perspectives", *Proceedings of the IEEE*, v.**38**, n.11, pp.1497–1529, Nov 1995

[52] H. D. Chiang, T. P. Conneen, A.J. Flueck, "Bifurcations and chaos in electric power systems: Numerical studies", *Journal of the Franklin Institute*, v.**331**, n.6, pp.1001–1036, Nov1994

[53] H. D. Chiang, L. Fekih-Ahmed, "Quasi-stability regions of nonlinear dynamical systems: optimal estimation", *IEEE Transactions on Circuits and Systems: I*, v.**43**, n.82, pp.636–642, Aug 1996

[54] H. D. Chiang, M. W. Hirsch, F. F. Wu, "Stability region of nonlinear autonomous dynamical systems", *IEEE Transactions on Automatic Control*, v.**33**, n.1, pp.16–27, Jan 1988

[55] H. D. Chiang, B. Y. Ku, J. S. Thorp, "A constructive method for direct analysis of transient stability", *IEEE Proc. 27th Conf. on Decision Control, Austin, TX*, pp.684–689 Dec 1988

[56] H. D. Chiang, J. Lee, Trust-Tech paradigm for computing high-quality optimal solutions: method and theory. In *Modern Heuristic Optimization Techniques: Theory and Applications to Power Systems*, K. Y. Lee and M. A. El-Sharkawi, Eds., pp.209–234, Wiley-IEEE Press, 2008

[57] H. D. Chiang, C. W. Liu, P. P. Varaiya, "Chaos in a simple power system", *IEEE Transactions on Power Systems*, v.**8**, n.4, pp.1407–1417, Nov 1993

[58] H. D. Chiang, A. K. Subramanian, "BCU dynamic security assessor for practical power system models", *IEEE PES Summer Meeting, Edmonton, Alberta, July 18–22*, pp.287–293, 1999

[59] H. D. Chiang, Y. Tada, H. Li, Power system on-line transient stability assessment (invited chapter), *Wiley Encyclopedia of Electrical and Electronics Engineering*, John Wiley & Sons, New York, 2007

[60] H. D. Chiang, J. S. Thorp, "Stability regions of nonlinear systems: a constructive methodology", *IEEE Transactions on Automatic Control*, v.**34**, n.12, pp.1229–1241, Dec 1989

[61] H. D. Chiang, T. Wang, Neighboring local-optimal solutions and its applications. In *Optimization in Science and Engineering*, Springer, pp.67–88, 2014

[62] H. D. Chiang, B. Wang, Q. Y. Jiang, Applications of Trust-Tech methodology in optimal power flow of power systems. In *Optimization in the Energy Industry*, Springer, pp.297–318, 2009

[63] H. D. Chiang, C. S. Wang, H. Li, "Development of BCU classifiers for on-line dynamic contingency screening of electric power systems", *IEEE Transactions on Power Systems*, v.**14**, n.2, pp.660–666, May 1999

[64] H. D. Chiang, F. F. Wu, P. P. Varaiya, "A BCU method for direct analysis of power system transient stability", *IEEE Transactions on Power Systems*, v.**8**, n.3, pp.1194–1208, Aug 1994

[65] H. D. Chiang, F. F. Wu, P. P. Varaya, "Foundations of direct methods for power system transient stability analysis", *IEEE Transactions on Circuits and Systems*, v.**34**, n.2, pp.160–173, Feb 1987

[66] H. D. Chiang, Y. Zheng, Y. Tada, H. Okamoto, K. Koyanagi, Y. C. Zhou, "Development of on-line BCU dynamic contingency classifiers for practical power systems", *14th Power System Computation Conference (PSCC), Spain, June 24–28*, 2002

[67] J. H. Chow, *Time-Scale Modeling of Dynamic Networks with Applications to Power Systems*, Springer-Verlag, Berlin, 1982

[68] C. C. Chu, H. D. Chiang, "Constructing analytical energy functions for network-preserving power system models", *Circuits Systems and Signal Processing*, v.**24**, n.4, pp.363–383, 2005

[69] L. O. Chua, A.-C. Deng, "Impasse points. Part I: numerical aspects", *International Journal of Circuit Theory and Applications*, v.**17**, n.2, pp.213–235, Apr 1989

[70] L. O. Chua, C. W. Wu, A. Huang, G. Zhong, "A universal circuit for studying and generating chaos – part I: route to chaos", *IEEE Transactions on Circuits and Systems*, v.**40**, n.10, pp.732–744, Oct 1993

[71] K. Ciliz, A. Harova, "Stability regions of recurrent type neural networks", *Electronic Letters*, v.**28**, n.11, pp.1022–1024, May 1992

[72] M. Cloosterman, N. van de Wouw, W. Heemels, H. Nijmeijer, "Stability of networked control systems with uncertain time-varying delays", *IEEE Transactions on Automatic Control*, v.**54**, n.7, pp.1575–1580, Jul 2009

[73] C. C. Conley, *Isolated Invariant Sets and the Morse Index*, American Mathematical Society, Providence, RI, 1978

[74] M. Corless, F. Garofalo, L. Glielmo, "New results on composite control of singularly perturbed uncertain linear systems", *Automatica*, v.**29**, n.2, pp.387–400, Mar 1993

[75] M. Corless, L. Glielmo, "On the exponential stability of singularly perturbed systems", *SIAM Journal on Control and Optimization*, v.**30**, n.6, pp.1338–1360, Nov 1992

[76] D. F. Coutinho, A. S. Bazanella, A. Trofino, A. S. Silva, "Stability analysis and control of a class of differential algebraic nonlinear systems", *International Journal of Robust and Nonlinear Control*, v.**14**, n.16, pp.1301–1326, 2004

[77] D. F. Coutinho, J. M. Gomes da Silva Jr., "Computing estimates of the region of attraction for rational control systems with saturating actuators", *IET Control Theory and Applications*, v.**4**, n.3, pp.315–325, 2008

[78] D. F. Coutinho, C. E. Souza, "Robust domain of attraction estimates for a class of uncertain discrete-time nonlinear systems", *8th IFAC Symposium on Nonlinear Control Systems, Bologna, Italy*, pp.185–190, 2010

[79] J. J. da Cruz, J. G. Geromel, "Decentralized control design for a class of nonlinear discrete time systems", *Proc. 25th Conf. on Decision and Control, Athens, Greece*, pp.1182–1183, 1986

[80] E. J. Davison, E. M. Kurak, "A computational method for determining quadratic Lyapunov functions for nonlinear systems", *Automatica*, v.**7**, pp.627, 1971

[81] R. J. Davy, I. A. Hiskens, "Lyapunov functions for multimachine power systems with dynamic loads", *IEEE Transactions on Circuits and Systems: I*, v.**44**, n.9, pp.796–812, Sept 1997

[82] J. Dieudonne, *Foundation of Modern Analysis*, 2nd edn, Academic Press,New York, 1969

[83] M. Djukanovic, D. Sobajic, Y.-H. Pao, "Neural-net based tangent hypersurfaces for transient security assessment of electric power systems", *International Journal of Electrical Power and Energy Systems*, v.**16**, n.6, pp.399–408, 1994

[84] Y. Dong, D. Cheng, H. Oin, "Applications of a Lyapunov function with a homogeneous derivative", *IEE Proceedings Control Theory and Applications*, v.**150**, n.3, pp.255–260, May 2003

[85] G. C. Ejebe, C. Jing, B. Gao, J. G. Waight G. Pieper, F. Jamshidian, P. Hirsch, "On-line implementation of dynamic security assessment at Northern States power company", *IEEE PES Summer Meeting, Edmonton, Alberta, July 18–22*, pp.270–272, 1999

[86] G. C. Ejebe, J. Tong, "Discussion of clarifications on the BCU method for transient stability analysis", *IEEE Transactions on Power Systems*, v.**10**, n.1, pp.218–219, Feb 1995

[87] A. M. Elaiw, X. Xia, "HIV dynamics: analysis and robust multirate MPC-based treatment schedules", *Journal of Mathematical Analysis and Applications*, v.**359**, n.1, pp.285–301, Nov 2009

[88] M. A. El-Kady, C. K. Tang, V. F. Carvalho, A. A. Fouad, V. Vittal, "Dynamic security assessment utilizing the transient energy function method", *IEEE Transactions on Power Systems*, v.**1**, n.3, pp.284–291, Aug 1986

[89] Electric Power Research Institute, *User's Manual for DIRECT 4.0, EPRI* TR-105886s, Electric Power Research Institute, Palo Alto, CA, Dec 1995

[90] D. Ernst, D. Ruiz-Vega, M. Pavella, P. Hirsch, D. Sobajic, "A unified approach to transient stability contingency filtering, ranking and assessment", *IEEE Transactions on Power Systems*, v.**16**, n.3, pp.435–443, Aug 2001

[91] V. Fairén, M. G. Velarde, "Dissipative structure in a nonlinear reaction–diffusion model with a forward inhibition; stability of secondary multiple steady states", *Reports on Mathematical Physics*, v.**16**, n.3, pp.421–432, 1979

[92] F. Fallside, M. R. Patel, "Step-response behaviour of a speed-control system with a back-e.m.f. nonlinearity", *Proceedings of the IEEE*, v.**112**, n.10, pp.1979–1984, Oct 1965

[93] N. Fenichel, "Geometric singular perturbation theory for ordinary differential equations", *Journal of Differential Equations*, v.**31**, pp.53–98, 1979

[94] C. A. Floudas, P. M. Pardalos, *Recent Advances in Global Optimization*, Princeton Series in Computer Science, Princeton, NJ, 1992

[95] A. A. Fouad, V. Vittal, *Power System Transient Stability Analysis: Using the Transient Energy Function Method*, Prentice-Hall, Englewood Cliffs, NJ, 1991

[96] J. H. Franks, *Homology and Dynamical Systems*, C.B.M.S. Regional Conf. Series in Mathematics, v.**49**, American Mathematical Society, Providence, RI, 1982

[97] R. Genesio, M. Tartaglia, A. Vicino, "On the estimation of asymptotic stability regions: state of the art and new proposals", *IEEE Transactions on Automatic Control*, v.**30**, pp.747–755, Aug 1985

[98] R. Genesio, A. Vicino, "Some results on the asymptotic stability of second-order nonlinear systems", *IEEE Transactions on Automatic Control*, v.**29**, n.9, pp.857–861, Sept 1984

[99] R. Genesio, A. Vicino, "New techniques for constructing asymptotic stability regions for nonlinear systems", *IEEE Transactions on Circuits and Systems*, v.**31**, pp.574–581, Jun 1984

[100] S. F. Glover, *Modeling and Stability Analysis of Power Electronics Based Systems*, PhD Thesis, Purdue University, 2003

[101] J. M. G. da Silva, S. Tarbouriech, "Anti-windup design with guaranteed region of stability: an LMI-based approach", *IEEE Transactions on Automatic Control*, v.**50**, n.1, pp.106–111, Jan 2005

[102] J. R. R. Jr. Gouveia, F. M. Amaral, L. F. C. Alberto, "Stability boundary characterization of nonlinear autonomous dynamical systems in the presence of a supercritical Hopf equilibrium point", *International Journal of Bifurcation and Chaos in Applied Sciences and Engineering*, v.**23**, n.12, 1350196, 2013

[103] G. Grammel, "Exponential stability of nonlinear singularly perturbed differential equations", *SIAM Journal on Control and Optimization*, v.**44**, n.5, pp.1712–1724, 2005

[104] J. W. Grizzle, J. M. Kang, "Discrete-time control design with positive semi-definite Lyapunov functions", *System and Control Letters*, v.**43**, pp.287–292, 2001

[105] L. T. Grujic, "Uniform asymptotic stability of nonlinear singularly perturbed and large scale systems", *International Journal of Control*, v.**33**, n.3, pp.481–504, 1981

[106] L. Grune, F. Wirth, "Computing control Lyapunov functions via a Zubov type algorithm", *Proc. 39th Conf. on Decision and Control, Sydney, Australia*, Dec 2000

[107] J. Guckenheimer, P. Holmes, *Nonlinear Oscillations, Dynamical Systems, and Bifurcations of Vector Fields*, Springer-Verlag, New York, 1993

[108] J. Guckeinheimer, P. Holmes, *Nonlinear Oscillations, Dynamical System, and Bifurcations of Vector Fields*, 1st edn., Springer-Verlag, New York, 1983

[109] V. Guillemin, A. Pollack, *Differential Topology*, Prentice-Hall, Englewood Cliffs, NJ, 1974

[110] N. Gunther, G. W. Hoffman, "Qualitative dynamics of a network model of regulation of the immune system: a rationale for the IgM to IgG switch", *Journal of Theoretical Biology*, v.**94**, pp.815–855, 1982

[111] W. Hahn, *Stability of Motion*, Springer-Verlag, New York, 1967

[112] J. K. Hale, *Ordinary Differential Equations*, Krieger, Huntington, NY, 1980

[113] J. K. Hale, *Introduction to Functional Differential Equations*, Applied Mathematical Sciences, v.**99**, Springer-Verlag, 1993

[114] J. K. Hale, H. Koçak, *Dynamics and Bifurcations*, Springer-Verlag, New York, 1991

[115] K. M. Halsey, K. Glover. "Analysis and synthesis on a generalized stability region", *IEEE Transactions on Automatic Control*, v.**50**, n.7, pp.997–1009, Jul 2005

[116] P. Hartman, *Ordinary Differential Equations*, Wiley, New York, 1973

[117] P. Hartman, *Ordinary Differential Equations*, 2nd edn., Classics in Applied Mathematics, SIAM, 1982

[118] M. A. Hassan, C. Storey, "Numerical determination of domains of attraction for electrical power systems using the method of Zubov analysis", *International Journal of Control*, v.**34**, pp.371–381, 1981

[119] G. Henkelman, G. Johannesson, H. Jonsson, Methods for finding saddle points and minimum energy paths. In *Progress on Theoretical Chemistry and Physics*, S. D. Schwartz, Ed., Kluwer Academic, pp.269–300, 2000

[120] D. Henrion, S. Tarbouriech, G. Garcia, "Output feedback robust stabilization of uncertain linear systems with saturating controls: an LMI approach", *IEEE Transactions on Automatic Control*, v.**44**, n.11, pp.2230–2237, Nov 1999

[121] D. J. Hill, I. M. Y. Mareels, "Stability theory for differential/algebraic systems with applications to power systems", *IEEE Transactions on Circuits and Systems*, v.**37**, n.11, pp.1416–1423, Nov 1990

[122] M. W. Hirsch. *Differential Topology*, Springer-Verlag, New York, 1976

[123] M. W. Hirsch, C. C. Pugh, M. Shub, "Invariant manifolds", *Bulletin of the American Mathematical Society*, v.**76**, n.5, pp.1015–1019, 1970

[124] M. W. Hirsch, S. Smale, *Differential Equations, Dynamical Systems and Linear Algebra*, Academic Press, New York, 1974

[125] I. Hiskens, R. Davy, "Lyapunov function analysis of power systems with dynamic loads", *Proceedings of the 35th IEEE Conference on Decision and Control*, v.**4**, pp.3870–3875, Dec 1996

[126] I. Hiskens, D. Hill, "Energy functions, transient stability and voltage behavior in power systems with nonlinear loads", *IEEE Transactions on Power Systems*, v.**4**, n.4, pp.1525–1533, Oct 1989

[127] J. J. Hopfield, "Neurons, dynamics and computation", *Physics Today*, v.**47**, pp.40–46, Feb 1994

[128] C. S. Hsu, *Cell-to-Cell Mapping: A Method of Global Analysis for Nonlinear Systems*, Applied Mathematical Sciences, v.**64**, Springer-Verlag, 1988

[129] P. Hsu, S. Sastry, "The effect of discretized feedback in a closed loop system", *Proc. 26th Conf. on Decision Control, Los Angeles, CA*, pp.1518–1523, 1987

[130] X. Hu, "Techniques in the stability of discrete systems", *Control and Dynamic Systems*, v.**66**, pp.153–216, 1994

[131] W. Hurewicz, H. Hallman, *Dimension Theory*, Princeton University Press, Princeton, NJ, 1948

[132] IEEE Committee Report, "Transient stability test systems for direct methods", *IEEE Transactions on Power Systems*, v.7, pp.37–43, Feb 1992.

[133] A. Jadbabaie, J. Hauser, "On the stability of recending horizon control with a general terminal cost", *IEEE Transactions on Automatic Control*, v.**50**, n.5, pp.674–678, May 2005

[134] J. L. Jardim, C. S. Neto, W. T. Kwasnicki, "Design features of a dynamic security assessment system", *IEEE Power System Conf. and Exhibition, New York, October 13–16*, 2004

[135] Z. Jing, Z. Jia, Y. Gao, "Research of the stability region in a power system", *IEEE Transactions on Circuits and Systems: I*, v.**50**, n.2, pp.298–304, Feb 2003

[136] L. B. Jocic, "On the attractivity of imbedded systems", *Automatica*, v.**17**, pp.853–860, 1980

[137] T. A. Johansen, "Computation of Lyapunov functions for smooth nonlinear systems using convex optimization", *Automatica*, v.**36**, n.11, pp.1617–1626, Nov 2000

[138] T. Kato, *Perturbation Theory for Linear Operators*, Springer-Verlag, Berlin, 1966

[139] M. Kaufman, R. Thomas, "Model analysis of the bases of multistationarity in the humoral immune response", *Journal of Theoretical Biology*, v.**129**, pp.141–162, 1987

[140] C. M. Kellet, A. R. Teel, "Smooth-Lyapunov functions and robustness of stability for difference inclusions", *System and Control Letters*, v.**52**, pp.395–405, 2004

[141] Y. Khait, A. Panin, A. Averyanov, "Search for stationary points of arbitrary index by augmented Hessian method", *International Journal of Quantum Chemistry*, v.**54**, n.6, pp.329–336, 1995

[142] H. K. Khalil, *Nonlinear Systems*, 3rd edn., Prentice Hall, 2002

[143] J. Kim, "On-line transient stability calculator", *Final Report* RP2206–1, EPRI, Palo Alto, CA, March 1994

[144] A. I. Klimushchev, N. N. Krasovskii, "Uniform asymptotic stability of systems of differential equations with a small parameter in the derivative terms", *Journal of Applied Mathematics and Mechanics*, v.**25**, n.4, pp.680–690, 1961

[145] D. E. Koditschek, "Exact robot navigation by means of potential functions: some topological considerations", *Proceedings of the IEEE International Conference on Robotics and Automation*, v.**4**, pp.1–6, 1987

[146] P. Kokotovic, *Singular Perturbation Techniques in Control Theory*, Lecture Notes in Control and Information Sciences, v.**90**, pp.1–55, Springer, 1987

[147] P. Kokotovic, R. Marino "On vanishing stability regions in nonlinear systems with high-gain feedback", *IEEE Transactions on Automatic Control*, v.**31**, n.10, pp.967–970, Oct 1986

[148] A. Korobeinikov, "Stability of ecosystems: global properties of a general predator–prey model", *Mathematical Medicine and Biology*, v.**26**, n.4, pp.309–321, 2009

[149] N. N. Krasovskii, *Problems of the Theory of Stability of Motion* (In Russian, 1959), English translation, Stanford University Press, Stanford, CA, 1963

[150] B. Krauskopf, H. M. Osinga, E. J. Doedel, M. E. Henderson, J. Guckenheimer, A. Vladimirsky, M. Dellnitz, O. Junge, "A survey of methods for computing (un)stable manifolds of vector fields", *International Journal of Bifurcation and Chaos*, v.**15**, n.3, pp.763–791, 2005

[151] P. Kundur, *Power System Stability and Control*, McGraw Hill, New York, 1994

[152] D.H. Kuo, A. Bose, "A generation rescheduling method to increase the dynamics security of power systems", *IEEE Transactions on Power Systems*, v.**10**, n.1, pp.68–76, 1995.

[153] Y. A. Kuznetsov, *Elements of Applied Bifurcation Theory*, Springer-Verlag, New York, 1995

[154] W. Langson, A. Allevne, "A stability result with application to nonlinear regulation", *Journal of Dynamic Systems Measurement and Control – Transactions of the ASME*, v.**124**, n.3, pp.452–445, Sept 2002

[155] D. S. Laila, M. Lovera, A. Astolfi, "A discrete-time observer design for spacecraft attitude determination using an orthogonality-preserving algorithm", *Automatica*, v.**47**, pp.975–980, 2011

[156] J. P. LaSalle, *Stability of Dynamical Systems*, SIAM, Philadelphia, PA, 1976

[157] J. P. LaSalle, *Stability Theory for Difference Equations*, Studies in Ordinary Differential Equations, Studies in Mathematics, v.**14**, pp.1–31, Mathematical Association of America, Washington, DC, 1977

[158] J. P. LaSalle, "Some extensions of Liapunov's second method", *IRE Transactions on Circuit Theory*, v.**7**, pp.520–527, 1960

[159] J. P. LaSalle, *The Stability and Control of Discrete Processes*, Springer-Verlag, New York, 1986

[160] J. P. La Salle, S. Lefschetz, *Stability by Lyapunov's Direct Method*, Academic Press, New York, 1961

[161] J. Lee, *Trajectory-based Methods for Global Optimization: Theory and Algorithms*, PhD Dissertation, Department of Electrical Engineering, Cornell University, Ithaca, NY, 1999

[162] J. Lee, H. D. Chiang, "A dynamical trajectory-based methodology for systematically computing multiple optimal solutions of general nonlinear programming problems", *IEEE Transactions on Automatic Control*, v.**49**, n.6, pp.888–899, 2004

[163] J. Lee, H. D. Chiang, "Theory of stability regions for a class of nonhyperbolic dynamical systems and its application to constraint satisfaction problems", *IEEE Transactions on Circuits and Systems: I*, v.**49**, n.2, pp.196–209, 2002

[164] A. Levin, "An analytical method of estimating the domain of attraction for polynomial differential equations", *IEEE Transactions on Automatic Control*, v.**39**, n.12, pp.2471–2475, Dec 1994

[165] A. Lewis, "An investigation of the stability in the large for an autonomous second-order two degree-of-freedom system", *International Journal of Non-Linear Mechanics*, v.**37**, n.2, pp.153–169, Mar 2002

[166] L. Liu, Y. Tian, X. Huang, *A Method to Estimate the Basin of Attraction of the System with Impulse Effects: Application to Biped Robots*, Intelligent Robotics and Applications, pp.953–962, Springer, 2008

[167] A. Llamas, J. De La R. Lopez, L. Mili, A. G. Phadke, J. S. Thorp, "Clarifications on the BCU method for transient stability analysis", *IEEE Transactions Power Systems*, v.**10**, n.1, pp.210–219, Feb 1995

[168] M. Loccufier, E. Noldus, "A new trajectory reversing method for estimating stability regions of autonomous nonlinear systems", *Nonlinear Dynamics*, v.**21**, pp.265–288, 2000

[169] L. A. Luxemburg, G. Huang, "On the number of unstable equilibria of a class of nonlinear systems", *26th IEEE Conf. on Decision Control*, pp.889–894, 1987.

[170] L. Luyckx, M. Loccufier, E. Noldus, "Computational methods in nonlinear stability analysis: stability boundary calculations", *Journal of Computational and Applied Mathematics*, v.**168**, n.1–2, pp.289–297, 2004

[171] Y. Mansour, E. Vaahedi, A. Y. Chang, B. R. Corns, B. W. Garrett, K. Demaree, T. Athay, K. Cheung, "B.C.Hydro's on-line transient stability assessment (TSA) model development, analysis and post-processing", *IEEE Transactions on Power Systems*, v.**10**, n.1, pp.241–253, Feb 1995

[172] P. C. Magnusson, "Transient energy method of calculating stability", *AIEE Transactions*, v.**66**, pp.747–755, 1947

[173] R. E. O'Malley, Jr. *Singular Perturbation Methods for Ordinary Differential Equations*, Applied Mathematical Series, v.**89**, Springer-Verlag, 1991

[174] C. M. Marcus, R. M. Westervelt, "Dynamics of iterated-map neural networks", *Physics Review A*, v.**40**, n.1, pp.501–504, 1989

[175] S. G. Margolis, W. G. Vogt, "Control engineering applications of V. I. Zubov's construction procedure for Lyapunov functions", *IEEE Transactions on Automatic Control*, v.**8**, pp.104–113, Apr 1963

[176] R. M. May, *Stability and Complexity in Model Ecosystems*, Princeton University Press, Princeton, NJ, 1973

[177] S. K. Mazumder, A. H. Nayfeh, D. Borojevic, "A nonlinear approach to the analysis of stability and dynamics of standalone and parallel dc-dc converters", *Proc. Applied Power Electronics Conference and Exposition*, pp.784–790, 2001

[178] F. Mercede, J. C. Chow, H. Yan, R. Fishl, "A framework to predict voltage collapse in power systems", *IEEE Transactions on Power Systems*, v.**3**, n.4, pp.1807–1813, Nov 1988

[179] J. L. Meza, V. Santibanez, R. Campa, "An estimate of the domain of attraction for the PID regulator of manipulators", *International Journal of Robotics and Automation*, v.**22**, n.3, pp.187–195, 2007

[180] A. N. Michel, R. K. Miller, *Qualitative Analysis of Large Scale Dynamical Systems*, Academic Press, New York, 1977

[181] A. N. Michel, R. K. Miller, B. H. Nam, "Stability analysis of interconnected systems using computer generated Lyapunov functions", *IEEE Transactions on Circuits and Systems*, v.**29**, pp.431–440, Jul 1982

[182] A. N. Michel, N. R. Sarabudla, R. K. Miller, "Stability analysis of complex dynamical systems: some computational methods", *Circuits, Systems and Signal Processing*, v.**1**, pp.171–202, 1982

[183] B. E. A. Milani, "Contractive polyhedra for discrete time linear systems with saturating controls", *Proceedings of the 38th IEEE Conference on Decision and Control*, pp.2039–2044, Dec 1999

[184] R. K. Miller, A. N. Michel, *Ordinary Differential Equations*, Academic Press, New York, 1982

[185] J. Milnor, "On the concept of attractor", *Communications in Mathematical Physics*, v.**99**, pp.177–195, 1985

[186] J. Milnor, *Topology from the Differential Viewpoint*, University of Virginia Press, 1965

[187] Y. Min, L. Chen, K. Hou, Y. Song, "The credible regions on the approximate stability boundaries of nonlinear dynamic systems", *IEEE Transactions on Automatic Control*, v.**52**, n.8, pp.1486–1491, 2007

[188] C. Mira, D. Fournier-Prunaret, L. Gardini, H. Kawakami, J. C. Cathala, "Basin bifurcations of two-dimensional noninvertible maps: fractalization of basins", *International Journal of Bifurcation and Chaos*, v.**4**, n.2, pp.343–381,1994

[189] R. Miron, K. Fichthorn, "The step and slide method for finding saddle points on multi-dimensional potential surfaces", *Journal of Chemical Physics*, v.**115**, n.19, pp.8742–8747, 2001

[190] S. Mokhtari, *et al.*, Analytical methods for contingency selection and ranking for dynamic security assessment, *Final Report* RP3103–3, EPRI, Palo Alto, CA, May 1994

[191] M. Morse, *The Calculus of Variations in the Large*, American Mathematical Society, 1934

[192] J. R. Munkres. *Topology – A First Course*, Prentice- Hall, Englewood Cliffs, NJ, 1975

[193] J. R. Munkres, *Topology*, 2nd edn., Prentice-Hall, 2000

[194] L. I. Nicolaescu, *An Invitation to Morse Theory*, Springer, 2007

[195] E. Noldus, J. Spriet, E. Verriest, A. Van Cauwenberghe, "A new Lyapunov technique for stability analysis of chemical reactors", *Automatica*, v.**10**, n.6, pp.675–680, Dec 1974

[196] R. Ortega, A. Loria, R. Kelly, "A semiglobally stable output feedback PI^2D regulator for robot manipulators", *IEEE Transactions on Automatic Control*, v.**40**, n.8, pp.1432–1436, Aug 1995

[197] F. Paganini, B.C. Lesieutre, A critical review of the theoretical foundations of the BCU method, *Technical Report* TR97-005, MIT Lab., Electromagnetic and Electrical Systems, July 1997

[198] F. Paganini, B.C. Lesieutre, "Generic properties, one-parameter deformations, and the BCU method", *IEEE Transactions on Circuits and Systems: I*, v.**46**, n.6, pp.760–763, Jun 1999

[199] M. A. Pai, *Energy Function Analysis for Power System Stability*, Kluwer Academic Publishers, Boston, MA, 1989

[200] J. Palis, "On Morse–Smale dynamical systems", *Topology*, v.**8**, pp.385–405, 1969.

[201] J. Palis, J. W. de Melo, *Geometric Theory of Dynamical Systems: An Introduction*, Springer-Verlag, New York, 1981

[202] J. Palis, W. Melo, *Introdução aos Sistemas Dinâmicos*, Edgard Blucher, 1978

[203] M. M. Peixoto, "On an approximation theorem of Kupka and Smale", *Journal of Differential Equations*, v.**3**, n.2, pp.214–227, Apr 1967

[204] G. Peponides, P. V. Kokotovic, J. H. Chow, "Singular perturbations and time scales in nonlinear models of power systems", *IEEE Transactions on Circuits and Systems*, v.**29**, n.11, pp. 758–767, Nov 1982

[205] L. Perko, *Differential Equations and Dynamical Systems*, Springer, New York, 1991

[206] R. M. Peterman, "A simple mechanism that causes collapsing stability regions in exploited salmonid populations", *Journal of the Fisheries Research Board of Canada*, v.**34**, pp.1130–1142, 1977

[207] K. Praprost, K. A. Loparo, "A stability theory for constrained dynamic systems with applications to electric power systems", *IEEE Transactions on Automatic Control*, v.**41**, n.11, pp.1605–1617, Nov 1996

[208] M. L. Psiaki, Y. P. Luh, "Nonlinear system stability boundary approximation by polytopes in state-space", *International Journal of Control*, v.**57**, n.1, pp.197–224, Jan 1993

[209] C. C. Pugh, "On a theorem of Hartman", *American Journal of Mathematics*, v.**91** n.2, pp.363–367, 1969

[210] W. Quapp, M. Hirsch, O. Imig, D. Heidrich, "Searching for saddle points of potential energy surfaces by following a reduced gradient", *Journal of Computational Chemistry*, v.**19**, pp.1087–1100, 1998

[211] M. Rabelo, L. F. C. Alberto, "An extension of the invariance principle for a class of differential equations with finite delay", *Advances in Difference Equations*, article ID 496936, pp.14, 2010

[212] F. A. Rahimi, M. G. Lauby, J. N. Wrubel, K. L. Lee, "Evaluation of the transient energy function method for on-line dynamic security assessment", *IEEE Transactions on Power Systems*, v.**8**, n.2, pp.497–507, May 1993

[213] C. Reddy, H. D. Chiang, "Finding Saddle points using stability boundaries", *Proc. 2005 ACM Symp. on Applied Computing*, pp.212–213, 2005

[214] C. Reddy, H. D. Chiang, "A stability boundary based method for finding saddle points on potential energy surfaces", *Journal of Computational Biology*, v.**13**, n.3, pp.745–766, Apr 2006

[215] L. Riverin, A. Valette, "Automation of security assessment for Hydro-Quebec's power system in short-term and real-time modes", *International Conference on Large High Voltage Electric Systems CIGRE*, pp.39–103, 1998

[216] C. Robinson, *Dynamic Systems. Stability, Symbolic Dynamics and Chaos*, 2nd edn., CRC Press, 1998

[217] H. Rodriguez, R. Ortega, G. Escobar, N. Barabanov, "A robustly stable output feedback saturated controller for the boost DC-to-DC converter", *Systems and Control Letters*, v.**40**, n.1, pp.1–8, 2000

[218] H. M. Rodrigues, L. F. C. Alberto, N. G. Bretas, "On the invariance principle: generalizations and applications to synchronization", *IEEE Transactions on Circuits and Systems: I*, v.**47**, n.5, pp.730–739, May 2000

[219] H. M. Rodrigues, J. H. Wu, L. R. A. Gabriel, "Uniform dissipativeness, robust synchronization and location of the attractor of parametrized nonautonomous discrete systems", *International Journal of Bifurcation and Chaos*, v.**21**, n.2, pp.513–526, 2011

[220] W. Rudin, *Principles of Mathematical Analysis*, International Series in Pure and Applied Mathematics, 3rd edn., McGraw-Hill, 1976

[221] A. Saberi, H. Khalil, "Quadratic-type Lyapunov functions for singularly perturbed systems", *IEEE Transactions on Automatic Control*, v.**29**, n.6, pp.542–550, Jun 1984

[222] M. Saeki, M. Araki, "A new estimate of the stability regions of large-scale systems", *International Journal of Control*, v.**32**, n.2, pp.257–269, 1980

[223] H. Sasaki, "An approximate incorporation of field flux decay into transient stability analysis of multimachine power systems by the second method of lyapunov", *IEEE Transactions on Power Apparatus and Systems*, v.**98**, n.2, pp.473–483, Mar–Apr 1979

[224] P. W. Sauer, M. A. Pai, *Power System Dynamics and Stability*, Prentice-Hall, Englewood Cliffs, NJ 1998

[225] L. A. Segel, "Multiple attractors in immunology: theory and experiment", *Biophysical Chemistry*, v.**72**, pp.223–230, 1998

[226] D. N. Shields, C. Storey, "The behavior of optimal Lyapunov functions", *International Journal of Control*, v.**21**, n.4, pp.561–573, 1975

[227] H. Shim, J. H. Seo, "Non-linear output feedback stabilization on a bounded region of attraction", *International Journal of Control*, v.**73**, n.5, pp.416–426, Mar 2000

[228] M. Shub, *Global Stability of Dynamical Systems*, Springer-Verlag, New York, 1987

[229] A. A. G. Siqueira, M. H. Terra, "Nonlinear *H*-infinity control applied to biped robots", *Proceedings of the 2006 IEEE International Conference on Control Applications, Munich, Germany*, Oct 2006

[230] K. K. Shyu, S. R. Chen, "Estimation of asymptotic stability region and sliding domain of uncertain variable structure systems with bounded controllers", *Automatica*, v.**32**, n.5, pp.797–800, 1996

[231] M. W. Siddiqee, "Transient stability of an a.c. generator by Lyapunov's direct method", *International Journal of Control*, v.**8**, n.2, pp.131–144, 1968

[232] D. D. Šiljak, *Large-Scale Dynamic Systems: Stability and Structure*, North-Holland, New York, 1978

[233] F. H. J. R. Silva, L. F. C. Alberto, J. B. A. London Jr., N. G. Bretas, "Smooth perturbation on a classical energy function for lossy power system stability analysis", *IEEE Transactions on Circuits and Systems: I*, v.**52**, n.1, pp.222–229, Jan 2005

[234] S. Smale, "Differential dynamical systems", *Bulletin of the American Mathematical Society*, v.**73**, pp.747–817, 1967

[235] K. T. Smith, *Primer of Modern Analysis*, Springer -Verlag, New York, 1983

[236] J. Sotomayor, Generic bifurcations of dynamical systems. In *Dynamical Systems*, M. M. Peixoto, Ed., pp.549–560, Academic Press, New York, 1973

[237] G. W. Steward, *Introduction to Matrix Computation*, Academic Press, New York, 1973

[238] B. Stott, "Power system dynamic response calculations", *Proceedings of the IEEE*, v.**67**, pp.219–241, 1979

[239] G. Strang, *Introduction to Linear Algebra*, Wellesley-Cambridge Press, 1994

[240] Y. Susuki, T. Hikihara, H. D. Chiang, "Stability boundaries analysis of electric power system with dc transmission based on differential-algebraic equation system", *IEICE Transactions on Fundamentals of Electronics, Communications and Computer Sciences*, v.**E87**-A, n.9, pp.2339–2346, Sept 2004

[241] Y. Tada, H. D. Chiang, "Design and implementation of on-line dynamic security assessment", *IEEJ Transactions on Electrical and Electronic Engineering*, v.**4**, n.3, pp.313–321, 2008

[242] Y. Tada, A. Kurita, Y. C. Zhou, K. Koyanagi, H. D. Chiang, Y. Zheng, "BCU-guided time-domain method for energy margin calculation to improve BCU-DSA system", *IEEE/PES Transmission and Distribution Conference and Exhibition*, 2002

[243] Y. Tada, T. Takazawa, H. D. Chiang, H. Li, J. Tong, "Transient stability evaluation of a 12,000-bus power system data using TEPCO-BCU", *15th Power System Computation Conference (PSCC), Belgium*, Aug 2005

[244] Y. Tada, A. Ono, A. Kurita, Y. Takahara, T. Shishido, K. Koyanagi, "Development of analysis function for separated power system data based on linear reduction techniques on integrated power system analysis package", *15th Conference of the Electric Power Supply, Shanghai, China*, Oct 2004

[245] M. Takegaki, S. Arimoto, "A new feedback method for dynamic control of manipulators", *Journal of of Dynamic Systems, Measurement, and Control*, v.**103**, n.2, pp.119–125, Jun 1981

[246] F. Takens, Constrained equations; a study of implicit differential equations and their discontinuous solutions. In *Structural Stability, the Theory of Catastrophes, and Applications in the Sciences*, Lecture Notes in Mathematics, v.**525**, pp.143–234, Springer, Berlin, 1976

[247] W. Tan, A. Packard, "Stability region analysis using polynomial and composite polynomial Lyapunov function and sum-of-squares programming", *IEEE Transactions on Automatic Control*, v.**53**, n.2, pp.565–571, Mar 2008

[248] R. Thom, *Structural Stability and Morphogenesis*, Benjamin, Reading, MA, 1975

[249] R. J. Thomas, J. S. Thorp, "Towards a direct test for large scale electric power system instabilities", *IEEE Proc. 24th Conf. on Decision and Control, Fort Lauderdale, FL*, pp.65–69, Dec 1985

[250] B. Tibken, "Estimation of the domain of attraction for polynomial systems via LMI's", *Proceedings of the 39th IEEE Conference on Decision and Control, Sydney, Australia*, Dec 2000

[251] J. Tong, H. D. Chiang, T. P. Conneen, "A sensitivity-based BCU method for fast derivation of stability limits in electric power systems", *IEEE Transactions on Power Systems*, v.**8**, n.4, pp.1418–1437, 1993

[252] U. Topcu, A. Packard, P. Seiler, T. Wheeler, "Stability region analysis using simulations and sum-of-squares programming", *Proceedings of the 2007 American Control Conference, New York*, Jul 2007

[253] G. J. Toussaint, T. Basar, "Achieving nonvanishing stability regions with high gain cheap control using H techniques: the second-order case", *Systems and Control Letters*, v.**44**, n.2, pp.79-89, Oct 2001

[254] N. Tsolas, A. Arapostathis, P. Varaiya, "A structure preserving energy function for power system transient stability analysis", *IEEE Transactions on Circuits and Systems*, v.**32**, n.10, pp.1041–1050, Oct 1985

[255] S. Ushiki, "Analytic expressions of the unstable manifolds", *Proceedings of the Japan Academy, Series A*, v.**56**, pp.239–243, 1980

[256] A. Vannelli, M. Vidyasagar, "Maximal Lyapunov functions and domains of attraction for autonomous nonlinear systems", *Automatica*, v.**21**, n.1, pp.69–80, 1985

[257] P. P. Varaiya, F. F. Wu, R. L. Chen, "Direct methods for transient stability analysis of power systems: recent results", *Proceedings of the IEEE*, v.**73**, pp.1703–1715, Dec 1985

[258] M. Varghese, J. S. Thorp, "An analysis of truncated fractal growths in the stability boundaries of three-node swing equations", *IEEE Transactions on Circuits and Systems*, v.**35**, pp.825–834, 1988

[259] A. B. Vasil'eva, V. F. Butuzov, *Asymptotical Expansions of the Solutions of Singularly Perturbed Systems*, Nauka, Moscow, 1973

[260] A. B. Vasil'eva, V. F. Butuzov, L. V. Kalachev, *The Boundary Function Method for Singular Perturbation Problems*, SIAM, Philadelphia, PA, 1995

[261] V. Veliov, "A generalization of the Tikhonov theorem for singularly perturbed differential inclusions", *Journal of Dynamical and Control Systems*, v.**3**, n.3, pp.291–319, 1997

[262] V. Venkatasubramanian, *A Taxonomy of the Dynamics of Large Differential-Algebraic Systems such as the Power Systems*, PhD Thesis, Washington University, Sever Institute of Technology, 1992

[263] V. Venkatasubramanian, W. Ji, "Coexistence of four different attractors in a fundamental power system model", *IEEE Transactions on Circuits and Systems: I*, v.**46**, n.3, pp.405–409, Mar 1999

[264] V. Venkatasubramanian, H. Schattler, J. Zaborsky, "Dynamic of large constrained nonlinear systems – a taxonomy theory", *Proceedings of the IEEE*, v.**83**, n.11, pp.1530–1560, Nov 1995

[265] M. Vidyasagar, *Nonlinear Systems Analysis*, Prentice-Hall, Englewood Cliffs, NJ, 1978

[266] M. Vidyasagar, "New directions of research in nonlinear system theory", *Proceedings of the IEEE*, v.**74**, pp.1060–1091, Aug 1986

[267] R. Villafuerte, S. Mondié, "Estimate of the region of attraction for a class of nonlinear time delay systems: a leukemia post-transplantation dynamics example", *Proc. 46th IEEE Conference on Decision and Control, New Orleans, LA*, pp.633–638, Dec 2007

[268] R. Villafuerte, S. Mondié, S. I. Niculescu, "Stability analysis and estimate of the region of attraction of a human respiratory model", *Proc. 47th IEEE Conference on Decison and Control, Mexico*, pp.2644–2649, 2008

[269] C. D. Vournas, N. Sakelladaridis, "Region of attraction in a power system with discrete ltcs", *IEEE Transactions on Circuits and Systems: I*, v.**53**, n.7, pp.1610–1618, Jul 2006

[270] N. Wada, *et al.*, "Model predictive tracking control using a state-dependent gain-schedule feedback", *Proc. 2010 Int. Conf. on Modelling, Identification and Control, Japan*, pp.418–423, 2010

[271] L. Wang, K. Morison, "Implementation of on-line security assessment", *IEEE Power and Energy Magazine*, v.**4**, n.5, Sept/Oct 2006

[272] W. X. Wang, Y. B. Zhang, C. Z. Liu, "Analysis of a discrete-time predator-prey system with allee effect", *Ecological Complexity*, v.**8**, n.1, pp.81–85, 2011

[273] S. Weissenberger, "Stability regions of large-scale systems", *Automatica*, v.**9**, pp.653–663, 1973

[274] E. R. Westervelt, J. W. Grizzle, C. C. de Wit, "Switching and PI control of walking motions of planar biped walkers", *IEEE Transactions on Automatic Control*, v.**48**, n.2, pp.308–312, Feb 2003

[275] E. R. Westervelt, J. W. Grizzle, D. E. Koditschek, "Hybrid zero dynamics of planar biped walkers", *IEEE Transactions on Automatic Control*, v.**48**, n.1, pp.42–56, Jan 2003

[276] S. Wiggins, *Global Bifurcations and Chaos: Analytical Methods*, Applied Mathematical Sciences, v.**73**, Springer-Verlag, New York, 1988

[277] S. Wiggins, *Introduction to Applied Nonlinear Dynamical Systems and Chaos*, Springer-Verlag, 1990

[278] H. K. Wimmer, "Inertia theorem for matrices, controllability, and linear vibration", *Linear Algebra and its Applications*, v.**8**, pp.337–343, 1974

[279] H. Xin, D. Gan, M. Huang, K. Wang, "Estimating the stability region of singular perturbation power systems with saturation nonlinearities: an LMI-based method", *IET Control Theory and Applications*, v.**4**, n.3, pp.351–361, 2010

[280] H. Xin, D. Gan, J. Qiu, Z. Qu, "Methods for estimating stability regions with application to power systems", *European Transactions on Electric Power*, v.**17**, n.2, pp.113–133, Mar/Apr 2007

[281] D. Xu, S. Li, Z. Pu, Q. Guo, "Domain of attraction of nonlinear discrete systems with delays", *Computers and Mathematics with Applications*, v.**38**, pp.155–162, 1999

[282] H. Yee, B. D. Spalding, "Transient stability analysis of multi-machine systems by the method of hyperplanes", *IEEE Transactions on Power Apparatus and Systems*, v.**96**, n.1, pp.276–284, Jan 1977

[283] T. Yu, J. Yu, H. Li, H. Lin, "Complex dynamics of industrial transferring in a credit-constrained economy", *Physics Procedia*, v.**3**, n.5, pp.1677–1685, 2010

[284] Y. Yu, K. Vongsuriya, "Nonlinear power system stability study by Lyapunov function and Zubov's method" *IEEE Transactions on Power Apparatus and Systems*, v.**86**, pp.1480–1485, 1967

[285] J. Zaborszky, G. Huang, B. Zheng, T. C. Leung, "On the phase portrait of a class of large nonlinear dynamic systems such as the power system", *IEEE Transactions on Automatic Control*, v.**33**, n.1, pp.4–15, Jan 1988

[286] L. A. Zadeh, C. A. Desoer, *Linear System Theory – The State Space Approach*, McGraw-Hill, New York, 1963

[287] J. Zhong, *et al.*, "New results on non-regular linearization of non-linear systems", *International Journal of Control*, v.**80**, n.10, pp.1651–1664, 2007

[288] Y. Zou, M. H. Yin, H. D. Chiang, "Theoretical foundation of the controlling UEP method for direct transient stability analysis of network-preserving power system models", *IEEE Transactions on Circuits and Systems*, v.**50**, n.10, pp.1324–1336, Oct 2003

Index

Lightning Source UK Ltd.
Milton Keynes UK
UKOW07n2316110716

278138UK00011B/143/P

9 781107 035409